Introductory Biomedical Imaging

Imaging is everywhere. We use our eyes to see and cameras to take pictures. Scientists use microscopes and telescopes to peer into cells and out to space. Doctors use ultrasound, X-rays, radioisotopes, and MRI to look inside our bodies. If you are curious about imaging, open this textbook to learn the fundamentals.

Imaging is a powerful tool in fundamental and applied scientific research and also plays a crucial role in medical diagnostics, treatment, and research. This undergraduate textbook introduces cutting-edge imaging techniques and the physics underlying them. Elementary concepts from electromagnetism, optics, and modern physics are used to explain prominent forms of light microscopy, as well as endoscopy, ultrasound, projection radiography and computed tomography, radionuclide imaging, and magnetic resonance imaging. This textbook also covers digital image processing and analysis. Theoretical principles are reinforced with illustrative homework problems, applications, activities, and experiments, and by emphasizing recurring themes, including the effects of resolution, contrast, and noise on image quality. Readers will learn imaging fundamentals, diagnostic capabilities, and strengths and weaknesses of techniques.

This textbook had its genesis, and has been vetted, in a "Biomedical Imaging" course at Lewis & Clark College in Portland, OR, and is designed to facilitate the teaching of similar courses at other institutions. It is unique in its coverage of both optical microscopy and medical imaging at an intermediate level, and exceptional in its coverage of material at several levels of sophistication.

Imaging in Medical Diagnosis and Therapy
Series Editors: Bruce R. Thomadsen, David W. Jordan

Beam's Eye View Imaging in Radiation Oncology
Ross I. Berbeco

Principles and Practice of Image-Guided Radiation Therapy of Lung Cancer
Jing Cai, Joe Y. Chang, Fang-Fang Yin

Radiochromic Film
Role and Applications in Radiation Dosimetry
Indra J. Das

Clinical 3D Dosimetry in Modern Radiation Therapy
Ben Mijnheer

Hybrid Imaging in Cardiovascular Medicine
Yi-Hwa Liu, Albert J. Sinusas

Observer Performance Methods for Diagnostic Imaging
Foundations, Modeling, and Applications with R-Based Examples
Dev P. Chakraborty

Ultrasound Imaging and Therapy
Aaron Fenster, James C. Lacefield

Dose, Benefit, and Risk in Medical Imaging
Lawrence T. Dauer, Bae P. Chu, Pat B. Zanzonico

Big Data in Radiation Oncology
Jun Deng, Lei Xing

Monte Carlo Techniques in Radiation Therapy
Introduction, Source Modelling, and Patient Dose Calculations, Second Edition
Frank Verhaegen, Joao Seco

Monte Carlo Techniques in Radiation Therapy
Applications to Dosimetry, Imaging, and Preclinical Radiotherapy, Second Edition
Joao Seco, Frank Verhaegen

Introductory Biomedical Imaging
Principles and Practice from Microscopy to MRI
Bethe A. Scalettar, James R. Abney

For more information about this series, please visit:
https://www.crcpress.com/Series-in-Optics-and-Optoelectronics/book-series/TFOPTICSOPT

Introductory Biomedical Imaging
Principles and Practice from Microscopy to MRI

Bethe A. Scalettar
James R. Abney

Illustrations by
Cyan Cowap

CRC Press is an imprint of the
Taylor & Francis Group, an **informa** business

First edition published 2023
by CRC Press
6000 Broken Sound Parkway NW, Suite 300, Boca Raton, FL 33487-2742

and by CRC Press
4 Park Square, Milton Park, Abingdon, Oxon, OX14 4RN

CRC Press is an imprint of Taylor & Francis Group, LLC

© 2023 Taylor & Francis Group, LLC

Reasonable efforts have been made to publish reliable data and information, but the authors and publisher cannot assume responsibility for the validity of all materials or the consequences of their use. The authors and publishers have attempted to trace the copyright holders of all material reproduced in this publication and apologize to copyright holders if permission to publish in this form has not been obtained. If any copyright material has not been acknowledged please write and let us know so we may rectify in any future reprint.

Except as permitted under U.S. Copyright Law, no part of this book may be reprinted, reproduced, transmitted, or utilized in any form by any electronic, mechanical, or other means, now known or hereafter invented, including photocopying, microfilming, and recording, or in any information storage or retrieval system, without written permission from the publishers.

For permission to photocopy or use material electronically from this work, access www.copyright.com or contact the Copyright Clearance Center, Inc. (CCC), 222 Rosewood Drive, Danvers, MA 01923, 978-750-8400. For works that are not available on CCC please contact mpkbookspermissions@tandf.co.uk

Trademark notice: Product or corporate names may be trademarks or registered trademarks and are used only for identification and explanation without intent to infringe.

ISBN: 978-1-032-32819-5 (hbk)
ISBN: 978-1-138-62668-3 (pbk)
ISBN: 978-1-315-22313-1 (ebk)

DOI: 10.1201/b22076

Typeset in Palatino
by KnowledgeWorks Global Ltd.

Access the Support Material: www.routledge.com/9781032328195

This book is dedicated to our parents:

Florence and Richard Scalettar

Helen and Horace Abney

We miss you.

Contents

List of Boxes

Preface

The seeds for this textbook were planted in 2008 when Lewis & Clark College was awarded a grant from the Howard Hughes Medical Institute (HHMI). The grant was focused on advancing interdisciplinary science at the College and included support for the development of a course that unites the life sciences and physics.

We ultimately decided to develop a course that covers prominent "biomedical" imaging techniques and the essential physics underlying these techniques. This option prevailed because imaging is a powerful, ubiquitous approach used in biology, medicine, and many other scientific disciplines, and because it has strong foundational ties to physics. Moreover, imaging arises frequently in everyday life in connection with the diagnosis and treatment of disease and injury, so students are interested in the topic and appreciate its importance.

A few years after the grant was awarded, Lewis & Clark College debuted "Biomedical Imaging," an upper-division physics course covering optical microscopy, several prominent medical imaging modalities, and (later on) image processing and analysis. The class now has the highest enrollment of the College's upper-division physics courses, in part due to strong enrollment by students majoring in the life sciences and chemistry.

The success of Biomedical Imaging at Lewis & Clark College suggested that other institutions might also profit from offering a similar course. Unfortunately, a major impediment for our institution was the absence of a textbook covering the wide-ranging material at an appropriate level. To eliminate this impediment, a follow-up, textbook-development project was initiated. Draft versions of this textbook were tested successfully for several years at Lewis & Clark College.

This textbook is targeted at undergraduate students with a fairly broad range of backgrounds in physics and mathematics. One year of general college physics and one year of college mathematics (ideally calculus) will suffice for nearly all of the material in the body of the text and some of the material in the optional "Boxes." Thus, most life science and chemistry students can readily use the text. This textbook also requires essentially no background in biology or chemistry and thus is appropriate for physics, engineering, and many math and computer science students. Supplementary material that requires a more advanced background in physics and/or mathematics (e.g., an involved derivation of a key equation) is presented in optional Boxes. The homework problems similarly are targeted at a broad audience; problems that involve more advanced physics and/or mathematics are labeled with an asterisk.

This entire textbook probably is well suited for a (fast-paced) two-quarter course. For a one-semester course, like the one at Lewis & Clark College, instructors can pick and choose the material that they want to cover. This textbook also can be used for a one-quarter course on optical microscopy or medical imaging.

One very popular component of Biomedical Imaging is the hands-on activities that allow students to explore some of the concepts (e.g., diffraction) and implement some of the techniques (e.g., phase-contrast microscopy, ultrasound, tomography, MRI, and image processing) that are covered in the lecture component of the course. These activities are described in this textbook at a reasonable level of detail, together with equipment and image-processing software suggestions. This textbook also includes in-class demonstrations that help to reinforce theoretical concepts introduced in lecture.

Imaging is a tremendously important topic with relevance to the sciences, medicine, technology, art, and "real life." We hope that instructors and students find this textbook useful for teaching and learning this amazing material.

Bethe A. Scalettar
Lewis & Clark College
Portland, OR

James R. Abney
Psi Star Intellectual Property LLC
Portland, OR

Acknowledgments

This textbook owes its existence to several generous organizations. Foremost among these is the Howard Hughes Medical Institute (HHMI), which provided monetary support to develop the course "Biomedical Imaging," which ultimately gave birth to this text. HHMI also funded the purchase of an ultrasound instrument and an optical computed tomography scanner used in hands-on course activities and, several years later, the Sherman Fairchild Foundation similarly contributed a desktop MRI system. Finally, Lewis & Clark College funded the conception and production of ~300 illustrations to enhance the educational value and visual appeal of the text.

A long list of organizations, and individuals, provided additional visual aids, contributing ~30 photographs and ~50 real biomedical images that appear in the text. We acknowledge these contributions individually in the figure legends.

Last, but certainly not least, many generous individuals contributed to the text by reading chapters in their area of expertise for accuracy, clarity, and emphasis. We extend our thanks to these individuals, and a few who contributed in other ways, in the (alphabetical) list below. In addition, we would like to extend a special thanks to Lou Chosen, who initially approached us about writing a textbook, to the many Lewis & Clark College students who piloted draft versions of the text, and to Daniel Axelrod, who read this entire textbook and offered many useful suggestions for its improvement.

Parvaneh Abbaspour
Lewis & Clark College

Steve Attinasi
Lewis & Clark College

Daniel Axelrod
University of Michigan

Thomas Barbara
Oregon Health & Science University

Lou Chosen
Taylor and Francis

Irene Duba
Rockefeller University

John Feleciano
Veterinary Diagnostic Imaging and Consultations

Edward Hsu
University of Utah

Nikolaus Loening
Lewis & Clark College

Edward Ritz
Lewis & Clark College

Richard Scalettar
University of California, Davis

About the Authors

Bethe A. Scalettar is a professor and chair of physics at Lewis & Clark College in Portland, OR. She joined the College in 1993 after receiving an undergraduate degree from the University of California at Irvine (majors, physics and mathematics), a PhD in biophysics from the University of California at Berkeley, and completing postdoctoral work at the University of California, San Francisco. Since arriving at Lewis & Clark College, Bethe's research and publications have focused primarily on elucidating the molecular basis of learning and memory, utilizing fluorescence microscopy as a primary tool. Bethe also has worked to enhance the interdisciplinary appeal of physics, most notably with the development of an undergraduate physics course entitled "Biomedical Imaging" and the writing of this supporting textbook.

James R. Abney is an intellectual property attorney at Psi Star Intellectual Property LLC in Portland, OR. He began practicing law in 1997 after receiving undergraduate degrees in physics and biology from the University of California at Irvine, a PhD in biophysics from the University of California at Berkeley, completing postdoctoral work at the University of California, San Francisco, and earning a J.D. from Lewis & Clark College School of Law. Jim's research and publications, which have continued during his legal career, have focused on the use of biophysical, notably fluorescence-based, techniques to answer questions about cell structure and function. Jim's legal work has spanned multiple technologies useful in science and medicine, including scientific instrumentation (especially light sources, detectors, and optics), medical devices, and biotechnology.

About the Illustrator

Cyan Cowap is an assistant supervisor for the YMCA's Expanded Learning Program, currently working at Empower Language Academy in San Diego, California. She joined the Academy in 2019 after receiving an undergraduate degree from Lewis & Clark College (major in studio art, minor in rhetoric and media studies) and studying for a semester at Queen Mary University of London. During her college years, Cyan held several long-term positions as an illustrator, most notably for the weekly Lewis & Clark College publication, The Pioneer Log. More recently, Cyan has collaborated with Bethe A. Scalettar and James R. Abney to create a textbook with hundreds of esthetically appealing illustrations that enhance understanding of key scientific and medical concepts.

1

Introduction

This chapter presents a brief overview of the history of biomedical imaging (augmented with some engaging anecdotes) and introduces the primary imaging modalities covered in the text. We start with light microscopy, which is the focus of the first section of the text, and finish with medical imaging.

1.1 Historical Overview of Light Microscopy

Light microscopy is very commonly used in both basic scientific research and industry. Light microscopy comprises a spectrum of approaches that use visible electromagnetic (EM) radiation to produce an image of an object (Fig. 1.1). Typically, the object is tiny (e.g., a cell), and the goal is to generate a high-magnification (500–1000×) image with excellent resolution and contrast. (Resolution refers to the ability to discriminate objects that are closely spaced; contrast refers to the extent of gray scale and/or color variation in the image.) The different approaches are distinguished by their methods of generating contrast and/or their resolution. A brief history of light microscopy is presented in Table 1.1.

1.2 Light Microscopic Modalities

This section is intended to provide the reader with a big picture view of the primary light microscopic modalities discussed in this text, without aiming for a thorough understanding. The techniques in Sections 1.2.1–1.2.4 differ primarily in their method of generating contrast. The techniques in Sections 1.2.5–1.2.7 use fluorescence (light emission) as a source of contrast and differ primarily in their degree of image blur and/or their resolution.

1.2.1 Brightfield

Brightfield is the most straightforward light microscopic approach. White light is transmitted through a sample, and contrast in the image is based on intensity (amplitude) modulations arising from differential absorption by the specimen, which appears dark or colored against a bright background (Fig. 1.2). Brightfield works well on intrinsically absorbing or stained specimens, which are designated "amplitude" specimens to emphasize their effect on light. Unfortunately, most cells and tissues are relatively transparent and

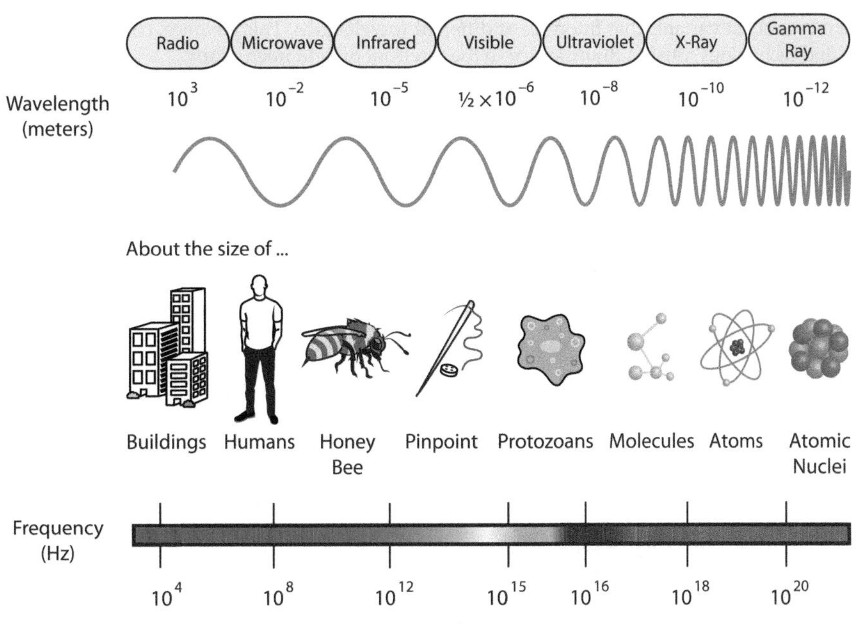

FIGURE 1.1
The EM spectrum and benchmark reference data. (Adapted from National Aeronautics and Space Administration [NASA].)

DOI: 10.1201/b22076-1

TABLE 1.1

Timeline of Some Key Developments in Light Microscopy

1600s	Galileo Galilei, Robert Hooke, and Antonie van Leeuwenhoek construct light microscopes. Hooke's observations with microscopes lead to the introduction of the word "cell," Hooke's term for the pores of cork. van Leeuwenhoek, a cloth merchant, constructs microscopes capable of achieving an amazing 200× magnification using a single lens, which functioned essentially as a magnifying glass. In addition to its superb magnification, van Leeuwenhoek's single-lens configuration was relatively free from the aberrations that plagued the two-lens microscopes of his contemporaries.
1800s	Ernst Abbe and Lord Rayleigh quantify the effects of diffraction on resolution. Abbe and Rayleigh independently predict that objects more closely spaced than ~200 nm will merge into a single unresolvable spot in the image produced by a light microscope. Around the same time, August Köhler develops a method of homogenous sample illumination that is still widely used today.
	Joseph von Gerlach, Camillo Golgi, Paul Ehrlich, and Hans Christian Gram pioneer the use of stains to enhance the contrast of cells and tissues, which intrinsically have low contrast because they essentially are transparent.
1900s	*1900s:* August Köhler builds the first ultraviolet (UV) microscope to achieve improved diffraction-limited resolution and notices that some objects fluoresce under UV illumination.
	1910s: Inspired by Köhler's observation, Oskar Heimstaedt builds the first fluorescence microscope. However, fluorescence microscopy becomes an important technique only in the latter part of the 20th century, after several key advances. These include the development of epi-fluorescence in the 1920s, the commercialization of the dichromatic mirror in the 1960s, and a surge in methods of fluorescence tagging that began in the 1980s and exploded in the 1990s.
	1930s: Frits Zernike develops phase-contrast microscopy, which provides high-contrast images of unstained living cells. Wilhelm Schmidt solidifies the potential of polarization microscopy, which has been in use for several years, when he visualizes the mitotic spindle in developing cells.
	1950s: Francis Smith and Georges Nomarski introduce differential interference contrast (DIC) microscopy, which also provides high-contrast images of unstained living cells. Marvin Minsky, a pioneer in the field of artificial intelligence, builds and applies for a patent on the first scanning confocal microscope, but the approach is not widely adopted for several decades.
	1960s: Osamu Shimomura isolates green fluorescent protein (GFP) from the jellyfish *Aequorea victoria* and observes that the protein fluoresces green upon irradiation with UV light. The full impact of this discovery would await advances achieved three decades later.
	1980s: David Agard and John Sedat develop deconvolution (software-based) methods of removing out-of-focus signal (a source of blur) from fluorescence images. Contemporaneously, the confocal microscope becomes another widely adopted (hardware-based) method of deblurring fluorescence images.
	Daniel Axelrod builds a total internal reflection fluorescence (TIRF) microscope that generates exceptionally blur-free images of structures, such as cell membranes, that adhere to (or are within ~100 nm of) a glass coverslip.
	1990s: Winfried Denk and co-workers, and Ernst Stelzer and co-workers, demonstrate that two-photon and light sheet microscopies are powerful fluorescence-based methods of imaging deep within intact tissues and live animals over long timescales.
	Martin Chalfie and co-workers demonstrate that native GFP can be used as a fluorescent reporter for gene expression and protein localization in living cells and organisms. Shortly thereafter, Roger Tsien and co-workers mutate native GFP and produce more robust variants that fluoresce at different wavelengths. This advance, coupled with other advances in labeling and technology, pushes fluorescence to the forefront of light microscopy.
2000s	Pioneering work by Stefan Hell, Eric Betzig, William Moerner, and their co-workers culminates in the introduction and popularization of stimulated emission depletion (STED) microscopy and photoactivated localization microscopy (PALM). These and related approaches significantly break the 200-nm limit on resolution in light microscopy. Contemporaneously, Mats Gustafsson and co-workers develop another super-resolution fluorescence technique, structured illumination microscopy (SIM).

must be fixed and stained for brightfield imaging. Thus, brightfield generally is not used to study living cells and tissues. This limitation was an impetus for the development of alternative approaches.

1.2.2 Darkfield

Darkfield microscopy is a simple variant on brightfield that bypasses the need for staining. Background illumination is excluded, so the field is dark. As a consequence, specimen contrast is greatly enhanced, similar to star visibility at night. Darkfield is an effective method of examining external details and small specimens that diffract or refract a small amount of light (Fig. 1.2). Disadvantages of darkfield include a need for intense illumination, because the signal is weak, and image contamination by scattering debris.

1.2.3 Phase-Contrast and Differential Interference Contrast

Phase-contrast microscopy is a sophisticated, Nobel Prize-winning technique that markedly improves the visibility of "phase" specimens, such as unstained living cells. Phase specimens primarily alter wave

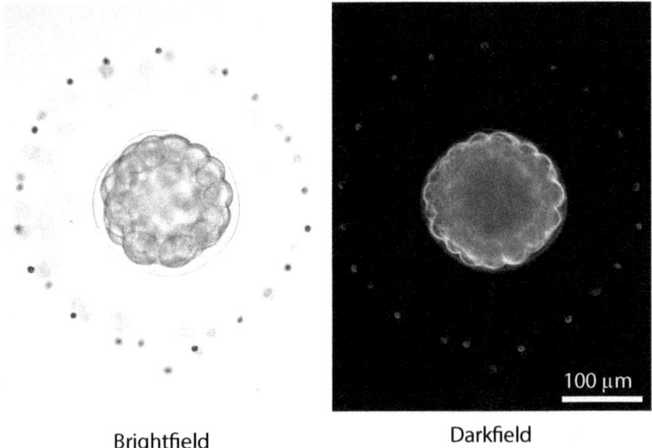

FIGURE 1.2
Comparison of brightfield and darkfield images of a sand dollar embryo. Contrast in the brightfield image was enhanced by partially closing the condenser aperture diaphragm, as discussed in Chapter 3. Bar = 100 µm. (Images courtesy of Michelangelo von Dassow, Ph.D.)

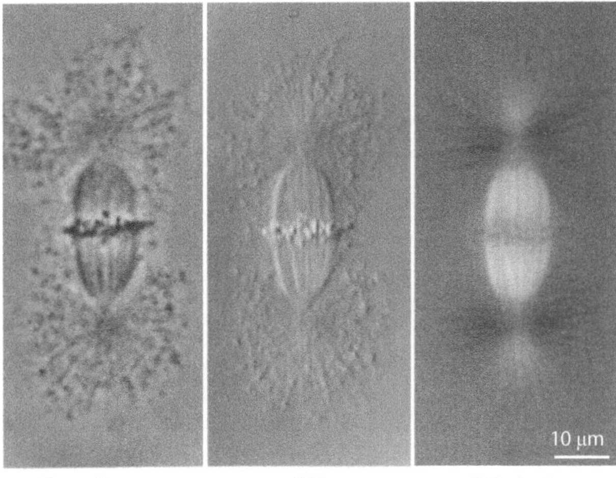

FIGURE 1.3
Comparison (left to right) of phase-contrast, DIC, and polarization microscopy images of an isolated mitotic spindle from sea urchin eggs at early metaphase. Bar = 10 µm. (Adapted with permission from Salmon, E.D., et al. 1980. Calcium-labile mitotic spindles isolated from sea urchin eggs (*Lytechinus variegatus*). *J. Cell Biol.* 86:355–365. https://doi.org/10.1083/jcb.86.2.355.)

phase (e.g., the location of crests) by slowing light propagation and are invisible under standard brightfield conditions because the eye is not sensitive to wave phase.

The goal of the phase-contrast method is to convert phase specimens into amplitude specimens (Section 1.2.1). To this end, the light waves that create the image of a phase specimen are manipulated to generate destructive interference. The outcome is a net image wave of reduced amplitude, and thus the phase specimen creates amplitude contrast (Fig. 1.3). Phase contrast is a powerful method of visualizing living cells but is subject to halo artifacts and has somewhat compromised resolution. The approach also works best for relatively thin, transparent specimens.

Differential interference contrast (DIC) is another sophisticated, interference-based modality that is well suited to imaging unstained living cells (Fig. 1.3). Advantages of DIC over phase contrast include higher resolution, better sectioning, and an absence of halos. On the other hand, DIC is not well suited to the study of cultured cells in plastic dishes, and it requires expensive, specialized optics.

1.2.4 Polarization Microscopy

Polarization microscopy is an additional approach that can be used to generate high-contrast images of living specimens. However, the method only works with specimens that are optically anisotropic. Crystals are one example, and so applications of polarization microscopy in geology are numerous. Biological specimens suited for polarization microscopy are less common but include very important structures, such as mitotic spindle fibers (Fig. 1.3). For suitable samples, polarization microscopy provides excellent image quality relative to the approaches described above, and it can be used to extract quantitative data. The images also can be beautifully colored.

1.2.5 Basic Fluorescence Microscopy

Fluorescence microscopy includes a broad spectrum of approaches. In all variants, the specimens fluoresce (emit radiation), primarily in the visible. Fluorescent specimens emit (emission) in response to energy input (excitation), which is supplied by shorter wavelength visible radiation. Because the input light is spectrally distinct, it can be blocked using filters. The result is a high-contrast image showing a fluorescence signal against a dark background.

Most biological samples are not intrinsically fluorescent, and thus samples must be labeled with fluorescent tags. The tags typically are designed to interact with specific constituents (e.g., in cells), so well-defined structures fluoresce (Fig. 1.4). Fluorescence microscopy dominates the biological sciences because the approach is sensitive, selective, and compatible with multi-color imaging of living specimens.

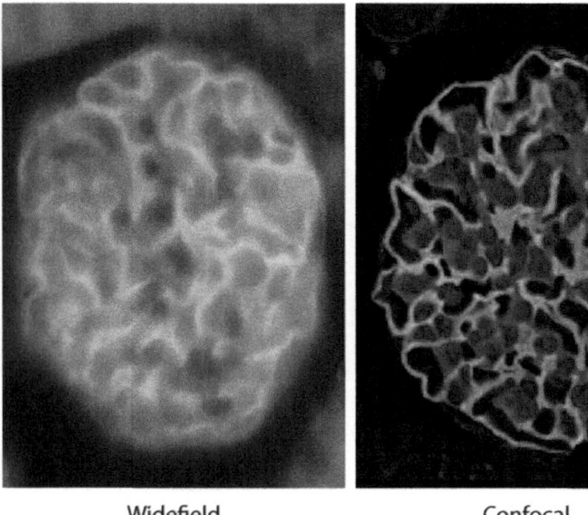

FIGURE 1.4
Comparison of widefield (left) and confocal (right) images of an ~20 μm-thick section of mouse kidney. Nuclei were stained with DAPI (blue), the cell membrane was stained with Alexa 488-wheat germ agglutinin (green), and actin filaments were stained with Alexa 568-phalloidin (red). Bar = 20 μm. (Adapted from Jonkman, J., et al. 2015. Any way you slice it – a comparison of confocal microscopy techniques. *J. Biomol. Tech.* 26:54–65. DOI: 10.7171/jbt.15-2602-003. With permission from Association of Biomolecular Resource Facilities.)

The main weakness of fluorescence microscopy is a need for intense excitation light, which causes bleaching and can also be phototoxic. In addition, images of thick specimens can be corrupted by out-of-focus blur.

There are two standard forms of fluorescence microscopy. The most basic form, termed "widefield" microscopy, generally illuminates all positions and all depths of the sample simultaneously (Fig. 1.4). An alternative form, termed "scanning" microscopy, illuminates the sample point-by-point via sequential scanning. A prominent form is laser-scanning confocal microscopy, which generates images with markedly reduced blur by scanning the sample with a focused illumination spot and then rejecting out-of-focus fluorescence using a pinhole filter (Fig. 1.4).

1.2.6 Axially Selective Illumination Methods

Both widefield and confocal microscopy involve illuminating the entire sample depth, with accompanying photobleaching and phototoxicity. There is another set of techniques, including total internal reflection fluorescence (TIRF), light sheet, and two-photon microscopies, that reduce blur by modifying the three-dimensional pattern of excitation light to avoid illumination of out-of-focus planes. These techniques can reduce both photobleaching and phototoxicity and, as a result, are better suited for long-term imaging of live specimens.

1.2.7 Super-Resolution Methods

Beginning in the 1990s, several fluorescence-based techniques were developed that surpass the diffraction limit on resolution. These include two sequential readout approaches, stimulated emission depletion (STED) microscopy and photoactivated localization microscopy (PALM), and a technique known as super-resolution structured illumination (SR-SIM).

A hallmark of the diffraction barrier is that simultaneously fluorescing molecules cannot be resolved if they are closer together than ~200 nm. This is because the images of closely spaced objects overlap and blur together. However, the overlap problem can be overcome if the images of closely spaced fluorophores are collected sequentially. This trick is the basis for STED and PALM, which can markedly beat the diffraction barrier. SR-SIM is based on a different trick and achieves more modest resolution enhancement. On the other hand, unlike STED and PALM, SR-SIM places relatively few special demands on fluorophores and thus is best suited among super-resolution techniques to multi-color imaging.

1.3 Historical Overview of Medical Imaging

Medical imaging comprises a spectrum of approaches that typically are used to assay for structural and/or functional changes associated with trauma or disease. What follows is a summary of key events in the development of four prominent medical imaging modalities that comprise the second section of the text. These are (1) ultrasound (US), which utilizes high-frequency sound waves, (2) projection radiography and computed tomography (CT), which utilize X-rays, (3) planar and tomographic nuclear imaging, which utilize gamma rays, and (4) magnetic resonance imaging (MRI), which utilizes radiofrequency radiation. Visible radiation does not penetrate the body well and thus plays a less prominent role in medical imaging. This reflects the fact that photon energies of visible light are well matched to energy spacings associated with electronic transitions, leading to strong absorption. The relative penetration capacities of different types of EM radiation, together with key interactions, are summarized in Table 1.2. A brief history of medical imaging is presented in Table 1.3.

TABLE 1.2

Penetration of the Body by EM Radiation

Type of Radiation	Extent of Absorption and Effect	Mechanism
Radio	Transparent; body is commonly penetrated by radio waves	Absence of matched energy spacings
Microwave	Largely transparent; tissue heating	Low absorption due to a match between photon energy and energy spacings associated with molecular rotation
Infrared	Penetrates a cm or more into skin/soft tissue; tissue heating	Greater absorption due to a match between photon energy and energy spacings associated with molecular vibration
Visible	Penetrates several mm into skin; tissue heating	Strong absorption due to a match between photon energy and energy spacings associated with electronic transitions of chromophores, including hemoglobin and melanin
Ultraviolet	Very low penetration of skin; shorter wavelength UV can ionize (knock electrons free)	Strong absorption arising from chromophores that include proteins and DNA
X-rays and Gamma rays	Relatively transparent; ionizing radiation	Photon energies too high to be matched to electronic transitions; interactions restricted to ionization

TABLE 1.3

Timeline of Some Key Developments in Medical Imaging

1800s	*1890s:* Wilhelm Röntgen discovers X-rays and sets the stage for the field of diagnostic radiography by generating the first radiographs of human bones. A notable example is a famous image that shows bones in the hand of Röntgen's wife and two rings on one of her fingers (Section 1.4.2). Röntgen did not patent his discoveries, which allowed a rapid surge in clinical, X-ray-based imaging. Unfortunately, the potential dangers of X-rays were not appreciated initially, and patients in some cases were exposed for hours, leading to radiation burns and other health problems.
	Inspired in part by Röntgen's work, Antoine Becquerel and Marie and Pierre Curie make groundbreaking advances on radioactivity that ultimately play a pivotal role in nuclear medicine.
1900s	*1910s/1920s:* William Coolidge invents an X-ray tube based on thermionic emission, which is the forerunner of all tubes still in use today. The Coolidge tube made it possible to control key beam parameters, such as intensity, and led to a marked decrease in exposure times. During this era, radiopaque barium is introduced as an agent to enhance contrast in (time-lapse) radiographs of the gastrointestinal tract.
	In response to the sinking of the Titanic, Paul Langevin invents an early sound transducer, which was used to detect submarines during World War I. This development helped set the stage for medical US.
	1930s: Carlo Perrier and Emilio Segre discover technetium-99*m*, which would become the dominant tracer in nuclear medicine following the development of the technetium-99*m* generator in the late 1950s.
	1940s: Edward Purcell and Felix Bloch discover nuclear magnetic resonance (NMR), setting the stage for MRI. Karl Dussik tries to detect brain tumors using US.
	1950s: Hal Anger develops a scintillation camera that still is utilized widely in nuclear imaging. Planar nuclear imaging blossoms and is applied to studies of the kidneys, skeleton, and heart.
	Ian Donald and Tom Brown pioneer clinical US, introducing the technique in obstetrics and gynecology. Donald and Brown also build a prototype instrument and thus sometimes are credited with inventing US.
	1960s: Basil Hirschowitz's work on the glass fiber endoscope in the 1950s leads to widespread production and use in the 1960s. Novel applications include real-time imaging of the interior of the stomach.
	David Kuhl and Roy Edwards introduce single-photon emission computed tomography (SPECT).
	1970s: Godfrey Hounsfield and Allan Cormack invent CT. Some credit also has been given to the Beatles because Hounsfield was an engineer at Electric and Musical Industries (EMI), the company that also signed the Beatles. This has led to (controversial) claims that revenue from Beatles albums in the 1960s helped Hounsfield pursue his work.
	Paul Lauterbur and Peter Mansfield invent MRI. Ironically, the journal Nature initially rejected Lauterbur's paper describing the invention, in part, because the images had low resolution. Even more ironically, the State University of New York failed to patent Lauterbur's invention out of concern that it would not generate sufficient revenue.
	Michael Phelps invents positron emission tomography (PET) and develops the first instrument in collaboration with Edward Hoffman. A few years later, Tatsuo Ido synthesizes ^{18}F-fluorodeoxyglucose, which would become the dominant PET tracer. However, the expense and somewhat limited resolution of PET hinder widespread use for a few decades.
	1980s: Willi Kalender develops spiral CT, which permits three-dimensional data acquisition in a breath hold. MRI and PET become clinical tools.
	1990s: Bruce Hasegawa develops SPECT/CT, and David Townsend and Ronald Nutt develop PET/CT. This fusion of functional and anatomical imaging was a major advance. Toshiba, Siemens, and General Electric introduce multi-slice CT detectors in the late 1990s. This innovation further accelerates data acquisition in CT.
2000s	Functional MRI, which was invented in the 1990s by Seiji Ogawa and co-workers, emerges as an important method of monitoring the activity of the human brain and obtaining valuable insight into how the mind works.
	Digital mammography is introduced. The approach produces images of better quality and with reduced radiation dose.
	Nuclear imaging experiences a renaissance due to its pivotal role in personalized medicine.

1.4 Medical Imaging Modalities

This section, like Section 1.2, contains a brief overview of the medical imaging modalities discussed in the text.

1.4.1 Ultrasound

US is the medical analog of sound navigation and ranging (SONAR). The approach is atypical in bypassing the use of EM radiation. Images are generated by directing a pulse of high-frequency sound into the body and detecting echoes that reflect or scatter from body interfaces and structures. Object depth is determined from the time that elapses between pulse initiation and echo detection. Object brightness (i.e., contrast) is determined by the strength of the echo, which depends on physical attributes of the object and its surroundings. US is preferred in obstetrics because it is extremely safe and is capable of real-time imaging (Fig. 1.5). US also is relatively inexpensive and portable and provides excellent images of soft tissues and organs. On the other hand, US is not well suited to the study of air-filled cavities or structures encased in bone, like the brain.

1.4.2 Conventional Radiography and Computed Tomography

Conventional radiography and CT are closely related techniques. Images are generated by detecting X-rays that traverse a patient positioned between a source and a detector. X-ray transmission, and thus the detected signal, is high for radiolucent regions of the body, which lack X-ray absorbing structures. In contrast, X-ray transmission and the detected signal are weak for radiopaque regions, such as bone. In images, radiolucent areas appear black, and radiopaque areas appear white (Fig. 1.6).

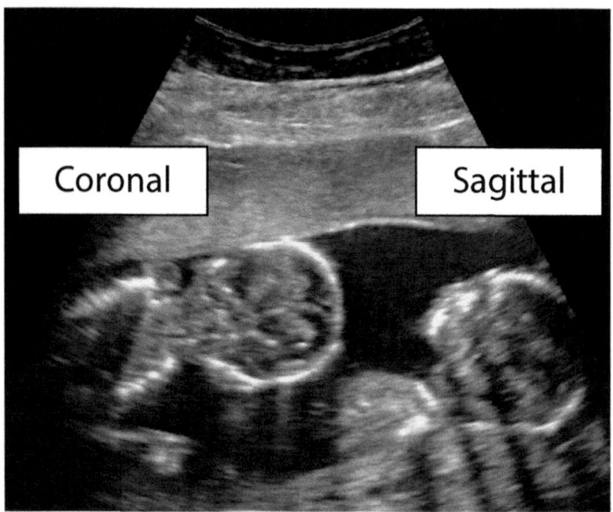

FIGURE 1.5
US image of identical twins at a gestational age of 15 weeks. (Adapted from Mikael Häggström. Creative Commons CC0 1.0 Universal Public Domain Dedication.)

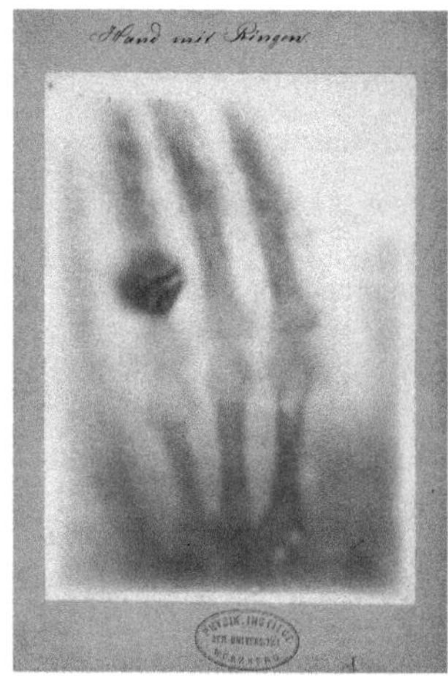
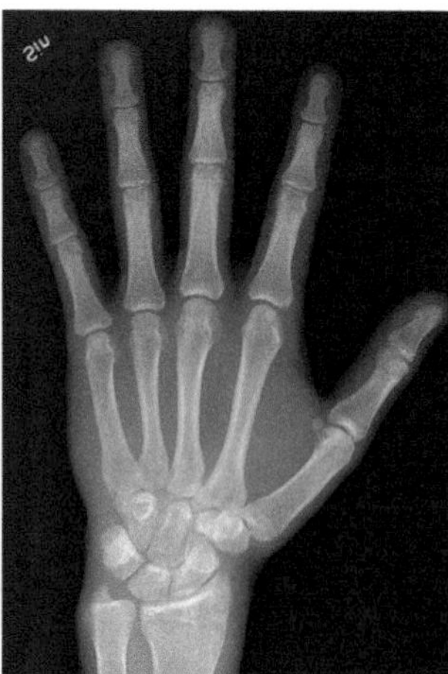

FIGURE 1.6
Comparison of planar radiographs of a hand from the late 19th and the early 21st centuries. The historic 1895 radiograph by Wilhelm Röntgen shows the hand of his wife, Anna Bertha Ludwig, revealing the bones of her hand and a ring on one finger. (Image courtesy of Wellcome Collection. Attribution-NonCommercial 4.0 International [CC BY-NC 4.0].) The modern radiograph shows the hand of a 31-year-old man. (Adapted from Mikael Häggström. Creative Commons CC0 1.0 Universal Public Domain Dedication.)

In conventional radiography, each image point reflects summed X-ray attenuation along an associated straight-line trajectory between the source and detector. The resulting "projection" images lack depth information and suffer from structural superposition. These weaknesses are overcome in CT. CT images are generated by collecting projections at many viewing angles and using the different views to reconstruct high-resolution, depth-preserving two-dimensional slices and three-dimensional images of anatomical structures in a patient. CT also has substantially improved ability to reveal soft tissues.

Despite its limitations, conventional radiography has many uses, including detecting broken bones and cancer of the lung and breast. CT is used for many of these same imaging purposes, and it has additional applications, including planning radiation therapy. CT also essentially has eliminated a once common practice: exploratory surgery. A primary weakness of projection radiography and CT is their use of ionizing radiation; another is a relatively limited ability to monitor function.

1.4.3 Planar and Tomographic Nuclear Imaging

Planar and tomographic nuclear imaging are emission analogs of projection radiography and CT, and they similarly utilize ionizing radiation. To generate an image, the patient is administered a radiopharmaceutical, which consists of a radioactive substance attached to a pharmaceutical. The radiopharmaceutical preferentially localizes to a particular site in the body, which becomes a source of gamma-ray photons generated by nuclear decay events. The gamma rays escape the body (ideally without interacting with body structures) and generate an image of the distribution of the radiopharmaceutical.

Detection of emitted, instead of transmitted, radiation is one of several key differences between nuclear and X-ray-based imaging. Another important difference is that nuclear imaging typically assays for disease-associated changes in function, which are manifest as altered radiopharmaceutical localization (Fig. 1.7). Common applications include monitoring increased metabolism of glucose analogs by cancerous cells and reduced metabolism in regions of the brain affected by cognitive impairment. Nuclear images generally have poorer resolution than X-ray images, but their contrast typically is very good.

1.4.4 Magnetic Resonance Imaging (MRI)

MRI is an imaging variant of NMR. MRI is renowned for its high-resolution, exquisite soft-tissue contrast, ability to monitor anatomy and function, and safety (Fig. 1.8).

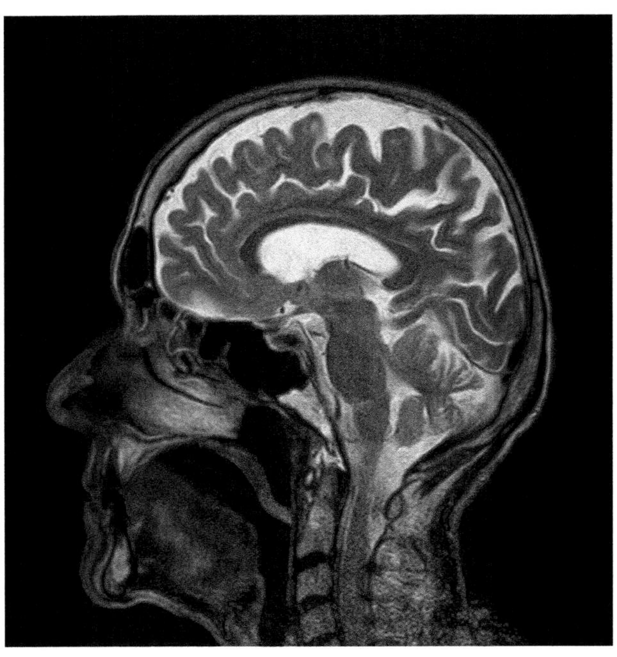

FIGURE 1.8
MRI image of a human head. (Image courtesy of Pixabay.)

MRI utilizes an unusual localization strategy to visualize protons in water in living tissue: signals arising from protons in different regions in the body are distinguished, and their distinct points of origin are identified, based on frequency. This approach, which is known as frequency encoding, is implemented by placing protons in a spatially varying magnetic field, which creates signals with frequencies that are determined by their location.

Similar to other approaches, image contrast in MRI is based on differences in signal amplitudes. One

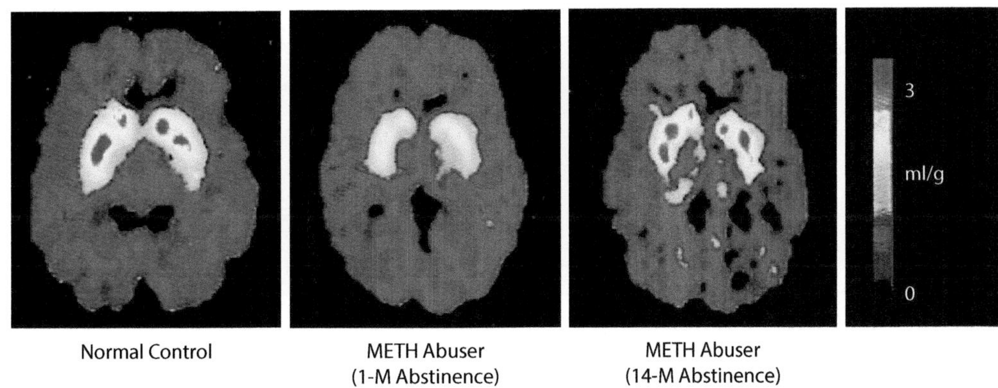

FIGURE 1.7
Emission tomography images of radioactive dopamine distribution in a normal brain and the brain of a recovering methamphetamine addict 1 and 14 months into abstinence. With time after quitting, dopamine signal is higher, indicating that dopamine transporters have regenerated. (Adapted with permission from Volkow, N.D., et al. 2001. Loss of dopamine transporters in methamphetamine abusers recovers with protracted abstinence. *J. Neurosci.* 21:9414–9418. DOI: https://doi.org/10.1523/JNEUROSCI.21-23-09414.2001. Copyright 2001 Society for Neuroscience.)

parameter that affects signal amplitude is proton density, which can be used as a source of contrast. However, soft tissues tend to have similar proton densities; thus "density-weighted" images are relatively low in contrast. The strength of MRI for soft-tissue imaging stems in part from the fact that the MRI signal amplitude can be made to depend on other, more variable parameters that generate much better contrast. Weaknesses of MRI include relatively slow data acquisition, high equipment cost, and incompatibility with metal implants.

Suggested Reading and Additional Resources

Bushberg, J.T., Seibert, J.A., Leidholdt, E.M., and Boone, J.M. 2012. *The essential physics of medical imaging (3rd Edition)*. Wolters Kluwer | Lippincott Williams & Wilkins, Philadelphia, PA. Chapter 1.

Freiherr, G. 2014. The eclectic history of medical imaging. Feature: Imaging Technology News. https://www.itnonline.com/article/eclectic-history-medical-imaging

Nature Publishing Group. 2009. Timeline: nature milestones | light microscopy. Available in Support Material.

2

Review of Essential Basics

Biomedical imaging techniques are founded extensively on interactions between waves, notably EM or sound waves, and matter, such as glass and tissue. We therefore begin with a brief review of basic properties of waves, including their mathematical description, and the most important consequences of interactions between waves and matter. Tailored specifics will be presented later, at appropriate points in the text.

2.1 Attributes of Waves

A traveling wave is a self-sustained disturbance that propagates from one location (e.g., its site of origin) to another. In the case of sound waves, the disturbance is a perturbation from the equilibrium positions of molecules in a solid medium (or alternatively in the associated pressure in a liquid or gas medium) created by a source, such as an instrument. In the case of EM waves, the disturbance is a changing EM field that might originate in an accelerating charge undergoing oscillatory motion.

Waves can exhibit many kinds of spatial variation. To begin, we focus on a simple, one-dimensional, traveling harmonic disturbance $y(x,t)$, where y could represent a pressure disturbance or the magnitude of an electric field (Box 2.1). An exemplary wave is depicted in Fig. 2.1. The corresponding mathematical description is

$$y(x,t) = y_0 \sin(kx \pm \omega t + \varphi) \tag{2.1}$$

The quantity y_0 is called the amplitude of the wave and determines its "height." The quantity $\theta = kx \pm \omega t + \varphi$ is called the phase of the wave and is measured in radians. The value of θ determines the magnitude of the disturbance at a given position and time. The quantity φ is a "phase shift" relative to a wave with phase $kx \pm \omega t$. The quantities $k = 2\pi/\lambda$ and $\omega = 2\pi\nu$ are the wave number and angular frequency, respectively, where λ and ν are the wavelength and frequency of the wave (Fig. 2.1). This wave is monochromatic (single λ) and propagates to the right when the minus sign is chosen and to the left when the plus sign is chosen. The wave propagates at speed v, where $\mathrm{v} = \lambda\nu$.

A phase shift between two EM waves can be generated when the two waves propagate in media of different refractive indices, n (Section 2.2), while traversing a common distance, Δx. The phase shift arises because the waves have different wavelengths in the two media, shifting the relative alignment of crests and troughs (Fig. 2.2). Specifically, the wavelength in each medium is $\lambda = \lambda_0/n$, where λ_0 is the vacuum wavelength (Section 2.2.1.5). Relatively simple arguments (Box 2.2) show that the phase shift, φ, is given by

$$\varphi = \frac{2\pi}{\lambda_0}(n_2 - n_1)\Delta x \tag{2.2}$$

where n_1 and n_2 are the two refractive indices. Equation (2.2) will figure prominently in our discussions of interference-based microscopy and polarization microscopy.

Waves can have more complicated spatial variation and dimensionality. Two very important examples, spherical and plane waves propagating in three dimensions, are shown in Fig. 2.3. Spherical waves are emitted by point sources; their amplitudes vary as $1/r$, where r is distance from the source, for $r \gg \lambda$. For closer

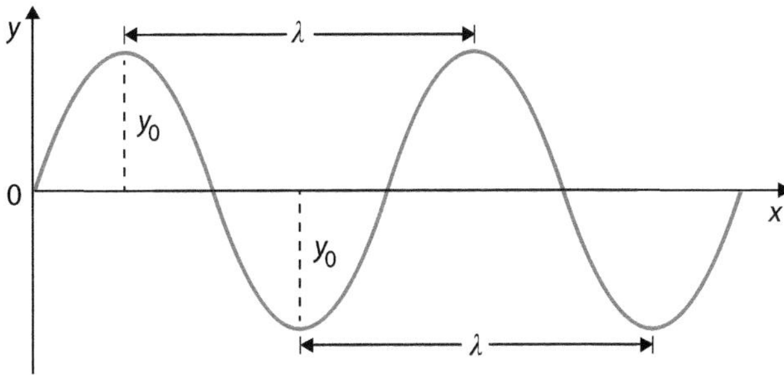

FIGURE 2.1
Snapshot of a traveling wave as a function of position for a fixed time, $t = 0$, and phase, $\varphi = 0$. The wavelength, λ, is the distance over which the wave's shape repeats, typically measured as a crest-to-crest or trough-to-trough distance. The period, T, (not shown) is the time it takes for one complete wave form to pass a given position. Finally, the frequency, ν, is the reciprocal of the period: the number of wave forms that pass any given position per unit time.

DOI: 10.1201/b22076-2

BOX 2.1 TRAVELING SOUND AND EM WAVES

A harmonic sound wave propagating along the x axis can be described in terms of spatial and temporal fluctuations in pressure $\Delta p(x,t)$ about equilibrium (e.g., atmospheric pressure ~10^5 Pa [Pascals]). These fluctuations are called compressions and rarefactions. The mathematical description is given below, where the fluctuation amplitude, Δp_{max}, typically is very small, ~10^{-2} Pa, for audible sounds. The speed of propagation is 343 m/s in air and 1540 m/s in water.

$$\Delta p(x,t) = \Delta p_{max} \sin(kx \pm \omega t + \varphi)$$

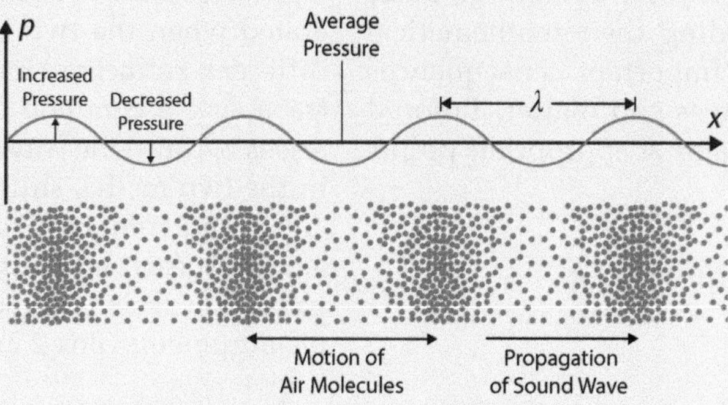

A harmonic plane EM wave propagating along the x axis has a more complex description. In particular, the EM wave consists of separate electric (E) and magnetic (B) fields. Moreover, these fields are both vectors that are perpendicular to the direction of propagation (here x) and to each other, so if the electric field is along y, the magnetic field is along z. (In subsequent sections, we typically will assume that the electric field propagates along z.) The mathematical description of these two disturbances and an associated graph are given below:

$$\vec{E}(x,t) = E_0 \sin(kx \pm \omega t + \varphi)\hat{y}$$

$$\vec{B}(x,t) = B_0 \sin(kx \pm \omega t + \varphi)\hat{z}$$

where $\hat{y}$ and $\hat{z}$ are unit vectors along the y and z axes, respectively. The direction of the electric field (here $\hat{y}$) is referred to as the direction of polarization. Polarization of EM radiation is discussed in Section 2.4 and will figure prominently in our discussions of polarization microscopy and DIC microscopy.

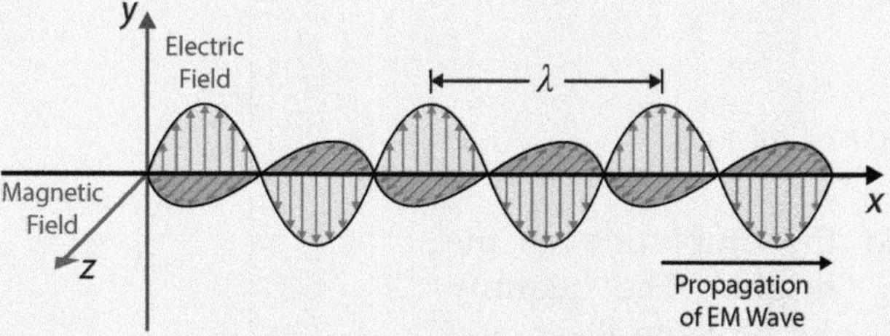

The speed of propagation in vacuum is $c = \lambda \nu = 3 \times 10^8$ m/s. If the EM wave instead propagates in a homogeneous, transparent medium, such as glass, its propagation speed and wavelength will be reduced by a parameter known as refractive index (also known as the index of refraction), as described in the next section.

Sound waves and EM waves are examples of "longitudinal" and "transverse" waves, respectively. The disturbance associated with a longitudinal wave occurs along the direction of propagation. In contrast, the disturbance associated with a transverse wave occurs perpendicular to the direction of propagation.

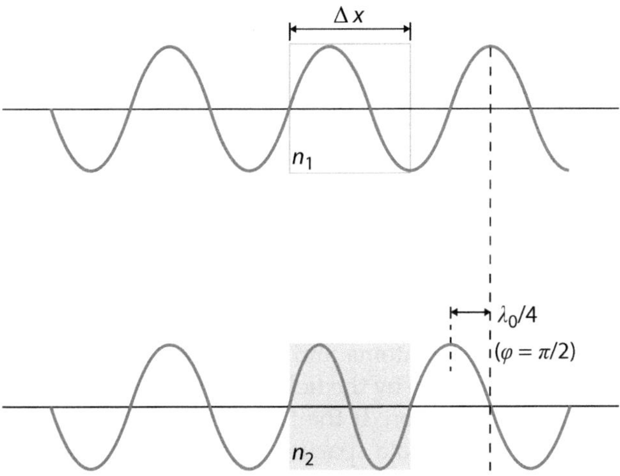

FIGURE 2.2
Phase shift between two waves induced by a disparity in refractive index over a distance of Δx. The wavelength of the lower wave is only 3/4 that of the upper wave in region Δx due to a higher refractive index in Δx. As a consequence, the two waves migrate out-of-phase by $\lambda_0/4$ or, equivalently, $\varphi = \pi/2$ rad, after traversing Δx.

distances, the amplitude includes significant $1/r^2$ and $1/r^3$ dependences. Plane waves can be generated by a wide stack of closely packed, parallel wires; their amplitude is constant. In addition, very far from a point source, spherical waves can be approximated as planar because their curvature becomes negligible (Fig. 2.3).

Fig. 2.3 also depicts the concepts of wavefronts and rays, which play key roles in graphical representations

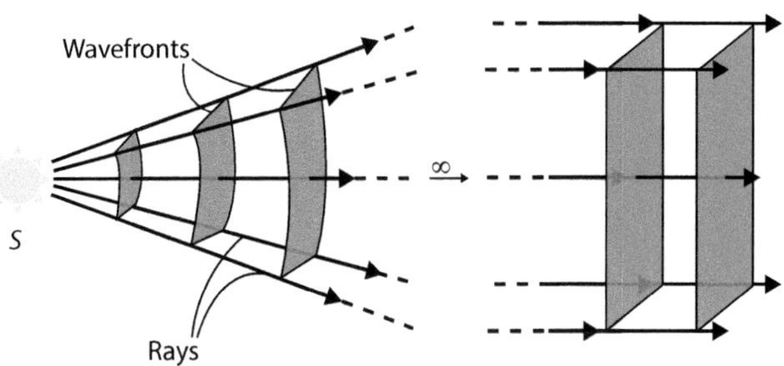

FIGURE 2.3
Schematic representations of diverging spherical waves and plane waves showing their wavefronts and rays.

of waves in two and three dimensions. Rays also are very important in geometrical optics. Wavefronts are surfaces of constant phase, θ; thus, the wave magnitude is the same at every point on a wavefront if the amplitude has the same fixed value at every point on the wavefront. This is true for both spherical and plane waves. Rays, in contrast, are arrows that point in the direction of energy propagation and are perpendicular to the wavefronts in isotropic media.

One of the most important attributes of waves is that they possess energy. For mechanical waves, like sound waves, which propagate in media, the energy has a kinetic (motion) component and a potential (deformation of the medium) component. In such cases, wave energy is described classically and can be quantified in terms of time-averaged intensity, I, which is energy per unit time per unit area (W/m^2).

BOX 2.2 PHASE SHIFTS AND OPTICAL PATH DIFFERENCE

Consider two simple one-dimensional waves, y_1 and y_2, that have different wavelengths, and thus different wave numbers, k_1 and k_2, over a distance Δx. We can write the waves as

$$y_1 = y_0 \sin(kx + k_1\Delta x - \omega t)$$

$$y_2 = y_0 \sin(kx + k_2\Delta x - \omega t)$$

The phase shift, φ, between the two waves is the difference in the arguments of the sine functions:

$$\varphi = (k_2 - k_1)\Delta x = \frac{2\pi}{\lambda_0}(n_2 - n_1)\Delta x$$

where we have $\lambda = \lambda_0/n$ (Section 2.2.1.5). Phase effects often are discussed using a quantity known as optical path difference (OPD), which lacks the constants that appear in the phase shift. OPD is thus defined as

$$\text{OPD} = (n_2 - n_1)\Delta x$$

Similarly, optical path length is defined as OPL = nx (i.e., the product of geometric distance and refractive index). Thus, in Fig. 2.2, the lower wave has a longer OPL in region Δx.

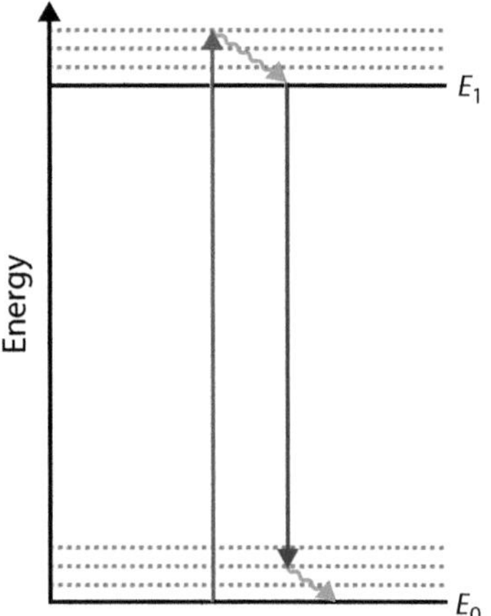

FIGURE 2.4
Schematic representation of quantized energy states and their involvement in fluorescence. An excited electronic energy state (E_1) and (often) excited vibrational energy levels (dotted lines) of a molecule serve as intermediates for fluorescence. Moreover, fluorescence emission (blue arrow) typically is preceded by nonradiative loss (orange arrow) of vibrational energy. Thus, the associated excitation (purple arrow) typically is of shorter wavelength than the resulting emission.

One important, generally valid attribute of intensity is its dependence on the square of the amplitude. For a harmonic wave as described by Eq. (2.1), this implies

$$I \propto y_0^2 \tag{2.3}$$

We will encounter some cases involving EM waves, notably X-ray-based imaging and MRI, where a classical analysis of energy is insufficient. To analyze these techniques adequately, it is necessary to invoke the fact that the energy states of atoms/molecules and nuclei are quantized, as shown in Fig. 2.4, as well as the fact that the energy in the EM field is quantized in "photon" units as

$$E_{\text{photon}} = h\nu \tag{2.4}$$

where $h = 6.63 \times 10^{-34}$ J.s is Planck's constant. Equation (2.4) shows that higher frequency, shorter wavelength photons have more energy than lower frequency, longer wavelength photons.

2.2 Interactions of Waves and Matter

2.2.1 Qualitative Description

When a wave interacts with matter, common outcomes include *absorption, emission, scattering, transmission,* and a *change in wave speed.*

2.2.1.1 Absorption

Absorption is the transfer of wave energy to matter. To get a feel for the physical origin of this effect, consider an ultrasound (US) wave propagating in a watery medium. The wave sets the water molecules into motion, and they rub past one another, transferring some of the energy in the US beam to the medium, where it is dissipated as thermal energy. This is referred to as (dissipative) *absorption* of the US beam.

In a similar spirit, when EM waves impinge on matter, electrons in atoms and molecules are set (classically) into motion by the field, resulting in a change in charge distribution. If the frequency of the radiation also is matched to a spacing ΔE between energy levels of an atom or molecule (Fig. 2.4), so that $\Delta E = h\nu$, the energy in the field can be absorbed, exciting the atom or molecule into a higher energy state. As with US, energy absorbed from an EM field usually is lost via dissipative collisions and converted into random kinetic energy of surrounding molecules.

2.2.1.2 Emission

Absorption of EM radiation can be followed by release of at least some energy radiatively, in the form of *emission*. One prominent example of emission is fluorescence (Fig. 2.4), which occurs when an electron in an excited state makes a downward radiative transition, typically ~1–10 ns after absorption. For molecules, the emitted photon always has a lower energy than the one that was absorbed; the energy is not "lost" but becomes spread out as vibrational and kinetic energy in the surroundings, i.e., as thermal energy.

2.2.1.3 Elastic Scattering

In addition to absorption, an incident EM wave also can induce a response from matter known as *scattering*. Scattering occurs when the frequency of incident EM radiation is not matched to that of a transition, and thus an atom or molecule responds to the field without excitation into a higher energy state. Instead, from a classical perspective, charge (in the ground state) is set into weak oscillation. The oscillating charge then immediately begins to emit radiation of the same frequency, which is called elastic scattering.

An important example is scattering by widely, randomly spaced molecules. This case arises in the earth's upper atmosphere and causes sunlight to undergo nearly omni-directional scattering. Lord Rayleigh demonstrated that the scattered intensity, *I*, is proportional to $1/\lambda^4$, helping to explain why the sky is blue. The shorter wavelength blue light undergoes more intense lateral scattering, so indirect light is blue.

2.2.1.4 *Transmission*

Transmission is forward propagation of the wave. In dense homogeneous transparent media, forward propagation of light occurs even in the absence of lateral and back scattering. Understanding the absence of these latter effects requires familiarity with the principle of superposition and interference and is discussed below.

2.2.1.5 *Change in Wave Speed*

The presence of a medium also affects *wave speed*. In the case of EM waves, the effect typically is a reduction in speed (phase velocity). The reduction in speed can be quantified in terms of a quantity known as refractive index, n, as follows:

$$\mathrm{v} = \frac{c}{n} \tag{2.5}$$

The associated wavelength reduction is

$$\lambda = \frac{\lambda_0}{n} \tag{2.6}$$

The frequency is unchanged. Our most important application will be the case of light propagation in glass with $n \sim 1.5$.

The properties of a medium that affect the speed of sound are medium elasticity/rigidity and density. For fixed density, the effect of increasing rigidity is to increase speed. In contrast, the effect of increasing density is to decrease speed, reflecting an effect of inertia. In general, sound travels more rapidly in solids than in liquids, and more rapidly in liquids than in gases. This is because the effect of elasticity typically dominates that of density. The effect of elasticity and density on speed can be quantified, for a fluid, in terms of bulk modulus, β, and density, ρ, as follows:

$$\mathrm{v}_{\mathrm{sound}} = \sqrt{\frac{\beta}{\rho}} \tag{2.7}$$

2.2.2 Reflection and Refraction at Boundaries

To this point, we have ignored the effects that boundaries have on waves. This section briefly reviews two of the most important of these effects, reflection and refraction.

At a mechanistic level, reflection and refraction of light are other manifestations of scattering and interference (Section 2.3). For our purposes, the most important attributes of these processes are two-fold. The first is that the reflected fraction depends on the degree of medium mismatch at the interface. This is quantifiable in terms of refractive index mismatch for EM waves and acoustic impedance mismatch for sound waves; the latter will be discussed in conjunction with US (Chapter 10). In the absence of any mismatch, the wave is entirely transmitted, similar to the case of no interface. The reflected component increases as mismatch increases. These qualitative observations can be quantified in terms of intensity reflection (R) and transmission (T) coefficients. For EM radiation that is normally incident on a boundary, these are

$$R = \frac{(n_1 - n_2)^2}{(n_1 + n_2)^2} \tag{2.8a}$$

$$T = \frac{4n_1 n_2}{(n_1 + n_2)^2} \tag{2.8b}$$

where n_1 and n_2 are the refractive indices of medium 1 and 2, respectively. In the expression for R, the effect of mismatch is embodied in the quantity $n_1 - n_2$. When $n_1 = n_2$, $R = 0$, and $T = 1$ (Demonstration 2.1). The reflection and transmission coefficients are much more complicated at other incidence angles and depend on both the angle and polarization of the incident light. The reflection and transmission of sound waves play a prominent role in US and are explored in Chapter 10.

SAMPLE PROBLEM 2.1 REFLECTION AND TRANSMISSION COEFFICIENTS

a. Show that light that is normally incident from air (n_1 = 1.0) onto glass (n_2 = 1.5) is ~4% reflected and ~96% transmitted.
b. How do these results change if the light is incident from water (n_1 = 1.33) onto glass? From oil (n_1 = 1.5) onto glass?
c. What happens if instead the light is traveling in the other direction (i.e., from glass onto air, glass onto water, or glass onto oil)?

SOLUTION

a. The reflected fraction when light is incident from air onto glass can be calculated from Eq. (2.8a):

$$R = \frac{(n_1 - n_2)^2}{(n_1 + n_2)^2} = \frac{(1.0 - 1.5)^2}{(1.0 + 1.5)^2} = \frac{0.25}{6.25} = 0.040 = 4\%$$

The transmitted fraction can be calculated from Eq. (2.8b) or, equivalently, by recognizing that the fraction of transmitted light is simply one minus the fraction of reflected light (because the light is either reflected or transmitted):

$$T = 1 - R = 0.096 = 96\%$$

b. The reflected fraction when light is incident from water onto glass and from oil onto glass can be calculated similarly using the appropriate refractive indices. The results for water onto glass are 0.36% reflected and 99.64% transmitted. The results for oil onto glass (no mismatch) are 0% reflected and 100% transmitted.

c. Examination of Eqs. (2.8a) and (2.8b) shows that the proportions of reflected and transmitted light are dependent only on the presence of the interface and not on whether the light is traveling from air, water, or oil onto glass (or vice versa). However, this simple result is valid *only* for light normally incident on an interface. Behavior at other angles is more complicated.

The second important attribute is that the transmitted wave refracts, or bends, at an interface between mismatched media, as shown in Fig. 2.5a. Refraction is at the heart of the function of lenses in microscopy. The degree of bending is governed by the Law of Refraction (also known as Snell's Law), which reads

$$n_1 \sin\theta_1 = n_2 \sin\theta_2 \tag{2.9}$$

The angles, θ_1 and θ_2, define the direction of incident and transmitted rays with respect to the normal to the interface, as shown. Important conceptual consequences of the law of refraction include the fact that light bends toward the normal when incident from a medium of lower refractive index onto a medium of higher refractive index, and away from the normal when incident from a medium of higher refractive index onto a medium of lower refractive index. In contrast, the Law of Reflection states that the reflected and incident rays make equal angles with respect to the normal (Fig. 2.5a).

One of the most familiar manifestations of refraction – the spectral separation of white light by a prism – reveals that blue light is bent more than red by a medium like glass (Fig. 2.5b). This, in turn, implies that refractive index is wavelength/frequency

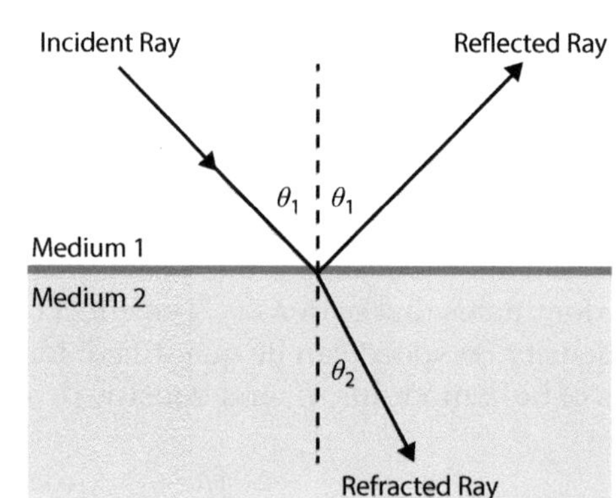

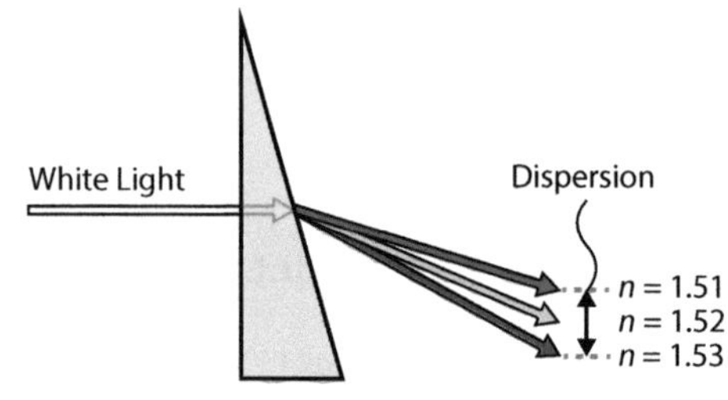

FIGURE 2.5
Pictorial representations of (a) reflection and refraction at an interface, and (b) separation of white light into its component colors by a prism, which is a manifestation of the dependence of refractive index on frequency/wavelength.

DEMONSTRATION 2.1 REFRACTIVE INDEX MATCHING

The effects of refractive index mismatch on reflection and refraction can be demonstrated by immersing glass rods in media of different refractive indices: air, water, and vegetable oil. The rod and oil have closely matched refractive indices, so reflection and refraction vanish, and the rod is invisible. (Photograph by the authors.)

SAMPLE PROBLEM 2.2 SNELL'S LAW/TRANSMISSION THROUGH A SLAB

Show that a ray incident on a parallel plate of thickness x (see picture) emerges from the plate parallel to its original direction of propagation but shifted laterally by the following amount:

$$\text{Shift} = \frac{x\sin(\theta_1 - \theta_2)}{\cos\theta_2}$$

We will invoke implications of this result in our discussion of thin lenses.

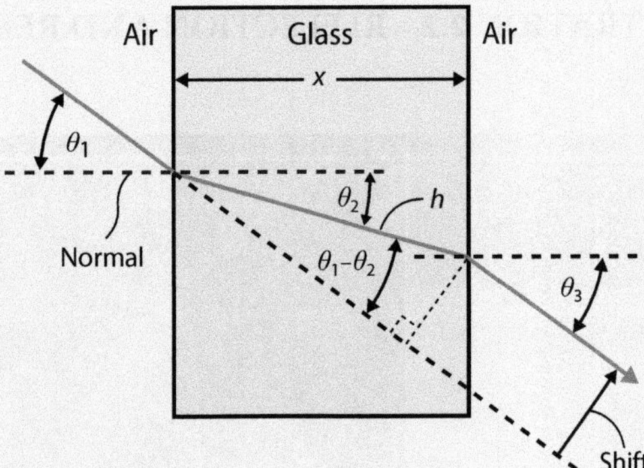

SOLUTION

The refraction of light as it passes from air to glass and from glass back to air is described in each case by Snell's Law:

$$n_1 \sin\theta_1 = n_2 \sin\theta_2 = n_3 \sin\theta_3$$

$$1.0\sin\theta_1 = 1.5\sin\theta_2 = 1.0\sin\theta_3$$

$$\sin\theta_1 = \sin\theta_3$$

$$\theta_1 = \theta_3$$

Hence, incident and exiting rays are parallel. The same result will apply whenever the same medium (e.g., air, water, oil) is found on both sides of the glass (or other material). Finally, the shift of the beam can be calculated as follows (using the definitions of sine and cosine as opposite over hypotenuse and adjacent over hypotenuse, respectively):

$$\sin(\theta_1 - \theta_2) = \frac{\text{Shift}}{h} = \frac{\text{Shift}}{x/\cos\theta_2}$$

This equation can be solved for the shift, yielding

$$\text{Shift} = \frac{x\sin(\theta_1 - \theta_2)}{\cos\theta_2}$$

Significantly, the ray will not be deflected if the plate is thin (negligibly thick), where x goes to zero, or if the plate has the same refractive index as the surrounding medium, because in this case $\theta_1 - \theta_2 = 0$.

dependent and increases with increasing frequency. This phenomenon, known as "dispersion," will be important later because it leads to aberrations in lens function.

Dispersion can be rigorously analyzed using the simple harmonic oscillator model. A simple conceptual explanation, applicable to a case like the prism, is as follows. Equation (2.5) reveals that refractive index differences are equivalent to speed differences, and that $n_{red} < n_{blue}$ is equivalent to $v_{red} > v_{blue}$. The speed difference arises because red light is not as well matched to atomic transitions in glass as blue light. Thus, red light produces a less significantly perturbed (i.e., less slowed) scattered wave and an associated less impeded net transmitted wave.

Another important consequence of refraction is the phenomenon of total internal reflection (TIR), which will figure prominently in our discussion of axially selective illumination techniques in Chapter 7. TIR can occur when EM radiation is incident from

DEMONSTRATION 2.2 REFLECTION AND REFRACTION

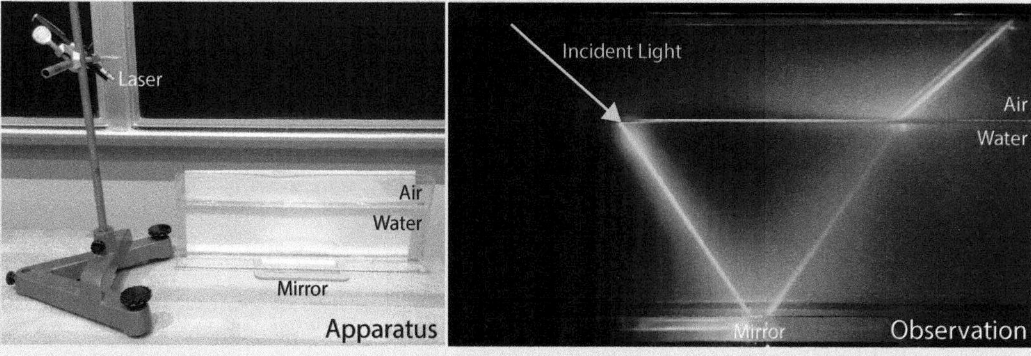

Reflection and refraction can easily be demonstrated using a laser pointer and a tank of water. In the pictured case, a green laser pointer is directed from left to right toward a partially filled tank. A hard-to-see portion of the beam is reflected back into the air. Another portion of the beam is refracted into the water. The angle of refraction is smaller than the angle of incidence, meaning that the light bends toward the normal, as described by Snell's Law. The beam bounces off the reflective bottom of the aquarium. The angle of reflection, relative to the normal, is equal to the angle of incidence, as described by the Law of Reflection. The final refracted beam bends away from the normal as it exits the water back into the air. (Photographs by the authors.)

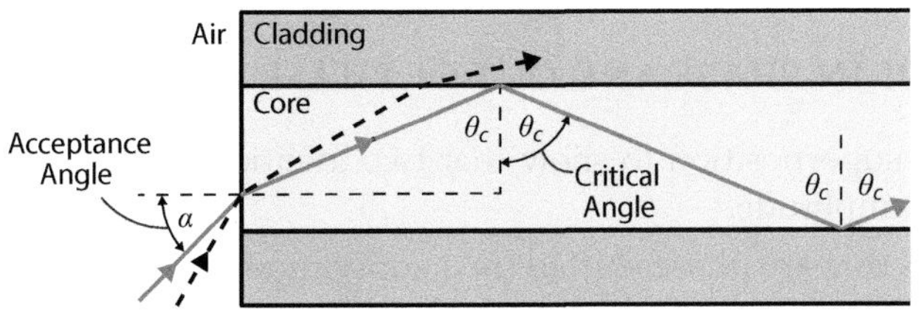

FIGURE 2.6
Propagation of light down a fiber optic cable occurs via TIR. The cable consists of a core with a higher refractive index than the surrounding cladding. Light that enters the cable within the acceptance angle undergoes TIR and does not exit the cable, leading to lossless transmission (apart from absorption in the core and cladding).

a region of higher refractive index onto a region of lower refractive index. In this case, the transmitted radiation bends away from the normal, and if the incidence angle is sufficiently high, the transmitted ray will be bent at an angle of 90° with respect to the normal. This means that there is no transmitted ray; the light is totally reflected (Fig. 2.6). The critical (minimum) incidence angle, θ_c, that produces TIR is given by invoking Snell's Law with $\theta_2 = 90°$:

$$\sin\theta_c = \frac{n_2}{n_1}$$

All larger incidence angles also produce TIR. Fiber optic cables use the phenomenon of TIR to ensure lossless transmission (Fig. 2.6).

2.3 Superposition of Waves

We now discuss one of the most important attributes of waves, which is their ability to superimpose and undergo constructive and destructive interference. Diffraction, the spreading of light when it encounters an obstacle and a fundamental limit on resolution in optical microscopy, may also be explained using the principle of superposition and Huygens's principle. This phenomenon will be dealt with extensively in Chapter 4.

When two or more waves propagate simultaneously in the same region, the net disturbance is the sum of the individual disturbances. This is known as the principle of superposition. The sum can be performed mathematically or graphically.

Equation (2.10) exemplifies the mathematical approach for two arbitrary waves:

$$y_{\text{total}}(x,t) = y_1(x,t) + y_2(x,t) \tag{2.10}$$

Sample Problem 2.3 implements the approach for two identical waves and for two equal-amplitude waves that are out of phase by π (i.e., by 180°). These two cases are examples of total constructive and destructive interference, respectively. Box 2.3 contains a more general mathematical treatment.

Fig. 2.7 exemplifies the graphical approach as applied to both total constructive and total destructive interference.

Constructive and destructive interference are the basis of many important phenomena and have many important applications, as we will see, including in optics and microscopy. As one example, we show how these effects can be used to understand the purely forward propagation/transmission of light in dense homogeneous transparent media. The connection between superposition and diffraction is explored in Chapter 4.

In completely homogeneous media, or in media where the inhomogeneities are spaced much closer together than a wavelength, no light is scattered sideways or backward because laterally and backscattered light exhibit destructive interference (Fig. 2.8). In contrast, forward scattered light exhibits constructive interference, and so the beam propagates forward essentially undiminished in energy. In media composed of random inhomogeneities spaced farther apart (such as the randomly positioned molecules in the atmosphere or water or glass), some light is scattered. In media with a regular geometrical arrangement of inhomogeneities with larger spacings (like crystals or manufactured gratings), some scattering occurs in highly discrete directions.

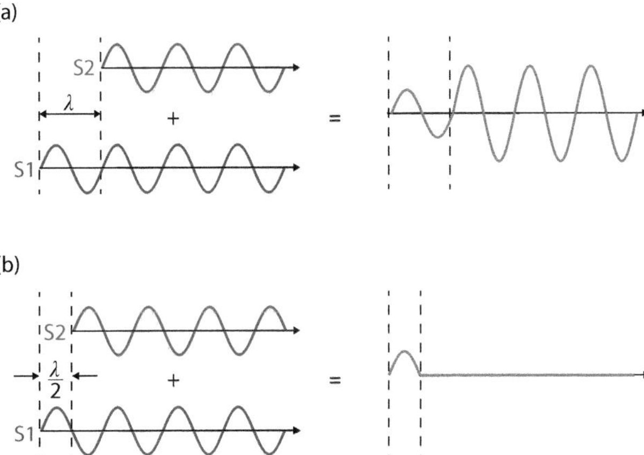

FIGURE 2.7
Graphical analysis of constructive and destructive interference. (a) When waves from two sources, S1 and S2, superimpose after traveling a distance that differs by an integral number of wavelengths, they travel crest on crest to produce a wave with twice the amplitude. (b) In contrast, when the same waves superimpose after traveling a distance that differs by an odd number of half wavelengths, they travel crest on trough to produce a wave with zero amplitude (HW 2.13).

SAMPLE PROBLEM 2.3 SUPERPOSITION OF IN-PHASE AND OUT-OF-PHASE WAVES

a. Use the mathematical form of the principle of superposition to show that two identical waves superimpose to produce a wave with twice the amplitude.
b. Similarly show that two equal-amplitude waves that are phase shifted by π superimpose to produce a wave with zero amplitude.

SOLUTION

a. The superposition principle, which states that the net disturbance is the sum of the individual disturbances, is given by Eq. (2.10). Therefore, the superposition of two identical (in-phase) waves with amplitude y_0 is

$$\begin{aligned} y_{\text{total}}(x,t) &= y_0 \sin(kx-\omega t+\varphi) + y_0 \sin(kx-\omega t+\varphi) \\ &= (y_0+y_0)\sin(kx-\omega t+\varphi) \\ &= 2y_0 \sin(kx-\omega t+\varphi) \end{aligned}$$

This is a wave with amplitude $2y_0$.

b. The superposition of two identical waves that are out of phase by π is

$$\begin{aligned} y_{\text{total}}(x,t) &= y_0 \sin(kx-\omega t+\varphi) + y_0 \sin(kx-\omega t+\varphi+\pi) \\ &= y_0 \sin(kx-\omega t+\varphi) - y_0 \sin(kx-\omega t+\varphi) \\ &= (y_0-y_0)\sin(kx-\omega t+\varphi) \\ &= 0 \end{aligned}$$

Here, we have made use of the trigonometric identity $\sin(\theta+\pi) = -\sin\theta$.

BOX 2.3 GENERAL TREATMENT OF SUPERPOSITION

Superposition of equal-amplitude harmonic disturbances that are phase shifted by an arbitrary angle φ can be readily calculated using trigonometric identities. Consider the superposition

$$y_{\text{total}}(x,t) = y_0 \sin(kx-\omega t) + y_0 \sin(kx-\omega t+\varphi)$$

This superposition can be rewritten in a more illustrative form as

$$y_{\text{total}}(x,t) = 2y_0 \cos\left(\frac{\varphi}{2}\right)\sin\left(kx-\omega t+\frac{\varphi}{2}\right) \tag{B2.3.1}$$

using the identity

$$\sin\theta_1 + \sin\theta_2 = 2\sin\left(\frac{\theta_1+\theta_2}{2}\right)\cos\left(\frac{\theta_1-\theta_2}{2}\right)$$

The amplitude of the wave in Eq. (B2.3.1) is $2y_0\cos(\varphi/2)$. Equation (B2.3.1) thus correctly predicts that the amplitudes associated with constructive ($\varphi = 0$) and destructive ($\varphi = \pi$) interference are $2y_0$ and zero, respectively.

Superposition of waves of non-equal amplitude can be addressed using phasor diagrams, as described in Hecht (2017) (HW 2.14).

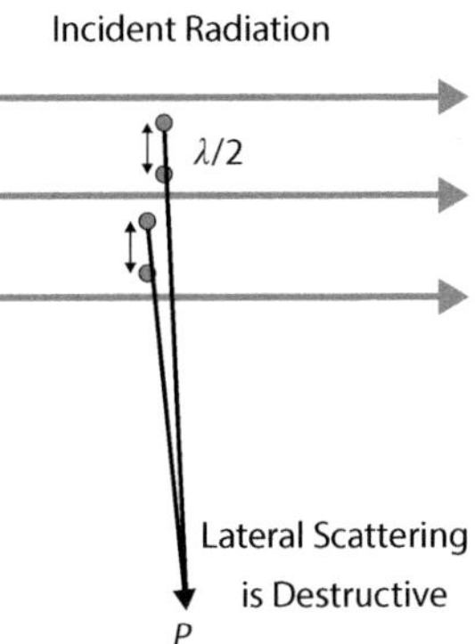

FIGURE 2.8
Schematic of the mechanism underlying the purely forward propagation of light in dense homogeneous media. In empty space, light propagates purely forward and is not visible from the side. In dense homogeneous media, scattering might lead to light that propagates sideways, but interference eliminates this possibility. For every atom in the medium, there is another $\lambda/2$ away. When scattered waves from two such sources, which have a lateral path difference $\Delta x = \lambda/2$, combine at a lateral point P, the result is destructive interference or an absence of laterally scattered light. The absence of backscattered light has similar origins.

2.4 Polarization of EM Radiation

This section contains a slightly lengthier and more technical review of material that is essential to understanding the polarization-sensitive microscopy techniques discussed in Chapter 5. If these techniques are not part of the course, the material can be omitted. Alternatively, the material can be read in closer alliance with coverage of these techniques.

EM waves are composed of electric, $\vec{E}$, and magnetic, $\vec{B}$, fields, which are vectors. Typically, $\vec{E}$ is emphasized due to its prominent effects on charge, and, when its direction is not random, the wave by convention is "polarized" along the direction of $\vec{E}$. The polarization direction discussed here is distinct from propagation direction, which usually is depicted as a ray arrow.

2.4.1 Types of Polarization

Polarization of light can be divided into three categories: random, linear, and elliptical, with circular polarization being a special case of elliptical (Fig. 2.9). Natural light is randomly polarized (unpolarized) because it is produced by many sources that emit light with rapidly and randomly changing polarization. Unpolarized light (Fig. 2.9a) can be represented as a superposition of two arbitrary, orthogonal, and equal-amplitude plane waves that are incoherent. Incoherence implies that the waves have rapidly and randomly fluctuating relative phase.

A monochromatic plane wave can be linearly polarized (Fig. 2.9b). In this case, the electric field has a fixed orientation, which is perpendicular to the direction of propagation and to the direction of the magnetic field (Box 2.1). However, the magnitude and sign of the field vary with time.

Superposition of orthogonal linearly polarized plane waves can produce elliptically and circularly polarized waves. One important example is superposition of two orthogonal plane waves of equal amplitude that differ in phase by 90°, which leads to circular polarization (Fig. 2.9c). The analogous unequal-amplitude superposition and superpositions with more general phase differences lead to elliptical polarization (Fig. 2.9d). For elliptical light, the tip of $\vec{E}$ describes an ellipse in any fixed plane orthogonal to the direction of propagation. Thus, both the direction and magnitude of the field change. These attributes will figure prominently in Chapter 5.

2.4.2 Generating and Manipulating Polarized Light

Common physical mechanisms that generate polarized radiation include selective absorption/dichroism, reflection, scattering, and birefringence (Fig. 2.10). Wire grid polarizers and Polaroid film operate via selective absorption. Polarization of sunlight is due to scattering. We will discuss birefringence in Chapter 5 in the context of polarization microscopy.

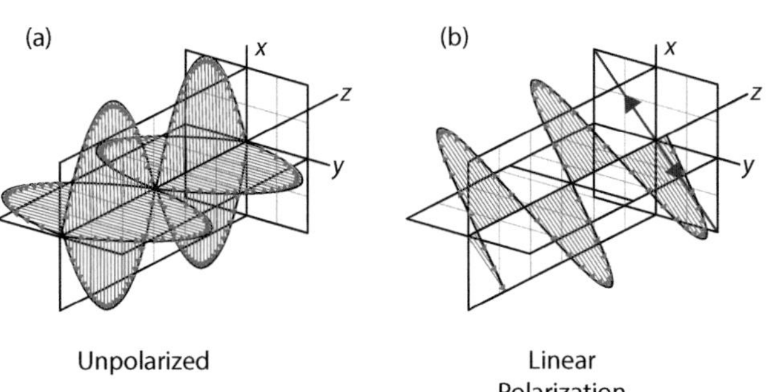

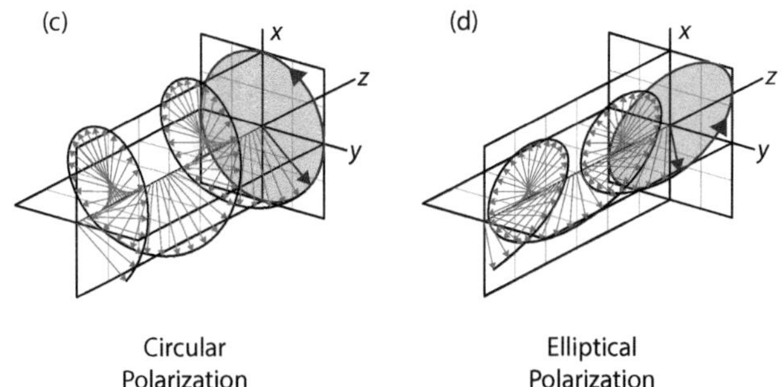

FIGURE 2.9
Snapshots of (a) unpolarized light and of three common types of polarized light: (b) linear, (c) circular, and (d) elliptical.

(a)

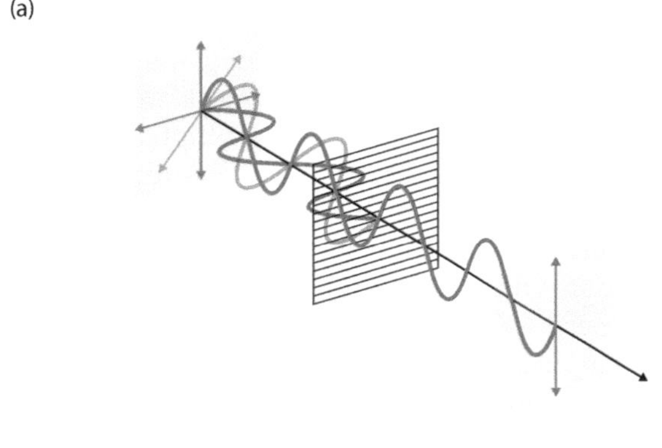

(b)

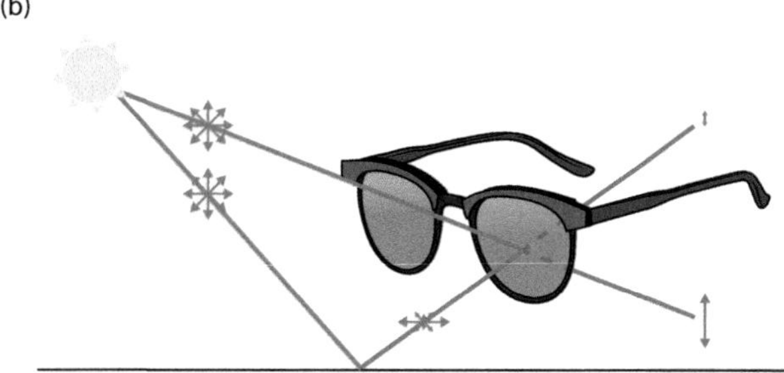

FIGURE 2.10
Exemplary mechanisms that produce polarized light. (a) Wire grid polarizers that produce linearly polarized microwave radiation can be fabricated by making the grid lines significantly smaller than the wavelength. These polarizers operate via a straightforward but perhaps counterintuitive mechanism. Electric fields polarized along the direction of the wires set electrons into motion and are absorbed as their energy is dissipated, whereas fields orthogonal to the wires are transmitted. In other words, the grid transmits radiation polarized perpendicular to the long axis of the grid. (b) Reflection from a surface is a common source of light that is partially or fully polarized parallel to the surface. Polarized glasses reduce glare by blocking the reflected polarized light. Polarization changes are depicted schematically.

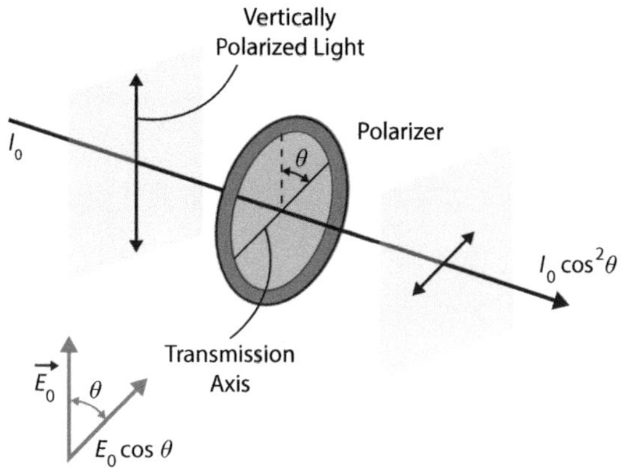

FIGURE 2.11
Effect of a linear polarizer on a linearly polarized input. The output field (right black arrow) is directed along the transmission axis. The output and input intensities are related by $\cos^2\theta$, where θ is the angle between the transmission axis and the direction of the input field.

Devices that output linearly polarized light (linear polarizers) play a key role in polarization-based microscopy, and we thus review their key attributes (Fig. 2.11). The first of these is the existence of a "transmission" axis, which dictates the direction and intensity of the output. Specifically, the output light (if nonzero) is polarized along the transmission axis. In addition, the output intensity depends on the extent of alignment between the input field and the transmission axis. This qualitative statement is quantified next.

First, consider an unpolarized input. The input can be viewed as an equal-amplitude superposition of light along, and orthogonal to, the transmission axis. Half of the light thus will be absorbed by the polarizer, and the output and input intensities, I_{out} and I_0, will be related by $I_{out} = \frac{1}{2} I_0$.

Second, consider a polarized input. In this case, the analysis is more complex. Assume that the polarized input wave has amplitude E_0 and is oriented at an angle θ with respect to the transmission axis. The output amplitude, E_{out}, is the component of the input along the transmission axis, so $E_{out} = E_0 \cos\theta$ (Fig. 2.11). Output and input intensities are proportional to the squares of their associated amplitudes (Eq. [2.3]) and thus are related by

$$I_{out} = I_0 \cos^2\theta \tag{2.11}$$

Equation (2.11) is known as Malus's law.

SAMPLE PROBLEM 2.4 MALUS'S LAW

Linearly polarized light of intensity I_0 is normally incident on a linear polarizer. How much light passes through the polarizer if the angle between the polarization of the incident light and the transmission axis of the polarizer is 0°? 30°? 45°? 60°? 90°?

SOLUTION

The incident light is polarized; therefore, Malus's law applies:

$$\frac{I_{out}}{I_0} = \cos^2\theta$$

Substituting the angles into this equation yields the following results:

Angle (°)	Output Intensity
0	1.00
30	0.75
45	0.50
60	0.25
90	0.00

Thus, the polarizer transmits all of the light when the angle between the polarization of the incident light and the transmission axis is 0° (i.e., they are parallel), half of the light when the angle is 45°, and none of the light when the angle is 90° (i.e., they are perpendicular).

Suggested Reading and Additional Resources

Born, M. and Wolf, E. 1999. *Principles of optics: Electromagnetic theory of propagation, interference, and diffraction of light.* Cambridge University Press, London. http://dx.doi.org/10.1017/CBO9781139644181

Hecht, E. 2017. *Optics (5th Edition).* Pearson Education Limited, Essex, England.

PhET Interactive Simulations. 2002–2022. https://phet.colorado.edu

Tipler, P.A. and Mosca, G. 2008. *Physics for scientists and engineers (6th Edition).* W.H. Freeman and Company, New York, NY.

HOMEWORK PROBLEMS

2.1 [Traveling Waves] A traveling sinusoidal wave, $y(x,t)$, has the following properties: $y(0,0)=0$, $A=10$ mm, $\nu=5$ Hz, and v = 1 m/s. Use this data to write an expression for the wave of the form specified in Eq. (2.1).

2.2 [EM Waves] An EM wave polarized along y and propagating along positive x has the following properties: $\nu=150$ MHz, $E(0,0)=E_0=2$ N/C (newtons per coulomb). Use this data to write expressions for the electric and magnetic fields. Hints: See Box 2.1; $B_0=E_0/c$.

2.3 [Effect of Medium on Wave Properties] A beam of green light (λ_0 = 500 nm) is used to interrogate a biological specimen. The light passes through air (n = 1.000), glass (n = 1.515), water (n = 1.335), and immersion oil (n = 1.515), often more than once. Calculate the speed, wavelength, and frequency of the light in each medium. Does the color of the light change from medium to medium?

2.4 *[Simple Harmonic Oscillator – Phase Shifts] The motion of an electron in the presence of an external electric field, $E_0e^{i\omega t}$, can be described using the damped, forced simple harmonic oscillator equation:

$$m\ddot{x}=-kx-b\dot{x}+qE_0e^{i\omega t} \quad \text{or}$$
$$m\ddot{x}+m\gamma\dot{x}+m\omega_0^2x=qE_0e^{i\omega t}$$

where k is stiffness, b is a damping constant, m is the electron mass, q is the electron charge (negative for an electron), $\gamma=b/m$, and $\omega_0^2=k/m$.

a. Assume that $x(t)=x_0e^{i(\omega t-\alpha)}$ and show that

$$x_0=\frac{qE_0}{m}\frac{1}{\sqrt{\left(\omega_0^2-\omega^2\right)^2+\gamma^2\omega^2}}$$

b. Use the result in (a) to shown that the phase lag, α, is given by

$$\tan\alpha=\frac{\gamma\omega}{\omega_0^2-\omega^2}$$

c. Compute the phase lag for the cases $\omega=0$, $\omega=\omega_0$, and $\omega=\infty$ radians.

2.5 [Reflection and Transmission Coefficients] A beam of light passes perpendicularly from a light source through a series of materials and interfaces and into a microscope objective. Calculate the percentage of light incident on the slide that passes into the front lens of the objective (ignoring secondary reflections). Hint: You can treat the specimen as water, ignore absorption, and avoid the most tedious calculations by using the numerical results in Sample Problem 2.1. How does the percentage change if the immersion oil between the coverslip and microscope objective is replaced with air?

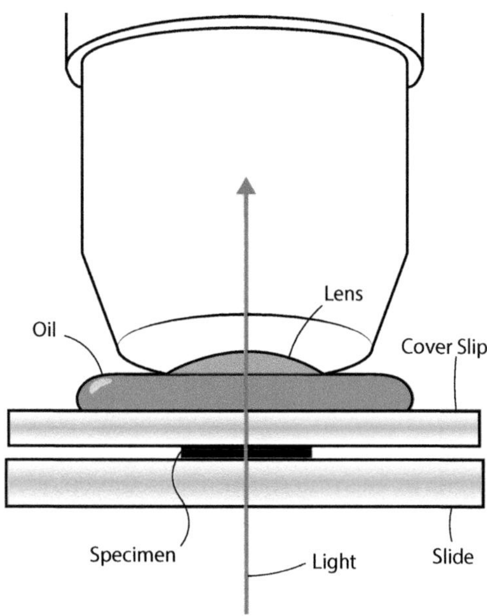

2.6 [Refractive Index – Sellmeier Equation] The wavelength dependence of the refractive index can be described empirically by the Sellmeier equation:

$$n^2(\lambda) = 1 + \frac{B_1\lambda^2}{\lambda^2 - C_1} + \frac{B_2\lambda^2}{\lambda^2 - C_2} + \frac{B_3\lambda^2}{\lambda^2 - C_3}$$

where λ is the wavelength, and B_i and C_i are experimentally measured parameters. The equation works quite well for a variety of transparent materials, where B_i and C_i are material dependent. The associated table shows these coefficients for borosilicate crown glass.

Coefficient	Value
B_1	1.03961212
B_2	0.231792344
B_3	1.01046945
C_1	$6.00069867 \times 10^{-3}$ μm²
C_2	$2.00179144 \times 10^{-2}$ μm²
C_3	1.03560653×10^{2} μm²

Using the Sellmeier equation, calculate the refractive index of glass for the following wavelengths: 400 nm, 550 nm, and 700 nm.

2.7 [Refractive Index – Sellmeier Equation (Numerical)] Write a program that calculates refractive index as a function of wavelength using the Sellmeier equation (HW 2.6). Plot $n(\lambda)$ for wavelengths between 400 nm and 700 nm. Note any trends.

2.8 [Reflection and Refraction] A beam of light traveling in air hits a glass surface, such as the front face of a microscope objective, at a 45° angle. (a) Calculate the angles of reflection and refraction for the reflected and refracted light. Assume that the refractive indices of air and glass are 1.00 and 1.515. Does the refracted ray bend toward or away from the normal? (b) Repeat the calculations for light traveling (i) through water (n = 1.33) before hitting the objective and (ii) through oil (n = 1.515) before hitting the objective. Compare the results.

2.9 [Law of Refraction] A monochromatic light ray passes through a cube. Which drawing best shows the direction of the ray as it exits the cube under the following conditions? (a) The cube is a glass prism surrounded by air. (b) The cube is an air inclusion surrounded by glass. (c) The cube is a glass prism surrounded by oil having the same refractive index as glass. Explain your choices.

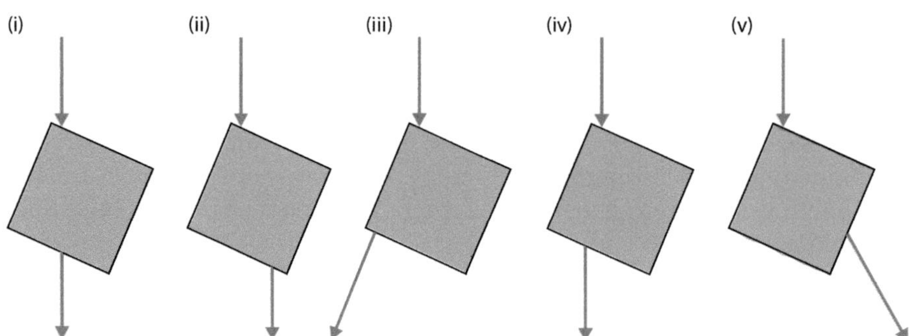

2.10 [TIR – Fiber Optic Acceptance Angle] The fiber optic acceptance angle, α, refers to the maximum incidence angle at which light incident on the face of a fiber optic cable will enter the cable and undergo TIR. (a) Show that the acceptance angle is given by the following expression, where n_{air}, n_{core}, and $n_{cladding}$ are the refractive indices of air, core, and cladding, respectively:

$$\alpha = \sin^{-1}\left(\frac{\sqrt{n_{core}^2 - n_{cladding}^2}}{n_{air}}\right)$$

Hint: See Fig. 2.6 and review the standard trigonometric identities in Appendix E. (b) A commercially available optical grade fiber optic cable has an acrylic polymer PMMA (poly[methyl methacrylate]) core ($n_{core} = 1.492$) and a fluorine polymer cladding ($n_{cladding} =$ 1.402). Calculate the acceptance angle of this cable, assuming as above that light enters the cable from air ($n_{air} = 1.000$).

2.11 *[Superposition – Mathematical] Compute the total wave generated from superposition of the following three waves:

$$y_1 = 0.10\sin(kx - \omega t + \pi/4)$$

$$y_2 = 0.20\sin(kx - \omega t)$$

$$y_3 = 0.10\sin(kx - \omega t - \pi/4)$$

Hint: First compute the superposition of y_1 and y_3 using the trigonometric identity in Box 2.3.

2.12 [Superposition – Graphical] Sketch the total wave generated from superposition of the two pulses shown below at time points $t = 1$ s, $t = 2$ s, and $t = 3$ s.

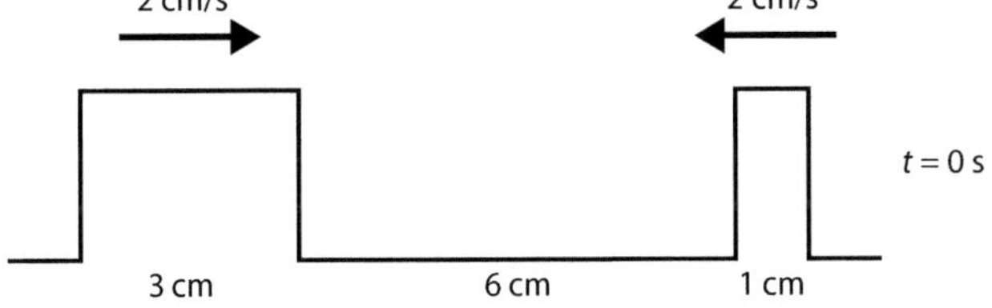

2.13 [Superposition – Acoustic Dead Spots] The principles of constructive and destructive interference apply to both EM and sound waves. Two loudspeakers are separated by 1.60 m. A listener is positioned 1.20 m in front of one of the loudspeakers (at a right angle to a line connecting the speakers). Assume that the speakers emit in phase. (a) Sketch the setup. (b) Write down the condition under which the listener will observe destructive interference. (c) For what frequencies is the sound at the listener a minimum? Assume that the speed of sound is 343 m/s.

2.14 *[Superposition – Phasors] Use the phasor approach to find the superposition of the following two waves:

$$E_1 = 4\sin(\omega t)$$

$$E_2 = 3\sin(\omega t + \pi/3)$$

2.15 [Phase Shift] Assume that the two waves shown in Fig. 2.2 are traveling through air, except where the bottom wave passes through the shaded box. (a) Calculate the difference in refractive index, $\Delta n = n_2 - n_1$, between the upper and lower boxes. Hint: Review Eq. (2.2). (b) Calculate the refractive index of the medium in the shaded box. What is a possible composition of the medium (based on materials discussed in this chapter)?

2.16 [Polarization] Prove that $I_{out} = \frac{1}{2} I_0$ for an unpolarized input by averaging Malus's Law over all angles.

2.17 [Polarization – Application] Gnaphosid spiders have several pairs of eyes and can detect polarized light using a pair of lensless eyes that have orthogonal transmission axes. Assume that linearly polarized light of intensity 1000 W/m^2 is incident on one of the two eyes at an angle of 60° with respect to the "transmission" axis. What is the angle of incidence with respect to the transmission axis of the other eye? What are the transmitted intensities for each eye?

Section I

Microscopy

3

Introduction to Image Formation by the Optical Microscope

The lens is the heart of a microscope. We therefore begin our study of microscopy with a conceptual discussion of how lenses form images by appropriately reconfiguring, or reshaping, a wavefront. For the moment, we will treat the topic by considering the rectilinear propagation of rays in homogeneous media and the bending of rays at interfaces, while neglecting the effects of diffraction and interference. This is the domain of geometrical optics, which is valid when the wavelength of EM radiation is negligible.

Geometrical optics will allow us to map an object into an image and predict key image properties, such as location and magnification. Wave optics, which describes interference and diffraction, will allow us to analyze image quality, such as resolution and contrast. Wave optics is discussed in Chapters 4 and 5.

3.1 Lens Function

Consider first an object, like the cat face in Fig. 3.1, that is viewed in reflected light. The object serves as a collection of scattering centers that generate diverging spherical waves with energy propagating along the direction of the rays. The purpose of a lens is to capture part of the wave that diverges from each source point and cause it to converge at a corresponding point to form an image of the object.

FIGURE 3.1
Spherical waves scattered from a face.

To accomplish this goal, *diverging* spherical waves are converted into *converging* spherical waves. There are two key ways to understand how a converging lens achieves this purpose. The first explanation is reshaping of the wavefront caused by lens-induced changes in wave speed. As shown in Fig. 3.2, a converging (convex) lens, which bulges in the middle, first reshapes a diverging spherical wave by intercepting and slowing the central, leading edge of the wave before intercepting and slowing the edges. As a consequence, the divergence of the spherical wavefront is reduced. The lens next reshapes the wave into a converging spherical wave by "releasing" the edges before releasing the center, so that the edges can move ahead.

The second explanation for the focusing effect of a lens is based on ray bending at an interface. As shown in Fig. 3.3, a ray incident on the upper half of a converging lens bends down (toward the normal) when it encounters the first interface, per Snell's Law, and it bends down again (but away from the normal) when it encounters the second interface. Rays incident on the lower half of the converging lens exhibit similar behavior but in the opposite direction. The result is that parallel, and moderately diverging, rays are made to converge and brought to focus. The opposite effect is observed with a diverging (concave) lens.

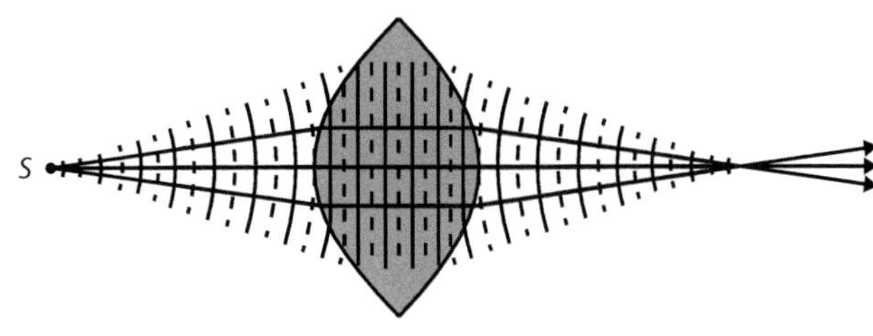

FIGURE 3.2
Reshaping of a spherical wavefront by a hyperbolic converging lens (Box 3.2).

DOI: 10.1201/b22076-4

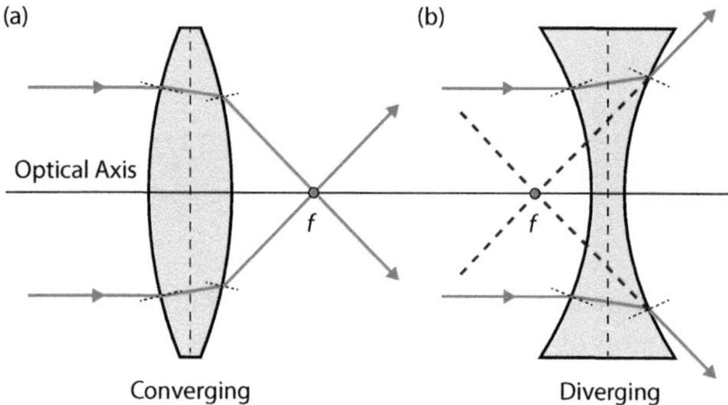

FIGURE 3.3
Refraction of rays by converging (a) and diverging (b) lenses. The normal to each interface at the point of ray incidence is shown using a dotted black line. Foci are designated using dots along the optical axis, which is a line that runs through the geometrical center of the lens and joins the two centers of curvature of its surfaces. Rays bend relative to the normal as dictated by Snell's Law. The result is that rays parallel to the optical axis are brought to focus on the optical axis after exiting a converging lens and appear to diverge from a focal point (Section 3.2) on the optical axis after exiting a diverging lens. In this diagram, and those that follow, lens radii of curvature and focal length are not necessarily to scale.

3.2 Optical Power, Focal Length, and Focal Plane

Two critical attributes of a lens are the location and size of the image that it produces. These attributes will be explored quantitatively in Section 3.4. Here we set the stage with a qualitative discussion and, in the process, introduce the important concepts of lens power, focal length, and focal plane.

Consider an idealized, thin (negligibly thick) converging lens. The lens has a limited power to refract rays and bring them to focus. Given its finite refractive ability, the location and existence of an image will depend on the degree of divergence of captured rays. Rays from a distant object, which essentially are parallel when they arrive at the lens, are relatively "easy" to bring to focus. The resulting image is thus close to the lens. However, as the object moves closer to the lens, the degree of divergence of captured rays increases, and the lens has more difficulty bringing the rays to focus. The image is thus pushed farther from the lens.

DEMONSTRATION 3.1 VISUALIZING RAY TRAJECTORIES THROUGH A LENS

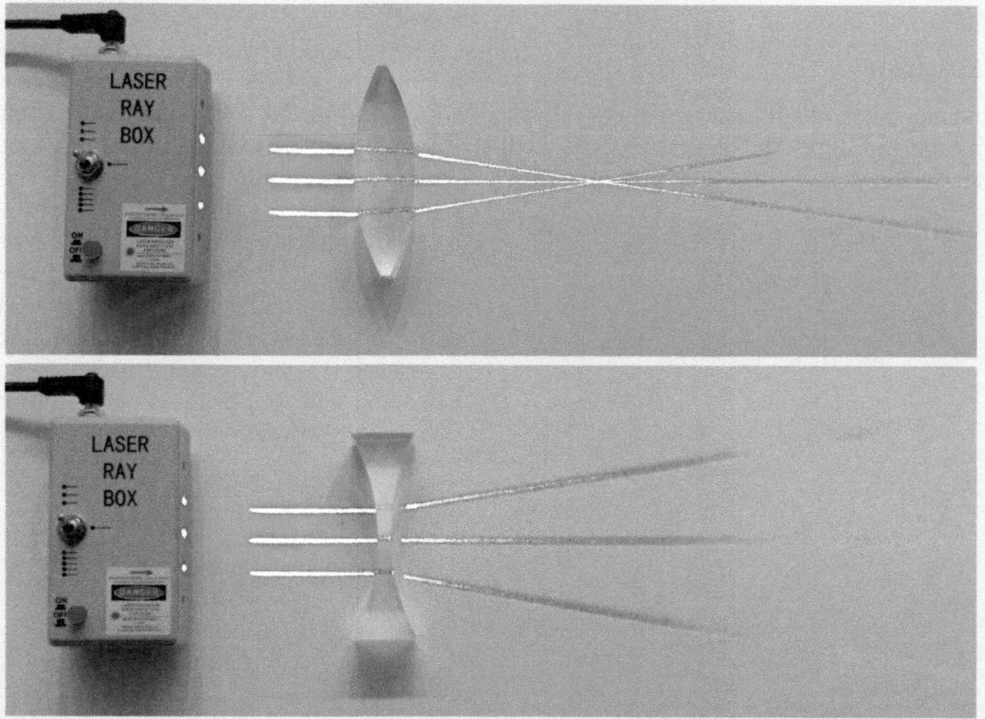

The trajectories of rays through lenses can be demonstrated in a visually compelling way using commercially available white board optics sets of modest cost. The set above, by American Scientific, shows the effects of converging and diverging lenses on three parallel laser beams, confirming the predictions in Fig. 3.3. (Photographs by the authors.)

At some point, the object gets too close to the lens, which is "overpowered" and unable to bring the rays to focus to form an image.

A converging lens brings parallel rays to focus at a distance f, known as the focal length, from the center of an idealized thin lens (Box 3.1). Lenses with short focal lengths are more powerful. Consistently, lens optical power, P, is defined to vary reciprocally with f:

$$P = \frac{1}{f} \tag{3.1}$$

The units of P are diopters (D) when f is expressed in meters. Thus, when $f = 5$ cm $= 0.05$ m, $P = 20$ D.

BOX 3.1 LENSMAKER'S FORMULA

The focal length of a lens is determined by its geometry and by the refractive indices of the lens and surrounding medium. For a spherical lens, the focal length is described by the Lensmaker's formula

$$P = \frac{1}{f} = \left(\frac{n_l}{n_m} - 1\right)\left(\frac{1}{R_1} - \frac{1}{R_2}\right) \tag{B3.1.1}$$

Here, n_l and n_m are the refractive indices of the lens and surrounding medium, respectively, R_1 is the radius of curvature of the first lens surface hit by light, and R_2 is the radius of curvature of the second lens surface hit by light. Equation (B3.1.1) is actually a simplified version of the Lensmaker's formula, valid for thin spherical lenses, which neglects a cross-term involving R_1, R_2, and the thickness of the lens. It provides a good approximation if the thickness of the lens is small relative to its radii of curvature. R_1 and R_2 can be independently positive or negative, depending on the shape of the lens and the direction of the light (Fig. B3.1.1).

Equation (B3.1.1) reveals several important properties of lens power (and thus focal length). First, lens power is dominated by the lens surface having the smallest radius of curvature, due to the reciprocal dependence of power on radius. Second, lens power is higher when the refractive index of the medium is smaller. In other words, lenses are stronger in air and weaker in water or oil. Finally, lens power (for a thin lens) is independent of the direction that light travels through the lens.

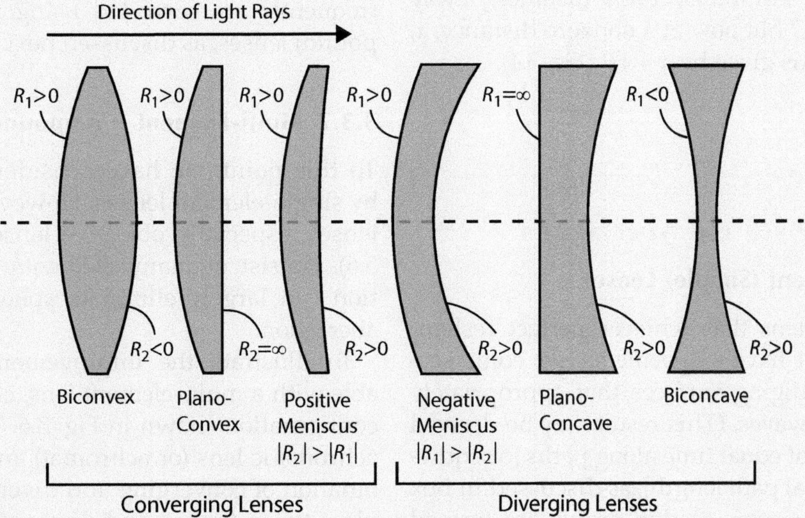

FIGURE B3.1.1
Types of spherical lenses and the associated sign conventions. The sign conventions are reversed when the direction of light rays is reversed. Converging lenses of different shapes are indistinguishable in terms of performance in the thin-lens, aberration-free approximation. This also is true of diverging lenses. However, lens shape and orientation are relevant in more realistic situations. For example, when the object is distant, a plano-convex lens with its convex side toward the object generates an image with minimal spherical aberration (Section 3.3). (For this scenario, the plano-convex lens is termed a "best form" lens.) The convex side needs to be oriented toward the object to ensure that bending occurs at both interfaces. In contrast, a biconvex lens provides the best performance when object and image distances are roughly similar.

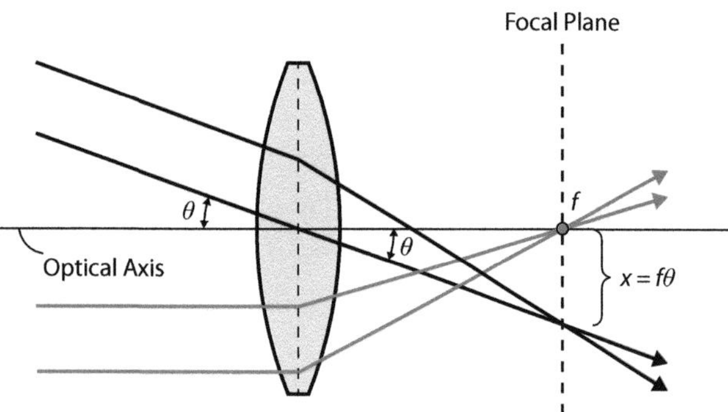

FIGURE 3.4
Parallel rays incident on a lens at a nonzero angle are brought to focus in the focal plane at a nonzero distance x from the optical axis, where in the small angle approximation $x = f\theta$. This result can be derived using the sine or the tangent. A subtlety associated with the choice of sine over tangent for the case of a microscope objective, which is designed to satisfy the Abbe sine condition, is discussed by Inoué and Spring (1997). In this and most subsequent figures, we have simplified the ray trace by showing ray refraction at the middle of the lens.

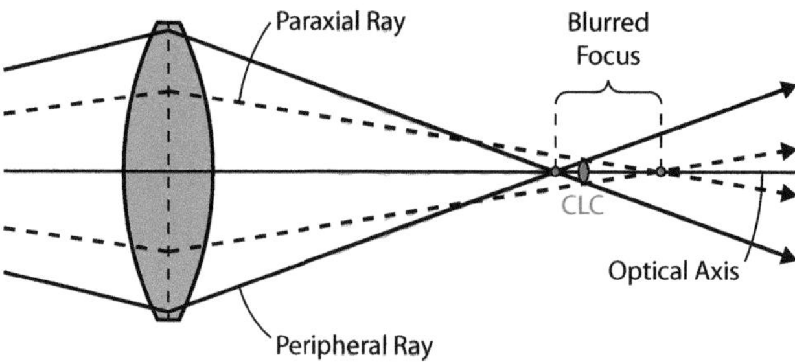

FIGURE 3.5
Non-ideal focusing caused by spherical aberration. Transverse spherical aberration and image blur are minimized at the circle of least confusion (CLC). In microscopy, spherical aberration also can arise from mismatches in refractive indices between the immersion medium and specimen.

Parallel rays can be incident on a lens at arbitrary angle, θ, with respect to the optical axis (Fig. 3.4). The rays in Fig. 3.3 are parallel to the optical axis. In this case ($\theta = 0$), their intersection point will be on this axis a distance f away from the lens center, known as the "image focus." (The analogous point on the object side is known as the "object focus.") In contrast, if the rays make a small (nonzero) angle with respect to the optical axis, they will still intersect at a distance f away in the "focal plane," but now at a nonzero distance, x, from the optical axis given by $x = f\theta$ (Fig. 3.4).

3.3 Types of Lenses and Aberrations

3.3.1 Single-Element (Simple) Lenses

A single-element lens that achieves perfect reshaping of waves must have a specific shape; conic sections create the bulging surfaces that appropriately reshape spherical waves. (This result may be derived using the concept of equal time along paths [or equivalently equal optical path length], as discussed in Box 3.2.) Lenses with a conic section profile are termed "aspheric." These lenses perform extremely well but are relatively difficult to manufacture. For this reason, spherical lenses often are used in their place. Unfortunately, spherical lenses suffer from a performance problem known as spherical aberration, which is depicted in Fig. 3.5. This type of aberration arises because more peripheral rays, which make larger angles with the optical axis, are refracted more than paraxial rays, which make smaller angles with, and travel close to, the optical axis. The result is a distribution of focal points, which blurs the image.

Even aspheric lenses suffer from a performance problem known as chromatic aberration, which arises because refractive index depends on frequency. In general, as shown in Fig. 2.5b, refractive index increases with increasing frequency/decreasing wavelength. Thus, for radiation in the visible, higher frequency radiation toward the blue end of the spectrum will be bent more by a converging lens than lower frequency radiation toward the red end (Figs. 2.5 and 3.6).

Spherical aberration and chromatic aberration most frequently are corrected using multi-element (compound) lenses, as discussed next.

3.3.2 Multi-Element (Compound) Lenses

To this point, we have considered wave reshaping by single-element lenses. However, most microscope lenses, especially objective lenses (Sections 3.5 and 3.6), consist of many elements, which in combination can largely eliminate spherical and chromatic aberration.

To illustrate the improvements that are achievable with a multi-element lens, consider the two-lens configuration shown in Fig. 3.6. This is known as an achromatic lens (or achromat) and consists of a combination of converging and diverging lenses made of glass types (crown and flint, respectively) with specific refractive indices. As shown in the ray diagram, the achromat corrects the extra convergence of blue light so that red and blue light focus at the same point. An achromat also is corrected for spherical aberration in the green. A more complicated configuration, known as an apochromat, is required to achieve good three-color correction.

BOX 3.2 ASPHERIC LENSES

Lens shape is dictated by the requirement that all paths from a point on the object to the corresponding point in the image contain the same number of vibrations, which is equivalent to all paths having the same transit time and OPL (Box 2.2).

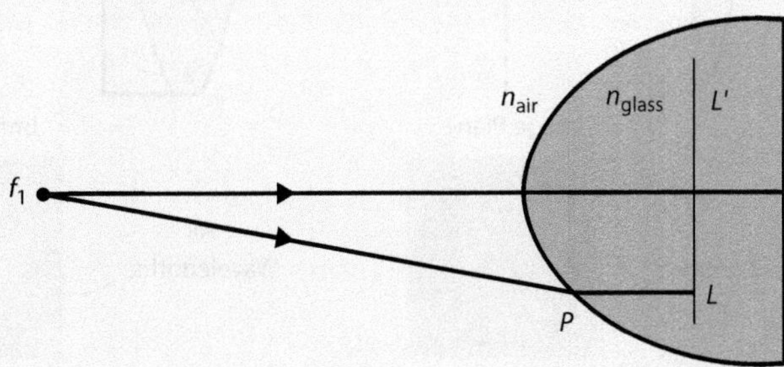

To determine the shape of the interface that will achieve this transformation, consider the lens geometry in the figure above. A point source located at f_1 emits a spherical wave that flattens into a surface of constant phase at the plane $\overline{LL'}$. In the notation of the figure, the requirement for a constant number of vibrations from the source point to $\overline{LL'}$ yields

$$\frac{\overline{Pf_1}}{\lambda_0/n_{air}} + \frac{\overline{PL}}{\lambda_0/n_{glass}} = \text{Constant} \tag{B3.2.1}$$

$$n_{air}\,\overline{Pf_1} + n_{glass}\,\overline{PL} = \text{Constant}' \text{ (a new constant)} \tag{B3.2.2}$$

Here, $\overline{Pf_1}$ and $\overline{PL}$ are distances, and n_{air} and n_{glass} are refractive indices of glass and air. In Eq. (B3.2.1), we have used Eq. (2.6) for the wavelength in a medium of refractive index *n*. Dividing both sides of Eq. (B3.2.2) by the speed of light in vacuum, *c*, and noting that $n_{air}/c = 1/\mathrm{v}_{air}$ and $n_{glass}/c = 1/\mathrm{v}_{glass}$, implies that the transit time is constant:

$$\frac{\overline{Pf_1}}{\mathrm{v}_{air}} + \frac{\overline{PL}}{\mathrm{v}_{glass}} = \text{time}_{air} + \text{time}_{glass} = \text{Constant}'/c = \text{Constant}'' \tag{B3.2.3}$$

Moreover, simple rearrangement of Eq. (B3.2.2) yields

$$\overline{Pf_1} + \frac{n_{glass}}{n_{air}}\overline{PL} = \text{Constant}''' \tag{B3.2.4}$$

This is the equation of a hyperbola with eccentricity $e = n_{glass}/n_{air} > 1$. Thus, when a point source is located at a focus, spherical waves will be transformed into plane waves inside the "lens." For the case where the ratio of refractive indices is less than one, the interface is ellipsoidal. To generate the lens depicted in Fig. 3.2, two "plano-hyperbolic" lenses can be fused on their planar interfaces.

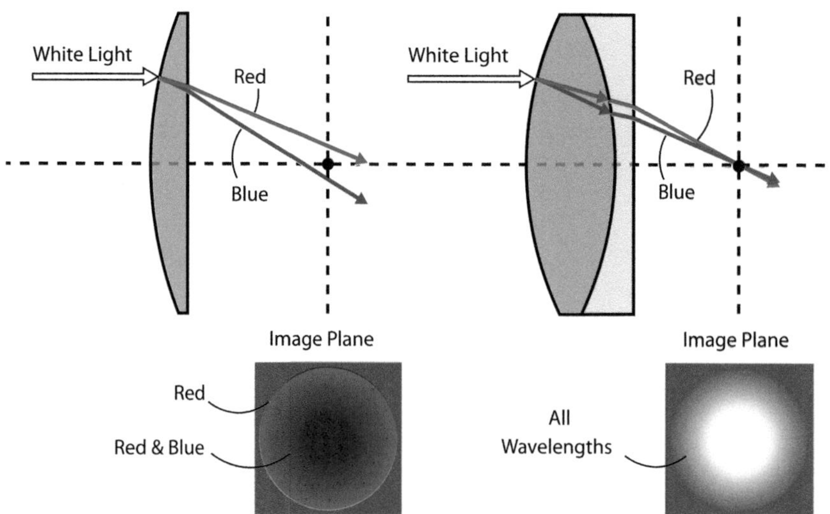

FIGURE 3.6
Schematic depiction of color correction by an achromatic doublet lens. The second (diverging) lens pulls the two colors back together because it has a much larger "chromatic dispersion" than the first lens. Thus, the second lens affects blue more than red. Moreover, the second lens is diverging and thus counteracts the effect of the first lens.

3.4 Objects and Images

3.4.1 Qualitative Analysis

The most obvious function of a microscope is to create highly magnified images of tiny objects. As a first step in understanding this function, we will use geometrical (ray) optics to quantify the size and location of images created by an idealized converging lens. We will then discuss how two such lenses can generate very high levels of magnification in a rudimentary compound microscope. Diverging lenses will play a subsidiary role in our discussion, but their behavior is explored in homework problems at the end of this chapter.

A converging lens creates two kinds of images – real and virtual. A real image is generated when rays exit the lens and converge to focus, creating an image on the opposite side of the lens. This will occur as long as the distance of the object from the lens, known as the object distance, o, exceeds the focal length (Fig. 3.7).

In contrast, if the object is positioned at a distance f from the lens, exiting rays will no longer converge. Instead, they will exit the lens parallel, at an angle dictated by their point of origin on the object. These statements can be verified by reversing the direction of the rays in Fig. 3.4.

Finally, if the object gets even closer to the lens ($o < f$), exiting rays will diverge. A virtual image will be generated in this case if a second, sufficiently powerful converging lens (such as the eye) collects the diverging rays and brings them to focus. When this occurs, the rays appear to emanate from a "virtual image," with position dictated by projecting the rays back to their apparent point of origin; simple magnifiers, and lenses that correct the nearsighted eye, work on this principle (Fig. 3.8). An image of this kind is virtual in the sense that light does not actually impinge on the location of the image.

The key qualitative attributes of image formation by a converging lens are summarized in Table 3.1. In the

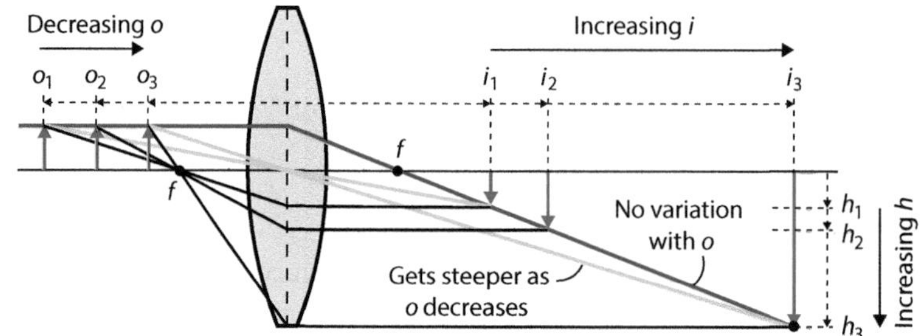

FIGURE 3.7
Example of ray tracing emphasizing the increase in the size of a real image as the object approaches the focal point of a converging lens. Object and image distances, along with f, are measured from the center of the lens.

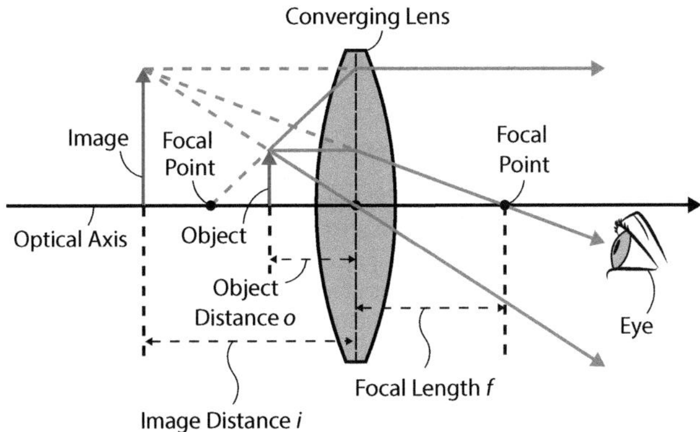

FIGURE 3.8
Example of ray tracing that generates a virtual image when the object is within one focal length of a converging lens.

table, and in our subsequent discussion, the symbols *o* and *i* denote the distances of the object and image from the center of an idealized thin lens. Upright, inverted, minified, and magnified image attributes are relative to the orientation and size of the object, as discussed in more detail below. Our goal now is to establish tools that can be used to confirm, and amplify on, the results in Table 3.1.

3.4.2 Quantitative Analysis

The results in Table 3.1 can be verified using two approaches: ray tracing and lens equations. These are discussed next.

3.4.2.1 Ray Tracing

The first approach to characterizing image formation, ray tracing, involves drawing a small, easily characterized subset of the myriad rays that propagate through the lens and using the point of intersection of these rays to determine the location and size of the image. In principle, only two rays are required to determine the intersection point, but typically three are used to enhance accuracy. In all cases, diagrams must be drawn to scale to obtain reliable results.

Fig. 3.7 shows a specific case, a thin converging lens with an object positioned at distances greater than f away. Three readily traced rays are shown in Fig. 3.7. Two of these rays (blue and black) propagate parallel to the optical axis on one side of the lens and pass through a focal point on the other side of the lens, as shown previously in Fig. 3.4. The third ray (green) passes through the center of the lens and exits undeflected. To understand this behavior, recall Sample Problem 2.2, which showed that a ray incident on a parallel plate exits parallel to its original direction of propagation but deflected a distance proportional to plate thickness. Rays incident on the center of a lens should behave similarly because the surfaces are roughly parallel. Moreover, if the lens is sufficiently thin, the deflection will be negligible. Hence, a central ray is undeflected.

Several important conclusions emerge from the ray tracing depicted in Fig. 3.7. First, the ray that emanates from the object parallel to the optical axis and passes through the "image focus" is unaltered in its orientation as the object changes position relative to the lens. In contrast, the central ray changes orientation and is directed at higher angles as the object approaches the lens. Thus, as object distance decreases, the point of intersection of the two rays and the location of the image are pushed farther from the lens. As a consequence, the image increases in size. Moreover, the image is inverted relative to the object. A careful examination of the figure also shows that the transition from minification (image smaller than object) to magnification (image larger than object) occurs when the object is located at $o = 2f$. In this case, $i = 2f$, and the object and image sizes are equal. We have thus confirmed the "real image" results appearing in Table 3.1.

If the object instead is positioned at a distance less than f from the lens, ray tracing needs to be modified slightly, as shown in Fig. 3.8. The three key rays are drawn as in Fig. 3.7, but now they diverge on exit from the lens, and so no real image will be generated. Thus,

TABLE 3.1
Summary of Image Information for Different Object Locations

Object Location	Image Type	Image Location	Image Orientation	Image Size / Object Size	Prominent Application[a]
$\infty > o > 2f$	Real	$f < i < 2f$	Inverted	< 1 (minified)	Eye/Camera
$o = 2f$	Real	$2f$	Inverted	= 1 (same size)	Photocopier
$2f > o > f$	Real	$2f < i < \infty$	Inverted	> 1 (magnified)	Objective
$o = f$		$\pm\infty$			Eyepiece/Magnifier/Collimator
$o \leq f$	Virtual	$o < \|i\|$	Upright	> 1 (magnified)	Magnifier

[a] Most of these applications are discussed in upcoming sections of this chapter. They are included here to underscore that all focal regimes are important.

to identify the location and size of the virtual image that would appear if the diverging rays were collected by a sufficiently powerful lens, like the eye, the rays are projected backward, as indicated by the dotted lines. Their point of intersection defines the virtual image, which is upright, magnified, and located farther from the lens than the object. We have thus confirmed the "virtual image" results appearing in Table 3.1. A summary of ray tracing techniques appears in Tutorial 3.1 at the end of this chapter.

3.4.2.2 Lens Equations

The second approach to characterizing image formation, lens equations, is based on implementation of lens and magnification equations together with appropriate sign conventions. We will focus on the thin lens and transverse magnification equations, as described below, but Newton's lens equation can also be used (Box 3.3).

The thin lens equation, also known as the Gaussian lens equation, can be derived from similar triangle arguments (HW 3.7). The thin lens equation reads

$$\frac{1}{i}+\frac{1}{o}=\frac{1}{f} \tag{3.2}$$

The associated transverse magnification equation, which determines the ratio of object height, h_o, to image height, h_i, also follows from similar triangle arguments and reads

$$M_T=\frac{h_i}{h_o}=-\frac{i}{o} \tag{3.3}$$

Table 3.2 summarizes sign conventions that must be employed when using these equations.

TABLE 3.2

Sign Conventions for Thin Lenses – Light is Assumed to be Incident From the Left

Quantity	Sign	Convention
Focal length (f)	+	converging (convex) lenses
	–	diverging (concave) lenses
Object distance (o)	+	real objects (on the left side of the lens/from which light diverges)
	–	virtual objects (on the right side of the lens/toward which light converges)
Image distance (i)[a]	+	real images (on the opposite [right] side of the lens from the object; light actually arrives at this location)
	–	virtual images (on the same [left] side of the lens as the object; light appears to diverge from this location)
Magnification (M_T)	+	image is upright (rightside up) with respect to object
	–	image is inverted (upside down) with respect to object

[a] We ignore images created by virtual objects, which arise in some multi-lens systems.

Several important conceptual results are apparent from the lens equations. The first is the "reciprocal" relationship between object and image distances (e.g., close objects imply distant images). This attribute is evident from Eq. (3.2) because the sum of the inverse of object and image distances is a constant ($1/f$). Therefore, if one distance increases, the other distance must decrease. The second, evident from Eq. (3.3), is the growth in size of real images with decreasing object distance and increasing image distance. A third attribute, evident from the minus sign in Eq. (3.3), is that real images are inverted.

SAMPLE PROBLEM 3.1 THIN LENS EQUATION

Use the thin lens equation to verify that a converging lens will form a real image, located in the focal plane, of an object located at infinity.

SOLUTION

The object distance, in this problem, is infinite, meaning that $o=\infty$. Substituting this value into the thin lens equation (Eq. [3.2]) yields the following result:

$$\frac{1}{i}+\frac{1}{o}=\frac{1}{f} \Rightarrow \frac{1}{i}+\frac{1}{\infty}=\frac{1}{f} \Rightarrow \frac{1}{i}+0=\frac{1}{f} \Rightarrow \frac{1}{i}=\frac{1}{f} \Rightarrow i=f$$

Thus, the image is real because f is positive, and the image is located in the focal plane because $i=f$.

This result provides an easy way to measure (or confirm) the focal length of a converging lens: use the lens to form an image of a distant object, such as a ceiling light, and measure the distance between the center of the lens and the image.

BOX 3.3 NEWTON'S LENS EQUATION

Newton's lens equation is an alternative description of the relationship between object distance, image distance, and focal length. The thin lens equation references object and image distances, o and i, respectively, relative to the center of a thin lens (see figure). Newton's lens equation, in contrast, references object and image distances relative to the focal points of the lens.

$$x_o x_i = f^2$$

Here, $x_o = o - f$ and $x_i = i - f$. The algebraic simplicity of Newton's lens equation emphasizes the reciprocal relationship between object and image distances. For example, because f^2 is a constant for a given lens, as x_o or x_i increases, the other must decrease, meaning that as the object gets nearer the lens, the image gets farther from the lens, and vice versa. Newton's lens equation is readily derived using arguments based on similar triangles (HW 3.9).

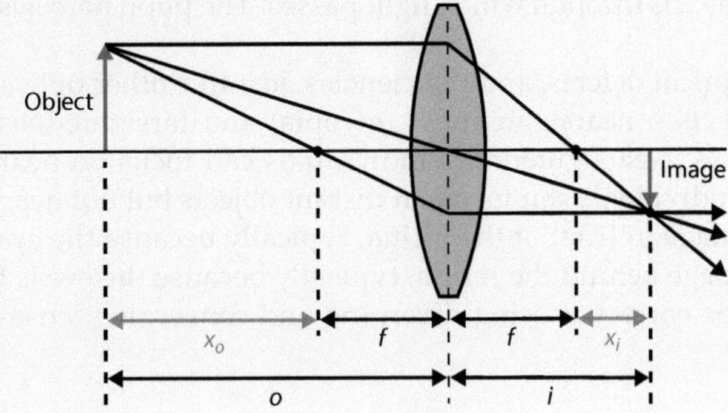

BOX 3.4 IMAGE FORMATION BY THE HUMAN EYE

The human eye is itself a sensitive and adaptive imaging system.

Fig. B3.4.1 shows an anatomic cross-section of a human eye and an associated ray diagram showing the path that light takes through the eye. Optically significant components of the eye are labeled. In brief, the cornea and lens bend and focus incident light, such that it forms a real image on the retina. The pupil,

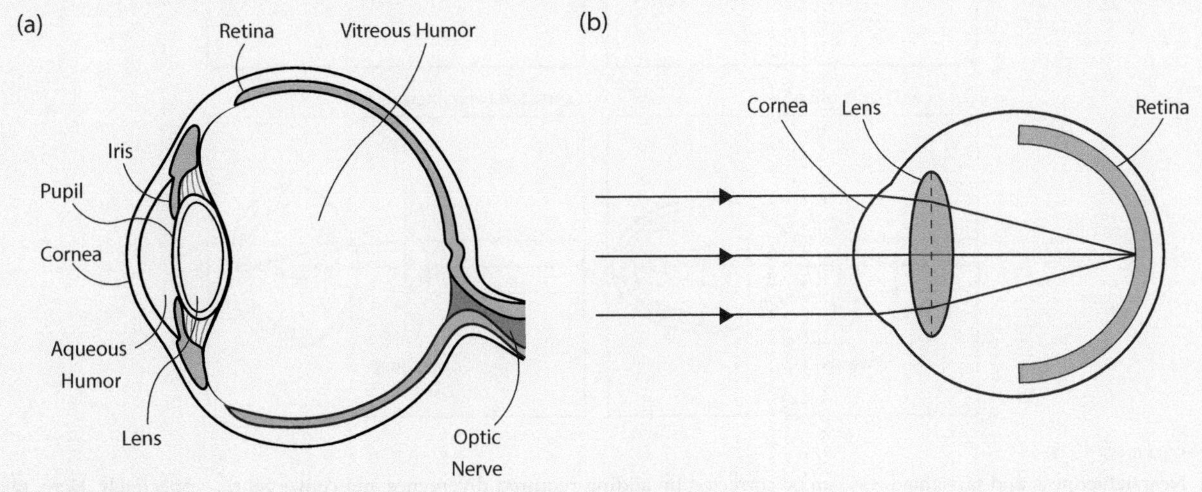

FIGURE B3.4.1
Cross-sectional views of (a) a human eye and (b) associated ray diagram.

positioned between the cornea and lens, determines the amount of light incident on the retina. The image is inverted: top and bottom, and left and right, are reversed. Rod and cone cells in the retina convert the image into electrochemical signals that are then transmitted to the visual cortex in the brain via the optic nerve. The visual cortex processes these signals, inverting and interpreting the image.

The lens and pupil are both adaptive. Specifically, special muscles control the shape of the lens, thickening it, and thereby increasing its ability to bend light, to focus on relatively close objects, and thinning it, and thereby decreasing its ability to bend light, to focus on relatively distant objects. This adaptation, termed "accommodation," is necessary because the distance between the lens and retina is fixed. In contrast, in fish eyes, the shape of the lens is constant, and focus is accomplished by changing the distance between lens and retina. The terms "near point" and "far point" denote the distances to the closest and farthest objects that can be brought to focus. These distances are, respectively, 25 cm and infinity, with infinity approximated as 6 m or 20 feet, for normal human eyes. The eyes lose their ability to accommodate as people age, which is why reading glasses are so common for people over forty.

The pupil is a hole in the iris through which light passes. The pupil narrows in bright light and widens in dim light.

The eye is subject to optical defects, and deficiencies, just like other optical systems. Fig. B3.4.2 illustrates two common defects – nearsightedness (myopia) and farsightedness (hyperopia) – and their correction with eyeglasses. Nearsightedness individuals can focus on nearby objects but not distant objects. Farsightedness individuals can focus on distant objects but not nearby objects. In nearsightedness, the eye forms an image in front of the retina, typically because the eye is too long. In farsightedness, the eye forms an image behind the retina, typically because the eye is too short. Nearsightedness and farsightedness can be corrected using diverging and converging lenses, respectively, as explored in HW 3.11 and 3.12.

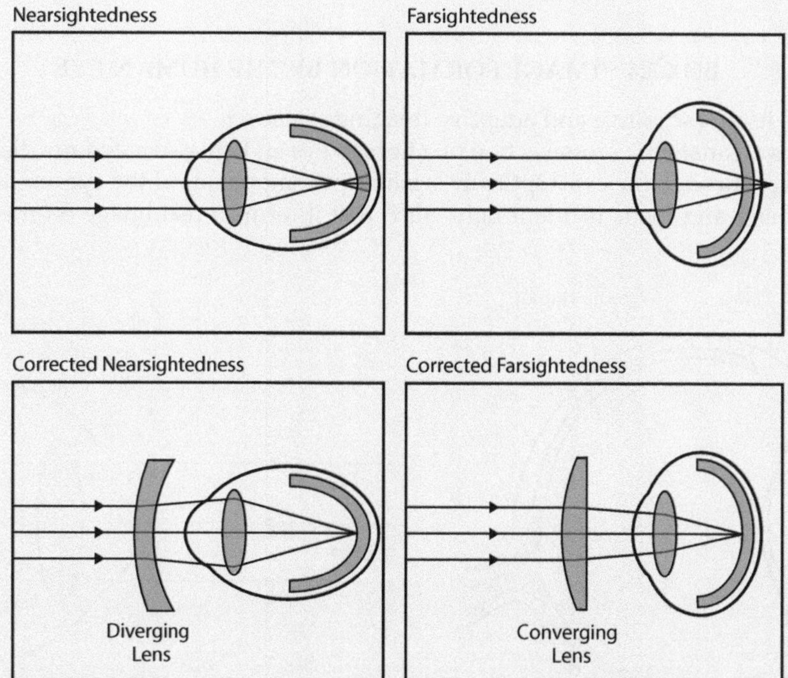

FIGURE B3.4.2
Nearsightedness and farsightedness can be corrected by adding required divergence and convergence, respectively. Here, ray bending by the cornea and lens have been combined.

3.4.2.3 Angular Magnification

In this section, we discuss simple magnifiers and angular magnification to facilitate our upcoming analysis of the microscope eyepiece, which functions as a magnifier (Section 3.5).

A magnifier is a simple convex lens that works with the eye to augment ray convergence. The magnifier makes an object appear enlarged because it can be brought closer to the eye, which views an enlarged virtual image produced by the magnifying lens (Fig. 3.9). In the case of the eyepiece, the object is at the lens focus, and thus the virtual image is "infinitely" large. Our previous definition of magnification loses meaning when the image has infinite height, and thus an angular analog of Eq. (3.3) is used instead to quantify the magnification of the eyepiece (and magnifiers generally).

The utility of angular magnification stems from the fact that the human brain evaluates object size based on the size of the associated image produced on the retina (Fig. 3.9a). Moreover, the size of a retinal image is determined by the angle that the object subtends. For example, as an object gets closer to the eye, it subtends a larger angle and appears enlarged. Thus, human perception of relative object size can be quantified using angle ratios.

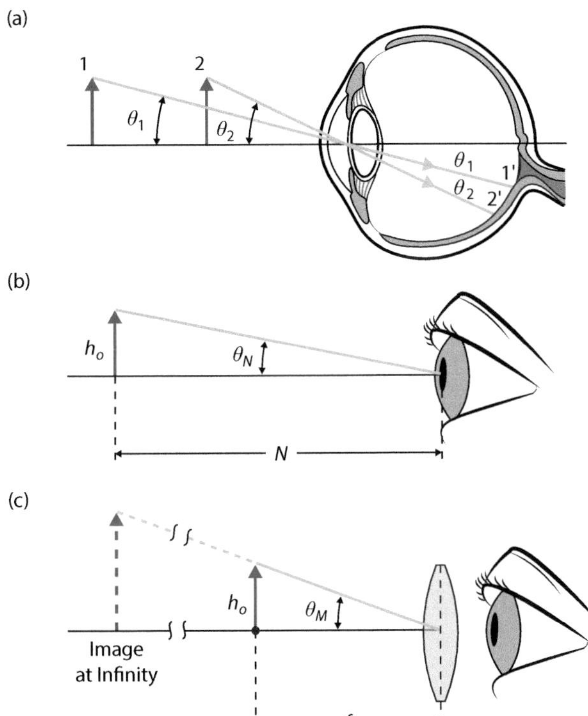

FIGURE 3.9
Attributes of angular magnification and simple magnifiers. (a) An object appears larger as it gets closer because it subtends a larger angle. Angular magnification is quantified using a ratio, where (b) the denominator is the angle subtended by the object at the near point, and (c) the numerator is the angle subtended when.

The angular magnification of an optical instrument is computed as the angle subtended by the object when the instrument is in use divided by a reference angular size. To implement this definition, we first consider an object of height h_o that is brought up to the near point, N, of an unaided eye (Box 3.4 and Fig. 3.9b). This configuration will create the largest "unaided" size on the retina that is free of blur and thus is used as the reference size. In the notation of Fig. 3.9, we have (using small angle approximations)

$$\tan\theta_N = \frac{h_o}{N} \approx \theta_N$$

As a follow-up, consider the same object viewed when a convex lens with $f < N$ is placed in front of the eye. To model the behavior of the eyepiece, we position the object at the focal point and the enlarged (virtual) image is at infinity. The angle subtended on the retina is given by

$$\tan\theta_M = \frac{h_o}{f} \approx \theta_M$$

Notably, $\theta_M > \theta_N$ because $f < N$, and the object is magnified. We now define the angular magnification of the eyepiece as

$$M_A = \frac{\theta_M}{\theta_N} = \frac{N}{f} \approx \frac{25}{f} \quad (3.4)$$

where we have inserted the near point, 25 cm, of a young adult with normal eyes (Box 3.4). The magnification predicted by Eq. (3.4) can be augmented slightly if the virtual image instead is positioned at the near point (at the expense of eye strain – HW 3.13).

3.5 The Rudimentary Compound Microscope

A rudimentary compound microscope is shown in Fig. 3.10. The microscope consists of two lenses. The first lens, known as the objective, creates a magnified real inverted "intermediate" image, which becomes the object for the second lens. The second lens, known as the eyepiece, acts as a magnifier and creates a further magnified virtual image, which is viewed by the eye.

The original object and intermediate image need to be positioned appropriately to achieve magnification of the intermediate image and to ensure that the eyepiece generates a virtual image that can be viewed by the eye. The results in Table 3.1, and the ray diagram in Fig. 3.7, reveal that the object should be positioned

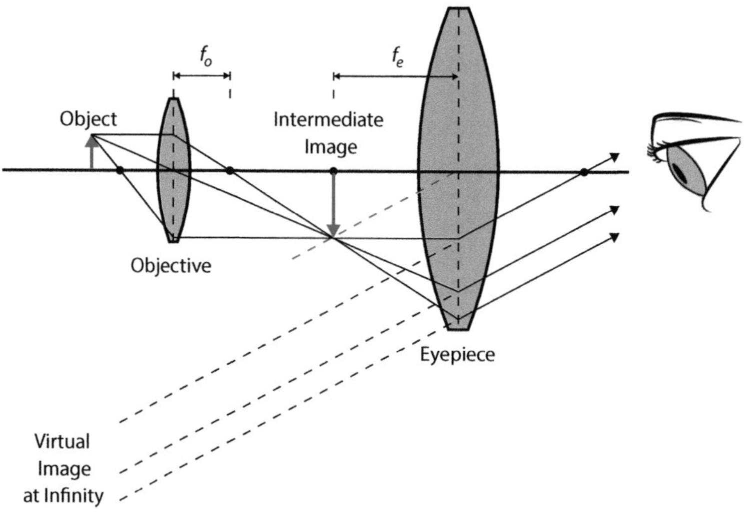

FIGURE 3.10
Schematic diagram of a rudimentary compound microscope. The objective generates a real, inverted, magnified intermediate image, which is located just at the eyepiece focal length and becomes the object for the eyepiece. The eyepiece generates a further magnified virtual image at infinity that is viewed by the relaxed eye. Magnifications, lenses, and eye are not to scale.

between $2f$ and f from the objective, and that object distances that just exceed the focal length will produce maximal magnification of the intermediate image. In addition, higher power objectives (which have shorter focal lengths) also will produce more magnified intermediate images.

The position of the intermediate image also is constrained to be in the focal plane of the eyepiece to ensure generation of a final virtual image at infinity. The total magnification, M_{total}, of the rudimentary microscope is $M_{total} = M_{obj} \times M_{eye}$, where M_{obj} and $M_{eye} \approx 25/f$ are the magnifications of the objective

SAMPLE PROBLEM 3.2 RUDIMENTARY TWO-LENS MICROSCOPE

Consider a two-lens microscope configuration like that shown in Fig. 3.10. The objective has a focal length $f_1 = 2$ cm, and the eyepiece has a focal length $f_2 = 10$ cm. The object is 2.1 cm from the objective. The virtual image created by the eyepiece is assumed to be at infinity. Determine the magnification of the final image and the objective/eyepiece separation.

SOLUTION

The key to solving this problem is recognizing that the image generated by the objective is the object of the eyepiece. The thin lens equation can be rearranged to yield an algebraic expression for an image position:

$$\frac{1}{i} + \frac{1}{o} = \frac{1}{f} \Rightarrow i = \frac{of}{o - f}$$

For the image created by the objective, i_o, we set $f_o = 2$ cm and $o_o = 2.1$ cm:

$$i_o = \frac{(2.1\text{ cm})(2\text{ cm})}{2.1\text{ cm} - 2\text{ cm}} = 42\text{ cm}$$

The associated magnification can be calculated using Eq. (3.3):

$$M_{T,o} = -\frac{42\text{ cm}}{2.1\text{ cm}} = -20$$

The magnification of the eyepiece can be computed using Eq. (3.4):

$$M_{T,e} = \frac{25 \text{ cm}}{10 \text{ cm}} = 2.5$$

The total magnification of the two-lens system is the product of the separate magnifications produced by each lens:

$$M_{T,\text{total}} = M_{T,o} \times M_{T,e} = -20 \times 2.5 = -50$$

To determine the lens separation, we note that the image produced by the objective must be at the eyepiece focal point. Thus, the lens separation is 42 cm + 10 cm = 52 cm.

and eyepiece, respectively. Activity A3.1 involves constructing a rudimentary compound microscope and exploring these issues.

3.6 The Research-Grade Optical Microscope

A research-grade microscope is much more sophisticated than its rudimentary counterpart because it is designed to produce a magnified image of exceptional quality. The way that this is accomplished will be addressed in the next few chapters as we discuss the theory of image formation and resolution and techniques for improving image contrast. This chapter closes with a discussion of the elements that comprise the research-grade microscope and how they are arranged to generate quality illumination to facilitate maximizing resolution and contrast (Fig. 3.11).

3.6.1 Köhler Illumination

To this point, we have largely ignored the fact that an image is generated via interactions between incident light and the specimen, and that there thus is a need for illumination. We tackle this issue now.

In the late 1800s, August Köhler developed a method of illumination that is still the standard for research-grade microscopy. Köhler's method has

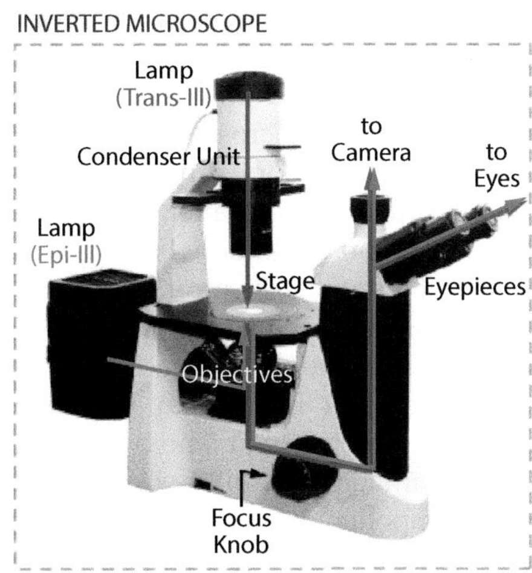

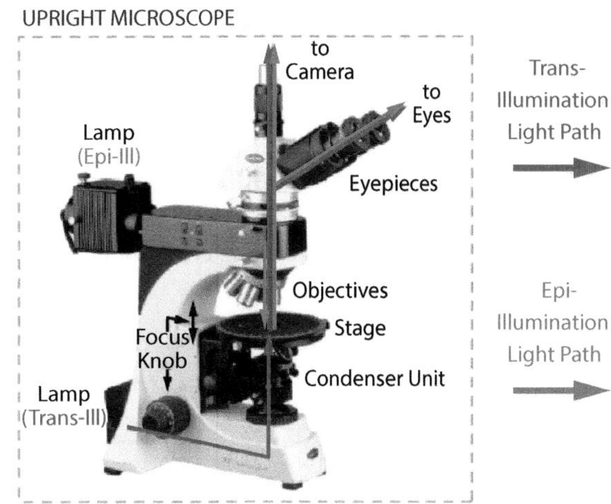

FIGURE 3.11
Photographs of two simple research-grade compound microscopes. In the inverted microscope (left), the objectives are positioned below the stage, which supports the sample, and the condenser is positioned above. In the upright microscope (right), the two components are flipped: the objectives are positioned above the stage, which again supports the sample, and the condenser is positioned below. The red and blue lines show light paths for trans- and epi-illumination. In trans-illumination, light passes from the condenser unit through the sample to the objective and then to the eyepieces or camera. Trans-illumination figures prominently in the contrast-enhancement techniques discussed in Chapter 5. In epi-illumination, the objective doubles as a condenser. Light passes through the objective to the sample and then back through the objective to the eyepieces or camera. Epi-illumination figures prominently in the fluorescence techniques discussed in Chapters 6–8. Both trans- and epi-illumination usually employ Köhler illumination. (Adapted with permission from Wiklund, M., et al. 2012. Acoustofluidics 18: Microscopy for acoustofluidic micro-devices. *Lab Chip* 12:3221–3234. DOI: 10.1039/c2lc40757d.)

several advantages. One is that the illumination is uniform, and thus variations in intensity in the image plane reflect specimen structure, and not illumination inhomogeneity. Another is that the illumination is confined to the portion of the specimen that is viewed, so there is no unnecessary specimen irradiation. Köhler's method also allows contrast to be enhanced by manipulating the illumination and/or by filtering information that encodes specimen details at appropriate sites in the illumination path (Section 4.4).

We start with a big picture view of Köhler's method. First, a collector lens and a condenser lens direct light from a source (e.g., a tungsten halogen lamp [Box 3.5]) onto the specimen plane as uniform (parallel/defocused) illumination. To this end, the collector lens focuses an image of the light source onto the front focal plane (FFP) of the condenser lens, which then directs the light as parallel rays onto the specimen (Figs. 3.4 and 3.12). After traversing the specimen, the unperturbed component of the illumination, and the light affected by the specimen, are collected by the objective lens to generate an image.

To implement Köhler illumination, and to understand how to optimize resolution and contrast, it is important to have a more detailed knowledge of the optical train in a research-grade microscope. This is our next topic.

3.6.2 The Optical Train

To understand light propagation through a research-grade microscope, it is useful to divide the system into two coincident paths. One – the light source path – focuses on propagation of the illumination, emphasizing planes of illumination focus and defocus. The other – the specimen/imaging path – similarly emphasizes planes of specimen focus and defocus (Fig. 3.13). It also is useful to introduce the concept of "conjugate"

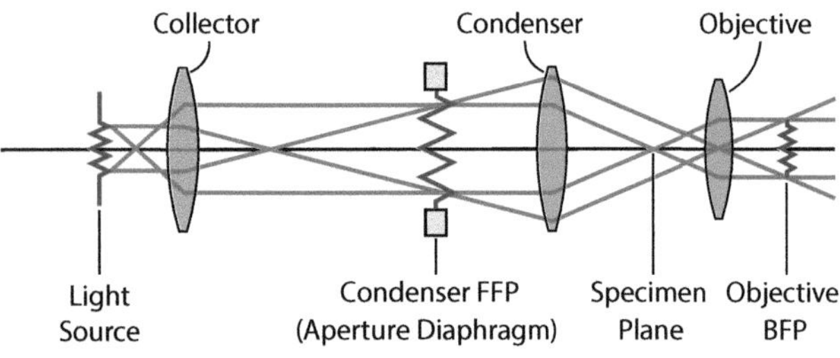

FIGURE 3.12
Ray diagram showing how the collector and condenser lenses generate uniform illumination at the specimen plane. A key step is the use of the collector lens to focus an image of the filament in the FFP of the condenser where the condenser aperture diaphragm is located. In an alternative method, critical illumination, the light is focused on the specimen plane (HW 3.17). Critical illumination is used when there is a need for higher intensity or structured (as opposed to uniform) illumination.

planes. Conjugate planes are simultaneously in focus and can be viewed superimposed.

Examination of Fig. 3.13 shows that the Köhler illumination optical train consists of two particularly useful sets of conjugate planes. One set of conjugate planes is simultaneously focused coincident with the light source. This set includes the location of the light source, the FFP of the condenser, the back focal plane (BFP) of the objective, and the exit pupil (where the eye pupil is placed). The figure also shows that the FFP of the condenser contains a condenser aperture diaphragm, which alters the cone of illumination to affect resolution, contrast, depth of field, and coherence (Fig. 3.14). The other set of conjugate planes is simultaneously focused coincident with the specimen. This set includes the field diaphragm, the specimen, the intermediate image plane, and the retina. As the name suggests, the field diaphragm affects the field of view (FOV) (Fig. 3.14).

Fig. 3.13 also shows that the two sets of conjugate planes are reciprocally related in the sense that light that is in focus in one set of conjugate planes is defocused (nearly parallel) in the other set of conjugate planes (and vice versa).

Fig. 3.13 shows the imaging path for two different microscope configurations: finite tube length and infinity corrected. Finite tube length microscopes were dominant until the 1980s. With this design, the specimen is in focus when it is a few microns beyond the FFP of the objective (for a typical high magnification lens), and rays exiting the objective converge toward the intermediate image plane, which is roughly 160 mm from the objective.

Finite tube length microscopes have the disadvantage that optical accessories that are inserted into the light path can effectively alter the tube length and introduce spherical aberration. For this reason, infinity corrected microscopes are the new standard. For this newer design, the specimen is positioned at the focal point of the objective, and thus light exits the objective as parallel rays. These then are brought to focus at the intermediate image plane by a "tube" lens. The primary advantage of infinity corrected optics is that auxiliary components can be introduced between the objective and the tube lens, where the light is parallel, with minimal introduction of aberration and minimal effect on focus.

There are two more subtle attributes of the optical train that affect resolution and contrast in optical microscopy, and thus the implementation of approaches like phase-contrast imaging, which will be discussed in Chapter 5. First, a diffraction pattern generated by the specimen appears in the BFP of the objective, as discussed in Section 4.4. Second, to enhance contrast, the illumination profile and the diffraction pattern can be altered at the FFP of the condenser and the BFP

BOX 3.5 LIGHT SOURCES

Arc lamps and lasers are the light sources that traditionally have dominated the field of optical microscopy. Here we discuss a few important attributes of arc lamps, light emitting diodes (LEDs), which are an emerging technology that is becoming increasingly prominent, and lasers. We omit a discussion of incandescent sources used for approaches like brightfield imaging (Section 5.1). (Data for graphs exported from University of Arizona Spectra Database [http://spectra.arizona.edu/].)

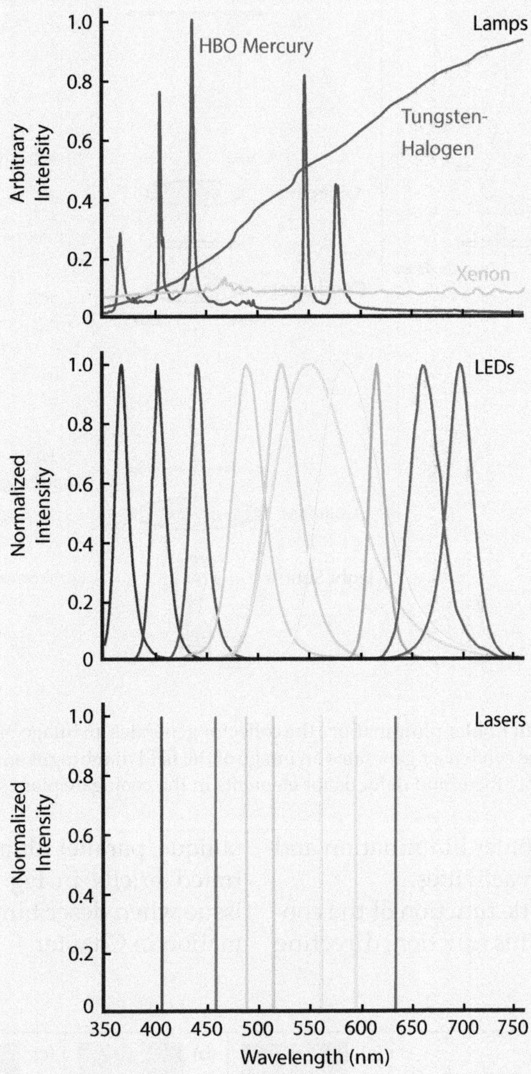

Arc lamps have figured prominently in optical microscopy, in part, because they are bright. Their illumination is spatially non-uniform but can be made homogenous at the specimen plane using the Köhler method. Arc output across the visible part of the spectrum varies markedly depending on the emitting element (see the figure). Notably, the spectrum is highly non-uniform for the commonly used mercury arc lamp, and fluorophores often are chosen based on having spectral properties that are well matched to the output of mercury.

LED-based light sources possess many desirable attributes that lamps lack and thus have largely replaced arc and incandescent lamps. A few key advantages of LEDs include low heat generation, low voltage DC operation, convenient on/off cycling, rapid wavelength switching, and very long lived, stable output. LED devices also offer diverse spectral output that is well matched to the excitation spectra of fluorophores that absorb in the UV, visible, and near-IR, and they are quite bright (see figure).

Lasers are a source of intense, collimated, monochromatic light. Some common laser lines are shown in the figure. Lasers will figure prominently in Chapters 6-8, which are devoted to fluorescence microscopy.

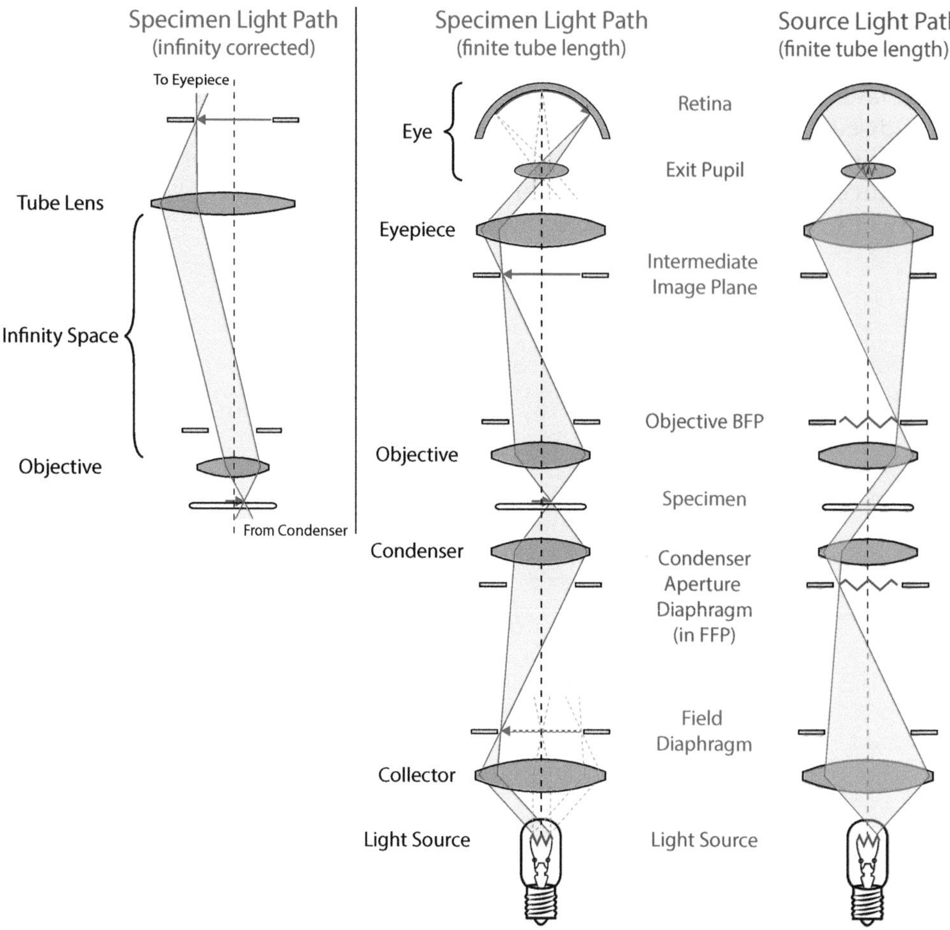

FIGURE 3.13
The Köhler illumination optical train. In Köhler illumination, the collector generates an image of the filament in the plane containing the condenser aperture diaphragm, and the condenser generates an image of the field diaphragm on the specimen plane. Representative rays are drawn that reveal relative sites of ray focus and defocus for elements in the conjugate planes.

of the objective, respectively. Köhler illumination and the optical train are explored in activities.

There also is a somewhat subtle function of the condenser that affects resolution. This function, directing oblique, parallel illumination at the specimen is illustrated briefly in Fig. 3.15. We will elaborate on this issue when describing the Abbe Theory of image formation in Chapter 4.

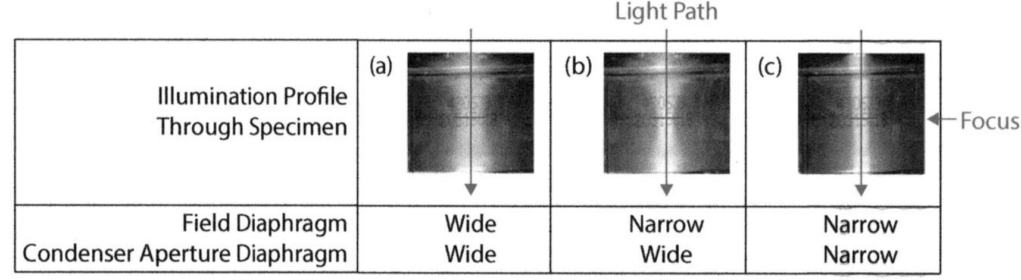

FIGURE 3.14
Effects of the aperture and field diaphragms on the illumination profile. Opening the aperture diaphragm increases the illumination cone, which enhances resolution and reduces contrast; these topics are the focus of Chapters 4 and 5. Given these antagonistic effects, a recommended compromise is to open the aperture diaphragm until ~75% of the full objective aperture is illuminated. Opening the field diaphragm increases the region of the specimen that is illuminated. The field diaphragm should be opened so that its edges are just outside the FOV. These effects of the two diaphragms are shown in photographs of the illumination profile passing through a very dilute milk solution. The photographs correspond to (a) a wide aperture and a wide field diaphragm opening, (b) a wide aperture and a narrow field diaphragm opening, and (c) a narrow aperture and narrow field diaphragm opening. Lettering on the container is irrelevant. (Photographs by Steven Attinasi and the authors.)

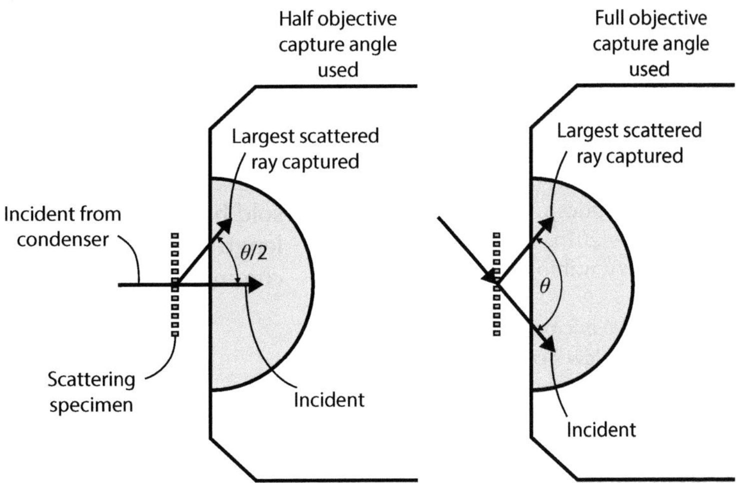

FIGURE 3.15
Oblique illumination increases the angular range of specimen-scattered radiation that is captured by the objective. Intuitively, this improves quality because more information from the specimen is transmitted to the objective and used to reconstruct the image. Abbe quantified the effect and demonstrated that it improves resolution by a factor of two, as shown in the next chapter. (Adapted from Scalettar, B.A., et al. 2015. Biomedical imaging in the undergraduate physics curriculum: module on optical microscopy. *Am. J. Phys.* 83:711–718. DOI: 10.1119/1.4921820. [CC BY 3.0])

TUTORIAL 3.1 RAY TRACING

The location and size of images formed by simple lenses can be determined by ray tracing. This process involves drawing three "principal" rays, all starting at the object, which show paths that light from the same point on the object takes through the lens:

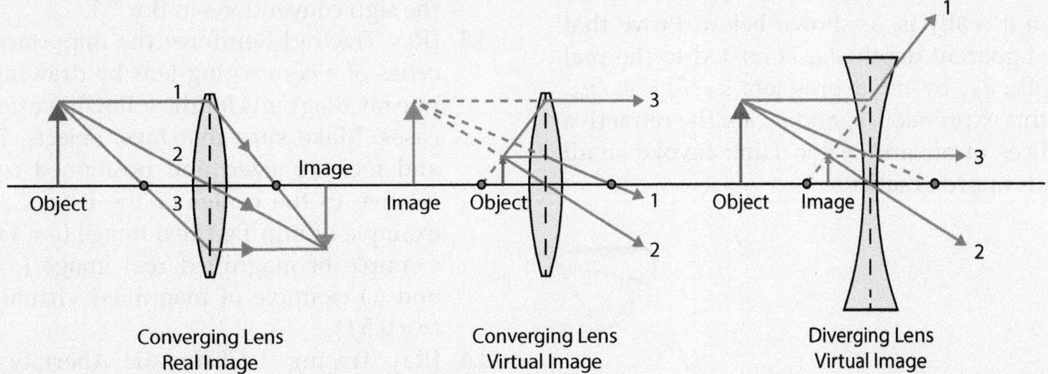

Ray 1 – a ray parallel to the centerline of the lens from the top of the object to the middle of the lens, then through the focal point on the far side of the lens (i.e., on the opposite side of the lens from the object). For a diverging lens, it will exit as if it came from the focal point on the near side of the lens.

Ray 2 – a ray through the center of the lens. This ray would, in fact, be offset as it passes through the lens (see Sample Problem 2.2); however, this offset is negligible for a thin lens.

Ray 3 – a ray through the focal point on the near side of the lens (i.e., on the same side of the lens as the object) to the middle of the lens, then parallel to the centerline of the lens. For a diverging lens, the ray that exits parallel to the centerline initially is directed at the focal point on the far side of the lens.

The rays converge at the image position, which will be on the far side of the lens (opposite the object) for real images, and on the near side of the lens (adjacent the object) for virtual images.

The location and size of the images can be determined using any two of these three rays.

Suggested Reading and Additional Resources

Hecht, E. 2017. *Optics (5th Edition)*. Pearson Education Limited, Essex, England. Chapter 5. (For a comprehensive discussion of geometrical optics).

iBiology: Microscopy Series. 2012 – 2015. *Refractive lenses and image formation*. https://www.ibiology.org/talks/refractive-lenses/

Inoue, S. and Spring, K.R. 1997. *Video microscopy: The fundamentals (2nd Edition)*. Springer, New York, NY.

Walker, J.S. 2016. *Physics (5th Edition)*. Pearson Prentice Hall, Upper Saddle Valley, NJ.

HOMEWORK PROBLEMS

3.1 [Wavefront Reshaping] Plane waves are normally incident on a thin converging lens. Sketch the wavefronts before and after the waves pass through the lens for a distance extending beyond the focal point on either side of the lens. Repeat for a thin diverging lens. Where do the wavefronts converge for the converging lens? From where do the wavefronts appear to diverge for the diverging lens? Hint: See Fig. 3.2.

3.2 [Refraction – Apparent Depth] An object below the surface of water appears higher than it really is, as shown below. Prove that the apparent depth, d_A, is related to the real depth, d_R, by the expression: $d_A/d_R = n_a/n_w$. In this expression, n_a and n_w are the refractive indices of air and water. Hint: Invoke small angle approximations.

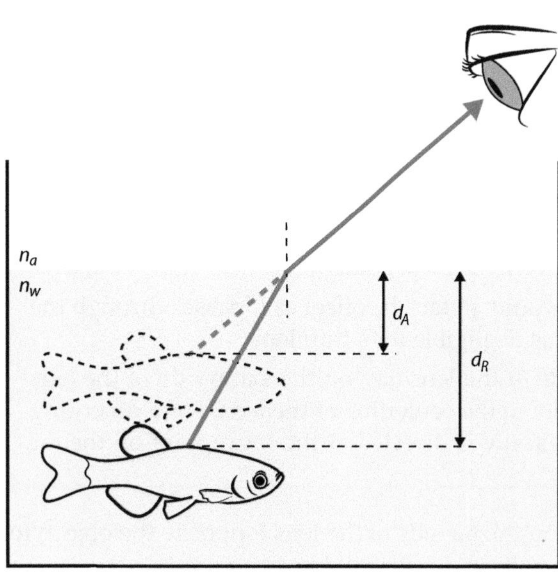

3.3 [Lens Power – Diopters/Reading Glasses] Reading glasses are converging lenses used to correct farsightedness. They usually can be purchased over the counter (although some states limit over-the-counter sales to weaker glasses). Reading glasses are typically sold based on lens power. Calculate the focal lengths (in cm) associated with the following common lens powers:

Lens Power	Focal Length?
1.00	
1.50	
2.00	
2.50	
3.00	

3.4 [Lensmaker's Formula – Chromatic Aberration] Consider a thin lens characterized by the following parameters: $R_1 = 25$ cm (radius of curvature of the lens surface closest to the light source), and $R_2 = 15$ cm (radius of curvature of the lens surface farthest from the light source). Sketch the lens. Calculate the focal lengths at the following wavelengths: 400 nm ($n_{400} = 1.531$), 550 nm ($n_{550} = 1.519$), and 700 nm ($n_{700} = 1.513$). What is the implication for image formation for a colored sample? Hint: Look at the sign conventions in Box 3.1.

3.5 [Ray Tracing] Reinforce the important properties of a converging lens by drawing accurate ray diagrams for the following exemplary cases. Make sure that rays, objects, images, and foci are sized and positioned correctly relative to the center of the lens. Cases: (a) example of minified real image ($o = 4.0f$); (b) example of magnified real image ($o = 1.5f$); and (c) example of magnified virtual image ($o = 0.5f$).

3.6 [Ray Tracing – Chromatic Aberration] The accompanying ray diagram traces green light as it passes through a thin converging lens. Sketch how you would need to modify this diagram for shorter wavelength, blue light and for longer wavelength, red light. Hint: Consider the relationship between refractive index and focal length.

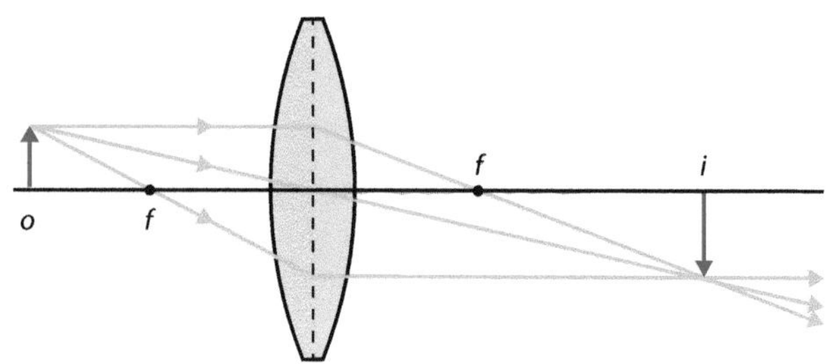

3.7 [Thin Lens Equation Derivation] Use the adjacent sketch and arguments based on similar triangles to prove the thin lens equation. Hint: The two blue triangles are similar, and the two yellow triangles are similar.

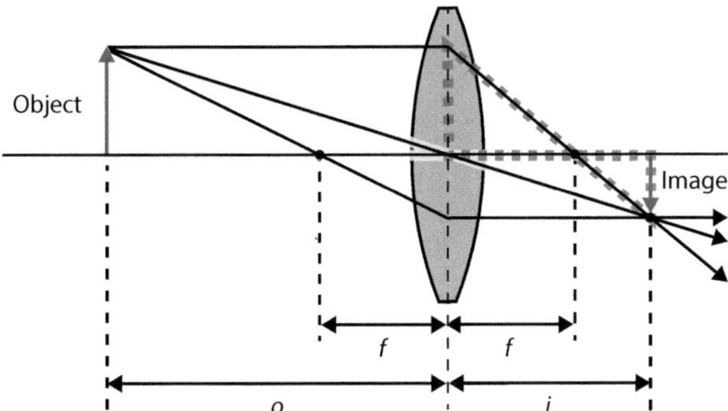

3.8 [Thin Lens Equation] Verify the results in Table 3.1 by completing the table of "special cases" below, using the thin lens equation and the associated sign conventions. Verify "image type" and "image location" in Table 3.1 by calculating, respectively, the sign and magnitude of i. Verify the associated "image orientation" and "image size/object size" by calculating, respectively, the sign and magnitude of M_T.

Case	Sign and Magnitude of i	Sign and Magnitude of M_T
$o = 100f$ $o = 2.1f$		
$o = 2f$		
$o = 1.9f$ $o = 1.1f$		
$o = f$		
$o = 0.9f$ $o = 0.1f$		

3.9 [Newton's Lens Equation Derivation] Use the adjacent sketch and arguments based on similar triangles to prove Newton's lens equation: $x_o x_i = f^2$. Hint: The two blue triangles are similar, and the two green triangles are similar.

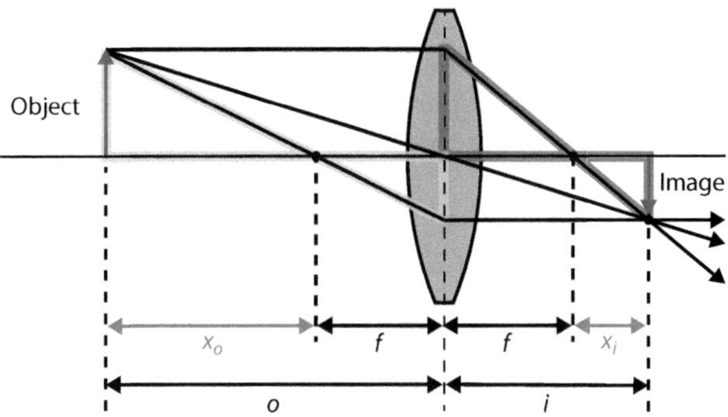

3.10 [Newton's Lens Equation] Newton's lens equation is useful for verifying the inverse correlation between object and image distances. To show this, calculate x_i/f, a measure of image distance, using Newton's equation for the following illustrative measures of object distances: (a) $x_o/f = 20$, and (b) $x_o/f = 1/20$.

3.11 [Vision Correction – Nearsightedness] Assume that a nearsighted patient has a far point of 300 cm. To correct the patient's vision, an optometrist designs glasses that, when placed 2 cm from the eye, create an image of an object at infinity at the far point, so that the object can be seen. What are the appropriate focal length and power for the glasses? Hint: The glasses create a virtual image of objects at infinity at the far point, so the relaxed eye can focus on distant objects.

3.12 [Vision Correction – Farsightedness] Assume that a farsighted patient has a near point of 60 cm. To correct the patient's vision, an optometrist designs glasses that, when placed 2 cm from the eye, enable the patient to read a book positioned at the standard 25-cm near point. What are the appropriate focal length and power for the glasses? Hint: The glasses need to create a virtual image of the book that is 60 cm from the eye.

3.13 [Magnifiers] The standard expression for angular magnification, Eq. (3.4), was derived in the text by assuming that the virtual image produced by the magnifier was positioned at infinity. It is possible to achieve higher magnifications, while still preserving image clarity, by moving the object closer to the lens until the virtual image is positioned at the near point, N. Show that the angular magnification under this condition is given by the following expression:

$$M_A = \frac{N}{f} + 1 = \frac{25}{f} + 1$$

3.14 [General Two-Lens Configuration] Consider two lenses (1 and 2) with focal lengths f_1 and f_2 that are separated by distance, d, as shown below. Show that the final image position, i_2, is given by

$$i_2 = \frac{f_2 d - \left[f_2 o_1 f_1 / (o_1 - f_1) \right]}{d - f_2 - \left[o_1 f_1 / (o_1 - f_1) \right]}$$

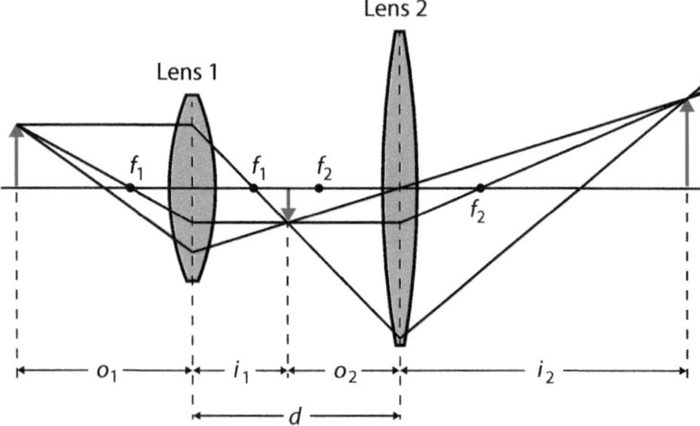

3.15 [Simple Two-Lens Configuration] Consider the two-lens configuration shown in HW 3.14 with $f_1 = 20$ cm, $f_2 = 10$ cm, $o_1 = 60$ cm, and $d = 60$ cm. Determine the location and magnification of the final image.

3.16 [Simple Two-Lens Configuration] Consider the following two-lens configuration:

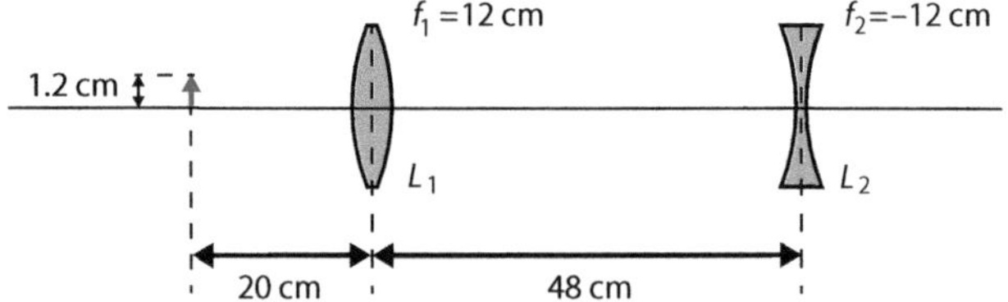

The setup superficially resembles the simple compound microscope in Sample Problem 3.2. However, in this case, lens 1 is a converging lens, and lens 2 is a diverging lens. (a) Use ray tracing to determine the approximate position, type (real or virtual, upright or inverted), and size of the image formed by lens 2. (b) Quantify these results using the thin lens equation.

3.17 [Critical Illumination] Draw a critical illumination ray trace diagram starting from the light source and continuing to the specimen plane. Hint: See Fig. 3.12.

3.18 [Proof – Finite Tube Length Microscope] Prove that the objective magnification, M_o, is given by $M_o = t/f_o$, where t is the tube length of the microscope and f_o is the focal length of the objective (see figure).

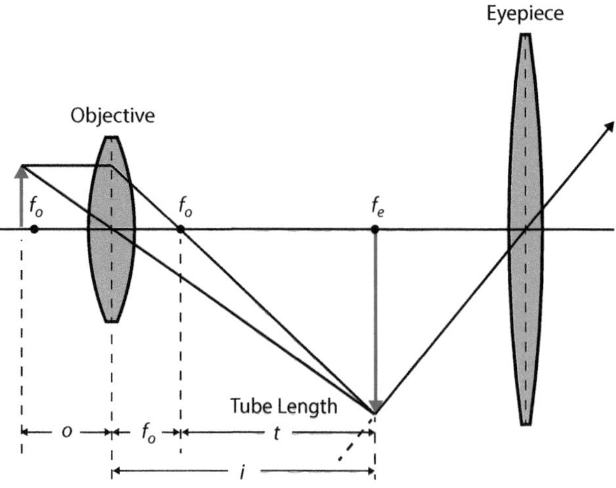

Hint: Invoke Eqs. (3.2) and (3.3).

3.19 [Proof – Finite Tube Length Microscope] Prove that $o = (M_o + 1) f_o / M_o$, where o is the object distance, and M_o and f_o are the objective magnification and focal length. Hint: Invoke Eqs. (3.2) and (3.3).

3.20 *[Proof – Magnification] Prove that $\frac{di}{do} = -M^2$, where M is the magnification. Hint: Differentiate Eq. (3.2), and remember that i is an implicit function of o. What does this expression imply about the relative sensitivity of image distance to changes in object distance at low versus high magnifications?

ACTIVITIES

A3.1. Construction of a simple compound microscope

Equipment: Optical rail
Lens holders
Converging lenses
Light source (e.g., an LED keychain flashlight)
Meter stick
Object
Index cards, microscope slide, and fine pen

Construct a simple compound microscope using the design and ideas discussed in this chapter. You will need to think about how to position the light source, condenser, object, objective, and eyepiece to create uniform illumination, a magnified intermediate image, and a magnified virtual image that you can visualize with your eye. (Hint: Look back at Fig. 3.10.) Make sure that you choose converging lenses (positive f).

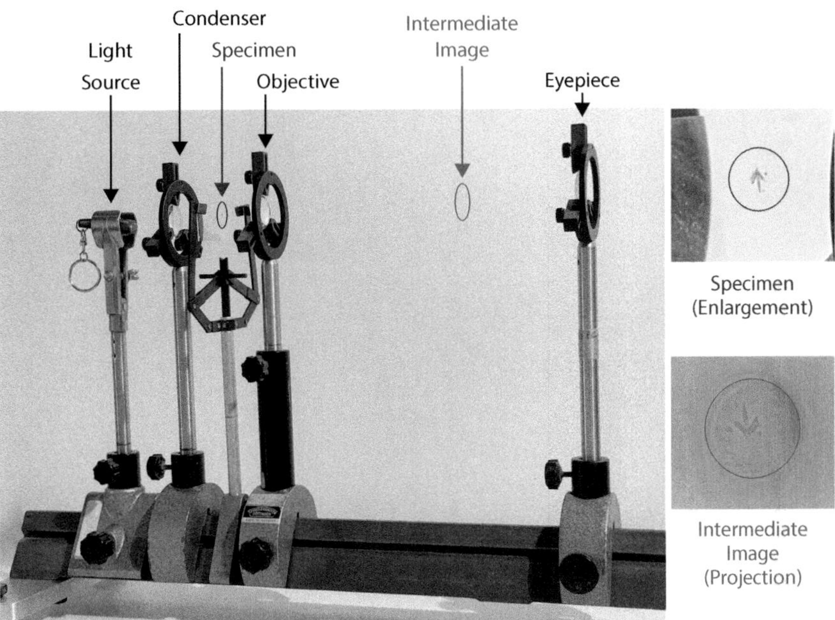

The following are just suggestions (see photographs by the authors). You can generate roughly uniform illumination by placing the flashlight at an appropriate distance from the condenser lens. A 5-cm lens will work for the condenser. A 2.5-cm lens works well as an objective, and a 10-cm lens works well as an eyepiece. Test the focal lengths using the $o = \infty$ approach described in Sample Problem 3.1, for example, by imaging a ceiling light onto a piece of paper placed on the floor. You can prepare a suitable object by drawing something small on a microscope slide with a fine tip pen.

Illuminate the object with the flashlight when you try to visualize the intermediate image (e.g., on an index card). The edge of an index card placed near the object can help define sharp focus. *Turn off the flashlight when you try to visualize the virtual image!*

Note the qualitative attributes of the intermediate and virtual images (e.g., upright/inverted and minified/magnified). How do your observations compare with the results in Table 3.1?

4

Wave Theory of Image Formation and Resolution

In geometrical optics, point objects create point images, and light from the object plane is focused on a well-defined image plane. However, both of these attributes of image formation are approximate. A more accurate picture is provided by wave optics, which incorporates diffraction and interference into image formation. Wave optics is the focus of the next two chapters.

The distinction between diffraction and interference is based largely on usage. Both effects result from superposition, but the term interference usually is restricted to superposition involving only a relatively small number of waves. In this chapter, we focus on the impact of diffraction on image formation in optical microscopy; interference will figure prominently in the next chapter.

4.1 Diffraction

Diffraction is a superposition effect that results in "spreading" of a wave when it encounters an obstacle (be it an obstruction or a hole). Thus, diffraction contributes to the ability of sound waves to spread, and be audible, around corners and the edges of doors. Diffraction also occurs when X-rays impinge on a crystal, generating a diffraction pattern that can be used to determine the crystal's structure. The effects of diffraction are important when the wavelength, λ, is not negligible relative to the size of the obstacle.

Diffraction in optical microscopy can have two distinct origins: (1) apertures, such as those in the objective, and (2) wavelength-sized structures in the specimen (Fig. 4.1). Apertures inevitably give rise to wave spreading that degrades image quality. Thus, when aberrations are negligible, resolution in optical microscopy is said to be "diffraction limited." Specimen details can give rise to diffraction "spots," or "orders," that encode information about the structure of the specimen, similar to the case of crystallography. We consider effects of apertures first and discuss specimen-derived diffraction in the context of Abbe Theory in Section 4.2.

4.1.1 Aperture-Induced Spreading

To cultivate intuition about diffraction by an aperture, consider first the case of a self-luminous, point-like object, such as a star or a fluorescent molecule, and the image generated using an aberration-free telescope or microscope. The image predicted by geometrical optics is a point – a perfect replica of the object. In contrast, the image predicted by wave optics is the far-field diffraction pattern created by a circular aperture (known as an Airy pattern), which is an imperfect, spatially spread out replica of the object (Fig. 4.2).

A rigorous derivation of the diffraction pattern created by a circular aperture is presented in many optics texts (e.g., Hecht 2017; Goodman 2017); the pattern consists of a bright central disk surrounded by alternating dark and bright rings (Fig. 4.2). A slit similarly creates a linear diffraction pattern, as shown in Demonstration 4.1 and discussed in Box 4.1.

Most of the intensity in a diffraction pattern is concentrated in its bright central maximum, whose size is thus a good measure of the degree of spread in the pattern. For a circular aperture, the full width of the central maximum that is projected on a distant screen (L away from an aperture of diameter D) is given by (Demonstration 4.1)

$$W_{\text{Airy}} = 2.44L\left(\frac{\lambda}{D}\right) \tag{4.1}$$

The theoretical foundations of this result are discussed in Section 4.1.3. In addition, the derivation of a very similar expression, valid for the full width of the central maximum in the diffraction pattern created by a slit, is presented in Box 4.1. Two key, resolution-related attributes of W_{Airy} are its reciprocal, or inverse, dependence on aperture diameter, which implies that narrow apertures create wide patterns, and its proportional dependence on wavelength. In particular, these attributes imply the following:

> Diffraction-induced spreading decreases, and thus resolution improves, with decreasing wavelength and increasing aperture size.

4.1.2 Quantifying Resolution

Before embarking on a quantitative analysis of resolution in optical microscopy, it is important to acknowledge that this is a complex topic, and that our discussion necessarily will be somewhat incomplete. For example, there are several ways to quantify resolution. Here we compare two prominent methods briefly, so that the

DOI: 10.1201/b22076-5

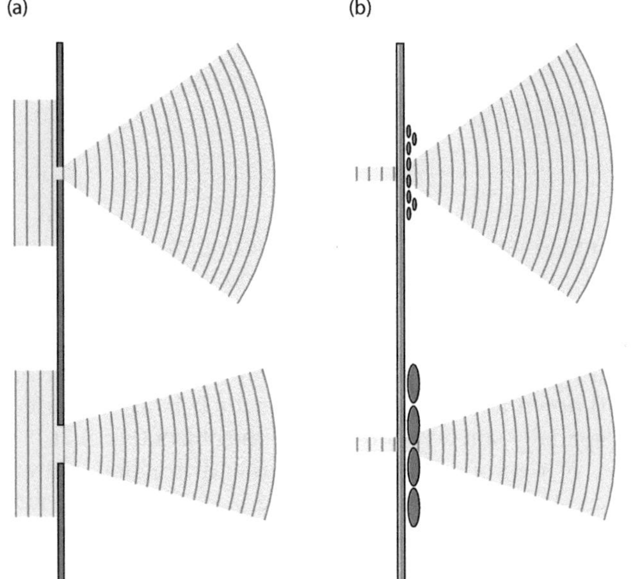

FIGURE 4.1
Diffraction of light by (a) apertures and by (b) particles with dimensions on the order of the wavelength of the light. Narrower apertures create wider patterns. Similarly, smaller particles (small ovals) diffract/scatter light through larger angles than larger particles (large ovals). The similar behavior of apertures and obstacles is the content of Babinet's principle.

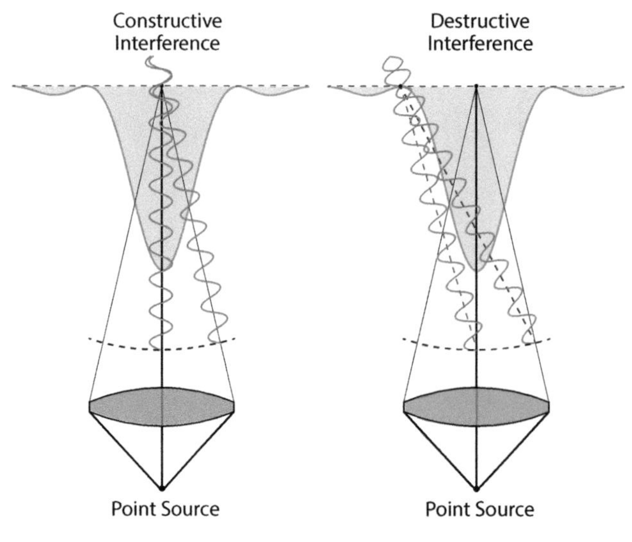

FIGURE 4.2
Schematic diagram describing the origin of the Airy pattern (shown shaded in light blue) in a microscope. The Airy pattern can be calculated by applying the Huygens-Fresnel principle to the dashed curve, which depicts a spherical wave consisting of EM fields exiting the objective lens and converging toward the geometrical image point. This principle states that every unobstructed point on a wavefront of frequency ν acts as a source of secondary spherical wavelets of the same frequency, and that the field beyond is the superposition of the secondary wavelets using their amplitudes and phases. The image point is surrounded by a bright spot created by constructive interference of the wavelets because all secondary sources are equidistant (one radius away) from the image point. In contrast, the dark ring that delimits the maximum arises from destructive interference of wavelets that have migrated π out of phase. This image pattern also can be calculated using a convolution operation, which is discussed in Box 6.2 of Chapter 6. (For simplicity, we have ignored the tube lens.)

reader is aware of the spectrum of possible approaches and their potential limitations. In subsequent sections, we focus on the simpler of the two methods.

One common way to quantify resolution is to use a single-parameter descriptor, such as the smallest lateral spacing (e.g., in μm) between two points in the specimen

DEMONSTRATION 4.1 DIFFRACTION BY APERTURES

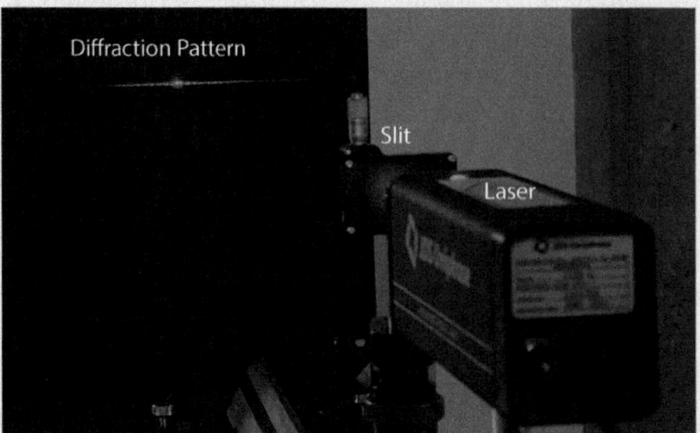

Diffraction by linear and circular apertures can easily be demonstrated using a laser and inexpensive slides containing holes or slits. Red and green lasers and holes or slits of different sizes can be used to demonstrate that the width of the diffraction pattern is proportional to the wavelength of the illuminating light and is inversely proportional to the size of the diffracting feature, as specified in Eq. (4.1). (Photograph by the authors.)

BOX 4.1 DIFFRACTION BY A SLIT

Diffraction-induced spreading is relatively easy to quantify for the case of EM radiation of wavelength λ impinging on a slit of width a (Fig. B4.1.1). Here we derive a pivotal feature of the resulting diffraction pattern, the full width of its bright central maximum, which is a key determinant of resolution in microscopy. Additional details are presented in most introductory physics texts.

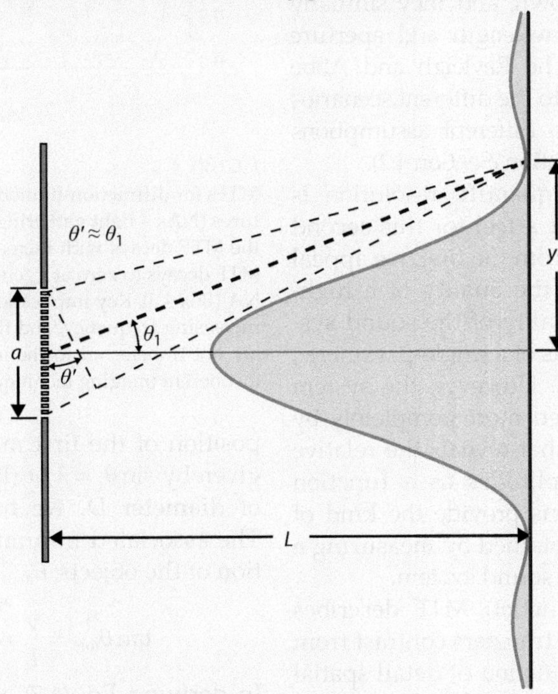

FIGURE B4.1.1
Huygens's sources (horizontal dashes) within a slit of width a generate a diffraction pattern of width $2y_1$ on a distant screen.

When a planar EM wavefront encounters a tiny slit, the wavefront is perturbed, and a new "diffracted" field is created. The diffracted field can be calculated by invoking the Huygens-Fresnel principle, which is described in the legend to Fig. 4.2. Fig. B4.1.1 shows the use of the Huygens-Fresnel principle to predict the location of bright and dark points in the intensity pattern created on a distant screen. The aperture contains many sources of secondary waves, which superimpose on the screen after traveling different path lengths. The center is a bright point created by constructive "interference" because waves from the top and bottom sources travel the same distance to the center, as do waves from the second from the top and second from the bottom sources, and so on. The bright central maximum is flanked by dark spots created by destructive interference. Simple geometry discussed in most introductory physics texts shows that the angular position, θ_1, of the first dark point is given by the expression

$$a \sin\theta_1 = \lambda \tag{B4.1.1}$$

This result also is subject to simple experimental verification. Starting with Eq. (B4.1.1), and invoking the small-angle approximation $\tan\theta \approx \sin\theta$, it is straightforward to derive the full width, W, of the bright central maximum (Fig. B4.1.1):

$$W = 2y_1 = 2L\tan\theta_1 \approx 2L\sin\theta_1 = 2L\left(\lambda/a\right) \tag{B4.1.2}$$

where W varies inversely with slit width and is proportional to wavelength. There are two conceptual ways to rationalize the first attribute, which often seems counterintuitive. First, when the slit is narrow, the sources within it are spatially close. Path differences thus tend to be small, and constructive interference persists over a large angle, producing a central maximum that is wide. Second, a narrow slit truncates the field significantly. This leads to the loss of the wavelets that helped to confine the beam and maintain forward propagation without spread (Section 2.3). As a consequence, the beam diffracts (spreads).

that can be detected by the microscope. As discussed below, under ideal conditions, this value is ~0.2 μm; the minimum spacing is nonzero because diffraction causes the images of the points to blur and overlap when they get too close. In upcoming sections, we discuss single-parameter descriptors proposed by Lord Rayleigh and Ernst Abbe. Both are well known, and they similarly reveal the critical role that wavelength and aperture play in dictating resolution. The Rayleigh and Abbe results are subtly different due to the different scenarios (models) analyzed and due to different assumptions about what is required for resolution (Section 4.2).

Another common way to quantify resolution is via a transfer function. To get a feel for this second approach, and how it differs from the first, we appeal to an analogy – quantifying the quality of a high-fidelity sound system. The quality of the sound system could be described in terms of a single-parameter, e.g., its high-frequency cutoff. However, the system quality also could be described more completely by measuring a response curve that reveals the relative input and output signal amplitudes as a function of frequency. Transfer functions provide the kind of detailed information that is obtained by measuring a frequency response curve of a sound system.

The modulation transfer function, MTF, describes how faithfully the microscope transfers contrast from the object to the image as a function of detail spatial frequency, while below the resolution limit. (High frequencies correspond to fine details.) In the context of the MTF, contrast measures the intensity variation about the mean divided by the mean, and thus reflects the detectability of intensity fluctuations against background (Chapter 5 and HW 5.1). An ideal MTF has a flat response and provides the most faithful reproduction of the object. In contrast, real MTFs tend to decay with increasing frequency (Fig. 4.3 and Box 4.3). Thus, contrast transfer is robust for coarser details and becomes increasingly poor for finer details. Single-parameter descriptors are very useful, but one limitation is their failure to highlight this important trend.

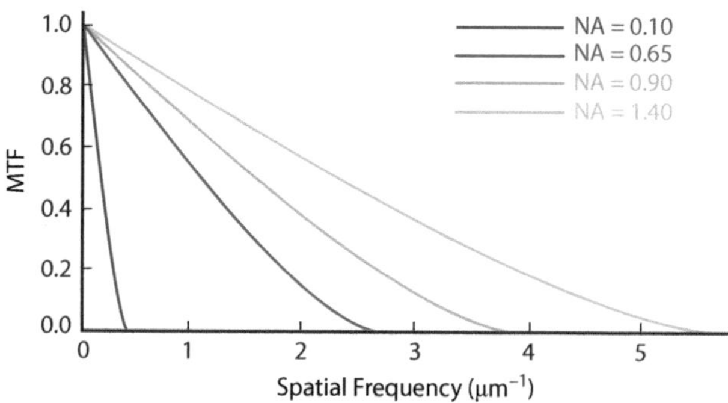

FIGURE 4.3
MTFs for diffraction-limited lenses with different numerical apertures (NAs – light gathering abilities – Section 4.1.3). In each case, the MTF decays with increasing spatial frequency. Moreover, the MTF decays to zero at a "cutoff" frequency that is proportional to NA (Box 4.3). Key implications are that contrast transfer falls with increasing frequency, and that frequencies higher than the cutoff are not transferred to the image. The plots shown here apply to incoherent imaging techniques like fluorescence (Section 4.5).

4.1.3 Rayleigh's Resolution Criterion

Resolution limits imposed by diffraction can be obtained approximately using relatively simple arguments, which lead to the famous Rayleigh resolution criterion.

To obtain Rayleigh's criterion, suppose that Sources 1 and 2 are two distant objects, such as stars. The objective lens of a telescope will image the objects as Airy disks, and as the angular separation of the objects decreases the Airy patterns will overlap increasingly until the patterns can no longer be resolved (Fig. 4.4).

Lord Rayleigh suggested that Airy patterns be regarded as just barely resolvable if the peak in one pattern falls on the first minimum of the second. The angular position of the first minimum of a slit of width a is given by $\sin\theta_1 = \lambda/a$ (Box 4.1). For a circular aperture of diameter D, we have similarly $\sin\theta_1 = 1.22\lambda/D$. The associated minimum resolvable angular separation of the objects, θ_{min}, is (Fig. 4.4)

$$\tan\theta_{min} = \frac{y}{L} \approx \sin\theta_{min} = \frac{1.22\lambda}{D} \approx \theta_{min} \quad (4.2)$$

In deriving Eq. (4.2), we have used the small angle approximation $\tan\theta \approx \sin\theta \approx \theta$ (in radians).

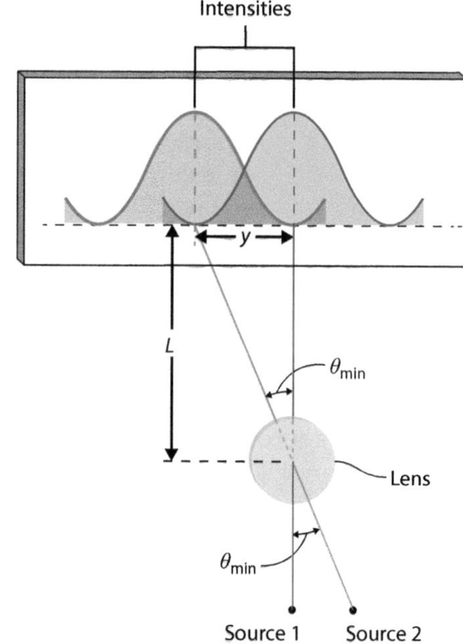

FIGURE 4.4
Diffraction patterns generated by two sources separated by θ_{min} are resolvable as two peaks; however, as the angular separation of the sources is reduced further, the patterns overlap and are not resolvable. For simplicity, only one ray from each source is shown.

SAMPLE PROBLEM 4.1 RESOLUTION OF AUTO HEADLIGHTS BY HUMAN EYE

The principles discussed in this section apply to many situations. For example, assume that the resolution of the human eye is limited by its pupil diameter. (a) Calculate the maximum distance at which the human eye can resolve automobile headlights as two separate objects. Assume that the nighttime pupil diameter is 5.0 mm, $\lambda_0 = 500$ nm, and the spacing between headlights is 1.1 m. (b) How does the distance change in the daytime? Assume that the daytime pupil diameter is 3.0 mm.

SOLUTION

a. We can determine the answer by calculating the angular resolution of the eye using Eq. (4.2) and then using the angle to calculate the distance. The angular resolution of the eye is

$$\theta_{\min} = \frac{1.22\lambda_0}{D} = \frac{1.22 \times 500\ \text{nm}}{5 \times 10^6\,\text{nm}} = 1.22 \times 10^{-4}\,\text{rad}$$

Therefore, the car's headlights will be resolvable as long as their angular separation exceeds this value. Invoking the content of Eq. (4.2), we have

$$\theta_{\min} = \frac{y}{L} \Rightarrow L = \frac{y}{\theta_{\min}} = \frac{1.1\ \text{m}}{1.22 \times 10^{-4}\,\text{rad}} = 9.0 \times 10^3\ \text{m} = 9.0\ \text{km}$$

Thus, the car's headlights will be distinguishable as two separate objects at distances up to 9 km. In real life, the intervening air, the finite size of the headlights, and other factors conspire to reduce this distance.

b. The distance at which the headlights are distinguishable scales with pupil size. In the daytime, when the pupil size is smaller, the distance shrinks to $3/5 \times 9 = 5.4$ km.

These results differ from our common experience that we "see better" in the daytime. While increased pupil size will increase resolution, the additional photons that hit the eye when illumination is bright improve resolution statistically and help to overcome dark noise. The subject of noise will be explored in Chapter 9.

The spatial version of Rayleigh's result, as applicable to a microscope, is much more difficult to derive. It involves invoking fairly sophisticated results from optics, notably Abbe's "sine condition." Born and Wolf (1999) presents the derivation, and in the interests of brevity, we refer the reader to their discussion. The result is

$$d_{\min} = \frac{0.61\lambda_0}{\text{NA}} \tag{4.3}$$

where λ_0 is the vacuum wavelength, and $d_{\min}$ is the minimum distance between two points that the microscope can resolve. In Eq. (4.3), the light-gathering ability of a lens is quantified in terms of numerical aperture:

$$\text{NA} = n \sin\theta_{\max} \tag{4.4}$$

where n is the refractive index of the medium in which the light propagates (e.g., the medium between the sample and objective), and $\theta_{\max}$ is half the angular aperture (collection angle) of the lens.

Equation (4.3) has several important implications:

1. Objects smaller than $d_{\min}$, although visible (if contrast is sufficient), will appear to have enlarged dimensions (an apparent radius) on the order of $d_{\min}$. This is most obvious as distortion in the image of a point object.
2. Objects more closely spaced than $d_{\min}$ will merge together and be unresolvable.
3. Resolution will improve with decreasing λ_0 and increasing NA.

Equation (4.3) often is quoted in an approximate form that is based on setting NA ~ 1.0 and replacing 0.61 by 0.50. This yields the approximation $d_{\min} \sim \lambda_0/2$. The resolution limit of optical microscopy (using visible radiation) thus often is stated to be ~200 nm, assuming the best-case scenario (i.e., violet light with $\lambda_0 \sim 400$ nm).

4.2 Abbe Theory

The effects of diffraction on resolution also were analyzed experimentally and theoretically in the latter half of the 19th century by another pioneer in optics,

Ernst Abbe. Unlike Rayleigh, Abbe analyzed nonluminous objects that were externally illuminated, and he concluded that diffraction can have a dual effect on image formation. First, a complex nonluminous object can generate diffraction "orders" when λ-sized structures in the object interact with the incident field. Second, the entrance aperture of the lens can limit the collection of specimen-generated diffraction orders. This second effect degrades the image that is encoded in the specimen-induced diffraction pattern.

In what follows, we describe Abbe's simple and elegant theory of the formation of images created by diffracting objects. We also derive Abbe's famous expression for diffraction-limited resolution, which, although mathematically similar to Rayleigh's criterion, has a more rigorous theoretical foundation and laid the foundations for the subject of Fourier optics (Box 4.3). Some additional experimental implications of Abbe theory are described in Section 4.4.

In developing his theory, Abbe studied an object known as a diffraction grating (Fig. 4.5), which consists of closely spaced lines or slits/sources equally separated by a distance d. It is easy to confirm (HW 4.9) that when a grating is illuminated *normally* (i.e., at a right angle – Fig. 4.5) with monochromatic light of wavelength λ, a set of bright diffraction spots is created, which arise from constructive interference. The relationship governing the angular position θ_m of the mth bright diffraction spot is given in most introductory physics texts and also is subject to simple experimental verification. It reads

$$d \sin \theta_m = m\lambda \quad m = 0, \pm 1, \pm 2, \ldots \tag{4.5}$$

An important conceptual prediction of Eq. (4.5) is that fine gratings (with small spacings, d) diffract over large angles, whereas coarse gratings (with large spacings) diffract over small angles. This concept will play a critical role in Abbe's interpretation of how a microscope resolves specimen features with different spacings, which Abbe modeled using gratings of different fineness.

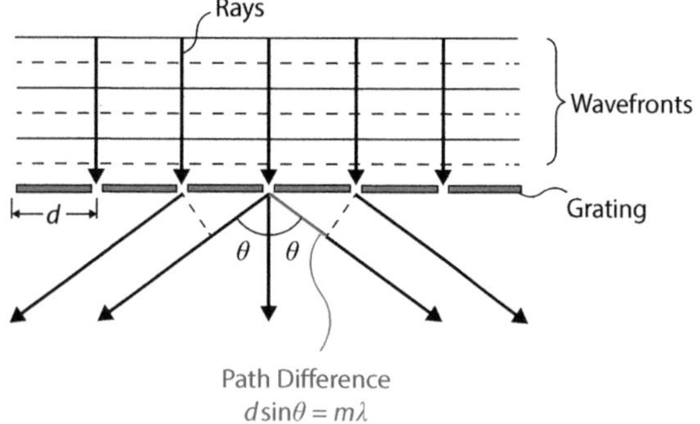

FIGURE 4.5
Schematic diagram of diffraction by a normally illuminated grating.

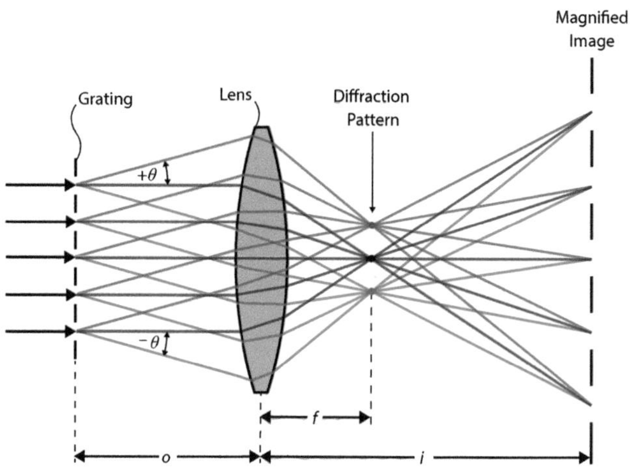

FIGURE 4.6
Diagram showing a ray view of the formation of a grating image in the image plane. Undeviated ($m = 0$) rays are black, and the $m = \pm 1$ diffraction orders are blue and red, respectively. The image is generated from interference of these three waves. The diagram also shows that the grating diffraction pattern appears in the BFP of the image-forming lens. This issue is discussed in Section 4.4.

Based on his studies of gratings, Abbe concluded that diffraction and interference play key roles in image formation. Specifically, some of the illumination is diffracted by small particles or features (size ~ λ) in the specimen, with smaller features, like finely spaced gratings, causing diffraction over larger angles (Fig. 4.1). Some of the illumination also traverses the specimen without interaction and is undeviated. Image formation then occurs via interference between at least two orders of light (e.g., the undeviated and first-order diffracted light) that are captured by the objective and brought to focus in the image plane (Fig. 4.6).

Abbe also concluded that a resolved grating image will be formed only if the objective captures at least the zeroth order and one of its nonzero (e.g., $m = \pm 1$) diffraction orders or, more generally, any two orders; capture of additional (higher) orders will enhance resolution. If only zeroth-order (undiffracted) light passes through, an unresolved image is formed because no specimen-specific information is collected (Fig. 4.7). These ideas imply that an image is inevitably an imperfect representation of the object, especially with respect to finer details, because the objective is unable to capture rays that diffract at large angles. In other words, a microscope has a limited resolving power.

Abbe used his theory of image formation to quantify the resolving power of an optical microscope. Specifically, Abbe derived a formula (engraved on his memorial) for the smallest detail $d_{\min}$ resolvable by a microscope, which states

$$d_{\min} = 0.5\lambda_0 / \mathrm{NA} \tag{4.6}$$

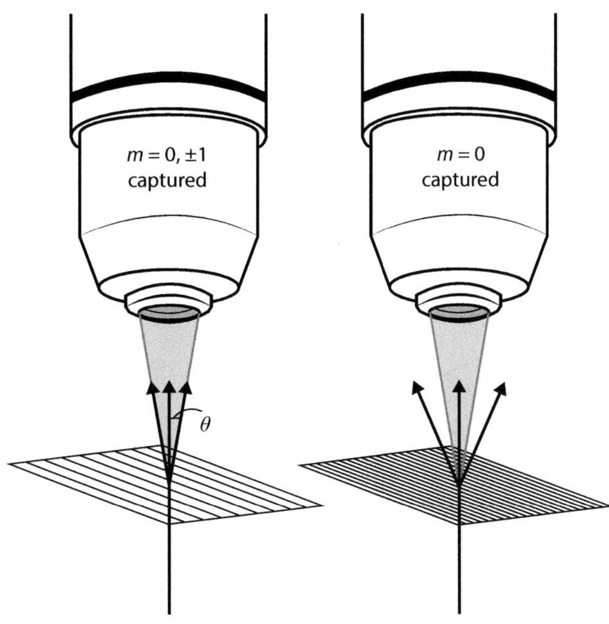

FIGURE 4.7
Schematic diagram of Abbe Theory. First-order spots are captured for a coarse grating (left), so a resolved image will be formed. In contrast, only zeroth-order (undiffracted) light is captured for a fine grating (right), which diffracts light through a greater angle, so an unresolved image will be formed.

Equation (4.6) is valid when the condenser and objective have the same numerical aperture ($NA_{cond} = NA_{obj}$), as shown in the derivation in Box 4.2. Comparison of Eqs. (4.3) and (4.6) shows that the Abbe resolution is *slightly* better than the Rayleigh resolution. This can be rationalized at least in part by the fact that Rayleigh's criterion demands some degree of contrast (minimally an ~20% intensity dip) for resolvability, whereas Abbe's criterion does not. (A less demanding criterion proposed by Sparrow gives a result that is nearly identical to that proposed by Abbe.) In addition, Rayleigh and Abbe analyzed incoherent and coherent imaging, respectively, and thus some disparity in results is not surprising (Sections 4.5 and 4.6).

Equation (4.6) can be derived for a special case, normal illumination (where Eq. [4.5] is valid), using very simple arguments. We do this now, emphasizing that the disparities with the full result highlight the importance of the condenser's role in generating oblique illumination.

Consider a normally illuminated diffraction grating with spacing d that acts as a specimen for an objective with capture half-angle θ_{max} that is immersed in oil with refractive index n. Imagine reducing the size of d. As a consequence, the first diffraction order spans a progressively larger angle. At some point, d becomes so small that the first order spans the objective half-angle and is just captured (Figs. 3.15 and 4.7). This is the smallest detail, d_{min}, or spacing that will be reproduced (i.e., resolved) in the image. Starting with Eq. (4.5), and using $m = 1$, yields

$$d_{min} \sin\theta_{max} = \frac{\lambda_0}{n}$$

$$d_{min} = \frac{\lambda_0}{n \sin\theta_{max}} = \frac{\lambda_0}{NA} \tag{4.7}$$

where we have substituted $NA = n \sin\theta_{max}$ and have invoked $\lambda = \lambda_0/n$ (Eq. [2.6]). Equations (4.6) and (4.7)

BOX 4.2 RESOLUTION WITH OBLIQUE ILLUMINATION

To derive Eq. (4.6) in full, again consider a specimen that is a grating with spacing d. In this case, under oblique illumination at angle θ_i the grating produces points of constructive interference for angles θ_i and θ_m satisfying a grating equation (for the rationale see the legend to Fig. B4.2.1):

$$dn_c \sin\theta_i + dn_o \sin\theta_m = m\lambda_0 \quad m = 0, \pm 1, \pm 2, \ldots \tag{B4.2.1}$$

In this equation, n_c is the refractive index of the medium between the condenser and specimen, n_o is the refractive index of the medium between the objective and specimen, λ_0 is the vacuum wavelength, and the angles satisfy standard sign conventions. According to Abbe, d_{min} corresponds to maximal separation between the $m = 0$ and $m = 1$ orders, consistent with capture by the objective. This situation occurs when maximally oblique illumination from the condenser generates an undiffracted spot on one side of the objective, and the grating generates a first-order spot on the opposite edge of the objective (Fig. 3.15), or, equivalently, when $n_o \sin\theta_m$ and $n_c \sin\theta_i$ take on their maximal values. Inserting these maxima, NA_{obj} and NA_{cond}, into Eq. (B4.2.1), we obtain

$$d_{min} = \frac{\lambda_0}{NA_{obj} + NA_{cond}} \tag{B4.2.2}$$

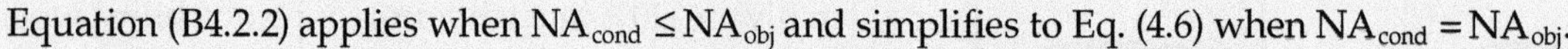
Equation (B4.2.2) applies when $NA_{cond} \leq NA_{obj}$ and simplifies to Eq. (4.6) when $NA_{cond} = NA_{obj}$.

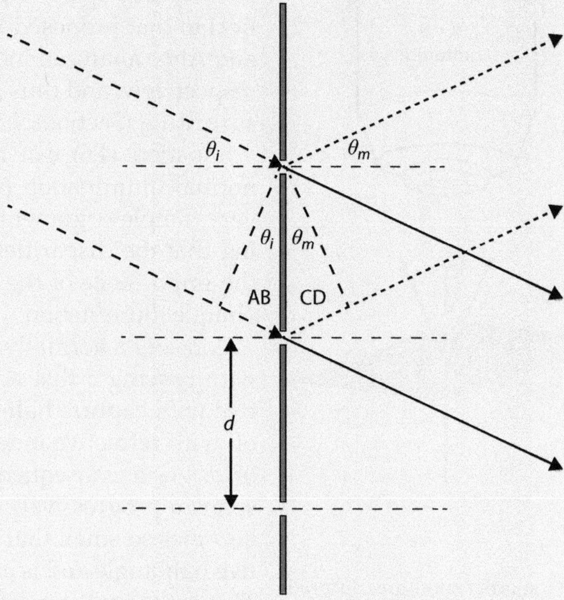

FIGURE B4.2.1
Geometry and derivation of the grating equation. Constructive interference will occur when the total path difference (AB + CD) contains an integral number of wavelengths (Fig. 2.7). Stated mathematically,

$$\frac{AB}{\lambda_0 / n_c} + \frac{CD}{\lambda_0 / n_o} = m \text{ where } m = 0, \pm1, \pm2\ldots$$

From the picture, we have

$$AB = d \sin\theta_i \text{ and } CD = d \sin\theta_m$$

which yields

$$dn_c \sin\theta_i + dn_o \sin\theta_m = m\lambda_0$$

where $m = 0, \pm1, \pm2\ldots$

are very similar, but the latter yields a resolution that is poorer by a factor of 2.

In Box 4.2, we show that the improved resolution predicted in Eq. (4.6) is achieved by using a high NA condenser that sends in oblique illumination. If the NA of the condenser is as high as that of the objective, the improvement is two-fold. The improvement can be understood conceptually by noticing that oblique illumination permits use of the entire NA of the objective to capture the first diffraction order (Fig. 3.15). Nonetheless, the full range of illumination is better than oblique alone.

4.3 The Significance of Numerical Aperture

At this juncture, it is worth discussing why NA is more relevant to resolution than θ_{max} alone. The essential point is shown in Fig. 4.8. In the absence of oil, rays bend away from the normal upon exiting the cover glass, and thus more oblique rays tend to miss the objective. Moreover, at angles that exceed ~40°, rays undergo TIR and do not enter the lens (Fig. 4.8a). The presence of oil reduces these effects, and resolution

SAMPLE PROBLEM 4.2 RESOLUTION OF MICROSCOPE OBJECTIVES

Calculate the Abbe resolution of the following commercially available microscope objectives, each of which has the same magnification, for $\lambda_0 = 400$ nm and $\lambda_0 = 700$ nm:

Objective	NA	Cost
63× Plan Achromat	0.65	$$
63× Plan Fluorite	0.75	$$$
63× Plan Apochromat	0.95	$$$$

Discuss any trends.

SOLUTION

The Abbe resolution is given by Eq. (4.6). Substituting values into the equation for the 63× plan achromat objective for 400 nm yields the following result:

$$d_{min} = \frac{\lambda_0}{2\text{NA}} = \frac{400 \text{ nm}}{2 \cdot 0.65} = 0.31\ \mu\text{m}$$

Repeating this process for the remaining objectives and wavelength completes the analysis.

Objective	NA	Resolution at 400 nm (µm)	Resolution at 700 nm (µm)
63× Plan Achromat	0.65	0.31	0.54
63× Plan Fluorite	0.75	0.27	0.47
63× Plan Apochromat	0.95	0.21	0.37

Resolution improves with decreasing wavelength, increasing numerical aperture, and (unfortunately!) increasing cost.

improves due to enhanced ray capture. Resolution is thus affected by a combination of n_{oil} and θ_{max}, as embodied in NA.

Numerical apertures of standard oil immersion objectives are bounded above by ~1.5, and a typical high NA is ~1.49. The upper bound arises because the refractive indices of oils used in optical microscopy are matched to those of glass (i.e., n_{oil} ~1.50), and the maximum value of the sine function is 1.00. Hence, $\text{NA}_{max} = n_{oil} \sin\theta_{max} = n_{oil} \times 1.00 \sim 1.50$ for conventional objectives. A select set of objectives use an oil with higher n.

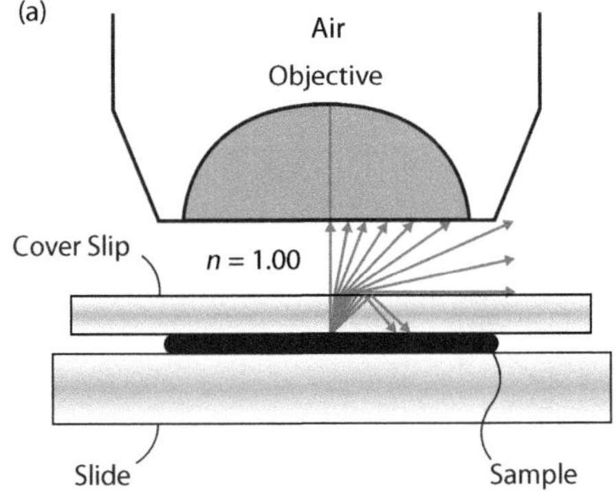

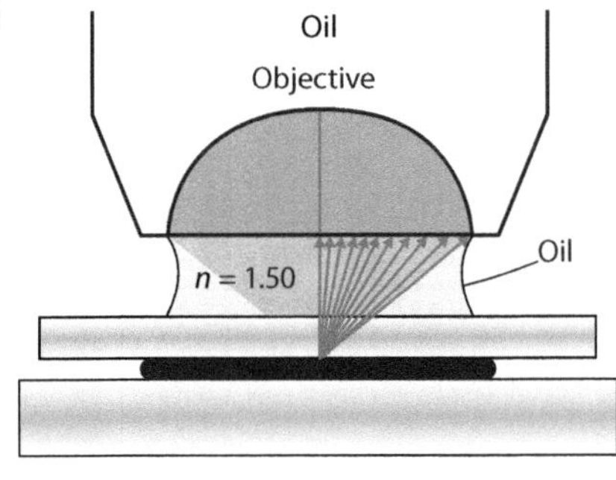

FIGURE 4.8
Effect of immersion oil on capture of rays exiting a sample. (a) Rays extensively miss the objective when air is present between the cover slip and objective. (b) Ray capture and resolution are markedly enhanced when oil with a refractive index that is roughly matched to that of glass is placed between the cover slip and the front element of an oil immersion objective.

4.4 Abbe's Experiments: Manipulating Specimen Diffraction Patterns

In the course of his experimental work, Abbe observed a pivotal, but relatively underappreciated, property of lenses, which we alluded to briefly in Fig. 4.6. Specifically, Abbe noticed that *the diffraction pattern of the specimen field distribution appears in the BFP of the objective, while the image simultaneously appears in the image plane.* Abbe also found that manipulating the diffraction pattern, which is known as "spatial filtering," alters the image (Fig. 4.9). In this section, we briefly discuss these issues in anticipation of their importance in Chapter 5. In Box 4.3, we connect Abbe's observations with Fourier analysis and Fourier optics.

The presence of the specimen diffraction pattern in the BFP of a lens can be derived rigorously, as discussed by Goodman (2017). This fact also is suggested by the alternating, reciprocally related (focused and defocused) field patterns in the Köhler illumination optical train (Fig. 3.13). Specifically, a specimen and its diffraction pattern are reciprocally related, which is consistent with their presence in reciprocally related conjugate planes in the Köhler train.

The presence of the specimen diffraction pattern in the BFP of a lens is relatively easy to verify experimentally. For example, consider the simple lens/laser setup shown in Fig. 4.10. The object is a mesh that generates an easily visualized diffraction pattern, and it is placed approximately in the FFP of the lens. With this setup, it is easy to confirm that the lens produces the object diffraction pattern in its BFP, as well as the image at infinity.

To demonstrate spatial filtering, place a rotatable slit aperture in the BFP of the lens, which can be used to "filter out" all but a line of spots in the diffraction pattern. Like Abbe, we find that the image of the mesh is altered in a predictable fashion. For example, filtering out horizontal spots has a dramatic, predictable effect on the image, as explored in Activity A4.2.

This simple discussion of spatial filtering shows that an image can be fundamentally altered by removing attributes that are encoded in the associated diffraction pattern. For our purposes, the primary application of this fact will be to alter an image to enhance contrast, which is the focus of Chapter 5.

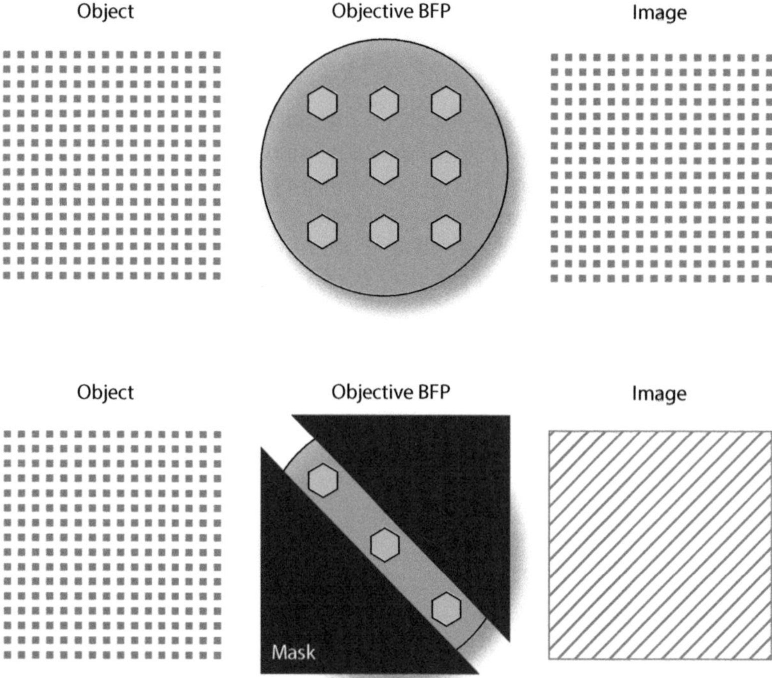

FIGURE 4.9
Effects of spatial filtering. An array of dots and its associated image are shown together with the pattern of diffraction orders in the BFP of the objective, which appears as replicas of the stopped down hexagonally shaped condenser aperture diaphragm. In the absence of spatial filtering (top panel), the image is a replica of the object. In contrast, in the presence of filtering (bottom panel) that leaves only diagonal diffraction orders, the image is markedly altered, consisting of a set of orthogonally oriented lines.

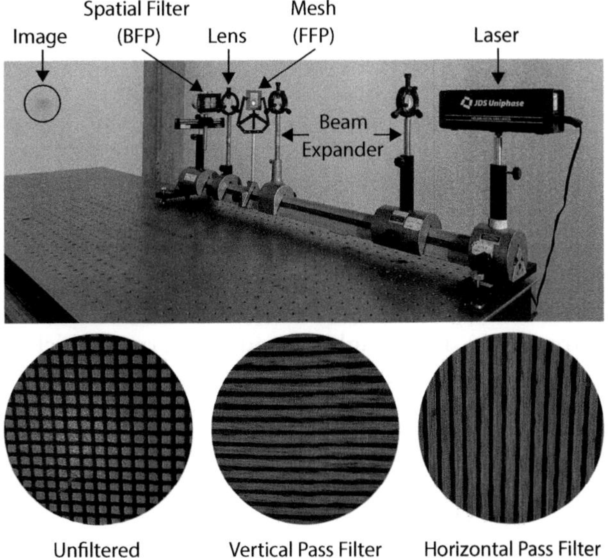

FIGURE 4.10
Diffraction and spatial filtering. The top panel shows an experimental setup for demonstrating the effects of spatial filtering in a simple and accessible manner. The bottom panels compare the unfiltered image of a mesh with filtered images prepared by placing a vertical slit (vertical pass filter) or horizontal slit (horizontal pass filter) in the BFP of the image-forming lens. The vertical and horizontal filters pass only diffraction spots in a vertical or horizontal line, respectively, centered at the origin. (Photographs by the authors.)

BOX 4.3 ABBE THEORY AND FOURIER OPTICS

Abbe theory is a description of image formation that is closely linked to Fourier analysis. Given the power and prevalence of Fourier analysis, here we provide a brief introduction to the methodology and establish the link between Fourier analysis and Abbe theory. Fourier analysis also will prove useful in our upcoming treatments of super-resolution imaging, image processing, and medical imaging. The reader interested in a more detailed discussion of Fourier analysis and its application to optics is referred to the treatments by Hecht (2017) and Goodman (2017).

We introduce Fourier analysis by considering its application to a periodic function in one dimension, $f(x)$, that has a spatial period λ. Fourier's theorem asserts that any periodic function can be written in the form

$$f(x) = \frac{A_0}{2} + \sum_{m=1}^{\infty} A_m \cos(kmx) + \sum_{m=1}^{\infty} B_m \sin(kmx) \tag{B4.3.1}$$

In Eq. (B4.3.1), $k = 2\pi/\lambda$. From a conceptual perspective, Eq. (B4.3.1) establishes that $f(x)$ can be synthesized as a superposition of sine and cosine functions with "spatial frequencies," $k_m = mk$, that are integral multiples (harmonics) of a lowest spatial frequency, k. The contribution of each frequency to the signal is determined by its associated amplitude, A_m or B_m. These amplitudes are given by the expressions (HW 4.10)

$$A_m = \frac{2}{\lambda} \int_0^{\lambda} f(x)\cos(kmx)\,dx \tag{B4.3.2a}$$

$$B_m = \frac{2}{\lambda} \int_0^{\lambda} f(x)\sin(kmx)\,dx \tag{B4.3.2b}$$

We illustrate some important attributes of Fourier synthesis in the context of a specific example, namely, the Fourier decomposition of an infinite periodic square wave, given over one period by

$$f(x) = \begin{cases} +1 & 0 < x < \lambda/2 \\ -1 & \lambda/2 < x < \lambda \end{cases}$$

The Fourier decomposition is (HW 4.11)

$$f(x) = \frac{4}{\pi}\left(\sin(kx) + \frac{1}{3}\sin(3kx) + \frac{1}{5}\sin(5kx) + \frac{1}{7}\sin(7kx) + \cdots\right) \quad \text{(B4.3.3)}$$

Fig. B4.3.1 shows plots of Eq. (B4.3.3) that include successively larger numbers of terms in the expansion. The plots reveal that the lowest frequency term captures the large-scale, overall spatial periodicity (detail) in the function. Higher frequency terms encode finer details, and with their addition the decomposition becomes a progressively more accurate representation of the square wave function.

Notably, Eq. (B.4.3.3) shows that the amplitudes, $B_m = 4/m\pi$, of the frequency contributions, $k_m = mk$, fall off with increasing spatial frequency. Thus, higher frequency terms, which encode finer details, make a progressively smaller contribution, and synthesis of the function from a sufficiently large subset of lower frequency terms leads to a reasonable approximation to the function.

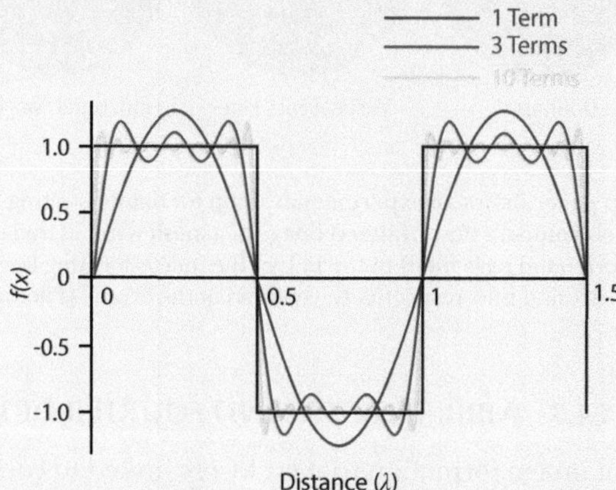

FIGURE B4.3.1
Graphical representation of the synthesis of portions of an infinite periodic square wave from its Fourier components. The lowest order term alone (shown in blue) is a sine wave that captures the overall periodicity of the function. The first three terms (in red) begin to show the square character of the function. The first ten terms (in green) represent the function quite well. The waveform continues to make discontinuous transitions between +1 and –1 for all x.

What is the connection between Fourier analysis and Abbe's analysis of diffraction patterns and image formation? The diffraction pattern is an intensity map, where intensity at each point can be shown to be proportional to the time average of the electric field squared or, equivalently, to the square of the Fourier amplitude at the associated spatial frequency. Given this, one connection is that every diffraction spot in the BFP in Abbe theory indicates the presence of a specific spatial frequency in the object Fourier transform, with spots at higher angles corresponding to Fourier components at higher frequencies. As a consequence, Abbe theory essentially is an analysis of the objective's finite ability to pass Fourier components and of the associated implications for image quality. For example, Abbe's conclusion that collection of more diffraction orders leads to an improved image is basically a statement that the Fourier synthesis becomes progressively more accurate with the inclusion of more Fourier components.

Another, more quantitative, connection is that Abbe theory yields a prediction for the period of the densest line spacing in a grating that is resolvable by an optical microscope. This spatial limit on resolution appears as a maximum spatial frequency in the Fourier spectrum of the image. Specifically, Abbe's expression for d_{min} corresponds to a reciprocally related maximum Fourier frequency passed by the objective, which is given by (Eq. [4.6])

$$k_{max} = \frac{1}{d_{min}} = \frac{2\text{NA}}{\lambda_0} \quad \text{(B4.3.4)}$$

One example of the role that k_{max} plays in Fourier optics was discussed in the legend to Fig. 4.3. Another example is illustrated in Fig. B4.3.2, which shows that contrast transfer falls off as object spatial frequency increases. Moreover, consistent with Eq. (B4.3.4), contrast transfer falls to zero as spatial frequency reaches k_{max}.

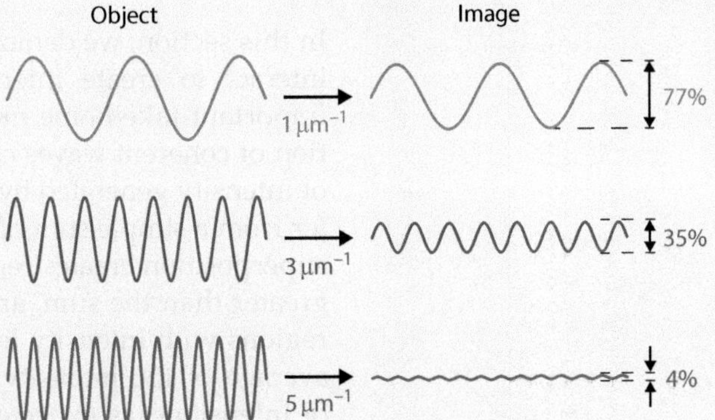

FIGURE B4.3.2
Schematic representation of the effects of diffraction on contrast transfer for an objective with NA = 1.4 and λ_0 = 500 nm. In the three cases shown, the object is a sinusoidal wave with a spatial frequency in inverse microns (μm^{-1}). The Abbe diffraction limit for this scenario (0.179 μm) corresponds to a frequency of 1/(0.179 μm) = 5.6 μm^{-1}. When the spatial frequency is relatively low, 1 μm^{-1}, contrast transfer is high (77%). At intermediate frequency, 3 μm^{-1}, contrast transfer is 35%, and at close to the Abbe limit, 5.0 μm^{-1}, contrast transfer is only 4%. Data are based on the contrast definition $C = (I_{max} - I_{min})/(I_{max} + I_{min})$ and the relationship $C^i(k) = \text{MTF}(k)C^o(k)$, where the superscripts refer to contrast in the image, *i*, and object, *o*. The relevant MTF is shown in Fig. 4.3.

We close this box with a model of a microscope that emphasizes a Fourier perspective (Fig. B4.3.3).

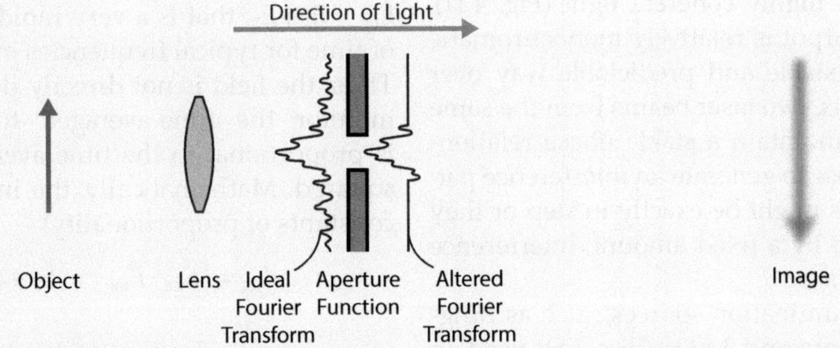

FIGURE B4.3.3
Fourier-based model of a microscope. The lens generates a Fourier transform of the object, which is close to the FFP, and then the aperture function fails to transmit the higher frequencies. The altered Fourier spectrum yields an image with lower resolution. Due to spatial constraints on this schematic, the size and position of the image are not to scale.

4.5 The Role of Coherence

To this point, we have ignored an important requirement for generating observable interference and diffraction patterns, which is that the sources have nearly the same frequency and be at least partially "coherent" (Fig. 4.11). Sources that are coherent maintain a constant relative phase. Here we briefly discuss coherence, and its implications for image quality, so that the reader appreciates the importance of coherence to the interference-based forms of optical microscopy discussed in the next chapter. Two examples are phase-contrast microscopy and differential interference contrast microscopy. In contrast, fluorescence imaging is incoherent.

Interference and diffraction rely on coherence because both effects arise when waves from two or more sources superpose to generate observable constructive and destructive interference. Moreover, interference is observable only if the waves from the sources superimpose stably over time. When sources

Coherent Source (Laser)

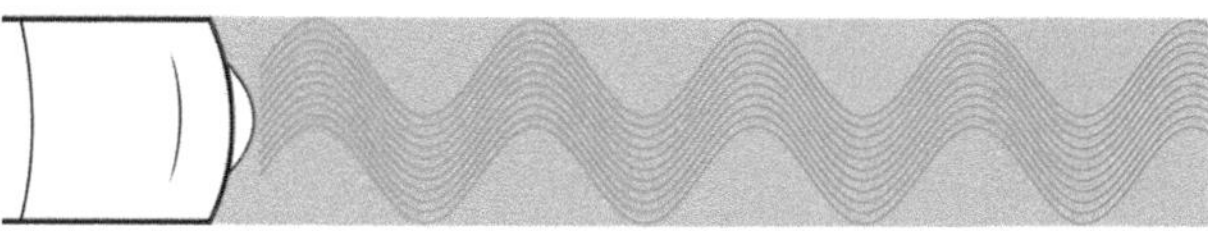

Incoherent Source (Bulb)

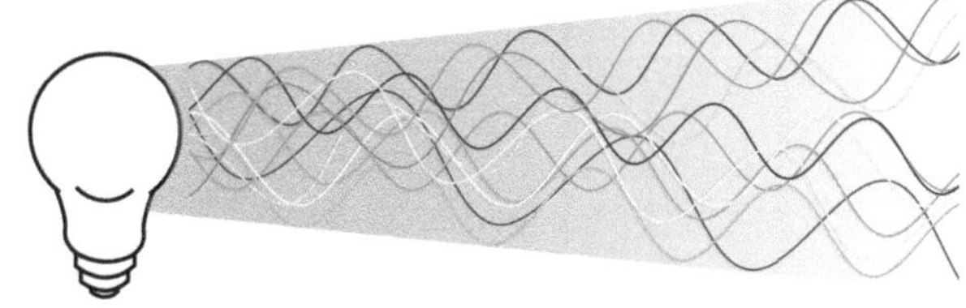

FIGURE 4.11
Waves emitted by a coherent source maintain a well-defined phase relationship, whereas waves emitted by an incoherent source do not.

are incoherent, the superposition at a given point oscillates rapidly between constructive and destructive, and this renders interference and diffraction effects unobservable.

Lasers are commonly used in optical microscopy and are a source of highly coherent light (Fig. 4.11). Specifically, their output is relatively monochromatic and oscillates in a stable and predictable way over long timescales. Thus, two laser beams from the same source can readily maintain a stable phase relationship and superimpose to generate an interference pattern. The two beams might be exactly in step or they might be out-of-step by a fixed amount; interference can occur in either case.

In contrast, the illumination sources, such as tungsten-halogen filaments and LEDs (Box 3.5), used in optical microscopy generate partially coherent illumination. The degree of coherence is far less than that of a laser but is sufficient for interference-based microscopy.

We close by addressing the relative image quality achieved for coherent and incoherent imaging. The incoherent transfer function (Fig. 4.3) has a cutoff frequency that is twice that of the coherent transfer function, and thus it sometimes is argued that the resolution of incoherent imaging is higher. However, direct comparison of transfer functions is problematic because incoherent imaging is linear in intensity, and coherent imaging is linear in field. Thus, the issue of relative resolution is complex. On the other hand, incoherent imaging is free of the "speckle" and "ringing" that frequently plague coherent imaging. In this latter respect, incoherent images often are of higher quality. The reader interested in more detail is referred to the text by Goodman (2017).

4.6 Mathematical Analysis of Interference from Two Coherent Sources [Optional]

In this section, we demonstrate how coherent sources interact to create interference mathematically. An important take-home message is that the superposition of coherent waves causes the spatial distribution of intensity generated by more than one source to differ from a simple sum of intensities. In particular, the superposition creates regions having intensity that is greater than the sum, and these are accompanied by regions with intensity less than the sum. The spatial average of the intensity is equal to that of the sum of intensities, as expected from energy conservation principles.

To illustrate these ideas mathematically, we will consider the simple superposition of two electric fields, $\vec{E}_1$ and $\vec{E}_2$, given by

$$\vec{E}_1(\vec{r},t) = \vec{E}_{01}\sin\left(\vec{k}_1\cdot\vec{r} - \omega t + \varphi_1\right)$$
$$\vec{E}_2(\vec{r},t) = \vec{E}_{02}\sin\left(\vec{k}_2\cdot\vec{r} - \omega t + \varphi_2\right)$$

where the symbol "·" denotes a dot (or scalar) product. These fields superpose to produce a total field, $\vec{E}_{tot} = \vec{E}_1 + \vec{E}_2$, that is a very rapidly changing function of time for typical frequencies in the visible (~10^{15} Hz). Thus, the field is not directly detectable. Instead, we monitor the time-averaged total intensity, which is proportional to the time average of the total field squared. Mathematically, the intensity is (neglecting constants of proportionality)

$$\begin{aligned} I_{tot} &= \left\langle \vec{E}_{tot}\cdot\vec{E}_{tot}\right\rangle = \left\langle \vec{E}_1^2 + \vec{E}_2^2 + 2\vec{E}_1\cdot\vec{E}_2\right\rangle \\ &= I_1 + I_2 + 2\left\langle \vec{E}_1\cdot\vec{E}_2\right\rangle \end{aligned} \tag{4.8}$$

where the brackets, $\langle\ \rangle$, denote a time average. The first two terms on the right-hand side of Eq. (4.8) are the individual intensities, I_1 and I_2. The last term is an interference term, which, if nonzero, will render intensity nonadditive.

To determine when the interference term is nonzero, we rewrite

$$\left\langle \vec{E}_1\cdot\vec{E}_2\right\rangle = \left\langle \vec{E}_{01}\cdot\vec{E}_{02}\sin\left(\vec{k}_1\cdot\vec{r} - \omega t + \varphi_1\right)\sin\left(\vec{k}_2\cdot\vec{r} - \omega t + \varphi_2\right)\right\rangle$$

using the trigonometric identity $\sin(\alpha-\beta) = \sin\alpha\cos\beta - \cos\alpha\sin\beta$. This yields

$$\begin{aligned} \left\langle \vec{E}_1\cdot\vec{E}_2\right\rangle &= \vec{E}_{01}\cdot\vec{E}_{02} \\ &\left\langle\left\{\sin\left(\vec{k}_1\cdot\vec{r}+\varphi_1\right)\cos(\omega t) - \cos\left(\vec{k}_1\cdot\vec{r}+\varphi_1\right)\sin(\omega t)\right\}\right. \\ &\left.\times\left\{\sin\left(\vec{k}_2\cdot\vec{r}+\varphi_2\right)\cos(\omega t) - \cos\left(\vec{k}_2\cdot\vec{r}+\varphi_2\right)\sin(\omega t)\right\}\right\rangle \end{aligned} \tag{4.9}$$

Equation (4.9) can be simplified considerably if we assume that φ_1 and φ_2 are stable (not functions of time) and use the results

$$\langle \sin^2(\omega t)\rangle = \langle \cos^2(\omega t)\rangle = \frac{1}{2}$$

$$\langle \sin(\omega t)\cos(\omega t)\rangle = 0$$

$$\cos(\alpha - \beta) = \cos\alpha\cos\beta + \sin\alpha\sin\beta$$

These identities yield

$$\left\langle \vec{E}_1 \cdot \vec{E}_2 \right\rangle = \frac{1}{2}\vec{E}_{01}\cdot\vec{E}_{02}\cos\left(\vec{k}_1\cdot\vec{r} + \varphi_1 - \vec{k}_2\cdot\vec{r} - \varphi_2\right) = \frac{1}{2}\vec{E}_{01}\cdot\vec{E}_{02}\cos\delta \quad (4.10)$$

where $\delta = \left(\vec{k}_1 - \vec{k}_2\right)\cdot\vec{r} + \varphi_1 - \varphi_2$ is a stable phase difference arising from the path length, $\vec{r}$, and any initial difference in phase angle, φ. Finally, from Eq. (4.8), we have

$$I_{\text{tot}} = I_1 + I_2 + \vec{E}_{01}\cdot\vec{E}_{02}\cos\delta \quad \text{(Coherent Sources)} \quad (4.11)$$

The last term in Eq. (4.11) gives rise to interference. Importantly, the interference term will vanish if the two fields are orthogonal, so that $\vec{E}_{01}\cdot\vec{E}_{02} = 0$. This point will prove important when we discuss polarization-based forms of optical microscopy. The interference term also will average to zero and vanish if the two sources are incoherent, and thus there is no stable phase relationship between the fields. For both of these cases, intensities are additive:

$$I_{\text{tot}} = I_1 + I_2 \quad \text{(Incoherent Sources)} \quad (4.12)$$

SAMPLE PROBLEM 4.3 TWO-SOURCE INTENSITIES IN COHERENT IMAGING

Show that, when their electric fields are parallel, the total intensity generated by two coherent sources can be greater than, less than, or equal to the total intensity that would apply for incoherent sources (i.e., $I_1 + I_2$).

SOLUTION

The total intensity generated by two coherent sources is given by Eq. (4.11):

$$I_{\text{tot}} = I_1 + I_2 + \vec{E}_{01}\cdot\vec{E}_{02}\cos\delta$$

This equation can be simplified when the electric fields are parallel. In particular, the dot product of $\vec{E}_{01}$ and $\vec{E}_{02}$ becomes a simple product of their magnitudes, E_{01} and E_{02}, yielding

$$I_{\text{tot}} = I_1 + I_2 + E_{01}E_{02}\cos\delta$$

This equation can be rewritten in terms of time-averaged intensities by noting that intensity is proportional to the square of the electric field:

$$I_1 \propto \left\langle E_1^2 \right\rangle = \frac{E_{01}^2}{2}$$

$$I_2 \propto \left\langle E_2^2 \right\rangle = \frac{E_{02}^2}{2}$$

This yields

$$I_{\text{tot}} = I_1 + I_2 + 2\sqrt{I_1 I_2}\cos\delta \quad \text{(SP 4.3.1)}$$

Equation (SP 4.3.1) shows that I_{tot} can be greater than, less than, or equal to $I_1 + I_2$ depending on the value of δ. Important examples of the first two cases occur when δ is an even integral multiple of π or an odd integral multiple of π, respectively. These cases correspond to maximum constructive and maximum destructive interference.

Suggested Reading and Additional Resources

BioDIP – Biopolis Dresden Imaging Platform. 2016. *Understanding the light microscope by Peter Evennett.* https://www.youtube.com/watch?v=60_jgZtyR6U

Born, M. and Wolf, E. 1999. *Principles of optics: Electromagnetic theory of propagation, interference, and diffraction of light.* Cambridge University Press, London. http://dx.doi.org/10.1017/CBO9781139644181

Goodman, J.W. 2017. *Introduction to Fourier optics (4th Edition).* Macmillan Learning, New York, NY.

Hecht, E. 2017. *Optics (5th Edition).* Pearson Education Limited, Essex, England.

iBiology: Microscopy Series. 2012–2015. *Microscopy: Abbe diffraction.* https://www.ibiology.org/talks/abbe-diffraction/

PhET Interactive Simulations. 2002–2022. *Physics simulation on wave interference.* https://phet.colorado.edu/en/simulation/wave-interference

Tipler, P.A. and Mosca, G. 2008. *Physics for scientists and engineers (6th Edition).* W.H. Freeman and Company, New York, NY. Chapter 33.

HOMEWORK PROBLEMS

4.1. [Diffraction Pattern Width] A helium neon laser emits red light with $\lambda_0 = 633$ nm. The light hits a slit of width 0.10 mm and creates a diffraction pattern on a wall 2.00 m away. What is the full width of the diffraction pattern? Hint: See Box 4.1.

4.2. [Resolving Power of the Eye at the Near Point] The "near point" of the human eye is the closest distance at which the human eye can bring an object into focus (Box 3.4). This distance typically is taken to be 25 cm. Show that the resolution of the eye at the near point is approximately 50 μm for green (500 nm) light and a daytime pupil diameter of 3 mm. Hint: Review Sample Problem 4.1.

4.3. [Resolving Power of the Eye – Pointillism] George Seurat created paintings consisting of closely spaced 2-mm diameter dots of pure color. A famous example is "A Sunday Afternoon on the Island of La Grande Jatte" (1884–1886). At a sufficient viewing distance, the colors in this painting appear to blend together due to diffraction. Determine the minimum required viewing distance to achieve the blending effect for all colors in the wavelength regime from 400 nm to 700 nm. Assume a daytime pupil diameter of 3 mm. How does this viewing distance compare with the size of the painting (approximately 2.1 m high by 3.1 m wide)? (Reprinted from Art Institute of Chicago. https://www.artic.edu/artworks/27992/a-sunday-on-la-grande-jatte-1884. CC0 Public Domain Designation.)

4.4. [Resolution – Useful Magnification] In HW 4.2, the minimum resolvable separation at the near point of the eye is stated to be on the order of 50 µm = 0.050 mm. In this problem, use a slightly larger (often quoted) value, which is 0.10 mm, to calculate a simple rule-of-thumb for the magnification, $M \sim 500\text{NA}$, that makes optimal use of the diffraction-limited resolution of an objective with a given NA. Hint: To derive the rule, magnify the smallest spacing that the objective can resolve (according to Abbe) onto the resolvable spacing of the eye at the near point. Assume $\lambda_0 = 400$ nm. Magnifications that markedly exceed ~500–1000 NA are termed "empty" and can introduce artifacts.

4.5. [Lateral and Axial Resolution] Rayleigh's estimate for lateral resolution, $0.61\lambda_0/\text{NA}$, is based on the separation between the center of the diffraction pattern and the first lateral minimum, as stated in the text. The separation between the center of the pattern and the first axial minimum, $2\lambda_0 n / \text{NA}^2$, can analogously be used to estimate axial resolution (i.e., resolution along, as opposed to perpendicular to, the optical axis). Calculate the lateral and axial resolution of the following commercially available microscope objectives:

Magnification	NA	Lateral?	Axial?
2×	0.10 (air)		
20×	0.75 (air)		
63×	0.95 (air)		
	1.40 (oil)		
100×	1.40 (oil)		

Notice that axial resolution is more sensitive than lateral resolution to NA. Assume $\lambda_0 = 550$ nm, $n_{\text{air}} = 1.0$, and $n_{\text{oil}} = 1.5$.

4.6. [Lateral Resolution – Telescope] The Mount Palomar telescope has a 508-cm-diameter objective mirror, which acts as a diffracting aperture.
   a. For the telescope, determine the minimum angular separation between two resolvable objects at a wavelength of 550 nm.
   b. Alpha Centauri A and B are stars that form a binary pair. The pair is ~4.35 light-years away from earth. What is the minimum separation of the stars if they are to be resolvable by the Palomar telescope? Hint: 1 light-year = 9.46×10^{12} km.

4.7. [Lateral Resolution – UV Microscope] The UV microscope was developed in the early 1900s to increase resolution beyond that available with visible light (Chapter 1). UV microscopes use short-wavelength UV illumination and quartz optics (because glass absorbs UV light). Images are focused and captured without direct viewing. A classic setup uses 275-nm UV light and a 1.25 NA quartz objective. What is the associated Abbe resolution? How does it compare to the best resolution (~0.2 μm) obtained using visible light? UV microscopes are still used to observe cellular components, such as DNA, that absorb in the UV. However, they have largely been replaced with other high-resolution techniques.

4.8. [Lateral Resolution – Electron Microscope] The electron microscope was developed in the 1930s, shortly after the introduction of quantum mechanics. Electron microscopes use high-energy electrons, instead of photons, to image specimens. This is possible because electrons, like other particles, exhibit both wave-like and particle-like properties (the "wave-particle duality"). The wave-like properties mean that electrons, like photons, undergo diffraction. The wavelength, λ, of an electron is given by the de Broglie equation:

$$\lambda = \frac{h}{p}$$

where h is Planck's constant, $p = mv$ is the momentum of the electron, v is the speed of the electron, and m is the mass of the electron. Calculate the electron wavelength in a standard 100 keV electron microscope (where 100 keV is the nonrelativistic kinetic energy, E_k, of the electron). (Hint: $E_k = mv^2/2$ and 1 keV = 1.6022×10^{-16} J.) What is the associated Abbe resolution? How does the resolution of an electron microscope compare with the best resolution obtained with the light microscope?

4.9. [Diffraction Grating] The bright spots produced by a normally illuminated diffraction grating occur at angles satisfying the equation $m\lambda = d \sin\theta$ (d = grating spacing).
   a. Use this equation to calculate the angles corresponding to bright spots for commercially available gratings with 100 lines/mm and 300 lines/mm under illumination by a laser emitting at 633 nm. You should find that the angular spread increases as grating spacing decreases. Hint: You first need to compute the grating spacing d from the number of lines/mm.

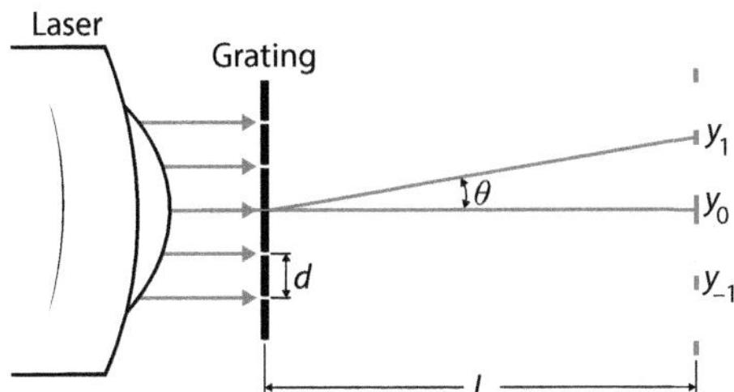

   b. For the coarser grating, the angles corresponding to maxima are not too large. Thus, the small angle approximation ($\tan\theta \approx \sin\theta \approx \theta$ [in radians]) can be used to compute the linear positions of bright spots. Use this approximation to prove that the linear positions y_m of the bright spots on a screen are given by $y_m = m\lambda(L/d)$ (see figure). This result reveals that the linear distances are inversely proportional to d. More significantly, it thus illustrates that small spacings, corresponding to fine details, are encoded by diffraction spots that are farther from the center.

4.10. *[Fourier Analysis] In Box 4.3, we stated that a periodic function $f(x)$ can be synthesized as follows (Eq. [B4.3.1]):

$$f(x) = \frac{A_0}{2} + \sum_{m=1}^{\infty} A_m \cos(kmx) + \sum_{m=1}^{\infty} B_m \sin(kmx)$$

where the amplitudes are given by

$$A_m = \frac{2}{\lambda} \int_0^{\lambda} f(x)\cos(kmx)\,dx$$

$$B_m = \frac{2}{\lambda} \int_0^{\lambda} f(x)\sin(kmx)\,dx$$

Prove these amplitude expressions by assuming (or proving) the validity of the famous orthonormality relationships for sine and cosine, which state (see the note below)

$$\frac{2}{\lambda}\int_0^{\lambda} \sin(kmx)\sin(knx)\,dx = \delta_{nm}$$

$$\frac{2}{\lambda}\int_0^{\lambda} \cos(kmx)\cos(knx)\,dx = \delta_{nm}$$

$$\frac{2}{\lambda}\int_0^{\lambda} \sin(kmx)\cos(knx)\,dx = 0$$

where the Kronecker delta, $\delta_{nm} \equiv \begin{cases} 0 & n \neq m \\ 1 & n = m \end{cases}$

Note: The cosine-cosine orthonormality relationship is not valid when $n = m = 0$. For this special case, it is easy to show that the integral is two instead of one. This anomaly is the origin of the factor of two that is associated with the A_0 term in the Fourier expansion.

4.11. *[Fourier Analysis] Use Eqs. (B4.3.1) and (B.4.3.2) to verify Eq. (B4.3.3).

4.12. [Fourier Analysis (Numerical)] Plot several progressively more complete versions of the sum in Eq. (B4.3.3), as exemplified in Fig. B4.3.1, to verify that the sum replicates the function. The plots easily can be generated using a program like Microsoft Excel®.

4.13. [Spatial Filtering] The left two sets of figures each show an object, its diffraction pattern in the BFP, and the associated image (neglecting finite lens effects). The right four sets of figures show the same objects and diffraction patterns, but with selected diffraction spots blocked (denoted by open squares). In each of the latter four cases, sketch the image that will result based on the spots that have been blocked.

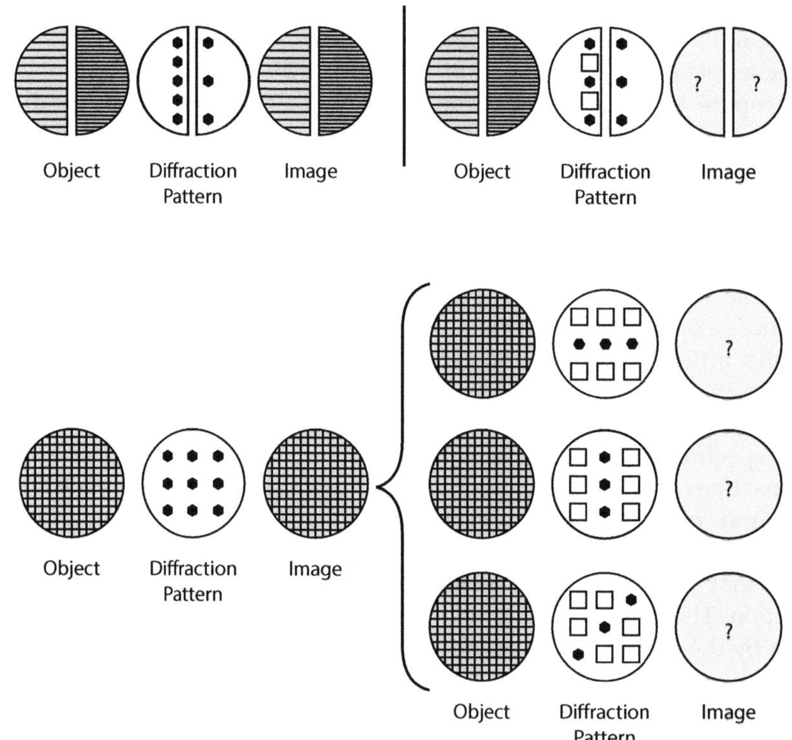

4.14. [Beam Expander] It sometimes is desirable to change the diameter of a light beam. A "beam expander" may be used to convert a collimated light beam of relatively smaller radius, r, into a collimated light beam of relatively larger radius, R (Activity A4.2). This problem explores the theory of a "Galilean beam expander."

a. Look up Galilean beam expander on the Internet and explain how it works using an appropriate ray diagram. Use the notation f_D for the focal length of the diverging lens and f_C for the focal length of the converging lens.

b. Use your ray diagram and simple geometrical arguments to prove that the beam is expanded according to the ratio: $R/r = -f_C/f_D$.

ACTIVITIES

A4.1 Generation of Simple Diffraction Patterns

1. Background: Solve HW 4.9b.
2. Equipment: Optical rail
3. Laser
4. Gratings
5. Optics holders (for laser and gratings)
6. Paper or screen
7. Ruler

In this activity, you will experimentally verify your theoretical analysis in HW 4.9b using an apparatus such as the one shown (photograph by the authors). The experiment is described for a 633-nm He-Ne laser and a three-panel (100, 300, and 600 lines/mm) Educational Diffraction Grating. However, any suitable laser and gratings may be used.

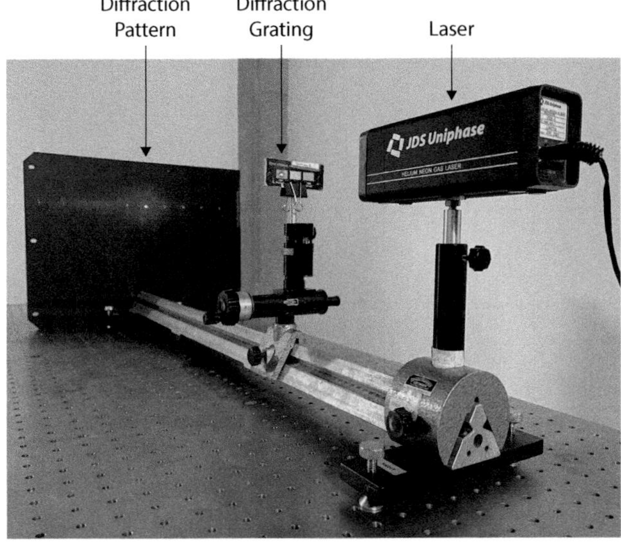

A. Illuminate the 100 lines/mm grating with the laser and project its diffraction pattern onto a piece of paper. Recopy the pattern to scale, mark the positions of the $m = 0$ and $m = \pm 1$ diffraction orders, and then measure the distances of the first-order spots from the central spot with a ruler. Also measure the distance, L, between the grating and the paper. Repeat this procedure for the 300 lines/mm grating. Compare your experimental data with the predictions of the theoretical result in HW 4.9b by calculating a percent difference.

B. Fill in the table, which will prove helpful in understanding spatial filtering.

Orientation of Lines in Grating	Orientation of Spots in Diffraction Pattern
Vertical	
Horizontal	

A4.2 Spatial Filtering

Explore the effects of spatial filtering using the equipment shown in Fig. 4.10.

A. Expand the laser beam to a spot of diameter ~1 cm when incident on a mesh having ~50–200 lines/in. To accomplish this, use diverging ($f_D = -2.5$ cm) and converging ($f_C = 40$ cm) lenses arranged in a Galilean beam expander configuration (HW 4.14). Place the mesh a little over one focal length from a second converging lens ($f = \sim 5$–10 cm) and place the variable aperture where the diffraction pattern created by the mesh is most crisply visible (roughly the BFP). With the aperture off the optical rail, or fully open, what do you observe on a distant screen?

B. Fill in the following table.

Manipulation	Orient the Aperture Vertically and Allow Only the Central Vertical Spots in the Diffraction Pattern Past the Aperture	Orient the Aperture Horizontally and Allow Only the Central Horizontal Spots in the Diffraction Pattern Past the Aperture	Orient the Aperture at 45° and Allow Only the Central 45° Angled Spots in the Diffraction Pattern Past the Aperture
Effect on the image of the mesh			

C. Explain these results. Hint: Review your observations in part (B) of Activity A4.1.

5

Contrast Enhancement in Optical Microscopy[1]

In addition to resolution, an image must possess contrast, which the eye detects as differences in luminance that arise from grayscale or color information in light emanating from a sample (Fig. 5.1). Contrast differs from resolution in that it is dictated by properties of the sample as well as by properties of the microscope.

Contrast generation is a rich topic when discussed in the context of optical microscopy because many commonly imaged specimens, such as cells and tissues, are intrinsically clear and colorless and thus tend (under standard illumination and detection conditions) to generate images with very poor contrast. One way to circumvent this problem is experimental and involves altering the sample by labeling cellular constituents with absorbing stains or fluorescent molecules that generate contrast via variations in the amplitude or color of the electric field (Box 5.1). A second approach is more optics based and gives rise to contrast-enhancing techniques such as darkfield, phase-contrast, polarization, and differential interference contrast (DIC) microscopies. We discuss these four techniques after introducing standard "brightfield" microscopy and highlighting the contrast problem that often plagues this approach. Fluorescence will be the focus of Chapters 6–8.

5.1 Brightfield Microscopy

In the absence of spatial filtering or deviation from standard Köhler illumination, an optical microscope is set up for "brightfield" imaging. In this mode, "amplitude" specimens stand out with high contrast set against a bright background, as described next (Fig. 5.2). Key advantages of brightfield imaging include high resolution, low cost, and ease of implementation.

5.1.1 Amplitude Specimens

Amplitude specimens modulate the amplitude (intensity) or color of EM radiation (Fig. 5.3) and thus can be distinguished from their surroundings by the eye under brightfield conditions.

Some biological specimens, such as chloroplasts, are intrinsically absorbing and thus are well suited for brightfield imaging. However, most cells and tissues are relatively transparent and need to be treated with absorbing stains to be visible via brightfield. Staining is a common practice in the fields of histology and microbiology, including cancer biopsies, but it is not suited to the study of living samples (Box 5.1).

5.1.2 Phase Specimens

Phase specimens primarily modulate the phase of EM radiation, without significantly affecting its amplitude (Fig. 5.3), and are very difficult to distinguish from their surroundings under brightfield conditions because the eye is not sensitive to wave phase.

5.2 Contrast-Enhancing Approaches

5.2.1 Common Themes

Several approaches are used to enhance the poor contrast generated by phase specimens in brightfield mode. One is to alter the illumination profile, typically in the FFP of the condenser, where the image of the light source is in focus in the Köhler train (Fig. 3.12). Another approach is to implement spatial filtering, as discussed briefly in Section 4.4, by manipulating the specimen diffraction pattern in the BFP of the objective.

Contrast-enhancing techniques also often use interference to convert invisible phase differences into visible amplitude differences. To illustrate the idea, consider two equal amplitude waves that are out of phase by half a wavelength (i.e., that have a phase difference of π). These waves are individually indistinguishable by the eye, but the phase difference between them can be detected via destructive interference that generates darkness (Figs. 2.7 and 5.4). Similarly, phase differences that are integral multiples of 2π are detectable via constructive interference that generates brightness (Fig. 5.4).

[1] This chapter is one of the more difficult in the text because it covers five different forms of microscopy that each could easily be the subject of its own chapter. Our aim here is to convey the essence of the different approaches.

DOI: 10.1201/b22076-6

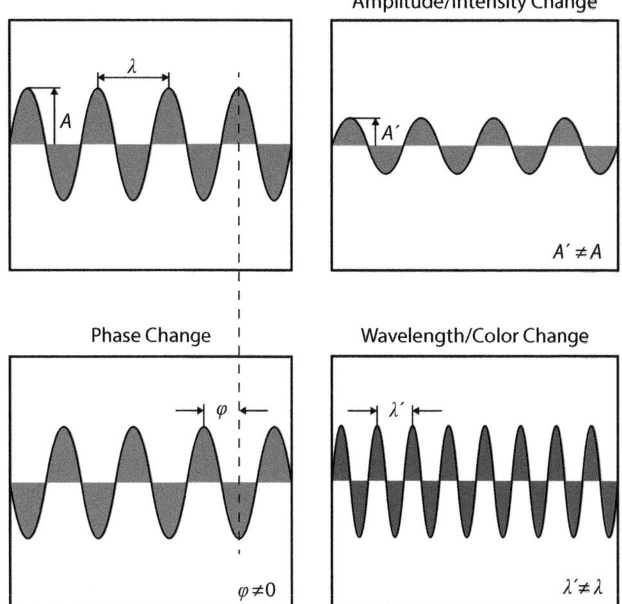

FIGURE 5.1
Schematic representation of important properties of light that are detectable by the eye (amplitude, A, and color, λ) and properties that are not (phase, φ).

The next several sections describe a spectrum of contrast-enhancing approaches that utilize these tools. We start with a conceptually simple method of making phase objects visible, which is to remove the illumination light so that weak specimen contrast is not obscured by a bright background.

5.2.2 Darkfield Microscopy

In this imaging mode, specimens become visible in the same way that stars become visible at night, by making the background dark (instead of bright). Advantages of darkfield include its low cost, its relatively simple

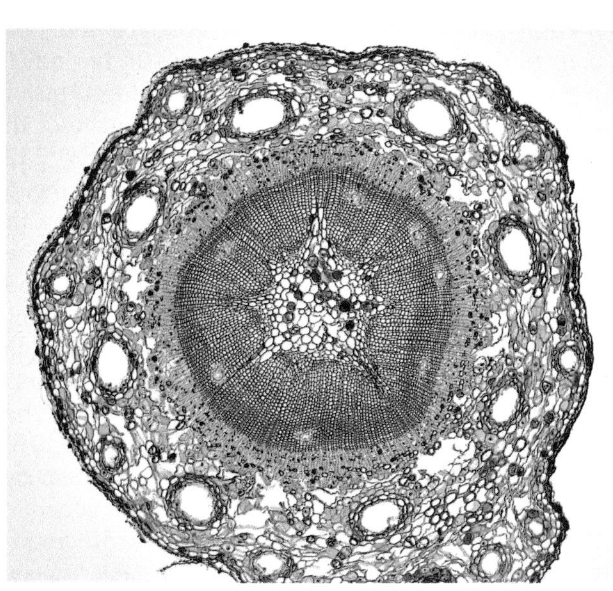

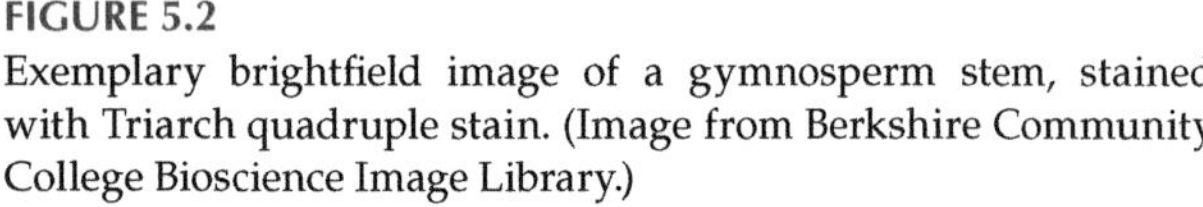

FIGURE 5.2
Exemplary brightfield image of a gymnosperm stem, stained with Triarch quadruple stain. (Image from Berkshire Community College Bioscience Image Library.)

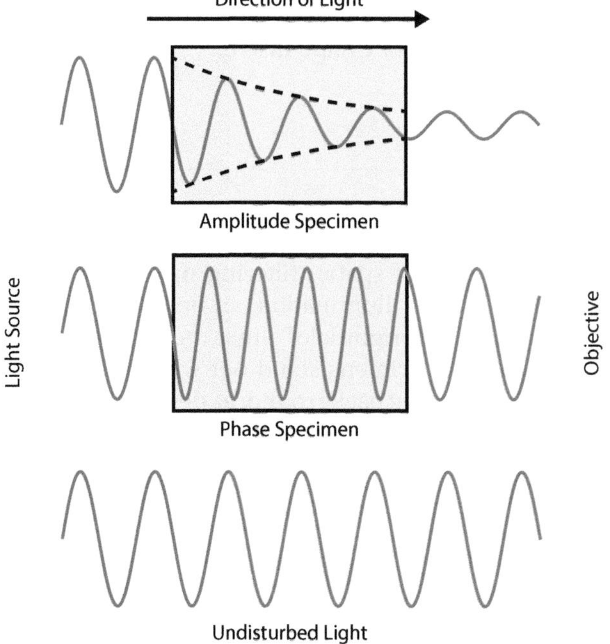

FIGURE 5.3
Schematic diagram showing the effects of amplitude and phase specimens on light.

BOX 5.1 HISTOLOGY

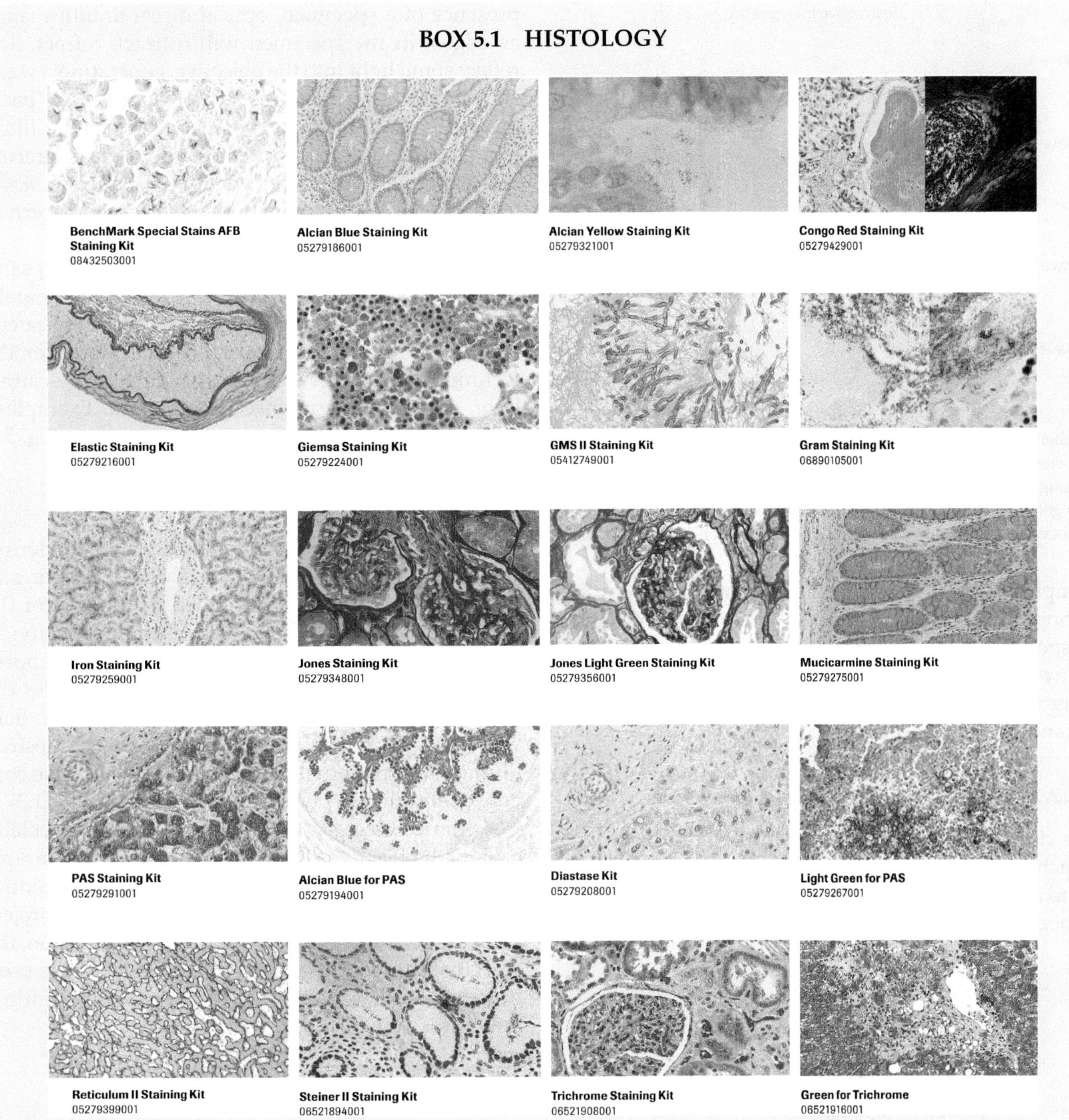

FIGURE B5.1.1
Exemplary commercially available staining kits. (Images and associated labels courtesy of Roche Corporation.)

Histology is the study of the microscopic structure of cells and tissues – structures that have intrinsically low contrast when imaged in brightfield mode. However, the brightfield contrast of cells and tissues can be markedly improved using absorbing stains (Fig. B5.1.1). Such stains can be used to distinguish different types of molecules and organelles within cells, different types of cells within tissues and organs, and native cells versus diseased or nonnative (e.g., bacterial or fungal) cells, among others. Several important commercially available stains are shown in the figure. In some cases (e.g., Alcian Blue and Congo Red), the stains consist of a single compound. In other cases (e.g., Giemsa Stain and Gram Stain), the stains consist of a mixture of compounds that differentially stain different features. Histology samples typically are prepared by fixing the sample with a crosslinker like paraformaldehyde to preserve sample morphology, embedding the fixed sample in a matrix such as paraffin to facilitate handling, sectioning the embedded sample into thin slices, and staining the slices for visualization under a microscope. The results can be visually striking.

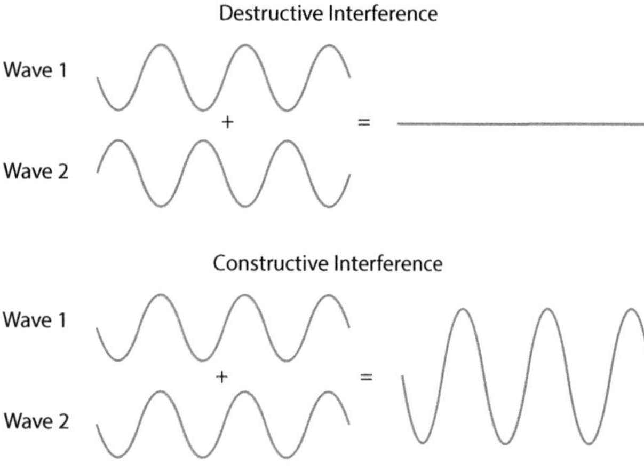

FIGURE 5.4
Schematic depiction of destructive and constructive interference of periodic waveforms. Two equal-amplitude waves that travel crest on trough (i.e., π out-of-phase) or crest on crest (i.e., in phase) add together to generate waves of zero amplitude (i.e., destructively interfere) and twice the amplitude (i.e., constructively interfere), respectively.

implementation, and its ability to markedly enhance contrast for thin unstained samples. Darkfield is especially useful for studying external details and tiny diffracting objects, such as bacteria and isolated organelles. A "darkfield microscopy examination" is a standard test for syphilis.

5.2.2.1 Mechanism of Contrast Generation

In darkfield microscopy, the illumination is oblique, such that the illuminating rays miss the objective in the absence of a specimen. Thus, in the absence of a specimen, the field is dark (Fig. 5.5). However, in the presence of a specimen, optical discontinuities (such as edges) in the specimen will diffract, refract, and reflect some light into the objective, generating a weak signal that nevertheless is visible because the background is dark. From the perspective of spatial filtering, darkfield microscopy involves high-pass filtering (by effectively removing the low-frequency $m = 0$ (undiffracted) component of the diffraction pattern in the BFP of the objective).

The strengths of darkfield, for the case of thin samples, are explored in Activity A5.1. Unfortunately, darkfield is not well suited to the study of thick specimens or for visualizing internal details. Moreover, the technique is sensitive to debris, which also scatters light, and requires intense illumination. Exemplary brightfield and darkfield images are shown in Fig. 5.6.

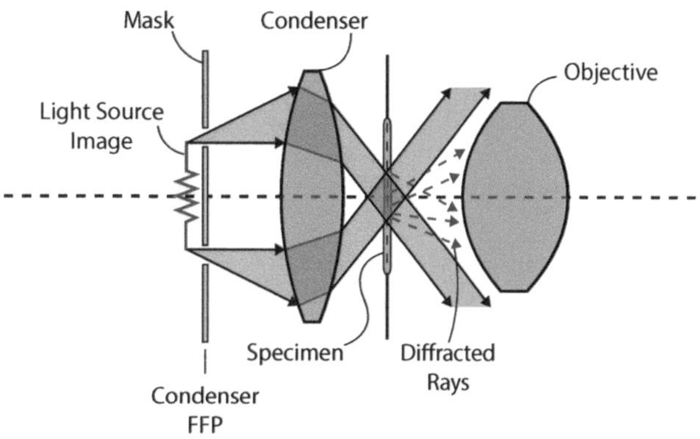

FIGURE 5.5
Schematic diagram that illustrates concepts underlying the implementation of darkfield microscopy. Oblique illumination and specimen-diffracted light are red and blue, respectively. The illumination misses the objective, and only specimen-scattered light is collected and used to generate an image. (Adapted from Scalettar, B.A., et al. 2015. Biomedical imaging in the undergraduate physics curriculum: module on optical microscopy. *Am. J. Phys.* 83:711–718. DOI: 10.1119/1.4921820. [CC BY 3.0].)

5.2.2.2 Experimental Implementation

Darkfield can be implemented by using a condenser with an NA that exceeds that of the objective and then using an annulus positioned in the FFP of the condenser to create a hollow cone of illumination in the specimen plane (Fig. 5.5). For an annulus of appropriate size, all of the illuminating rays will miss the objective in the absence of a specimen, and the field will be dark. A simple way to achieve the desired illumination pattern is to use an oversized phase condenser annulus (Section 5.2.3.3 and Activity A5.1.3).

A more elegant method employs a specially designed objective and condenser. In the example shown in Fig. 5.7, the condenser contains a central mask (not shown) and mirrors that are used to project a hollow cone of light onto the sample. In addition, the objective contains an iris diaphragm that can be used to reduce its NA to ensure that the oblique illumination misses the objective.

5.2.3 Phase-Contrast Microscopy

In this imaging mode, *spatially varying* phase shifts in light that are induced by a transparent specimen are converted into amplitude variations and thereby into contrast. The method was developed by Frits Zernike, earning him a Nobel Prize in Physics in 1953, and involves manipulation of the illumination profile in the FFP of the condenser, spatial filtering in the BFP of the objective, and interference. Advantages of phase-contrast imaging include its relatively low cost, its ease of implementation, and its ability to produce high-quality images of "phase" objects, most notably living, unstained cells. Unlike DIC, phase is not subject to orientation-dependent artifacts, and it is insensitive to polarization and birefringence effects, permitting examination of cultured cells grown in plastic dishes.

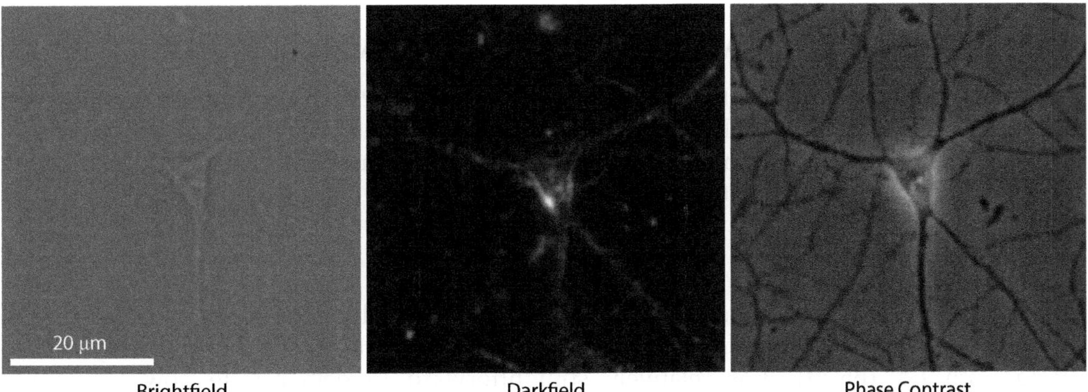

FIGURE 5.6
Brightfield (left), darkfield (middle), and phase-contrast (right) images of the same hippocampal neuron. Contrast in the brightfield image was enhanced by partially closing the condenser aperture diaphragm. The cell body in the phase image shows the typical halo artifact (Section 5.2.3.2). Bar = 20 μm. (Adapted from Scalettar, B.A., et al. 2015. Biomedical imaging in the undergraduate physics curriculum: module on optical microscopy. *Am. J. Phys.* 83:711–718. DOI: 10.1119/1.4921820. [CC BY 3.0].)

5.2.3.1 Vector Analysis of Phase Specimens

The phase-contrast method can be illustrated using wave and vector diagrams, which depict how electric fields combine to create an image. In the main text, we focus on these visually oriented approaches. The method also can be demonstrated analytically, and the reader is encouraged to study this alternative approach, which is relatively straightforward (Box 5.2).

It is useful to start with a big picture view of the effect of a thin, phase specimen on an incident wave. To this end, consider the scenario in Fig. 5.8.

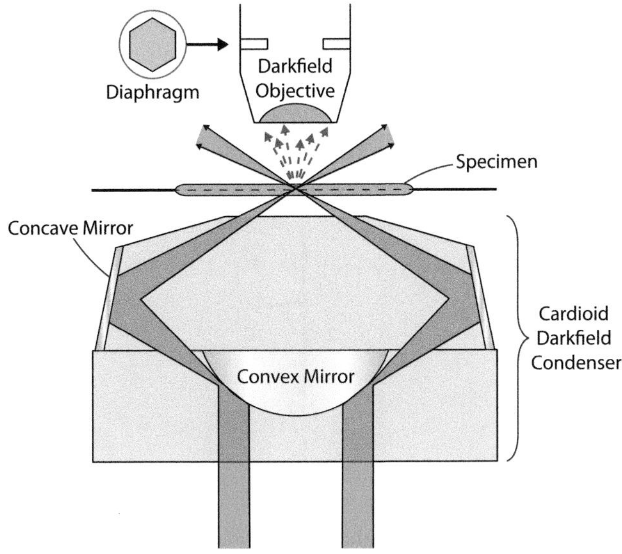

FIGURE 5.7
Schematic of an exemplary, commercially designed darkfield condenser. A mirror in the center of the condenser deflects light onto another mirror resembling a cardioid of revolution. The output is a hollow cone of light that is free from spherical and chromatic aberration. A cardioid darkfield condenser can have an NA as high as ~1.30.

The specimen introduces a localized region of phase retardation into a plane wave, so the outgoing wave is indented instead of planar.

Specimen effects such as those depicted in Fig. 5.8, and their manifestation in images, can be illustrated and analyzed conveniently using vector diagrams. In this approach, the image of the specimen is regarded as arising from the superposition (interference) of an undiffracted zeroth-order, Z, wave and a specimen-diffracted, D, wave, respectively passing through or emanating from the same spot on the sample. This is the Abbe view. In addition, each wave is represented using a vector with length proportional to the associated wave amplitude and with an orientation that reflects the wave's phase shift relative to Z (Fig. 5.9). The amplitude and phase of the total, T, wave at an image point can be determined from vector addition

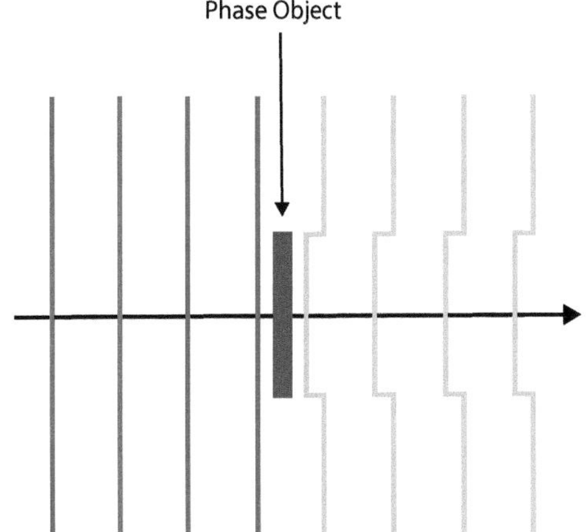

FIGURE 5.8
Effect of a rectangular phase specimen on a plane wave (red). The outgoing wave (green) contains an indentation (phase lag) caused by the specimen.

BOX 5.2 ANALYTICAL ANALYSIS OF PHASE CONTRAST

The phase-contrast effect also can be demonstrated using analytical arguments. We start by assuming that a specimen in the xy-plane is illuminated with a monochromatic wave given in the specimen plane by $E(z=0,t)=E_0\sin(\omega t)$. The specimen induces a location-dependent phase shift $\varphi(x,y)$ in the illuminating wave. As a consequence, a constant-amplitude wave given by

$$E(\vec{r},t)|_{z=0}=E_0\sin\left[\omega t+\varphi(x,y)\right] \tag{B5.2.1}$$

exits the specimen plane. We also use Eq. (B5.2.1) to describe the field in the image plane, which amounts to ignoring magnification and orientation effects and assuming that the lens performs very well.

Starting with Eq. (B5.2.1), it is relatively straightforward to demonstrate how to enhance the contrast from a phase specimen. First, Eq. (B5.2.1) is expanded using the trigonometric identity $\sin(\alpha+\beta)=\sin\alpha\cos\beta+\cos\alpha\sin\beta$ and limited to the case of very small phase shifts (where $\cos\varphi=1$ and $\sin\varphi=\varphi$):

$$E(x,y,t)=E_0\left[\sin(\omega t)\cos\varphi+\cos(\omega t)\sin\varphi\right]\approx E_0\left[\sin(\omega t)+\varphi\cos(\omega t)\right] \tag{B5.2.2}$$

This equation shows that in the very small shift approximation the wave in the image plane is the sum of two terms that are $\pi/2$ out of phase. Next, the wave in the image plane is converted into an amplitude-modulated wave by introducing an additional phase shift of $\pi/2$ between the two terms (e.g., transforming a sine into a cosine) to yield a wave of the form:

$$E(x,y,t)=E_0\left[1+\varphi(x,y)\right]\cos(\omega t) \tag{B5.2.3}$$

As desired, this wave is amplitude modulated, but the effect of the object on the amplitude (as embodied in φ) is relatively small. To rectify this, the undiffracted wave also is attenuated. Mathematically, this is represented by replacing the 1 in Eq. (B5.2.3) with a number $\alpha<1$, which accounts for the attenuation of the zero-order term. Ideally, α should be chosen to be around the same size as the average φ to achieve the maximum sensitivity to $\varphi(x,y)$.

of the fields arising at the conjugate object point, i.e., from $\vec{T}=\vec{Z}+\vec{D}$. Nonzero specimen contrast requires amplitude modulation; thus, contrast appears as a varying difference in length between $\vec{T}$ and $\vec{Z}$ that depends on location in the image.

5.2.3.1.1 Phase Specimens Using Brightfield

We start our vector analysis of the phase-contrast method with standard brightfield imaging of a phase specimen (Fig. 5.9). The brightfield vector diagram is relatively simple and is key to understanding Zernike's phase-contrast approach.

To generate the brightfield vector diagram, we invoke the fact that a pure phase specimen does not alter wave amplitude (Fig. 5.3). Thus, $\vec{T}$ and $\vec{Z}$ have the same length. However, $\vec{T}$ is rotated with respect to $\vec{Z}$ by a small angle, φ, which represents the specimen-induced phase shift (typically a lag). To determine the length and direction of $\vec{D}$ in the diagram, we invoke the relationship $\vec{T}=\vec{Z}+\vec{D}$; this shows that $\vec{D}$ is small and nearly orthogonal to (i.e., 90° out-of-phase with) $\vec{Z}$ (Fig. 5.9). These results also are proven in Box 5.2.

The orthogonality of $\vec{Z}$ and $\vec{D}$ plays a pivotal role in the contrast problem that plagues brightfield imaging of phase specimens. We turn now to Zernike's method of altering this 90° relationship so the Z

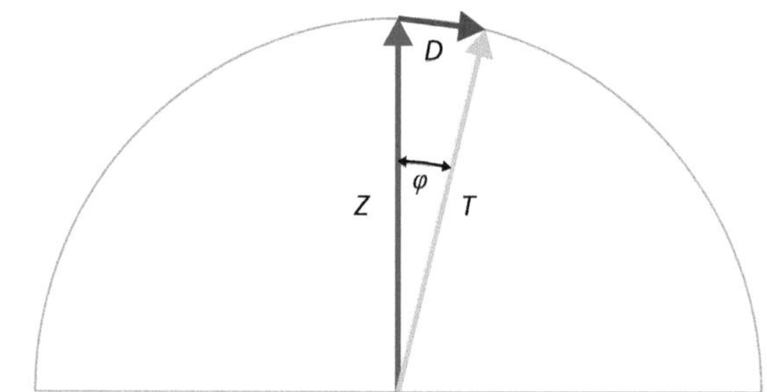

FIGURE 5.9
Contrast problems in brightfield imaging of a thin phase specimen depicted using a vector diagram. The vector diagram shows a small phase lag φ (presumably different at different points in the image) that arises from effects of specimen optical path length (HW 5.3). Phase lags are represented by clockwise rotations. For a pure phase specimen, $|\vec{Z}|=|\vec{T}|$ everywhere in the image, and thus contrast in brightfield imaging is absent. Here, the difference between the zeroth-order Z wave and the total T wave is denoted as the "diffracted wave" D. Z and D are out of phase by $\pi/2$.

and D waves instead superpose to generate "phase contrast."

5.2.3.1.2 The Phase-Contrast Approach

Zernike's method is based on using spatial filtering to create a 180° phase relationship between the Z and D waves. The important consequence is that Z and D will then undergo (partial) destructive interference and create a T wave with reduced amplitude (Figs. 2.7 and 5.4). Thus, T is distinguishable from Z, and the phase specimen has amplitude contrast.

To understand Zernike's method from a theoretical perspective, refer back to Fig. 5.9 and envision selectively advancing $\vec{Z}$ by 90°. In the altered diagram, $\vec{Z}$ is directed horizontally to the left, while $\vec{D}$ continues to point horizontally to the right (Fig. 5.10). Thus, 90° advancement generates a 180° (anti-parallel) phase relationship between $\vec{Z}$ and $\vec{D}$, which in turn creates destructive interference with the length of $\vec{T}$ less than that of $\vec{Z}$. Destructive interference, coupled with the spatial variation of $\vec{D}$, creates a grayscale image with good contrast.

How did Zernike implement selective advancement of Z experimentally, without altering the phase of D? To accomplish this goal, Zernike restricted the illumination, which otherwise fills the BFP of the objective, so that Z could be altered selectively in the BFP. One possible approach involves illumination that is generated by a point source (Fig. 5.10). Point-like illumination can be generated by stopping down the condenser aperture diaphragm. In this case, the Z wave is confined to a conjugate *point* in the BFP of the objective. In contrast, the D wave is spread out in the BFP (Fig. 4.6). Thus, Z can be advanced, with only a minimal effect on D, by etching a small, appropriately deep, spot in the center of a transparent phase plate located in the BFP (Sample Problem 5.1). An alternative approach that yields brighter, and thus preferable, illumination, as well as higher resolution, is described in Section 5.2.3.3.

Zernike's method is known as positive phase contrast because Z is advanced. Z also can be retarded by 90° to generate contrast via constructive interference (negative phase contrast), as shown in Fig. 5.10. In both approaches, Z typically also is deliberately attenuated by slightly darkening the phase plate in the objective's BFP where Z passes through it (in this case, at the center) without affecting the amplitude of the D wave, which bypasses the attenuation region as

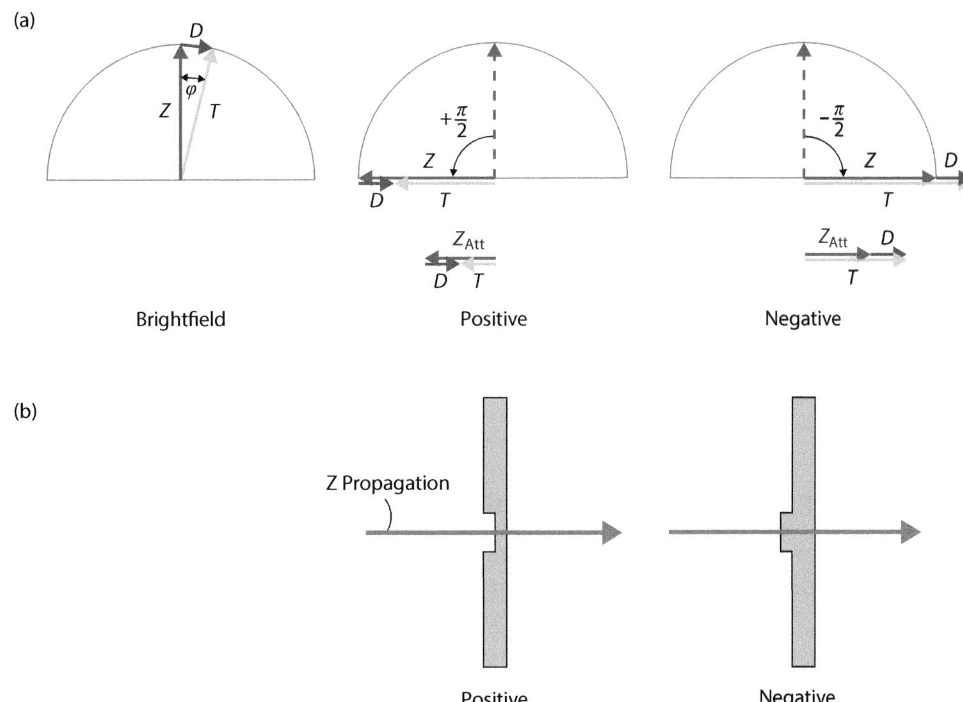

FIGURE 5.10
(a) Vector diagrams showing the amplitude and phase relationships arising in positive and negative phase contrast in the absence (top) and presence (bottom) of attenuation. Attenuating Z makes $|\vec{Z}|$ more comparable to $|\vec{D}|$ and thereby increases the relative difference between $|\vec{T}|$ and $|\vec{Z}|$. The brightfield vector diagram is shown for reference. (b) Basic mechanism of advancing or retarding the Z wave after it has emerged from the sample. In this simplified example, the illumination is assumed to emanate from a point, so Z maps to a point in the BFP of the objective. Z can be advanced (positive phase contrast) by etching a region in its path. Alternatively, Z can be retarded (negative phase contrast) by inserting a bump in its path. Unlike Z, D is relatively unaltered because it spreads out in the BFP and largely misses the advancing or retarding filter (Fig. 5.12).

it travels through the rest of the BFP. This deliberate attenuation increases the relative difference between the amplitudes of Z and T (Fig. 5.10), further increasing contrast between different spots in the image.

5.2.3.2 Specimen Appearance in Positive Phase Contrast

Specimen appearance in phase contrast depends on the sign and magnitude of the specimen-induced phase shift. A thin specimen with a refractive index that is higher than that of the surrounding medium will appear darker gray against a lighter gray background because the length of $\vec{T}$ is less than that of $\vec{Z}$ (Fig. 5.10). In contrast, a thin specimen with a refractive index that is lower than that of the surrounding medium, or one that generates a large phase shift (Fig. 5.11), generally will appear bright.

Weaknesses of phase contrast include a failure to work well for thick specimens, and the existence of "halos" in images (Fig. 5.6). Halos arise in part because some of the light diffracted by coarse features passes through the filter in the BFP and is phase shifted, when this should be the fate only of the undiffracted wave (Figs. 5.10 and 5.12). The presence of an annulus in the condenser also reduces NA and thus resolution. Exemplary brightfield and phase-contrast images are shown in Fig. 5.6.

5.2.3.3 Experimental Implementation

To implement phase-contrast microscopy, the undiffracted and diffracted waves must be separated so that Z can be manipulated (shifted and attenuated) separately in the BFP of the objective using a phase filter, as explored in Activity A5.1.2. In Fig. 5.10, we illustrated one possible approach using illumination that emanates from a point. In real applications, an extended source is used to achieve brighter illumination and higher resolution.

The extended light source is created by passing the illumination through an annulus positioned in the FFP of the condenser; this generates illumination at the specimen that emanates from a ring (Fig. 5.12). After impinging on the specimen plane, Z waves, which are not diffracted, are collected by the objective, and they then pass through an annular "phase plate" in the objective BFP (a plane conjugate to that containing the condenser annulus). The phase plate phase shifts and attenuates the Z wave, similar to the etch and bump described in Fig. 5.10. In contrast, waves that are diffracted pass through all regions of the BFP and are predominantly unaltered.

5.2.4 Polarization Microscopy

In this imaging mode, optical anisotropies in a (birefringent) sample are used to create contrast. Polarization microscopy generates useful images for a restricted, but important, set of biologically relevant structures, such as spindle and collagen fibers, gout crystals, and (normal and malaria-infected) red blood cells. Applications in geology abound due to the importance and prevalence of crystalline samples.

To understand contrast generation in polarization microscopy, we briefly review the key properties of birefringent specimens and of polarized light and polarizers. In this way, we also set the stage for understanding DIC microscopy, which is more ubiquitously applicable in the biological sciences. Additional background on polarized light and polarizers is presented in Section 2.4.

5.2.4.1 Birefringence

In our past discussion of reflection and refraction (transmission), we implicitly were dealing with an isotropic material (like glass), which possesses one

SAMPLE PROBLEM 5.1 FABRICATION OF RING FOR POSITIVE PHASE CONTRAST

Design a 90° ($\pi/2$) advancing etch for the phase-contrast Z wave. Assume that the plate is fabricated from glass and that the illumination has vacuum wavelength 500 nm. Hint: Review Fig. 5.10.

SOLUTION

The phase shift created when light passes through media with different refractive indices reflects the different number of times the wave oscillates within each medium and is given by Eq. (2.2) (Box 2.2). In this problem, we want to create a phase shift of $\pi/2$ between waves that pass between glass and the etched region and solve for the etch thickness, Δx. Invoking Eq. (2.2) with $n_g = 1.50$ and $n_a = 1.00$ yields

$$\varphi = \frac{2\pi}{\lambda_0}\left(n_g - n_a\right)\Delta x = \frac{2\pi}{500 \text{ nm}}(1.50 - 1.00)\Delta x = \frac{\pi}{2}$$

We conclude that $\Delta x = 250$ nm.

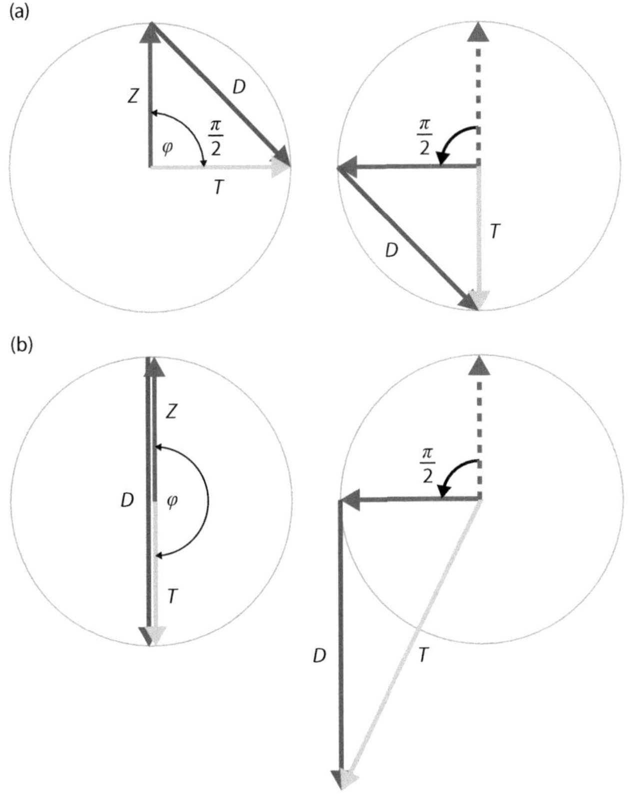

FIGURE 5.11
Effect of large specimen-induced phase shifts on object brightness, using positive phase contrast. Left panels depict vector relationships in the absence of Zernike phase advancement of Z, for two particularly large values of φ, either $\pi/2$ (a) or π (b). In both cases, $|\vec{Z}|=|\vec{T}|$, and there is no possibility of contrast. Right panels show the effects of a Zernike phase advancement of positive $\pi/2$ (for simplicity here in the absence of attenuation of Z) for these two cases of φ. For $\varphi=\pi/2$ (top right), $|\vec{Z}|$ still equals $|\vec{T}|$, and there is no alteration in total intensity. But for $\varphi=\pi$ (lower right), a large increase in total intensity will appear. For example, for specimens that induce a phase shift of π (b), $|\vec{T}|=\sqrt{5}\,|\vec{Z}|$ (HW 5.5). So, in summary, for positive phase contrast, specimens that induce small phase shifts will appear darker, specimens that induce $\varphi=\pi/2$ shifts will appear unaltered in intensity, and specimens that induce $\varphi=\pi$ shifts will appear brighter. Deliberate attenuation of the Z wave (not shown here) will alter the contrast inversion profile.

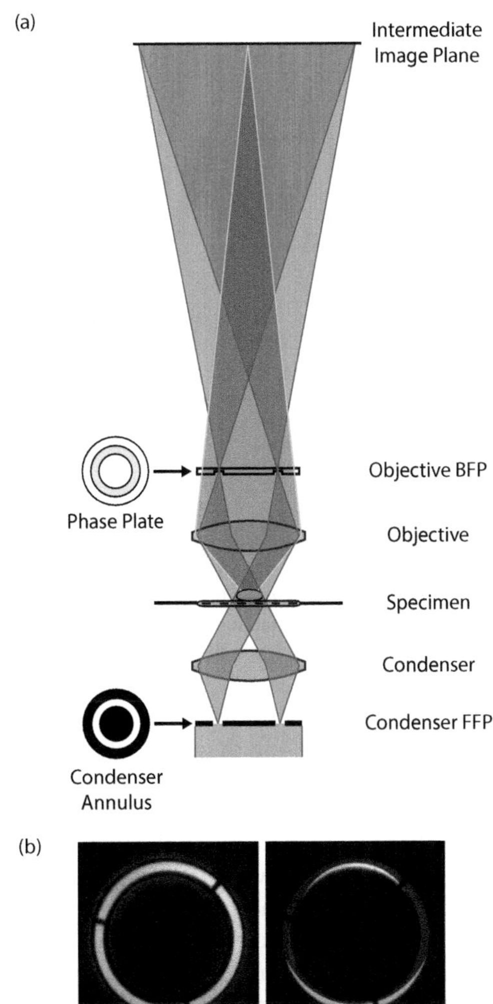

FIGURE 5.12
Implementation of phase-contrast imaging. Panel (a) is a schematic that illustrates key features of the light path in phase-contrast imaging. Zeroth-order and diffracted light are shown in red and blue, respectively. The undiffracted light passes through the phase plate (small indented rectangles) in the objective BFP and is shifted and attenuated. Panel (b) shows the condenser annulus (bright) in the condenser FFP and the phase plate (dark) in the objective BFP in aligned and misaligned configurations that produce good and poor phase contrast, respectively. Use of a green filter optimizes the phase-contrast effect because the phase plate is designed to phase shift green light by $\pi/2$. (Panel (a) adapted from Scalettar, B.A., et al. 2015. Biomedical imaging in the undergraduate physics curriculum: module on optical microscopy. *Am. J. Phys.* 83:711–718. DOI: 10.1119/1.4921820 [CC BY 3.0]. Photographs in Panel (b) by authors.)

refractive index that is valid for all polarizations and directions of ray incidence. Light that impinges on an interface between two isotropic materials is split into one reflected ray and one refracted ray.

In contrast, for an optically anisotropic material (a birefringent specimen), refractive index depends on the direction of ray incidence and light polarization, and refraction can be complex. In the main text, we restrict our discussion to a relatively simple set of biologically relevant birefringent specimens and a simple, frequently encountered illumination/specimen geometry. A more general treatment is presented in Hecht (2017).

Consider a specimen, such as a fiber, that has a nonrandom orientation of molecules along its long axis. We assume that when the fiber is viewed under a microscope, it is oriented with its long axis orthogonal to the direction of propagation of the incident light, as shown in Fig. 5.13. Because of the structural asymmetry, light polarized along the long axis of the

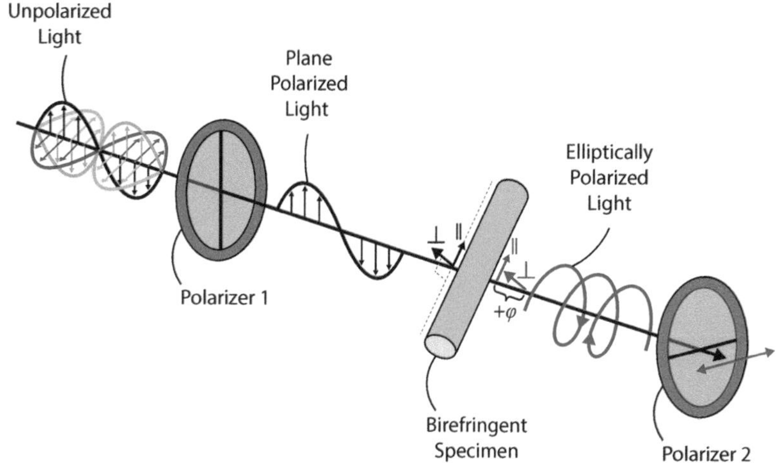

FIGURE 5.13
Schematic depiction of the birefringence of a fiber. In the schematic, light incident on the fiber is vertically polarized (its electric field is oriented along the vertical) after passing through Polarizer 1. The incident light can be decomposed into two waves. One wave, designated ||, is polarized parallel to the long axis of the fiber; the other wave, designated ⊥, is polarized perpendicular to the long axis. These waves encounter different refractive indices and thus exit the fiber out of phase by φ. In most cases, the two waves superimpose to generate elliptically polarized light, which has a nonzero component along Polarizer 2 (an analyzer) that generates a signal.

specimen will encounter a different refractive index than light polarized along the short axis, and the two polarizations will propagate with distinct speeds, since $v = c/n$ (Eq. [2.5]). In the case shown in Fig. 5.13, the parallel light encounters a higher refractive index and is preferentially slowed down. Birefringence is defined to be the difference in the refractive indices.

In what follows, we discuss in some detail how birefringence generates an image in polarization microscopy. However, the essence of the idea is shown in Fig. 5.13. The sample is illuminated with polarized light, which typically has components along both the long and short axes of the fiber. The two component fields get out-of-phase as they propagate through the fiber because they travel at different speeds. As a consequence, the net field exiting the fiber has an altered (generally elliptical) polarization. A "crossed" analyzer blocks light with the incident polarization, but the elliptical light in part can pass the analyzer to create signal (Demonstration 5.1).

DEMONSTRATION 5.1 EXTINCTION AND SIGNAL GENERATION USING CROSSED POLARIZERS

Two Parallel Polarizers	Two Crossed Polarizers	Two Crossed Polarizers (w/ Intervening Angled Polarizer)
Light Passes	Light is Blocked	Light Passes

Extinction by crossed polarizers, and signal generation by a third, intervening polarizer, can be demonstrated easily using a light source, such as an out-of-work overhead projector, and a set of linear polarizers. The intervening polarizer (like the birefringent fiber) rotates the plane of polarization and allows some light to pass through the final polarizer. (Photographs by the authors.)

In our simplified scenario, the interaction of polarized light with the specimen produces two orthogonally polarized waves (parallel and perpendicular to the long axis) that follow identical trajectories. The creation of two orthogonal, non-diverging transmitted waves is a simple example of birefringence. The crystal calcite can provide a more dramatic example, where the two orthogonal waves also diverge as they traverse the material because they refract (bend) differently. This phenomenon is known as double refraction and gives rise to a double image when calcite is appropriately oriented over printed material (Fig. 5.14). Divergent waves will arise in our discussion of DIC but are not considered here.

5.2.4.2 Mechanism of Contrast Generation

In biological polarization microscopy, the birefringent specimen is placed between two polarizers with

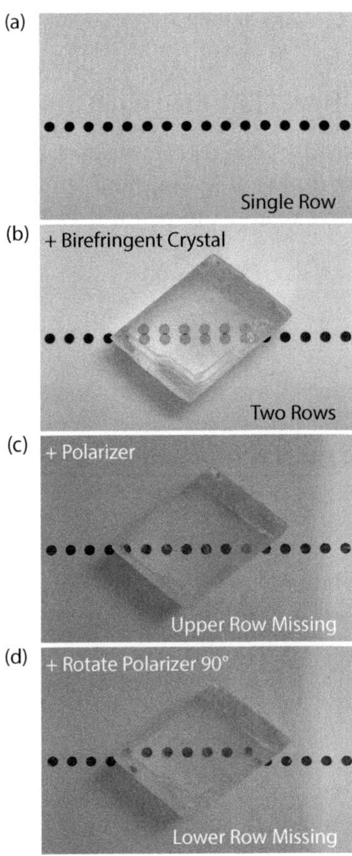

FIGURE 5.14
Double refraction by a calcite rhomb. The lighter images in (a) and (b) lack a polarizer and show the rhomb generating two images of a single line of dots via double refraction. The darker images in (c) and (d) each have a polarizer, but the polarizer transmission axes are orthogonally oriented. Notice that a different set of dots is eliminated in (c) and (d), demonstrating that the waves that generate the two images are orthogonally polarized. The tinting in Panels (c) and (d) is due to the polarizer, which blocks half of the unpolarized light. (Photographs by the authors.)

transmission axes that are orthogonally oriented; these are known as crossed polarizers. The first polarizer generates linearly polarized light that the specimen splits (Fig. 5.13). The two waves follow identical paths through the specimen, but they encounter different refractive indices, $n_{\parallel}$ and $n_{\perp}$, and thus acquire a phase difference given by

$$\varphi = \frac{2\pi}{\lambda_0}\left(n_{\parallel} - n_{\perp}\right)t \tag{5.1}$$

Here t is the specimen thickness traversed. Equation (5.1) is a simple rewrite of Eq. (2.2).

In this text, we discuss contrast generation by emphasizing the role that the two phase-shifted waves play in creating light with altered polarization (Fig. 5.13 and Section 2.4.1). This viewpoint also will prove useful when we discuss DIC microscopy. In Box 5.3, we augment this somewhat qualitative discussion with a rigorous theoretical analysis.

Consider first two crossed polarizers with no intervening specimen. The first polarizer, when illuminated with unpolarized light, produces light that is linearly polarized along the direction of its transmission axis (Figs. 2.11 and 5.13). This light impinges on the second polarizer, known as an analyzer, with its orthogonally oriented transmission axis, and no light is transmitted, as predicted by Malus's Law (Eq. [2.11]). The field is dark, an effect known as extinction (Demonstration 5.1).

Notably, introducing an intervening third polarizer can rotate the plane of light polarization and thereby generate a signal against the dark background (Demonstration 5.1 and Sample Problem 5.2). The key point is that birefringent specimens also have the ability to alter light polarization and thereby restore signal throughput. Specifically, for most specimen-induced phase shifts an intervening birefringent specimen will create elliptical light, which has a nonzero component along the analyzer that creates signal (Figs. 5.13 and 5.15).

A theoretical expression that rigorously describes the analyzer-generated intensity pattern, I_A, as a function of specimen-induced phase shift, φ, is derived in Box 5.3. The result is

$$I_A = I_P \sin^2(2\theta)\sin^2(\varphi/2) \tag{5.2}$$

In Eq. (5.2), I_P is the intensity incident from the polarizer, and θ is the angle between the specimen's long axis and the transmission axis of the polarizer.

Equation (5.2) makes several useful predictions about the effects of specimen-induced phase shift. The output intensity is zero when $\varphi = 0$ and there is no specimen effect. The output increases with increasing φ and is maximal when $\varphi = \pi$. In this case, the light

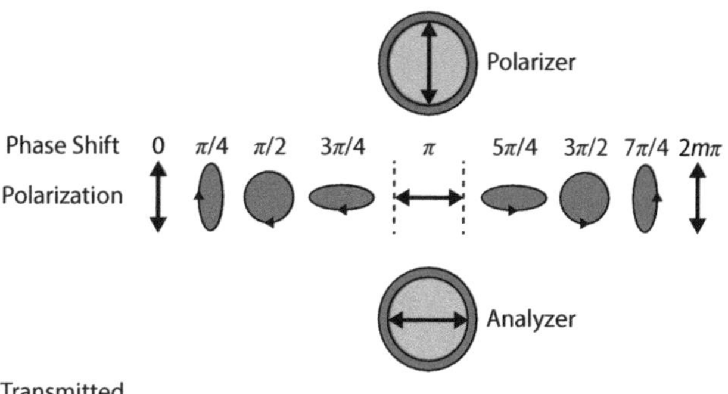

FIGURE 5.15
Exemplary output polarizations (for an equal split along the specimen axes), and the associated intensities of light transmitted through the analyzer for different phase shifts. There is a transmitted signal except when the phase shift is a multiple of 2π. In this case, there is no signal because the specimen outputs the entering polarization. The specimen axis is assumed to be at 45°, as shown in Fig. 5.13.

output by the specimen is linearly polarized maximally along the analyzer (Fig. 5.15). For larger shifts, the intensity falls, reaching zero whenever $\varphi = 2m\pi$ (m integral) (Fig. 5.15 and HW 5.11). This last case will arise in our discussion of interference colors in HW 5.12 and Activity A5.2.

Equation (5.2) also leads to several useful predictions related to specimen orientation (Fig. 5.13 and HW 5.11). One is that the specimen will be dark when its long axis makes an angle of 0° or 90° with respect to the transmission axis of the polarizer. The absence of signal has a relatively simple interpretation. For example, when $\theta = 0°$, the long axis in Fig. 5.13 is parallel to the polarizer, and the parallel field receives all of the energy. Because the perpendicular field receives no energy, nothing passes the analyzer. Another useful prediction is that the specimen will be maximally bright when rotated by 45° with respect to the orientations at which it vanishes (Fig. 5.16).

SAMPLE PROBLEM 5.2 CROSSED POLARIZERS WITH INTERVENING POLARIZER

A linear polarizer is placed at a 45° angle between the two crossed polarizers in a polarization microscope. (a) Calculate the fraction of initially unpolarized light that makes it through all three polarizers. Hint: Review the discussion of polarization in Section 2.4. (b) How would the result differ if the linear polarizer were placed *after* the crossed polarizers?

SOLUTION

a. The fraction of light passing through all three polarizers is the ratio of the final intensity, I_3, (after the third polarizer) to the initial intensity, I_0 (before the first polarizer). The effect of the first polarizer is to polarize the initially unpolarized light and to reduce its intensity by 50%:

$$I_1 = \tfrac{1}{2} I_0$$

The effects of the second and third polarizers are given by Malus's Law (Eq. [2.11]), since the light incident on each polarizer already is polarized:

$$I_2 = I_1 \cos^2\theta_{21} = \tfrac{1}{2} I_0 \cos^2\theta_{21}$$

$$I_3 = I_2 \cos^2\theta_{32} = \tfrac{1}{2} I_0 \cos^2\theta_{21} \cos^2\theta_{32}$$

Here, θ_{21} is the angle between the first and second polarizers, and θ_{32} is the angle between the second and third polarizers. The ratio of the final to initial intensities can now be calculated as the ratio of I_3 to I_0 (where $\theta_{21} = \theta_{32} = 45°$).

$$\text{Ratio} = \frac{I_3}{I_0} = \frac{\tfrac{1}{2} I_0 \cos^2\theta_{21} \cos^2\theta_{32}}{I_0} = \tfrac{1}{2} \cos^2\theta_{21} \cos^2\theta_{32} = \tfrac{1}{2}\left(\frac{\sqrt{2}}{2}\right)^2 \left(\frac{\sqrt{2}}{2}\right)^2 = \frac{1}{8}$$

b. The fraction of light passing through all three polarizers is zero if the additional linear polarizer is placed *after* the crossed polarizers because no light makes it past the crossed polarizers to reach the additional polarizer.

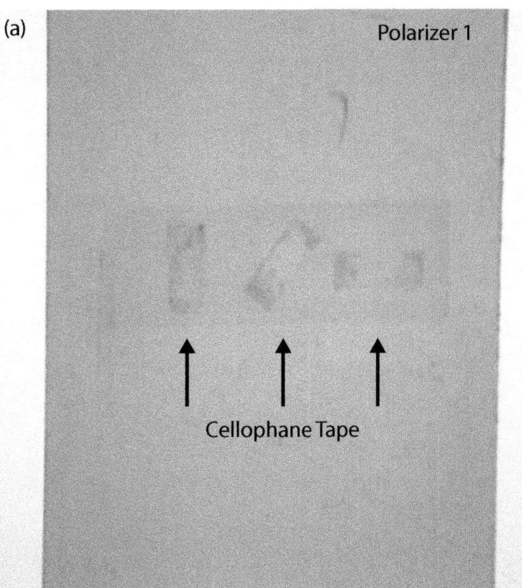

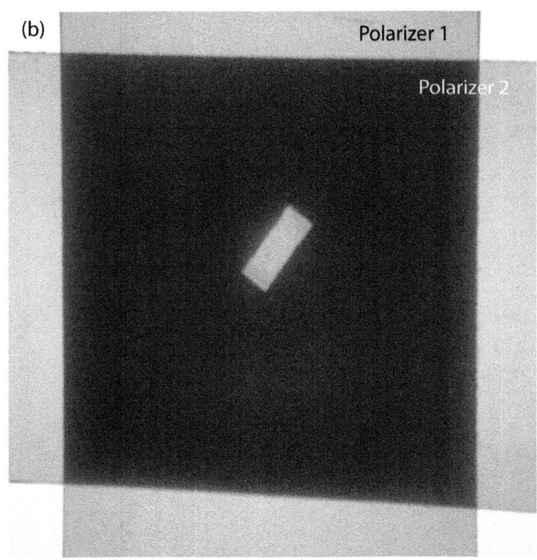

FIGURE 5.16
Birefringent cellophane tape between crossed polarizers. (a) The tape is shown in three orientations, differing in 45-degrees increments, placed atop a single polarizer. (b) When a crossed polarizer is placed on top of the configuration in (a), only the tape at 45 degrees is visible. Tape that works well can be obtained from Educational Innovations, Inc. (Photographs by the authors.)

BOX 5.3 DERIVATION OF EQ. (5.2)

Here we sketch a derivation of Eq. (5.2). The material in this box, and several upcoming boxes, represents a wave using a complex exponential function instead of a cosine or sine function (Box 2.1 and Eq. [2.1]). The utility of the complex formalism is discussed in Hecht (2017). In brief, complex exponentials often are used in lieu of cosine or sine because mathematical manipulations in general are easier with an exponential. Moreover, a cosine or sine representation can be extracted from a complex exponential representation by taking the real or imaginary part of the complex exponential, respectively. This follows from Euler's equation, which reads: $e^{i\theta} = \cos\theta + i\sin\theta$.

Our goals are to compute the field that exits a birefringent specimen (fiber) positioned between crossed polarizers and its component along the analyzer in terms of specimen orientation and specimen-induced phase shift.

To implement the derivation, we assume the geometry in Fig. B5.3.1. The polarizer generates a field with amplitude A, which we decompose into fields along, and orthogonal to, the long axis of the fiber. The angle between the long axis of the fiber and the transmission axis of the polarizer is θ, and the x and y axes are as shown.

The field that impinges on the specimen can be expressed in terms of the unit vectors $\hat{x}$ and $\hat{y}$ as

$$\vec{E}_1 = A\sin\theta\,\hat{x} + A\cos\theta\,\hat{y}$$

The specimen will induce a phase shift φ between the fields along $\hat{x}$ and $\hat{y}$, and thus the field after exiting the specimen is altered and becomes

$$\vec{E}_2 = A\sin\theta\, e^{i\varphi}\hat{x} + A\cos\theta\,\hat{y}$$

The term $e^{i\varphi}$ describes the phase shift.

Finally, this field impinges on the analyzer, which has a transmission axis oriented at θ with respect to the x axis. Thus, the total component along the analyzer is

$$E_A = A\sin\theta\cos\theta e^{i\varphi} - A\cos\theta\sin\theta = A\sin\theta\cos\theta\,(e^{i\varphi} - 1)$$

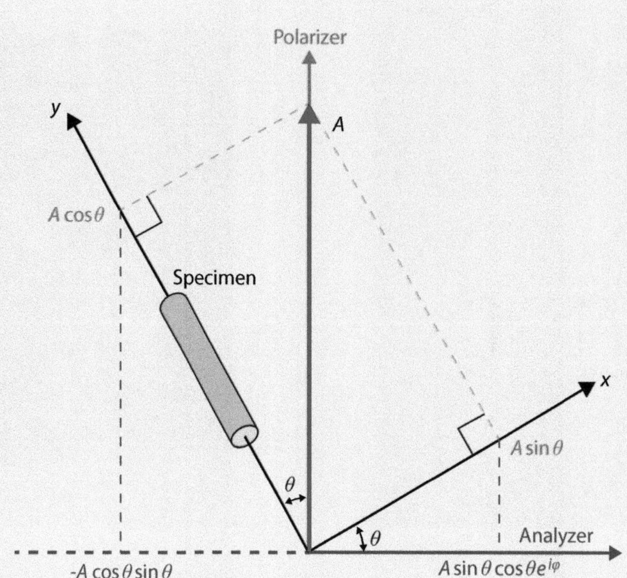

FIGURE B5.3.1
Schematic representation of the mathematics of light throughput by a birefringent specimen between crossed polarizers.

The associated intensity passed by the analyzer, I_A, is proportional to the absolute square of E_A; thus, we have (invoking several prominent trigonometric identities (Appendix E) and using * to denote the complex conjugate):

$$I_A \propto E_A E_A^* \propto A^2 \sin^2\theta \cos^2\theta (e^{i\varphi} - 1)(e^{-i\varphi} - 1)$$

$$I_A \propto A^2 \frac{\sin^2(2\theta)}{4}(1 - e^{i\varphi} - e^{-i\varphi} + 1)$$

$$I_A \propto A^2 \frac{\sin^2(2\theta)}{4}(2 - 2\cos\varphi) = A^2 \sin^2(2\theta)\sin^2\left(\frac{\varphi}{2}\right)$$

$$I_A = I_P \sin^2(2\theta)\sin^2\left(\frac{\varphi}{2}\right)$$

Here, I_P is the intensity incident from the polarizer, as defined in the text.

5.2.4.3 Experimental Implementation

A polarization microscope includes two essential components not present in a microscope designed for standard brightfield microscopy (Fig. 5.17). These are a first polarizer, positioned before the specimen, and a second orthogonally oriented polarizer, the analyzer, positioned between the BFP of the objective and the detector. Additional useful components include a strain-free condenser, strain-free objectives, and a compensator/retardation plate. Strain-free lenses avoid undesirable birefringence from the optics that can obscure weak birefringence from specimens. Compensators and retardation plates introduce additional path differences and thereby alter image contrast and color (Section 5.2.5.1 and Activity A5.2). For example, contrast improves somewhat if a compensator is used to increase the background slightly relative to total darkness.

Gut granules in the worm *Caenorhabditis elegans* are intrinsically birefringent and thus can be studied using polarization microscopy. Fig. 5.18a shows a representative polarization microscopy image of gut granules in wild type worms.

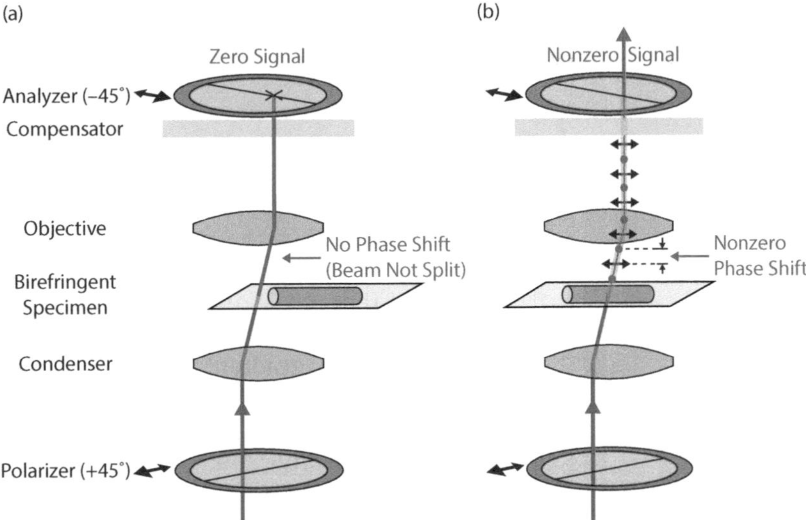

FIGURE 5.17
Schematic of a polarization microscope. (a) In the absence of a specimen, the crossed polarizers extinguish the signal, as in Demonstration 5.1. (b) In the presence of a birefringent specimen, the incident plane-polarized light is split into || and ⊥ rays. These typically get out of phase, as indicated by the vertical shift between the arrow and dot, and combine to produce a wave that has a nonzero component along the analyzer. The polarizer and analyzer are shown at ±45° but usually are oriented north/south and east/west in real microscopes.

5.2.5 Differential Interference Contrast

In this imaging mode, gradients in the phase lag introduced by the specimen are converted into contrast. DIC images have a pseudo-three-dimensional appearance and emphasize edges; they also can be colored. Advantages of DIC over phase contrast include the absence of halos, the absence of annuli that reduce NA, and improved axial resolution, which allows thicker specimens to be imaged. On the other hand, DIC requires expensive, specialized optics and is more complex than phase imaging. Thus, DIC is a somewhat less commonly used approach.

5.2.5.1 Mechanism of Contrast Generation

DIC is in some respects similar to polarization microscopy. In both approaches, signal is created when two orthogonally polarized beams superimpose to generate a nonzero component along an analyzer. One fundamental difference between the approaches is that in DIC the specimen is not birefringent. Thus, the two beams are not generated by the specimen but instead by birefringent optical components (prisms).

The image-generating strategy in DIC is somewhat unusual, and so it is worthwhile beginning with the essence of the approach (Fig. 5.19). The essential

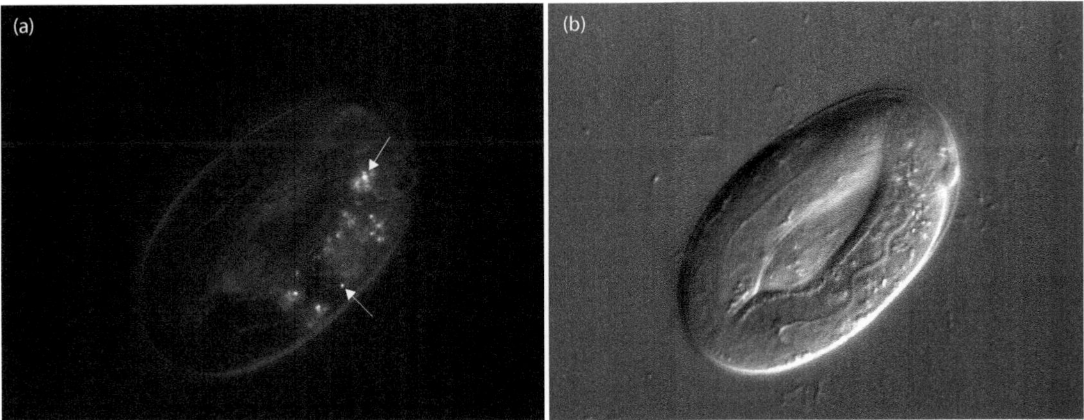

FIGURE 5.18
Comparative (a) polarization and (b) DIC images of a *Caenorhabditis elegans* pretzel stage embryo. The polarization image shows birefringent material (arrows) in intestinal cells. (Images courtesy of Holly Hale and Professor Greg Hermann [Lewis & Clark College].)

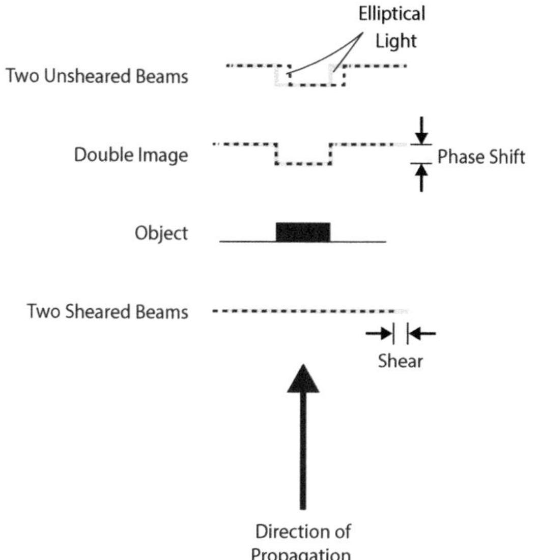

FIGURE 5.19
Essence of image formation in DIC. The diagram depicts the effects of prism-induced lateral shear and specimen-induced phase shifts on wavefronts propagating along the optical axis. Lateral shear between waves is depicted as a left/right translation, and a phase difference between waves is depicted as a vertical displacement. The rectangular object introduces a rectangular phase retardation into the sheared (green and black dotted) wavefronts. This is depicted here in plots of phase versus position. Unshearing creates phase differences at the edges of the object that create elliptical light. The orthogonal polarizations of the two wavefronts ensure that they can independently encode an image and will interfere only when they are unsheared and then combined into the same polarization by an analyzer.

idea is to generate two independent, slightly laterally shifted (sheared) waveforms that can be used to create a "double image." Specifically, each waveform traverses the specimen and produces an image, which is an imprint of the phase profile (Figs. 5.8 and 5.19). After traversing the specimen, the two waveforms are reverse laterally shifted (unsheared), and thus the two images superimpose slightly out-of-register. The critical consequence of unshearing is that phase differences between the waveforms arise in regions of the specimen containing refractive index gradients, such as edges. Thus, these regions generate elliptical light that can pass the analyzer to create signal, similar to polarization microscopy (Fig. 5.19).

Fig. 5.20 illustrates key aspects of implementation together with a ray view of image formation. The two sheared beams are generated by a beam-splitting prism composed of a birefringent material such as calcite, which exhibits double refraction (Fig. 5.14). Specifically, this prism, in conjunction with the condenser, creates pairs of equal amplitude, closely spaced, and laterally sheared parallel beams, which traverse each point in the sample. A second beam-combining prism, in conjunction with the objective, then realigns the beams. In regions where there is no gradient in the specimen-induced phase lag, the two beams reunite to create the original linear polarization, which is blocked by the analyzer. In contrast, in regions containing gradients,

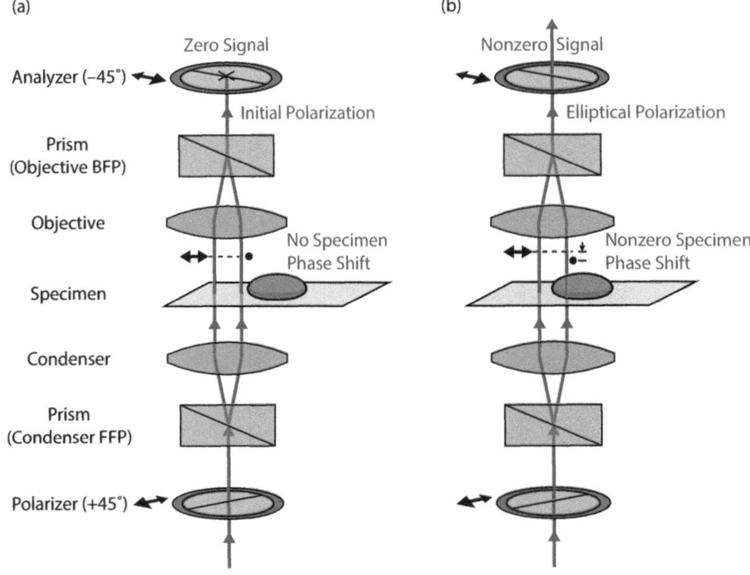

FIGURE 5.20
Schematic of DIC implementation and contrast generation using rays. A first DIC prism generates two equal-energy beams that are polarized at ±45° with respect to the polarizer transmission axis. The beams are laterally sheared by less than the resolution limit, typically 100–150 nm. A second DIC prism recombines the beams. (a) In the absence of a gradient, the orthogonal beams recombine to produce light with the original polarization, which is blocked by the analyzer. (b) In the presence of a gradient, the beams recombine to produce elliptically polarized light, a component of which passes through the analyzer. The shear of the beams is greatly exaggerated.

the beams reunite to create elliptically polarized light that passes the analyzer to create signal. Signal strength is determined by the degree of "ellipticity" (Fig. 5.15).

A mathematical expression that rigorously describes the analyzer-generated intensity pattern, $I_A(x)$, as a function of position, x, along the direction of shear in a DIC image is derived in Box 5.4. The result is

$$\begin{aligned} I_A(x) &= \tfrac{1}{2} I_P \left(1 - \cos\left[\varphi'(x)\Delta x + r\right]\right) \\ &= I_P \sin^2\left[\left(\varphi'(x)\Delta x + r\right)/2\right] \end{aligned} \tag{5.3}$$

Equation (5.3) shows that the intensity depends on the gradient of phase lag as embodied in the term involving the spatial derivative φ' and on the magnitude of beam shear, Δx. $I_A(x)$ also depends on a "retardation" parameter, r, which is a phase shift introduced between the orthogonal beams. We address the function of r next and leave its experimental implementation for Section 5.2.5.2.

To illustrate the utility of retardation, we first examine contrast in its absence. When $r = 0$, Eq. (5.3) becomes $I_A(x) = I_P \sin^2(\varphi'\Delta x/2)$, and intensity variations are determined by values of the sine-squared function near zero (because Δx is small). The sine-squared function varies slowly near zero, so sensitivity to the gradient in phase is weak. Introducing a nonzero r will move the function to a region of more rapid variation and thus help to rectify this problem.

Introduction of retardation also converts images that have a dark background into images that have a gray background and a three-dimensional, "shadow cast" appearance (Fig. 5.21). The origin of this effect, which causes cells to appear bright on one side and dark on the other along the direction of shear (Figs. 5.18b and 5.21c), is described in Fig. 5.21. Finally, the role of retardation in DIC's ability to produce color images is explored in Box 5.5.

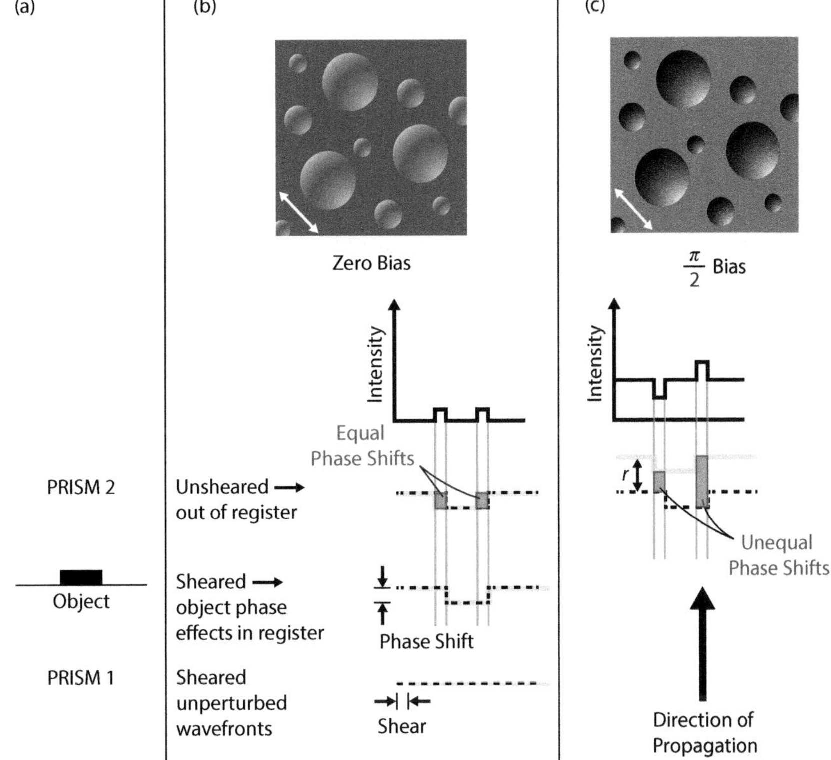

FIGURE 5.21
Schematic description of the effects of introducing bias retardation. Lateral shear and phase shifts are depicted as in Fig. 5.19. The specimen in Panel (a) introduces a rectangular phase retardation (indentation) into the sheared (green and black dotted) wavefronts in Panels (b) and (c). In Panel (b), the sheared wavefronts are unsheared by the second prism and objective in the absence of bias retardation, and there is no phase difference except at the specimen edges. The phase difference, which in general creates elliptical light, is equal in magnitude at the two edges and thus creates equal intensity signals at the two edges. The resulting intensity pattern shows bright edges superimposed on a black background, as exemplified in the image at the top of Panel (b). In Panel (c), the wavefronts are unsheared in the presence of nonzero (e.g., $r = \pi/2$) retardation, which is depicted as an overall vertical shift between the green and black wavefronts. In this case, the edges are associated with phase shifts of different magnitude. At one edge, the phase shift is less than $\pi/2$. This edge will appear darker than the background; see the transmission patterns in Fig. 5.15. At the other edge, the phase shift will be larger than $\pi/2$, and tending toward a maximal transmission shift of π. This edge will appear brighter than the background. Thus, in the presence of quarter-wave retardation, a specimen will have a shadow cast appearance. Double-headed white arrows in the images indicate the direction of wave shear.

BOX 5.4 DERIVATION OF EQ. (5.3)

Here we sketch a derivation of Eq. (5.3). The arguments are similar to those presented in Box 5.3, so this discussion is slightly more succinct.

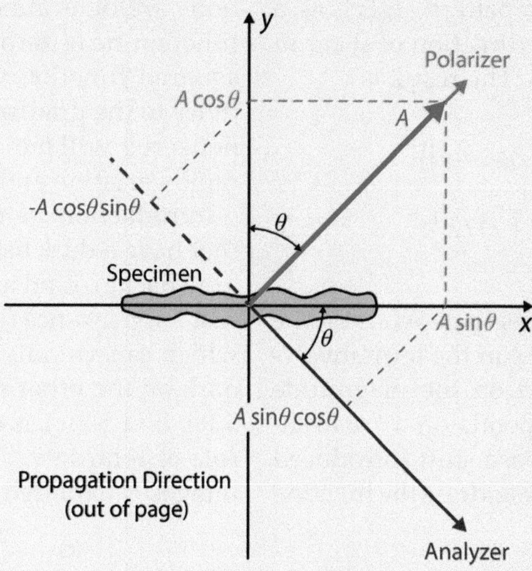

FIGURE B5.4.1
Schematic representation of the geometry used in deriving Eq. (5.3). Optical elements are shown compressed along the direction of light propagation (z) to emphasize the effects of orientation. The transmission axis of the polarizer is oriented at $\theta = 45°$ with respect to the x and y axes, which specify the polarization directions of the two orthogonal beams generated by the first DIC prism. The analyzer and polarizer are crossed.

Our goal is to model the effects of the beam-splitting and beam-recombining prisms as well as the effect of the analyzer. To this end, we assume the geometry in Fig. B5.4.1. The direction of light propagation is z, and, for simplicity, the specimen has a phase profile, $\varphi(x)$, that varies only along x, which also is taken to be the direction of wave shear.

The polarizer generates a field with amplitude A, which the beam-splitting DIC prism decomposes into two equal-amplitude, orthogonally polarized, and laterally sheared fields. Specifically, the total field that impinges on the specimen can be expressed in terms of the unit vectors $\hat{x}$ and $\hat{y}$ as

$$\vec{E}_1 = A\sin\theta\,\hat{x} + A\cos\theta\,\hat{y}$$

The field encounters the specimen and acquires a phase shift (Figs. 5.19 and 5.20). Subsequently, the second DIC prism realigns (unshears) the two orthogonal fields, thereby generating a superposition of fields that have slightly different shifts, $\varphi(x)$ and $\varphi(x+\Delta x)$. As a consequence, the field has the form

$$\vec{E}_2 = Ae^{i[\varphi(x+\Delta x)+r]}\sin\theta\,\hat{x} + Ae^{i\varphi(x)}\cos\theta\,\hat{y}$$

In this expression for $\vec{E}_2$, we have attributed the entire lateral shift, Δx, to one beam and have included the possibility that the second prism adds bias retardation, r, which is also attributed to the same beam.

To proceed further, we invoke Taylor's theorem (for small Δx) to write

$$\varphi(x+\Delta x) \cong \varphi(x) + \varphi'(x)\Delta x$$

In this limit, the field that exits the second prism is given by

$$\vec{E}_2 = Ae^{i[\varphi(x)+\varphi'(x)\Delta x+r]}\sin\theta\,\hat{x} + Ae^{i\varphi(x)}\cos\theta\,\hat{y} = Ae^{i\varphi(x)}\left(e^{i[\varphi'(x)\Delta x+r]}\sin\theta\,\hat{x} + \cos\theta\,\hat{y}\right)$$

Finally, this field impinges on the analyzer. In standard DIC, the polarizer and analyzer are crossed and $\theta = 45°$. Given this, the total component along the analyzer is

$$E_A = Ae^{i\varphi(x)}\left(e^{i[\varphi'(x)\Delta x + r]}\sin\theta\ \cos\theta - \cos\theta\ \sin\theta\right) = \tfrac{1}{2}Ae^{i\varphi(x)}\left(-1 + e^{i[\varphi'(x)\Delta x + r]}\right)$$

The associated intensity is

$$I_A(x) = E_A E_A^* = \tfrac{1}{4}A^2\left(-1 + e^{i[\varphi'(x)\Delta x + r]}\right)\left(-1 + e^{-i[\varphi'(x)\Delta x + r]}\right)$$

$$I_A(x) = \tfrac{1}{2}I_P\left(1 - \cos[\varphi'(x)\Delta x + r]\right) = I_P \sin^2\left[(\varphi'(x)\Delta x + r)/2\right]$$

5.2.5.2 *Experimental Implementation*

The transmitted light DIC optical train, which is relatively complex, is shown schematically in Figs. 5.20 and 5.22. In brief, a first prism and the condenser decompose the incident light into two parallel beams that are orthogonally polarized and laterally sheared by a small amount. These beams impinge on the specimen and then enter the objective, which together with a second prism, refocuses and reverses (in the absence of retardation) what the condenser and first prism did, generating linearly or elliptically polarized light depending on the absence or the presence of a gradient in the specimen-induced phase lag and/or bias. The analyzer transmits the gradient-dependent intensity onto the image plane.

Weaknesses of DIC include expense and complexity. In addition, polymers, such as those in plastic dishes that typically are used to grow cultured cells, need to be avoided, and birefringent specimens can create confusing images. The three-dimensional appearance of images also must be interpreted with caution because it typically bears little relationship to actual topography.

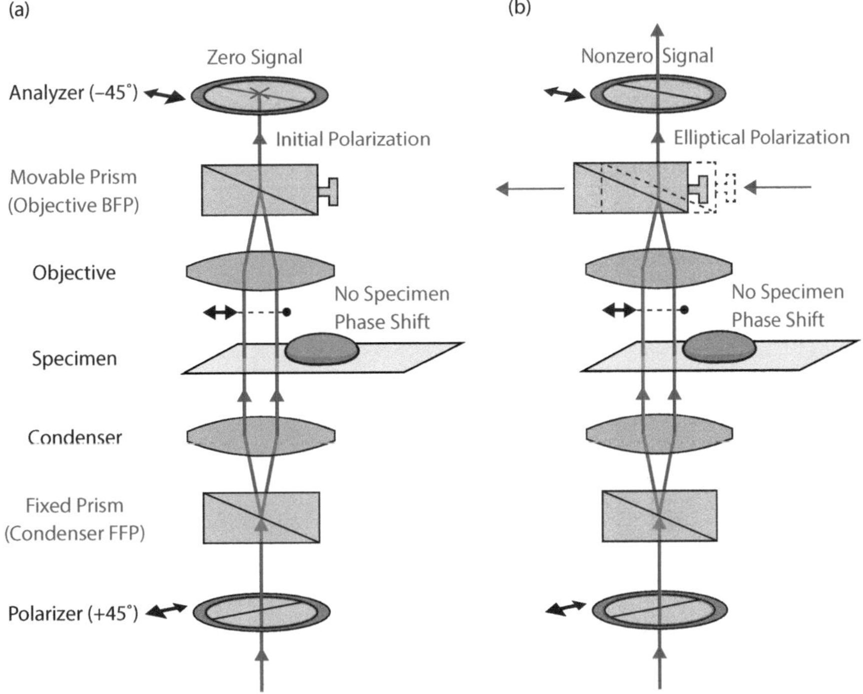

FIGURE 5.22
The DIC optical train emphasizing implementation of bias retardation and its effects on the output waveform for the background (in the absence of a specimen-induced phase shift). (a) In the absence of bias, the two prisms are aligned so that the second prism reverses the effects of the first, and the background is dark. (b) Introduction of bias by translating the second prism leads to a phase shift between the orthogonal background beams and a gray background. The shear of the beams is again greatly exaggerated.

BOX 5.5 COLOR IN DIC IMAGES

DIC has an ability to produce images that are spectacularly colored (Fig. B5.5.1). In this box, we briefly sketch the theoretical basis of this effect using the intensity expressions derived in Box 5.4.

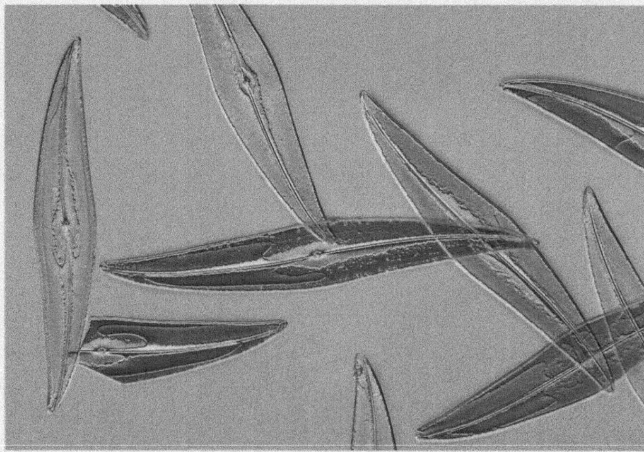

FIGURE B5.5.1
Color DIC image of the diatom *Pleurosigma angulatum*. This organism is popular as a test specimen for microscopy. (Image courtesy of Frank Fox [mikro-foto.de].)

The first of the two equivalent expressions for DIC intensity in Eq. (5.3) and Box 5.4 reads

$$I(x) = 1 - \cos\left[\varphi'(x)\Delta x + r\right] = 1 - \cos\beta$$

Here we have ignored overall multiplicative factors. Fig. B5.5.2 shows a plot of this intensity pattern as a function of $\beta = \varphi'(x)\Delta x + r$. The colored bars (for a fixed bias r) reveal the effect of the sample term $\varphi'(x)\Delta x$ on intensity for each of the primary colors red, green, and blue. At $r = 0$, the intensity curve is relatively flat. This implies that DIC will be relatively insensitive to sample phase variations because they have little effect on intensity. In contrast, for $r = \pi/2$, the intensity curve is steep, and thus DIC will be sensitive to sample variations.

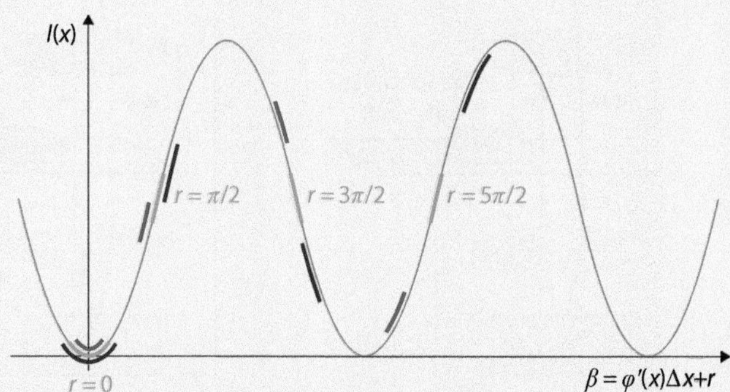

FIGURE B5.5.2
Intensity variations in DIC due to the effect of the sample and bias. The bias shown corresponds to the color green.

The graph also shows that different colors differ significantly in intensity for nonzero r. This variation arises from the dependence of β on wavelength through the effect of wavelength on r, refractive index, and phase shift (Eq. [2.2] and Box 2.2). The brightness differences associated with red, green, and blue generate an image that is colored (Fig. B5.5.1).

TABLE 5.1

Comparison of Optical Techniques (All Techniques Employ Köhler Illumination)

Technique	Optical Basis	Implementation	Image	Applications	Live Cells?
Brightfield	Absorption	Standard Köhler illumination	Specimen is dark (or colored) on a bright background	Naturally colored specimens; stained specimens, especially bacteria and tissues	No[a]
Darkfield	Only specimen perturbed (e.g., diffracted/ scattered) light	$NA_{condenser} > NA_{objective}$ Special condenser/ objective pair or annulus in FFP of condenser, equivalent to blocking zeroth-order diffraction spot in BFP of objective	Specimen is bright on a dark background	Thin samples, notably tiny unstained biological specimens, such as bacteria and diatoms	Yes
Phase contrast	Interference between *Z* and *D* waves, contrast determined by differences in the magnitude of the phase shift (OPL)	Special phase-contrast condenser and phase-contrast objective with phase rings	Specimen is gray scale, halo artifacts	Thin samples, notably unstained cells in tissue culture dishes	Yes
Polarization	Birefringence	Polarizers, before and after specimen, at 90° relative to one another	Specimen is bright (including colored) on a darker or colored background	Birefringent samples; mineralogy	Yes
DIC	Interference between sheared beams, contrast determined by *gradients* in phase (OPL)	Polarizers, as in polarization microscopy, and birefringent prisms	Specimen is gray scale or colored, image has shadow cast, pseudo-three-dimensional appearance, no halos	High-contrast imaging and optical sectioning of cells, thick samples	Yes

[a] Brightfield microscopy can be used to study naturally colored living samples, but such samples (e.g., chloroplasts) are uncommon. Instead, it typically is used to study dead fixed and stained samples.

5.3 Comparison of Approaches

In this chapter, we have discussed five prominent optical microscopy techniques. To facilitate comparison and mastery of the approaches, we end with a concise comparative summary of their important attributes, including mechanisms of contrast generation, implementation, and applications (Table 5.1).

Suggested Reading and Additional Resources

Hecht, E. 2017. *Optics (5th Edition)*. Pearson Education Limited, Essex, England.

iBiology: Microscopy Series. 2012 – 2015. Contrast Generation for Transmitted Light. https://www.ibiology.org/online-biology-courses/microscopy-series/contrast-generation-transmitted-light/; https://www.ibiology.org/talks/microscope-imaging-koehler-illumination/; https://www.ibiology.org/talks/set-up-koehler-illumination/

Inoue, S. and Spring, K.R. 1997. *Video microscopy: The fundamentals (2nd Edition)*. Springer, New York, NY. ISBN 0-306-45531-5.

Murphy, D.B. and Davidson, M.W. 2013. *Fundamentals of light microscopy and electronic imaging (2nd Edition)*. Wiley-Blackwell, Hoboken, NJ.

Spencer, M. 1982. *Fundamentals of light microscopy*. Cambridge University Press, Cambridge, England.

Wayne, R. 2019. *Light and video microscopy (3rd Edition)*. Academic Press, Burlington, MA.

HOMEWORK PROBLEMS

5.1 [Contrast Metrics and Effects of Background] Various metrics have been developed to quantify contrast. Two common metrics are listed below:

Metric	Definition	Application
Weber	$C_W = \frac{I_{max} - I_{min}}{I_{min}}$	Ratio of the difference in the maximum and minimum intensities to the minimum intensity. Typically used to describe images of small objects on an otherwise uniform field. In this case, the intensity of the uniform field will be approximately equal to the average intensity. Range: 0–∞.

Metric	Definition	Application
Michelson	$C_M = \frac{I_{max} - I_{min}}{I_{max} + I_{min}}$	Ratio of the difference in the maximum and minimum intensities to an "average" intensity expressed as a sum of the maximum and minimum intensities. Typically used to describe periodic images, such as interference patterns, in which lighter and darker regions occupy approximately equal areas. Range: 0–1.

The best metric depends on the nature of the image and the use to which the metric is being put. The metrics can be used to study the effect of an undesired background intensity, or noise, on image contrast. Assume for simplicity that the background is constant (i.e., that it contributes the same amount to both minimum and maximum intensities in an image). Show that a constant background will always reduce contrast in the image, whether measured by the Weber or Michelson metric. Hint: Add the same constant intensity background, I_B, to both I_{max} and I_{min} in the definitions of contrast.

5.2 [Neutral Density Filters] Microscopes use a variety of filters to control the quantity and nature of light. Neutral density (ND) filters are used to reduce the amount of light incident on the sample and/or detector. They are called ND filters because they are spectrally neutral: they reduce light equally across all visible wavelengths. ND filters are typically characterized by a single number called "optical density" (OD) that describes the amount of light that the filter transmits (the transmittance, T):

$$T = 10^{-\text{OD}}$$

$$\text{OD} = -\log T$$

a. Calculate the amount of light transmitted through ND filters having the following common ODs: 0.1, 0.2, 0.3, 0.5, 1.0, 1.5, and 2.0.
b. Most ND filters are absorptive: they block light by absorbing it. What happens to the energy from the absorbed light?
c. Optical densities, unlike transmittance, are additive. Thus, when multiple filters are used, the total OD is simply the sum of the individual ODs. Calculate the OD necessary to *block* 75% of the incident light. Which combination(s) of filters from part (a) can be used?

5.3 [Brightfield Microscopy – Phase Relationships for a Phase Specimen] The figure shows a typical brightfield vector diagram. In the main text, we assumed that φ is very small. In this case, the angle between Z and D is $\pm\pi/2$. Show that the angle between the Z and D waves, Φ, is more generally given by $\Phi = \pi/2 + \varphi/2$. Hint: The vector addition process generates an isosceles triangle because $|\vec{Z}| = |\vec{T}|$.

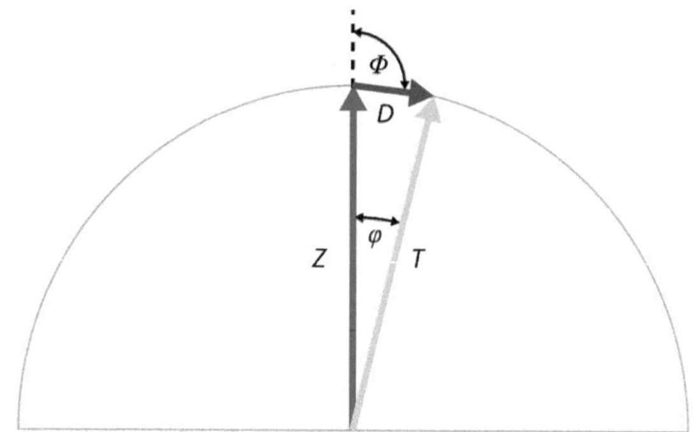

5.4 [Darkfield Microscopy – Vector Diagram] Draw a vector diagram for darkfield microscopy showing the relative amplitudes and phases of the Z, D, and T waves. Hint: Review the wave diagram for phase-contrast microscopy, paying particular attention to the definitions of these waves.

5.5 [Phase-Contrast Microscopy – Vector Diagram] Verify the assertion made in the legend to Fig. 5.11, which is that a phase shift of π between the Z and T waves implies that $T = \sqrt{5}Z$.

5.6 *[Phase-Contrast Microscopy – Optimal Attenuation] In phase-contrast imaging, the undiffracted beam is both phase shifted and attenuated to improve contrast. To verify the importance of attenuation, implement the following analysis.

a. Write the image field, $T(x,y)$, as the sum of a spatially invariant, unscattered field, $|Z|e^{i\omega t}$, and a spatially varying scattered component, $|D|e^{i(\omega t+\varphi)}$, which is out of phase with the unscattered field by $\varphi(x,y)$ due to the effects of the sample. Starting with this expression, show that the image intensity $I(x,y) = |T(x,y)|^2$ is given by an interference pattern of the form

$$I(x,y) = |Z|^2 + |D|^2 + 2|Z||D|\cos[\varphi(x,y)]$$

b. Contrast, $C(x,y)$, may be defined in terms of the minimum and maximum intensities, I_{min} and I_{max}, using the Michelson metric:

$$C(x,y)=\frac{I_{max}-I_{min}}{I_{max}+I_{min}}$$

Use your result from part (a) to demonstrate that the contrast of the interference pattern is given by

$$C(x,y)=\frac{2r(x,y)}{1+r(x,y)^2}, \text{ where } r(x,y)=\frac{|D(x,y)|}{|Z|}$$

c. Determine the value of $r(x,y)$ that maximizes the contrast using methods from calculus and confirm your result with a plot of $C(x,y)$ versus $r(x,y)$.

d. Use your result from part (c) to explain the importance of attenuating the Z wave, which typically is much more intense than the D wave. Interestingly, in the limit where the attenuation is 100%, phase contrast becomes darkfield.

5.7 [Phase-Contrast Microscopy – Optimal Attenuation (Numerical)] Write a program that adds two harmonic waves of different phase and amplitude. Study the mechanism underlying phase contrast by generating plots of Z, D, and T waves for positive and negative phase contrast. Hint: Think about the appropriate phase shift to use between Z and D in each case.

5.8 [Phase-Contrast Microscopy – Negative Phase Contrast] Sample Problem 5.1 described the construction of a $\pi/2$ advancing etch for use in positive phase-contrast microscopy. An alternative is negative contrast where the Z waves travel through more material and therefore are retarded in phase. Calculate the thickness of material, Δx, that would have to be added to a glass slab relative to air to retard Z by $\pi/2$. Assume that $n_a = 1.00$, $n_g = 1.50$, and that the illumination has a vacuum wavelength of 500 nm.

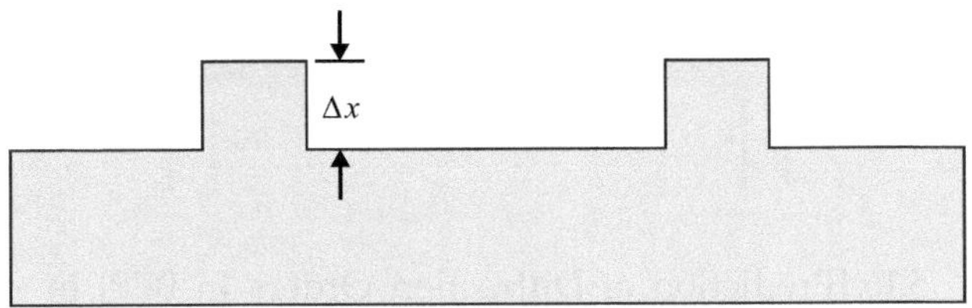

5.9 [Polarizers – Reduction of Glare] A fisherman is wearing polarized sunglasses, which have a vertically oriented transmission axis. Assume, for simplicity, that the light reflected from the water is purely horizontally polarized (Fig. 2.10b). While looking for fish, the fisherman tilts his head at an angle of 30° with respect to the vertical. What percent of the reflected light will be transmitted by the glasses?

5.10 [Polarimetry] The concentration of glucose in urine – an important marker for diabetes – can be measured using polarimetry. Polarimetry exploits the fact that glucose is optically active, which, for our purposes, means that a glucose sample will allow some light to be transmitted through crossed polarizers by rotating the plane of polarization, as shown below. The angle of rotation is proportional to the thickness and concentration of the glucose sample. Thus, glucose concentration can be determined by measuring the amount of light transmitted and comparing the result to a calibration chart.

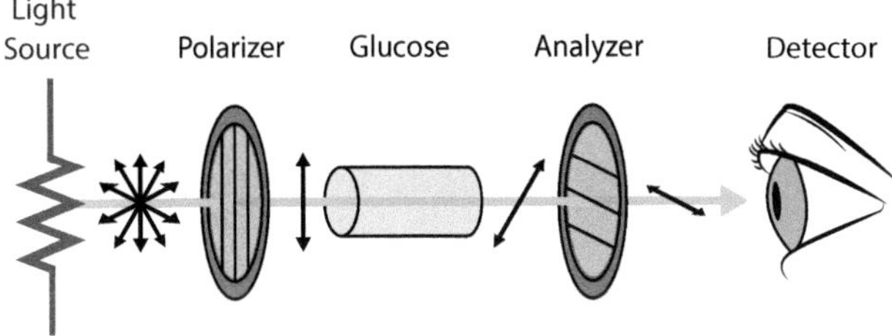

a. In a polarimeter, what fraction of the unpolarized incident light is transmitted by the first polarizer?

b. When a glucose specimen is placed in a polarimeter, 30% of the incident light is transmitted by the second polarizer. What was the angle of rotation?

c. Determine the percent transmission by a sample with one quarter the concentration of that in part (b).

5.11 [Polarization Microscopy] Use Eq. (5.2) to verify that the specimen will be dark when its symmetry axis is oriented at 0° or 90° with respect to the transmission axis of the polarizer and when the specimen-induced phase shift is given by $\varphi = 2m\pi$ (m integral). Also use Eq. (5.2) to verify that the specimen will be maximally bright when the specimen-induced phase shift is given by $\varphi = m\pi$ (m odd integral) (Fig. 5.15).

5.12 [Polarization Microscopy – Interference Colors] If a crumpled piece of cellophane is placed between crossed polarizers, it will generate a highly multi-colored image, which arises when interference leads to removal of particular visible wavelengths from the incident white light, while other colors are partially transmitted. Visible wavelengths can be removed if the cellophane is sufficiently thick. If this is not the case, the cellophane will look bright but colorless (Fig. 5.16). (Image courtesy of Ed Knepley [https://edknepleyphoto.photos/].)

a. To illustrate the origin of interference colors, suppose that white light is incident on a small region of the cellophane where the thickness and birefringence have values that result in a phase shift of 2π for blue light ($\lambda_0 = 450$ nm). Refer to Fig. 5.15 to explain why blue light will not be transmitted by the analyzer. Notably, the removal of blue light will typically lead to a yellow interference color because red and green are partially transmitted.
b. What will be the interference color from this region if the cellophane instead is placed between parallel polarizers?

The important take-home message from this problem is that different visible wavelengths will be removed from regions of the cellophane of different thicknesses, producing an image that is variably colored.

5.13 *[Circularly Polarized Light] Verify that equal amplitude, orthogonal electric fields that are out-of-phase by $\pi/2$ generate circularly polarized light. To this end, consider the vector sum of two orthogonal fields that are propagating along z and are given by

$$\vec{E}_x(z,t) = E_0\cos(kz - \omega t)\hat{x}$$

$$\vec{E}_y(z,t) = E_0\cos\left(kz - \omega t - \frac{\pi}{2}\right)\hat{y}$$

a. Show that the resultant field is given by $\vec{E}(z,t) = E_0\left[\cos(kz - \omega t)\hat{x} + \sin(kz - \omega t)\hat{y}\right]$.
b. Prove that the resultant field has a constant scalar magnitude, E_0, which is consistent with circular polarization.
c. Convince yourself that at a given point z the field executes clockwise circular motion at angular frequency ω (as seen by an observer toward whom the field is moving). For example, draw the field at the point $z = 0$ for the time points $t = 0, \pi/2\omega, \pi/\omega, 3\pi/2\omega$. This is known as a right-circularly polarized field.

5.14 [Birefringence – Application] Birefringent crystals are used to construct quarter-wave retardation plates, which can be used to convert plane-polarized light into circularly polarized light. The plate has a geometry similar to that discussed in Section 5.2.4.1 and introduces a phase shift of $\pi/2$ between two orthogonal, equal amplitude waves; this generates circularly polarized light (HW 5.13). Assume that light of vacuum wavelength 550 nm is incident on a plate composed of quartz. In this case, the two orthogonal waves experience refractive indices of 1.544 and 1.553. What is the (minimum) possible thickness of the plate?

5.15 [DIC – Signal Strength] The cartoon below shows two object point scenarios in a DIC experiment in the absence of bias. For each case, indicate if the associated image point will be bright or dark. Justify your answer! Hint: Review Figs. 5.19 and 5.20.

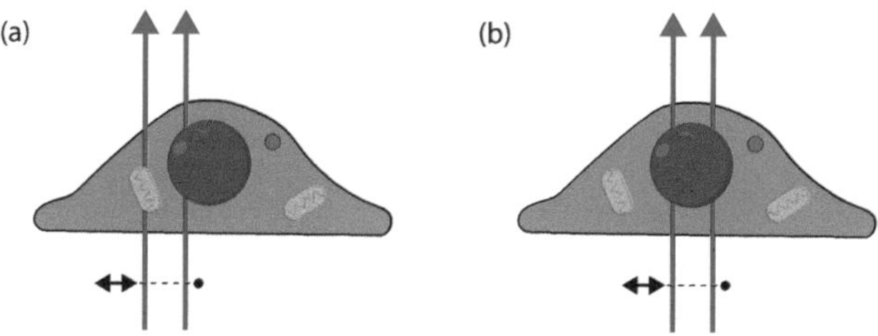

5.16. [Prediction of Diffraction Orders in BFP] In Activity A5.1, you will verify some aspects of the Abbe Theory of image formation using a real microscope and a grating as a specimen. In preparation for the experiment, carry out the following calculations.

a. Your specimens will be gratings with 500 lines/mm and 1000 lines/mm illuminated by green light with λ ~550 nm. You will use two air/dry (not oil immersion) objectives. One is a 10× objective with an NA = 0.3; the other is a 40× with an NA = 0.7. Use the definition of NA to calculate the half angle over which rays are captured by each objective, where the intervening medium is air with $n = 1.0$.
b. Now use the grating equation $m\lambda = d \sin\theta$ to predict the highest order diffraction spot (largest m) that will be captured by each objective for each of the two gratings.

Lens	Grating = 500 lines/mm	Grating = 1000 lines/mm
10×		
40×		

ACTIVITIES

A5.1 Tour of Brightfield, Darkfield, and Phase-Contrast Microscopies

This activity is written under the assumption that there is access to an older inverted microscope, such as the Olympus IMT-2.

1. [Köhler Illumination] Follow the steps below to set up Köhler illumination, which produces uniform partially coherent light in the specimen plane. There are several key features of Köhler illumination, which are illustrated in the schematic in Fig. 3.13. Look at the schematic before beginning this exercise, so the manipulations make sense. In this activity, we assume that the lamp has been correctly focused on the FFP of the condenser and will simply confirm this. The reader interested in learning how to focus the lamp on the condenser FFP is referred to the lecture on Köhler illumination in the iBiology microscopy series.
   A. Turn on the transmitted light.
   B. Open the field diaphragm completely.
   C. Open the condenser aperture diaphragm completely.
   D. Focus on an amplitude specimen, such as fixed cells, using a low magnification objective (e.g., 10×).
   E. With the specimen in focus, close the field diaphragm and focus the field diaphragm onto the specimen plane using the condenser translation.
   F. Center the field diaphragm with the condenser lens centration knobs and then open the field diaphragm just to the edge of the FOV. The specimen and field diaphragm are now in conjugate planes.
   G. Remove the diffusor and insert the Bertrand lens into the light path. This will allow you to look in the BFP of the objective.
   H. Verify that the lamp is correctly focused on the condenser FFP by viewing its (focused) image in the objective BFP. Close down the condenser aperture diaphragm and verify that it is in focus in the objective BFP. Explain why looking in the objective BFP is an appropriate way to view the condenser FFP.
   I. Reinsert the diffusor and switch back to the sample plane. Open the condenser aperture diaphragm all of the way and then stop it down. Notice that the contrast improves. In brightfield mode, the condenser aperture diaphragm typically is set to fill ~75% of the objective BFP to enhance contrast without excessive loss of resolution.
2. [Brightfield and Phase-Contrast Microscopies] Follow the steps below to set up phase contrast and to compare brightfield and phase-contrast images of transparent specimens.
   A. Choose a phase-contrast objective and focus on the specimen in brightfield mode. Use the field and condenser aperture diaphragms as an aid because your specimen is low contrast.
   B. Choose a matching phase annulus by rotating the condenser turret.
   C. Use the Bertrand lens to look in the objective BFP. You should see two annuli – the illumination annulus in the condenser FFP and the phase plate in the objective BFP. Use the condenser annulus centering knobs to align the two annuli.
   D. Switch back to the specimen plane. When focused, the specimen should have nice contrast.
   E. Switch the condenser annulus back to brightfield.
   F. What happens to the contrast when you switch to brightfield imaging?

3. [Darkfield Microscopy] In this exercise, you will investigate the effects of darkfield imaging and see how easily it can be implemented. This activity is written under the assumption that an objective with a sufficiently low NA, and a phase annulus of sufficient size, are available to meet the conditions specified in Section 5.2.2.2.
   A. Make sure that the condenser aperture diaphragm is open. Focus on the specimen with the condenser set on brightfield (no annulus in the condenser). After the specimen is in focus, make sure that the field diaphragm is focused and centered on the specimen, and that it is opened just to the edge of the field of view.
   B. An intentionally oversized phase annulus can be used as a mask. Your task is to find the one that works. Begin with the smallest annulus. Does it create a darkfield effect? If not, successively rotate through the annuli until you find one that works. What do the cells look like?
   C. With the annulus in place, and no sample on the stage, look through the objective. Describe the appearance of the field. Explain.
   D. Suppose your condenser has an NA = 0.55. Would you expect darkfield imaging to work with the 40× objective (NA = 0.70) analyzed in HW 5.16? Justify your answer!

4. [Abbe's Experiments] In this exercise you will repeat some of Abbe's pioneering experiments using specimens consisting of pieces of diffraction grating attached to (mounted on) a cover slip. This activity is written as a complement to HW 5.16 and thus the lenses and gratings have NAs and spacings consistent with HW 5.16.
   A. Place the 500 lines/mm grating on the stage and focus on the grating using the 10× objective in brightfield mode. It may help to close down the condenser aperture diaphragm because this increases contrast, and the grating is hard to see. Also, you can stop down the field diaphragm and focus on it because the field diaphragm and specimen should be in focus simultaneously. Reopen the field diaphragm when you are in focus.
   B. Use the Bertrand lens to look in the objective BFP. Stop down the condenser aperture diaphragm (if you did not already do so) and bring it into focus using the focus for the BFP. You should see replicas of the condenser aperture diaphragm. The number of replicas corresponds to the number of diffraction orders "m" that the objective captures. Fill out the table below, including data for the 1000 lines/mm grating.
   C. Repeat parts (A) and (B) using the 40×.

Lens	# of spectra observed (500 lines/mm)	Agreement with HW 5.16?	Grating visible?	# of spectra observed (1000 lines/mm)	Grating visible?	Agreement with HW 5.16?
10×						
40×						

   D. Notice that the diaphragm images (except for $m = 0$) display a gradient of colors. Explain using the relationship between λ and diffraction angle.
   E. If you remove the grids, are there any replicas in the BFP? How did Abbe interpret this?

A5.2 Introduction to Polarization Microscopy

This activity is written under the assumption that there is access to a polarizing microscope with attributes similar to the Steindorff monocular polarizing microscope, which has a mica full-wave plate in addition to a polarizer and analyzer. The sample must be birefringent. Cellophane tape mounted to a microscope slide is one option.

1. Focus on the birefringent specimen without the analyzer or wave plate in the light path.
2. Answer and implement the following:
   A. What is the appearance of the field with the analyzer out?
   B. Insert the analyzer into the light path. What happens to the field? Explain briefly!
   C. Locate a birefringent specimen. What does it look like? Explain briefly!
   D. Rotate the stage through 360°. What happens to the appearance of your birefringent specimen?
   E. Insert the wave plate into the light path. What happens to the field? Explain briefly!
   F. With the plate in, what happens to the appearance of your birefringent specimen? Explain briefly!

6

Fluorescence Microscopy

Fluorescence microscopy differs from the techniques discussed in Chapter 5, which use specimen-induced phase effects to create images of clear and colorless, unstained samples. Fluorescence, in contrast, uses the absorption and re-emission of light to generate an image. The molecule that generates the emission is known as a "fluorophore." Most commonly, specific components of a cell are tagged with an extrinsic fluorophore that absorbs and emits in the visible part of the EM spectrum (Fig. 6.1). Ultraviolet radiation is avoided because it is much more damaging.

Fluorescence microscopy dominates the biological sciences. We will discuss the reasons throughout this chapter, but, in brief, this is because the technique is

1. *Sensitive*: Signal is viewed against a dark background, similar to darkfield and polarization microscopies (Fig. 6.1).
2. *Selective*: Structures and locations of interest are specifically tagged with fluorescent probes, while native molecules typically are nonfluorescent.
3. *Versatile*: Multiple structures within the same specimen can be imaged using spectrally distinguishable fluorophores (Fig. 6.1). Moreover, living samples can be studied.

Disadvantages of fluorescence include the need for intense illumination, which can damage cells and lead to photobleaching (loss of fluorescence), and an almost universal need to introduce extrinsic fluorophores, which can perturb physiology.

6.1 Attributes of Fluorophores and Fluorescence

Fluorophores and the physics of fluorescence will play major roles in our upcoming discussion. We therefore briefly review these important topics.

6.1.1 Fluorophore Structure

The fluorophores used in fluorescence microscopy generally are organic molecules having an aromatic structure formally consisting of alternating single and double bonds (Fig. 6.2). Aromatic structures are useful because they have delocalized, mobile electrons that undergo transitions between states having energy differences that are matched to visible radiation. Rare-earth metal complexes, notably lanthanide chelates, which exhibit a long-lived "luminescence" (microsecond to millisecond), also are useful probes.

6.1.2 Fluorescence Excitation and Emission Spectra

The absorption (excitation) and emission attributes of a molecular fluorophore can be described using a Jablonski diagram, which shows the (quantized) energy levels associated with the fluorophore's electronic and vibrational states (Figs. 6.3 and 6.4). Before excitation, the molecule typically resides in its lowest (ground) electronic and vibrational energy states. After absorbing light with energy matched to that of a quantum transition, the molecule is promoted into any of the possible vibrational energy states associated with an excited electronic state. The existence of multiple possible excited states means that absorption occurs over a range of light frequencies, termed an absorption band, having a nonzero width (Figs. 6.3 and 6.4).

Fluorophores generally spend several nanoseconds in the excited electronic state before relaxing back to the ground electronic state and possibly emitting radiation (Fig. 6.3). This timescale is known as the "fluorescence lifetime" and depends on both the fluorophore and the environment. Before emission occurs, excited vibrational energy is lost nonradiatively (typically within ~10^{-12} seconds) to the surroundings in the form of heat. Similarly, if a fluorophore relaxes into an excited vibrational state associated with the ground electronic state, the vibrational energy also is dissipated nonradiatively. These facts imply that emitted light is of lower energy than excitation light or, equivalently, that $\lambda_{\text{emission}} > \lambda_{\text{excitation}}$. Sir George Stokes discovered this wavelength relationship, termed the "Stokes shift," while observing fluorescence emission from quinine (Fig. 6.5 and Demonstration 6.1). The existence of multiple possible decay states means that the emission band also has a finite width (Figs. 6.3–6.5).

DOI: 10.1201/b22076-7

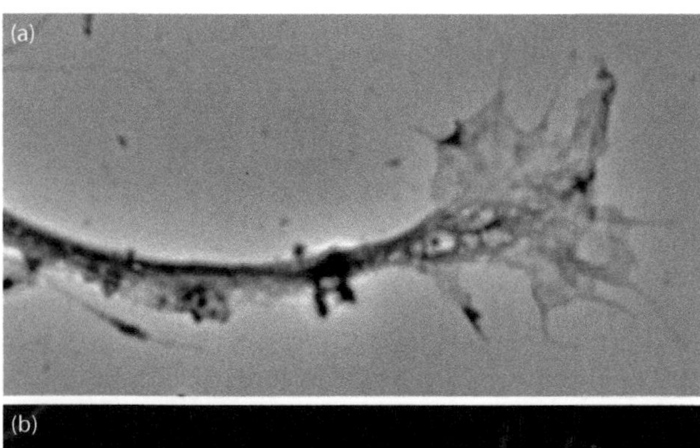

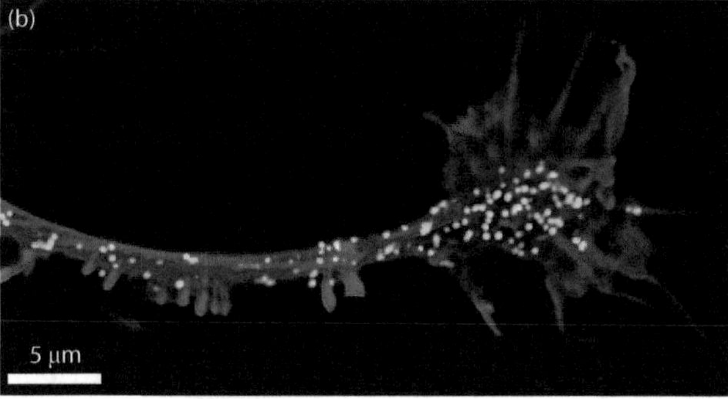

FIGURE 6.1
(a) Comparative phase-contrast and (b) three-color fluorescence images of a growth cone in a cultured hippocampal neuron. Unlike the phase image, the fluorescence image shows the distribution of three specific subcellular components. These are microtubules (blue), actin filaments (red), and dense-core vesicles (DCVs) (green). Scale bar = 5 μm. (Adapted with permission from Silverman, M. A., et al. 2005. Mechanisms of transport and exocytosis of dense-core granules containing tissue plasminogen activator in developing hippocampal neurons. *J. Neurosci.* 25:3095–3106. https://doi.org/10.1523/JNEUROSCI.4694-04.2005. Copyright 2005 Society for Neuroscience.)

Fluorescein
BODIPY
Cy5
AMC
Cascade Blue Hydrazide

FIGURE 6.2
Structures of some common fluorophores. The prevalence of alternating single and double bonds (conjugation) is evident. The names are color coded to match the emission wavelengths.

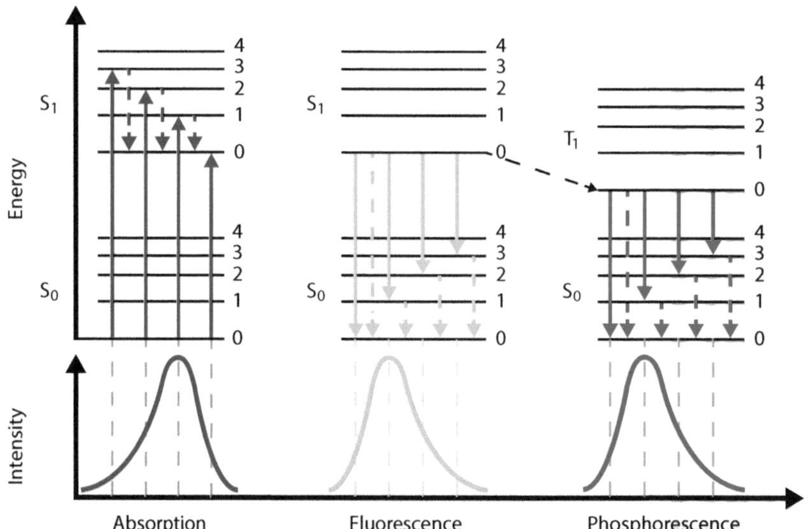

FIGURE 6.3
Jablonski Diagram. Electronic states are labeled using the letter S (for a singlet spin state), and vibrational states are labeled with numbers. The lowest electronic and vibrational states are denoted S_0 and 0, respectively. Solid lines denote radiative transitions (absorption: blue and emission: green/red), and dashed lines denote nonradiative transitions, such as vibrational relaxation or internal conversion between electronic states with the same spin. The letter T signifies a long-lived (triplet spin) state that leads to phosphorescence.

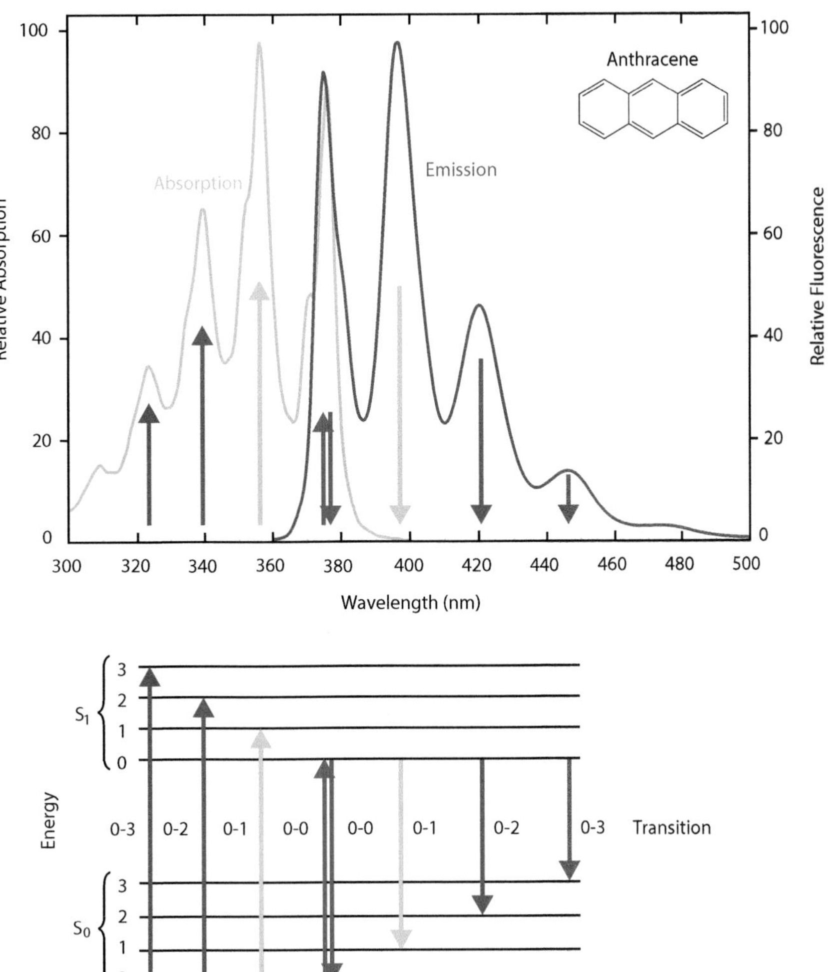

FIGURE 6.4
Absorption and emission spectra for anthracene. The spectrum of anthracene shows unusually well-defined peaks associated with transitions into different vibrational states (HW 6.3). These peaks are identified using colored arrows. (Data for graph exported from University of Arizona Spectra Database [http://spectra.arizona.edu/].)

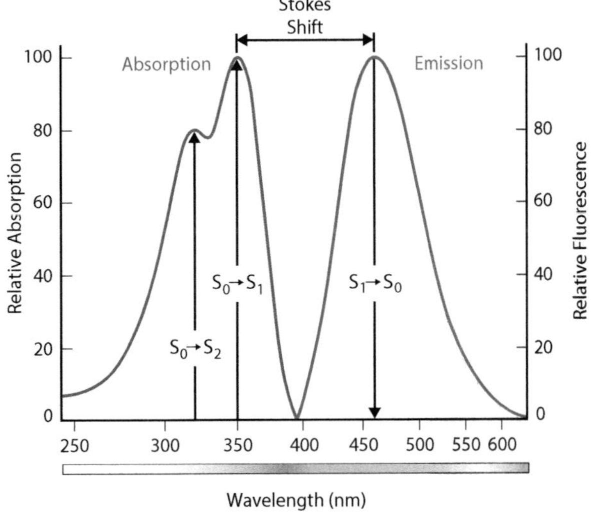

FIGURE 6.5
Absorption and emission spectra for quinine, showing the Stokes shift. (Data for graph exported from University of Arizona Spectra Database [http://spectra.arizona.edu/].)

DEMONSTRATION 6.1 UV EXCITABLE FLUORESCENT MINERAL KIT

The Stokes shift and the rapid decay of the fluorescent state can be demonstrated easily and inexpensively using a UV light source and fluorescent minerals. The minerals fluoresce in the visible under UV excitation and "immediately" cease fluorescing when the excitation light is removed. (Photographs by the authors.)

SAMPLE PROBLEM 6.1 STOKES SHIFT

Use the data in Fig. 6.5 to calculate (a) the approximate energy difference between the ground (S_0) and first excited (S_1) singlet states of quinine, and (b) the approximate energy loss that generates the Stokes shift.

SOLUTION

a. The energy difference between S_0 and S_1 can be calculated using the expression (Eq. [6.2]):

$$E_1 - E_0 = \frac{hc}{\lambda}$$

We can estimate $\lambda = 350$ nm from the location of the arrow in Fig. 6.5. Inserting the relevant numerical values yields

$$E_1 - E_0 = \frac{\left(6.63 \times 10^{-34}\right) \times \left(3.00 \times 10^{8}\right)}{350 \times 10^{-9}} = 5.68 \times 10^{-19} \text{ J} = 3.55 \text{ eV}$$

(This result can be obtained more rapidly using the result $hc = 1240$ eV·nm, which sometimes is quoted as the mnemonic 1234 eV·nm.)

b. The energy loss that generates the Stokes shift can similarly be estimated from the difference in energy between the excitation peak (E_{EX}) and emission peak (E_{EM}). We thus have

$$E_{EX} - E_{EM} = \frac{1240}{350} - \frac{1240}{460} = 0.84 \text{ eV}$$

The existence of a Stokes shift is pivotally important to fluorescence microscopy. In particular, the shift allows the fluorescence signal to be viewed superimposed on a dark background, which markedly enhances the sensitivity of fluorescence microscopy (Fig. 6.1). A dark background is generated by excluding the shorter wavelength excitation from the detector using a filter (Section 6.3.2).

Fluorophore emission spectra are a complex function of structure. However, it is approximately true that emission wavelength increases with increasing fluorophore size. This observation can be rationalized using a simple quantum model – an electron of mass m in an impenetrable box of length L. The quantized energy levels available to an electron in an impenetrable box are given by (HW 6.6 and 6.7)

$$E_j = \frac{j^2 h^2}{8mL^2} \quad j = 1, 2, 3\ldots \tag{6.1}$$

where h is Planck's constant and j is a quantum number. Thus, when an electron in an impenetrable box makes a transition from a higher energy excited state into a lower energy state (with quantum numbers j_2 and j_1), energy conservation dictates that the wavelength of the emitted photon is given by

$$h\frac{c}{\lambda} = E_2 - E_1 = \frac{h^2}{8mL^2}\left(j_2^2 - j_1^2\right) \tag{6.2}$$

Equation (6.2) shows that $\lambda \propto L^2$ and thus that emission wavelength tends to increase with increasing fluorophore size (HW 6.8).

6.2 Fluorescence Labeling Techniques

Most biological specimens have low intrinsic fluorescence (in the visible). Therefore, to obtain a fluorescence signal, specimens must be labeled. Fluorescence labeling is achieved using many, continually evolving approaches. Here we highlight a subset of possible approaches based on (1) conventional stains and (2) genetically encoded fluorescent reporters. Our emphasis is on labeling biomolecules in cells.

6.2.1 Conventional Staining

Below we divide conventional staining into common subcategories, although these of necessity overlap.

6.2.1.1 Covalent Labeling

In this approach, a fluorophore is covalently attached (chemically bonded) to a biomolecule of interest. This is accomplished by engineering the fluorophore to include a functional group that will react with a preexisting target on the biomolecule. A standard target for proteins and nucleic acids is a primary amine group (NH_2). Amines are attractive because they are charged at physiological pH and thus generally are surface exposed and chemically accessible. Moreover, amines also are highly nucleophilic and thus readily react with a spectrum of functional groups, including *N*-hydroxysuccinimide (NHS) esters and isothiocyanates. The concept is illustrated in Fig. 6.6.

6.2.1.2 Immunofluorescence Labeling

In this approach, a "primary" antibody is generated against an antigen, which frequently is a subcellular structure. The antibody is introduced into the sample, where ideally it recognizes and binds predominantly to the antigen. Unfortunately, in studies of subcellular components, cells are fixed to cross-link proteins and permeabilized using a detergent to facilitate antibody passage. Thus, immunofluorescence is not commonly applied to living cells.

Primary antibodies produced by "host" animals are not intrinsically fluorescent and therefore need to be tagged with fluorophores. There are two common tagging methods. The first is "direct labeling" of the primary antibody, which involves chemically linking the primary to a fluorophore (e.g., using methods like those described in Section 6.2.1.1). The product is a fluorescently tagged primary that can be visualized using fluorescence microscopy (Fig. 6.7).

The second method is "indirect labeling." This alternative involves generating a fluorescently tagged secondary antibody that recognizes and binds to the untagged primary (Fig. 6.7 for more detail). Binding of the secondary yields a fluorescently tagged primary that again can be visualized using fluorescence microscopy.

Each method has its strengths. Strengths of direct immunofluorescence include reduced labeling time and complexity as well as reduced species cross-reactivity and nonspecific binding. Strengths of indirect immunofluorescence include reduced cost, increased flexibility, and amplified signal due to binding of multiple secondary antibodies to the primary (Fig. 6.7).

6.2.1.3 Fluorescent Chemical Reagents

This approach uses small chemicals that are intrinsically fluorescent and that associate preferentially with specific subcellular structures or molecules.

FIGURE 6.6
Exemplary covalent labeling scheme showing a reaction between an NHS ester on a fluorophore, Fl, and an amine group on a protein of interest to yield a fluorescently labeled protein. The fluorophore and protein are joined by an amide bond –(CO)-(NH)–.

Prominent examples include the nucleic acid labels 4′,6-diamidino-2-phenylindole (DAPI) and Hoechst 33258 and the organelle labels MitoTracker® and LysoTracker® (Fig. 6.8).

Binding mechanisms are dye specific. DAPI binds to the minor groove of DNA with a resulting ~20-fold enhancement in fluorescence. MitoTracker® dyes are cationic fluorophores that accumulate in active mitochondria. In some cases, small chemical fluorophores are membrane permeant and thus can be used to label living cells.

6.2.2 Genetically Encoded Fluorescent Reporters

In this approach, DNA encoding a fluorescent reporter is introduced into cells, and the cellular protein synthesis machinery then generates the fluorescent probe. Strengths of genetically encoded fluorescent reporters include the fact that they can be introduced into, and targeted within, cells using minimally disruptive methods. The probes also are exceptionally versatile; they can be designed to visualize a broad spectrum of subcellular processes, including molecular distribution and dynamics, and biochemical reactions. Disadvantages of genetically encoded reporters include their relatively large size and "overexpression," which have the potential to alter the expression, targeting, and activity of both exogenous and endogenous proteins.

Enhanced green fluorescent protein (EGFP) fusion proteins are a prominent example of a genetically encoded fluorescent reporter (Fig. 6.9). HaloTag®

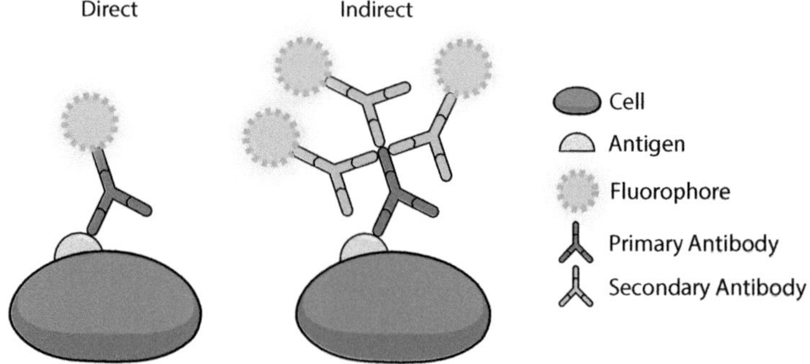

FIGURE 6.7
Cartoon of direct and indirect immunofluorescence labeling. In the indirect method, primary and secondary antibodies are generated using different host animals. For example, the secondary antibody might be raised in goat against a primary antibody produced in rabbit, which acts as the antigen in the goat.

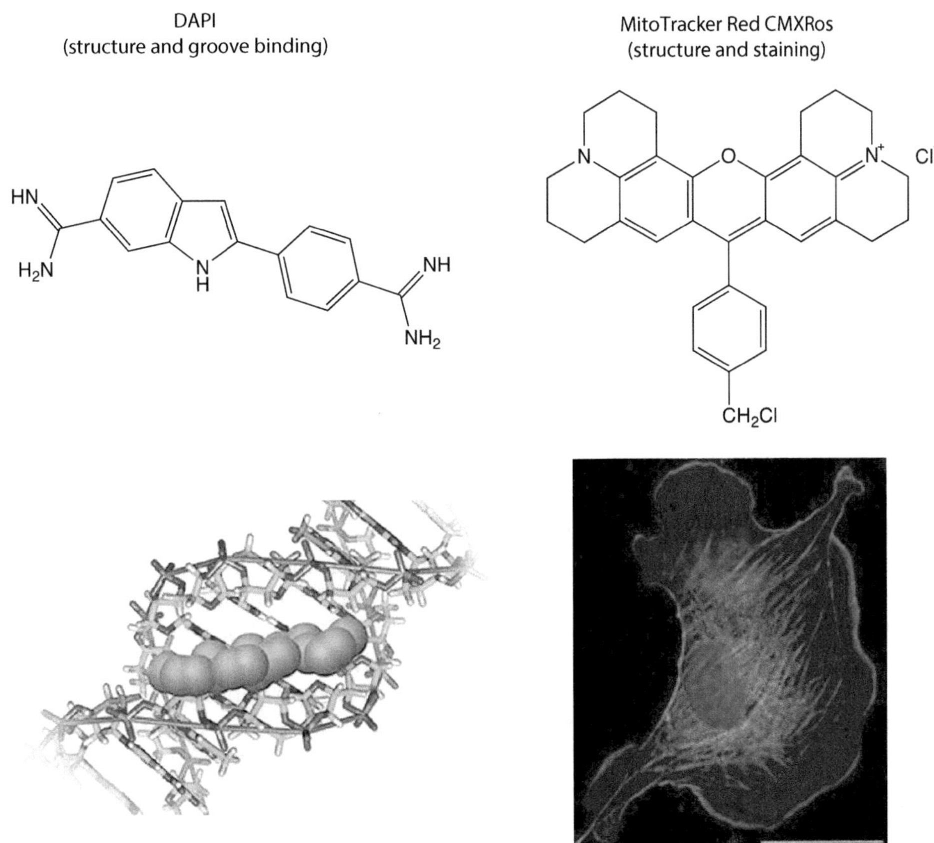

FIGURE 6.8
Exemplary fluorescent chemical reagents. Upper panels show the chemical structures of DAPI and MitoTracker Red CMXRos®. The lower left panel shows DAPI (blue spheres) binding to the minor groove of DNA. The lower right panel shows a live cultured epithelial cell labeled with DAPI, Alexa 488-phalloidin (an actin filament stain), and MitoTracker Red (a mitochondrial stain). Scale bar = 20 μm. (DAPI image reprinted with permission from Bio-Rad Laboratories, Inc. Cell image adapted from Jonkman, J., et al. 2015. Any way you slice it – a comparison of confocal microscopy techniques. *J. Biomol. Technol.* 26:54–65. DOI: 10.7171/jbt.15-2602-003. With permission from Association of Biomolecular Resource Facilities.)

fusions are another example (Fig. 6.10). Unlike EGFP fusions, HaloTag fusions are not intrinsically fluorescent. Instead, they are rendered fluorescent when an exogenous ligand linked to a fluorophore attaches covalently to HaloTag. Advantages of HaloTag fusions include versatility and brightness. Disadvantages include additional steps, such as incubation of cells with the ligand followed by washes.

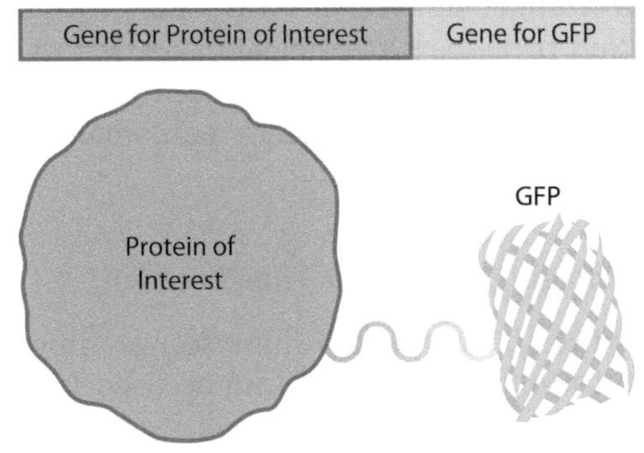

FIGURE 6.9
Schematic illustration of a gene coding for a GFP/protein of interest fusion (top) and the associated fluorescent gene product (bottom).

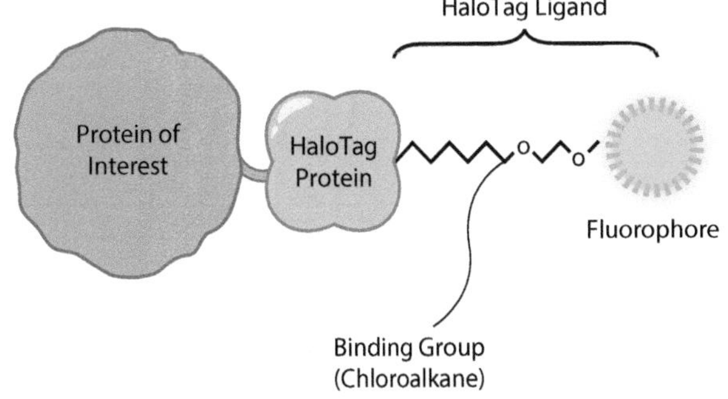

FIGURE 6.10
Schematic of HaloTag labeling. HaloTag fusions bind ligands containing a reactive chloroalkane group, which binds covalently to HaloTag, and a reporter group, which can be any of multiple fluorophores.

6.3 The Widefield Fluorescence Microscope

The preferred optical configuration for fluorescence microscopes differs from the transmitted light configurations discussed at the end of Chapter 3. In addition, fluorescence microscopes include several components that we have not previously encountered. We discuss these issues in turn.

6.3.1 The Epi-Fluorescence Optical Train

The discussion in Chapters 3–5 focused on transmitted light optical microscopy techniques. These are implemented using a "trans" illumination scheme in which a separate condenser and objective deliver illumination, and collect light, that is transmitted through the specimen (Figs. 3.11 and 3.12).

In contrast, fluorescence microscopy overwhelmingly is implemented as a reflected light technique using "epi" illumination (Fig. 6.11). In epi mode, the objective also acts as the condenser, directing reflected excitation light onto the specimen and collecting that portion of the fluorescence that is reemitted back toward the objective and sending it to a detector. As with transmitted light microscopy, the Köhler method is used to achieve homogeneous illumination.

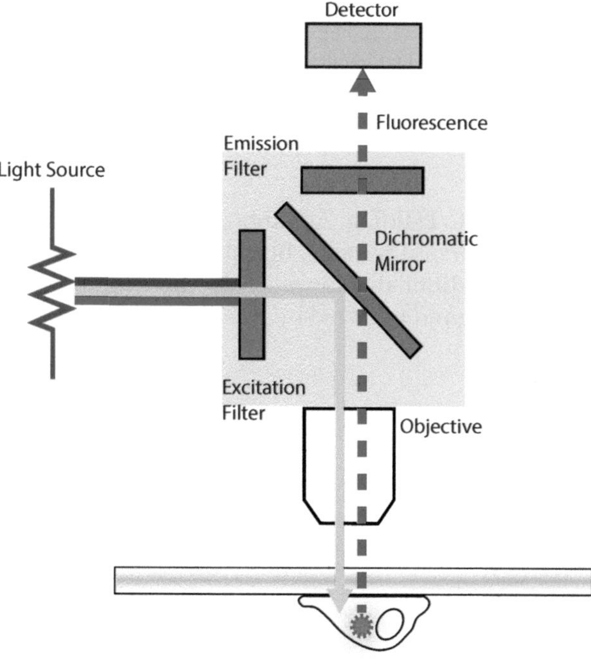

FIGURE 6.11
Schematic of the epi-illumination light path. The diagram emphasizes the behavior of light with different wavelengths. Key components of a filter cube are enclosed in the blue box.

Epi-illumination has several advantages over trans-illumination. These include automatic alignment of the objective and condenser. In addition, separation of excitation and emission light is improved in two ways. First, most of the excitation travels away from the objective. Second, epi-illumination employs a dichromatic mirror that markedly improves discrimination of excitation and emission light (Fig. 6.11). Dichromatic mirrors frequently are packaged with other components into a filter cube, which is our next topic.

6.3.2 Filter Cubes

Filter cubes are used to achieve selective excitation of a specific fluorophore and selective transmission of the fluorophore's emission to a detector. To these ends, filter cubes contain three key components: an excitation filter, a dichromatic mirror, and an emission filter (Fig. 6.11). The excitation filter functions to transmit shorter wavelengths that effectively excite a particular fluorophore while blocking longer wavelengths that might be confused with emission light. Suitable excitation filters include "shortpass" filters, which pass wavelengths below some cutoff, and, more commonly, "bandpass" filters, which filter light both below and above the most suitable excitation wavelengths (Fig. 6.12).

The dichromatic mirror (often called a dichroic mirror) aids in separating excitation and emission light. It has a complex reflection/transmission profile, reflecting the shorter wavelength excitation light onto the sample and transmitting the longer wavelength emission light onto the detector. If the mirror were perfect, this would suffice, but imperfections necessitate an additional (emission) filter. This final component functions to transmit the fluorophore's emission and block any residual excitation light. Emission filters also often are bandpass, but in some cases, such as single-color imaging, "longpass" filters are used; these pass all wavelengths exceeding some value and thus yield a stronger signal (Fig. 6.12). Filters with intricate dual- and triple-bandpass capabilities that reduce filter changes and facilitate simultaneous multi-color imaging also are available from many manufacturers (Section 9.4.1.1.6).

6.3.3 Imaging Modes for Thin Samples

Conventional, widefield fluorescence microscopy is a versatile technique that can be used to generate several types of images. Here we focus on imaging modes relevant for thin samples (Fig. 6.13) and thus temporarily forgo discussion of three-dimensional

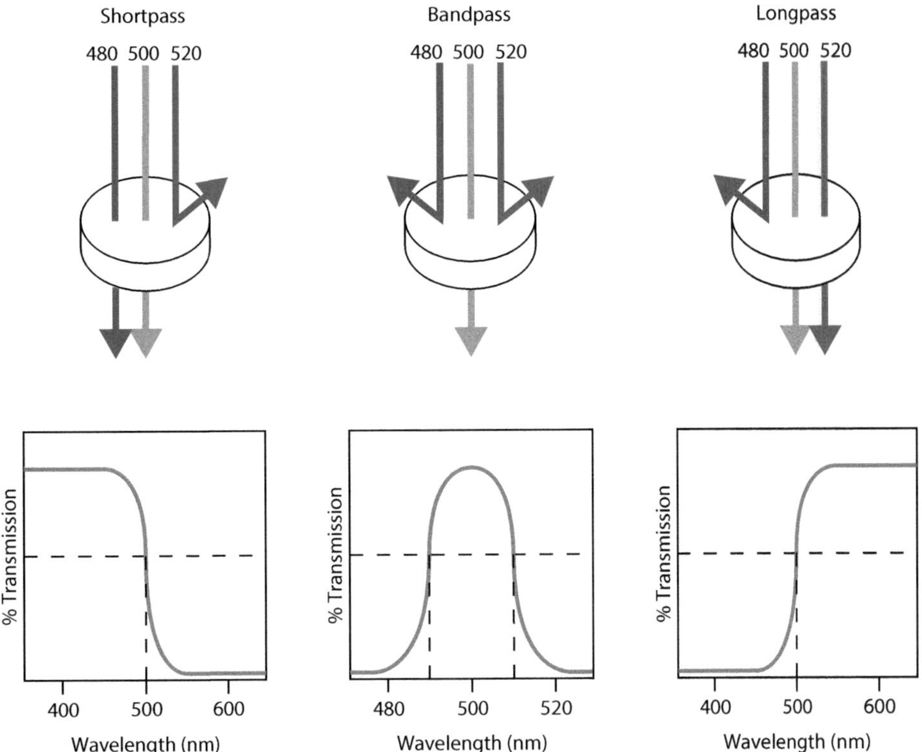

FIGURE 6.12
Schematic comparison of the transmission profiles of shortpass, bandpass, and longpass filters.

(multi-plane) imaging as well as approaches used to deal with blur from thicker samples (Section 6.3.4).

Images of thin specimens can be single-color, multi-color, or time-lapse (for the case of living specimens). Multi-color and time-lapse images are derived from two or more images that (usually) are collected sequentially and from the same focal plane (Fig. 6.13). Multi-color images typically are generated by sequentially capturing pictures of subcellular constituents that are labeled with spectrally distinguishable fluorophores (Sample Problem 6.2); this involves switching filter cubes to achieve selective imaging of each fluorophore (assuming single bandpass filter cubes are used) and then overlaying the images (Fig. 6.13). Time-lapse (single-color) images are generated by sequentially capturing pictures at specified time intervals and then combining the pictures into a movie.

In most cases, collection of fluorescence images is implemented by a highly automated computer program that directs rotation of filter cubes into the light path, implements excitation for a specified time interval, and collects images at specified time intervals, etc. Thus, in large part, image acquisition amounts to mastering the specifics of the program that drives the microscope.

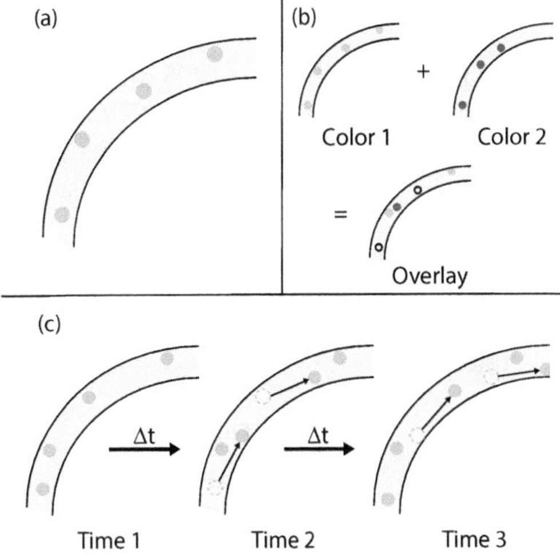

FIGURE 6.13
Cartoon of imaging modes. (a) The protocol used to generate a single-color image involves selecting the filter cube for the color and collecting signal for a specified time interval. (b) Generating a dual-color image involves selecting the filter cube for color 1, collecting signal for a specified time interval, repeating for color 2, and overlaying images. Here, green circles denote color 1 fluorescent features, red circles denote color 2 fluorescent features, and yellow circles denote overlap between color 1 and color 2 in the overlay. (c) Generating a single-color, time-lapse movie involves selecting the filter cube, collecting signal, waiting a time Δt, repeating this basic process the appropriate number of times, and constructing the movie.

SAMPLE PROBLEM 6.2 MULTI-COLOR IMAGING

Successful multi-color imaging requires generating a robust signal from each fluorophore and avoiding significant contamination of a signal from one fluorophore by signals from other fluorophores. In the case of laser-based illumination, this involves choosing appropriate laser lines and emission filters and choosing between options that include simultaneous and sequential imaging.

Consider a sample labeled with the three fluorophores DAPI, EGFP, and Texas Red. The excitation (dashed lines/unfilled areas) and emission (solid lines/filled areas) spectra for these three fluorophores are shown below. (Data for graph exported from ThermoFisher Scientific's Fluorescence SpectraViewer [www.lifetechnologies.com/handbook/spectraviewer].) To generate robust signals, you want to choose a laser line that overlaps well with each fluorophore's excitation spectrum. Similarly, you want to choose an emission filter that will pass a significant fraction of each fluorophore's emission with (hopefully) minimal signal contamination from other fluorophores.

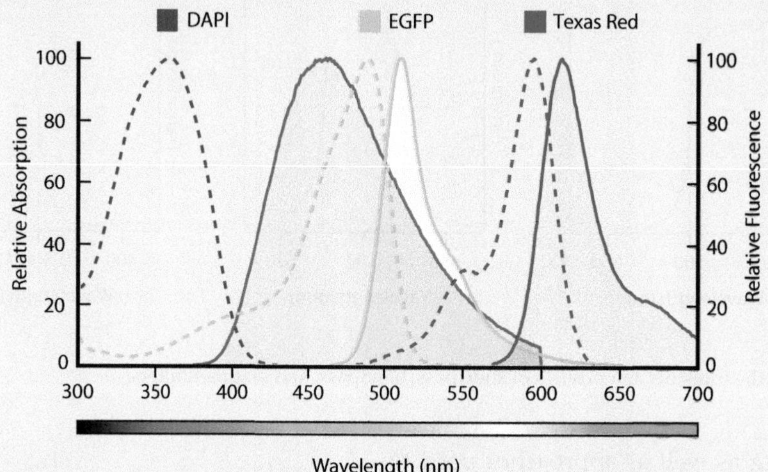

Your laser line options are 405 nm, 458 nm, 488 nm, 514 nm, 561 nm, and 633 nm.

Your emission filter options are 350/50 nm, 450/50 nm, 490/50 nm, 525/50 nm, 590/50 nm, and 625/50 nm.

Note: The notation 350/50 nm means that the filter passes EM radiation with wavelength ranging from 325 nm to 375 nm (i.e., 25 nm on either side of 350 nm).

a. Choose the best excitation line for each of the three fluorophores and enter your choice into the table. Note: DAPI is the hardest to excite with these laser lines, but one choice will work with low efficiency. The use of monochromatic laser light for excitation obviates the need for an excitation filter.

Fluorophore	Excitation Line	Emission Bandpass
DAPI		
EGFP		
Texas Red		

b. Choose the best emission filter for each fluorophore and enter your results into the table.

c. In general, fluorophores in multi-color samples are imaged sequentially (Fig. 6.13). Sequential imaging involves directing only one laser line onto the sample at a time, collecting the resulting fluorophore emission, and then repeating this process for the other fluorophores. For living samples, it can be preferable to image fluorophores "simultaneously," which involves using multi-laser excitation and then collecting the emission simultaneously using a different detector for each fluorophore. Avoiding significant contamination of a signal from a fluorophore by signals from other fluorophores is much more difficult to achieve with simultaneous imaging, which is one reason this approach is less commonly used.

For each fluorophore combination below, determine if simultaneous and/or sequential imaging is viable with your filters and place a "check mark" in the table when the method is an option.

Fluorophore	Simultaneous	Sequential
DAPI/EGFP		
DAPI/Texas Red		
EGFP/Texas Red		
DAPI/EGFP/Texas Red		

SOLUTION

a. Examination of the excitation spectra shows that the 405 nm, 488 nm, and 561 nm laser lines will overlap best for DAPI, EGFP, and Texas Red, respectively.

b. Examination of the emission spectra similarly shows that the results listed in the following table produce the best overlap.

Fluorophore	Excitation Line	Emission Bandpass
DAPI	405 nm	450/50
EGFP	488 nm	525/50
Texas Red	561 nm	625/50

c. Examination of the spectra shows that the DAPI emission signal will significantly contaminate the signal that EGFP produces in its detector in a simultaneous imaging experiment. This eliminates simultaneous imaging options involving DAPI and EGFP, as shown in the table. Other options are viable.

Fluorophore	Simultaneous	Sequential
DAPI/EGFP		√
DAPI/Texas Red	√	√
EGFP/Texas Red	√	√
DAPI/EGFP/Texas Red		√

6.3.4 The Blur Problem for "Thick" Samples

When sample thickness exceeds several μm, conventional, widefield fluorescence images are plagued by blur because the objective collects both in-focus and out-of-focus light. One way to illustrate the origin of blur is to examine the three-dimensional image of a (fluorescent) point source. This image is known as a point spread function (PSF), which, for a well-corrected (aberration-free) system, is a three-dimensional diffraction pattern with radial and axial symmetry. More generally (e.g., for aberrated systems), the PSF has asymmetry and greater spatial extent.

Fig. 6.14a shows an axial view of the diffraction pattern generated by a 100× objective (NA = 1.4) under conditions in which aberrations were negligible. The pattern was obtained experimentally by imaging a tiny fluorescent bead, which is essentially a point source of light. The figure clearly shows that the image of a point source is not a point. In particular, the axial profile is highly elongated.

Panel (b) quantifies the effect, showing a plot of the integrated (summed) intensity in transverse (x/y) planes separated by 0.10 μm along the axial (z) direction (Activity A6.1). The plot also underscores the fact that the intensity pattern extends over μm-length distance scales in the image plane. This means that out-of-focus intensity extensively contaminates intensity from the focal plane. The net effect is to create extensive blur in fluorescence images of thick samples.

There are two common approaches to alleviate the blur problem: confocal microscopy and deconvolution. These are discussed in the next two sections. More specialized techniques, such as total internal reflection fluorescence microscopy and multi-photon microscopy, are discussed in Chapter 7.

6.4 Confocal Microscopy

Confocal microscopy is a popular method of obtaining relatively blur-free fluorescence images. There are several variants of the technique. We focus initially on

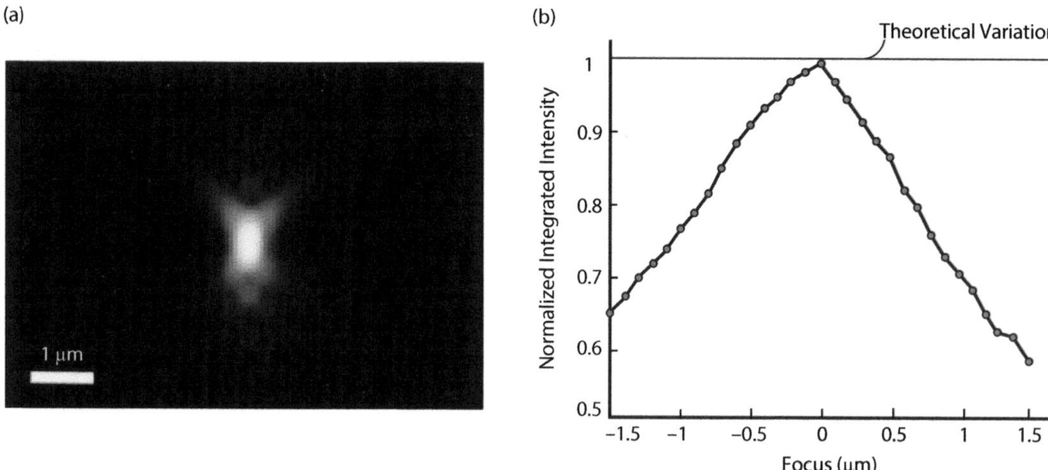

FIGURE 6.14
Axial view (a) of a PSF generated by a Nikon 100x objective and an associated plot (b) of the measured integrated intensity in each of 15 lateral planes on either side of optimal focus (0). The axial view was generated from a three-dimensional image that was obtained by imaging a fluorescent bead over a distance of four microns on either side of the point of optimal focus (Section 6.4.5). Theoretically, the integrated intensity should be constant (horizontal line in Panel [b]), but restricting the field size, as done here, causes a falloff in intensity with distance from optimal focus. Bar = 1 μm. (Adapted with permission from Scalettar, B. A., et al. 1995. Dispersion, aberration and deconvolution in multi-wavelength fluorescence images. *J. Microsc.* 182:50–60. DOI: 10.1046/j.1365-2818.1996.122402.x.)

laser scanning confocal microscopy (LSCM), which is the most common embodiment. We then discuss spinning disk confocal microscopy (SDCM), which is better suited to the study of living samples. Both techniques use pinholes to block out-of-focus light.

6.4.1 Key Optical Principles of LSCM

Two key features of LSCM are shown schematically in Fig. 6.15. These features involve illuminating small areas sequentially and reducing the detection of out-of-focus emission. Together these reduce blur.

The illuminated region on the specimen is reduced by focusing laser-derived excitation light to a diffraction-limited spot, which typically has in-focus dimensions on the order of 0.5 μm. The illumination also has an "hourglass" (i.e., hyperbolic) profile around focus (Box 7.2). Spot illumination, in conjunction with a small pinhole located in a specimen-conjugate plane in front of the detector, yields preferential detection of in-focus signal and improved lateral resolution, as discussed below. On the other hand, spot illumination does not directly generate an image of an extended sample. Instead, LSCM systems must build an image

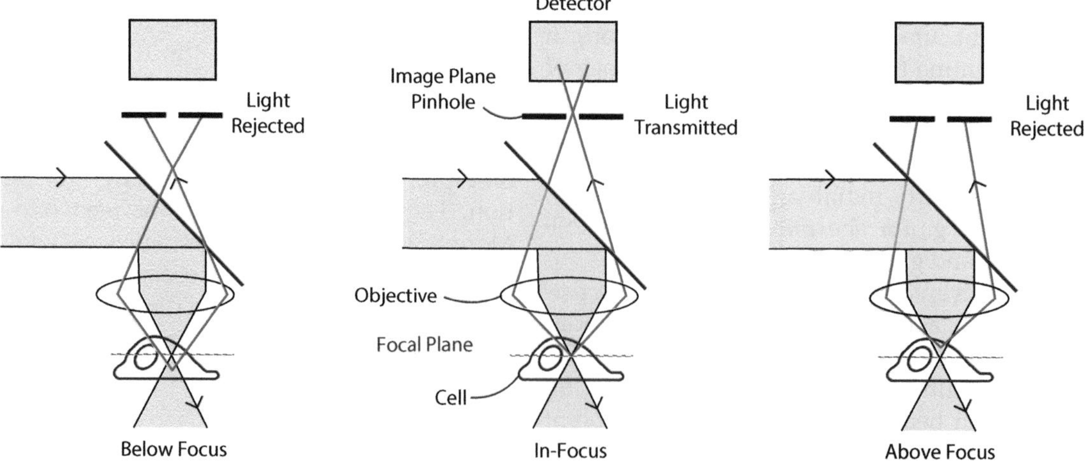

FIGURE 6.15
Attributes of laser-based illumination and light rejection in LSCM. Laser-based illumination is focused to a small spot on the specimen. The beam spreads above and below focus and excites fluorophores both in and away from the focal plane (Box 7.2). Light rejection by a confocal pinhole leads to exclusion of the out-of-focus signal.

by scanning the spot over the sample, as discussed in Section 6.4.2.

Out-of-focus contamination is reduced via spatial filtering using a pinhole positioned in a conjugate image plane (Fig. 6.15). Pinhole filtering is a ubiquitous feature of the various embodiments of confocal microscopy. The basic idea is that light emitted by a point in the focal plane is focused to a tight spot in the conjugate image plane. The size of this spot is matched to that of a "confocal" (conjugate) pinhole (HW 6.15). Thus, in-focus light passes through the pinhole and onto the detector. In contrast, light from points above and below the focal plane generates a large blur spot, which mostly misses the pinhole. The result is marked rejection of out-of-focus fluorescence and an associated reduction in image blur.

Lateral discrimination in confocal microscopy also is somewhat improved relative to widefield microscopy. One reason is that points excited by the periphery of the (Gaussian) scanning beam generate a weaker signal (Box 7.2). Another reason is that peripheral points are not centered on the pinhole, and thus their signal is rejected preferentially.

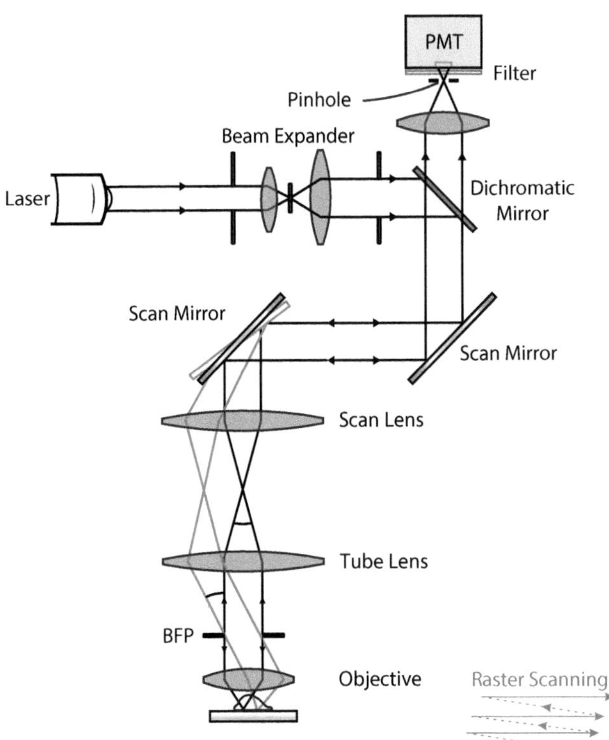

FIGURE 6.16
Simplified ray diagram for LSCM. To minimize the spot size on the sample, the laser beam is expanded to fill the BFP of the objective. (This is a typical "reciprocal" Fourier behavior.) Rectangular "raster scanning" of the specimen is achieved using two mirrors that deflect the beam along horizontal and vertical directions in the specimen plane. New line scans are preceded by flyback during which no data typically are collected. Despite the movement of the excitation spot, "descanned" emission arrives back at the fixed location of the pinhole in front of the detector. This is depicted for two (black and red) trajectories.

6.4.2 Instrumentation for LSCM

The key components of an LSCM are shown in Fig. 6.16. Epi-illumination is used, with excitation light reflected by a dichromatic mirror onto scanning mirrors, which oscillate and thereby raster the excitation light over the specimen. The objective both directs the excitation light onto the sample and collects the emission, which follows the same path as the excitation (due to the nearly instantaneous propagation of light). The emission ultimately arrives back at the dichromatic mirror and is transmitted to the detector after passing through an emission filter.

Importantly, the emission light is focused to a fixed image point despite the fact that the excitation light oscillates over the specimen (Fig. 6.16). Thus, a fixed emission pinhole positioned at this image point can be used to reject out-of-focus fluorescence.

In LSCM, the signal that passes through the pinhole usually is quantified using a photomultiplier tube (PMT) (Fig. 6.16). The signal fluctuates in time due to variations in specimen fluorescence intensity, and the PMT voltage fluctuates accordingly. This analog signal is periodically digitized and used to reconstruct an image from stored signals.

6.4.3 Key Optical Principles of SDCM

One of the major weaknesses of LSCM is that images are acquired relatively slowly due to limitations imposed by the scanning mirrors and typical signal strength. Generally, an LSCM system will scan ~1 pixel/μs. (Pixels, or "picture elements," will be discussed in detail in Chapter 9.) This scan rate limits acquisition times to ~1 second for images with dimensions on the order of 1000×1000 pixels^2; thus, faster subcellular processes cannot be monitored. Much more rapid image acquisition is obtained with SDCM. This method, which is the confocal method of choice for living samples, is our current topic.

In SDCM systems, the temporal limitations on point-scanning systems are circumvented via multi-beam scanning using a Nipkow disk, which contains tens of thousands of pinholes. Similar to LSCM, the disk and its pinholes are located in a conjugate image plane. A subset of these (~1000) is illuminated, and the objective projects their minified images onto the sample to achieve multi-beam illumination (Fig. 6.17). The specimen can be scanned in less than a millisecond during a partial rotation of the disk.

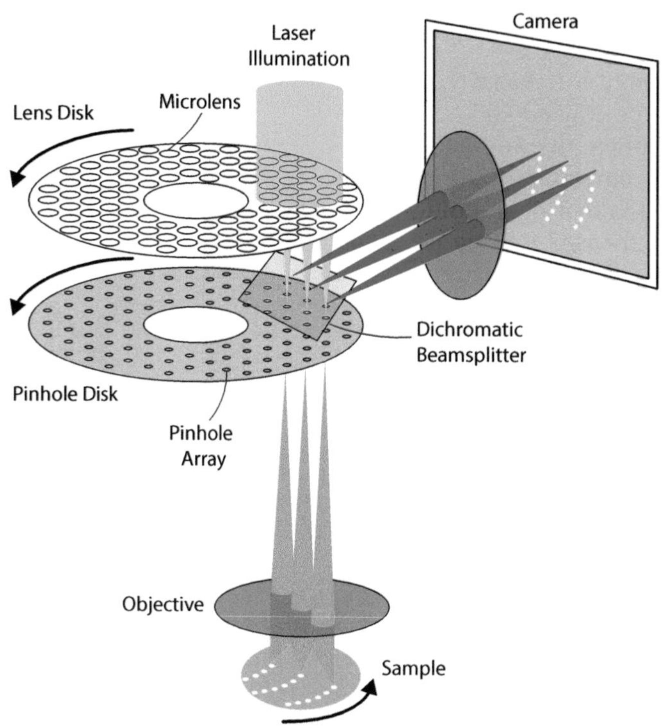

FIGURE 6.17
Schematic of a Yokogawa SDCM system. The dichromatic mirror has a reflection/transmission profile that is the inverse of that in a traditional epi-fluorescence microscope. Rapid acquisition is achieved using pinholes arranged in sets of nested spirals that illuminate the specimen uniformly and generate an image each 30° rotation of the disk. (Adapted with permission from Gräf, R., et al. 2005. Live cell spinning disk microscopy. *Adv. Biochem. Engin./Biotech.* 95:57–75. DOI: 10.1007/b102210.)

6.4.4 Instrumentation for SDCM

Fig. 6.17 shows key details of a state-of-the-art SDCM system manufactured by Yokogawa Electric Corporation. A key innovation in this system is the introduction of a second (upper) spinning disk, which contains a set of Fresnel microlenses that focus the excitation light onto aligned pinholes in the lower Nipkow disk. The use of microlenses in the excitation path enhances light transmission ~10-fold and yields transmission as high as ~60%, relative to systems that lack microlens arrays. The result is markedly improved image brightness.

After excitation, emission generated by each Nipkow pinhole passes back through the same pinhole, leading to preferential rejection of out-of-focus fluorescence signal. Unlike the excitation, the emission bypasses the microlens array and is projected onto a very sensitive detection device, such as an electron-multiplying charge-coupled device (EMCCD) or a high quantum efficiency back-illuminated scientific complementary metal-oxide semiconductor (sCMOS) device (Chapter 9).

The Yokogawa system can, in principle, capture up to 2000 frames/s, which is markedly superior to an LSCM. In reality, other limitations typically lead to vastly reduced acquisition rates. One example is the need to collect an acceptably strong signal from a dim sample, which often places an upper bound of 10 frames/s on acquisition rates.

Additional advantages of SDCM relative to LSCM include reduced sample photobleaching and the use of novel ultrasensitive detectors that facilitate imaging of dim samples. In principle, SDCM also can be implemented with arc lamps, and thus excitation is not necessarily limited to laser lines. Disadvantages of SDCM relative to LSCM include somewhat poorer out-of-focus rejection, especially for thick or heavily stained specimens, which is caused by pinhole crosstalk (Fig. 6.18). Spinning disk instruments also are not well suited to implementing complementary useful approaches, such as fluorescence recovery after photobleaching, that require focused, high-intensity illumination (Box 6.1).

6.4.5 Imaging Modes

Confocal microscopy can be used to implement each of the imaging modes discussed in Section 6.3.3. In addition, confocal microscopy frequently is used to conduct three-dimensional spatial imaging of thick fixed or living specimens (Fig. 6.19a–c). Three-dimensional spatial imaging is implemented by collecting a series of single-plane images (optical sections) with the objective focused at different depths

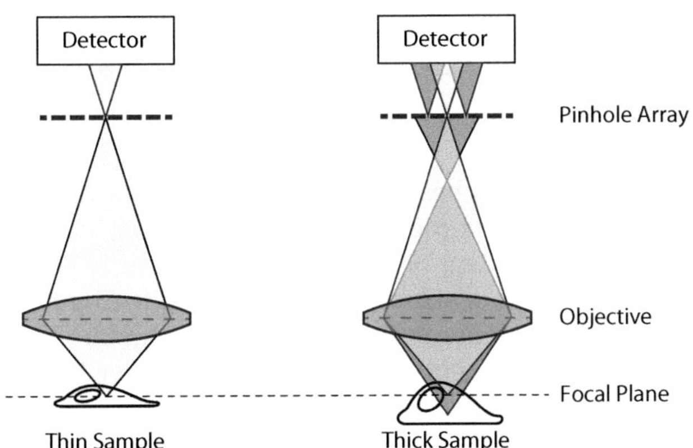

FIGURE 6.18
Schematic of pinhole crosstalk. Out-of-focus light that would be blocked in LSCM can reach the detector via adjacent pinholes in SDCM.

within the specimen. Focus can be changed using a microstepper motor under computer control. A series of three-dimensional images collected sequentially as a function of time generates a four-dimensional (x,y,z,t) image (Fig. 6.19d).

6.5 Deconvolution

Deconvolution is another popular method of removing blur. It is based on well-established input-output relationships that describe distortion induced by a device like a microscope. Deconvolution differs from confocal microscopy in that it is a software, versus hardware, "fix" to the blurring problem.

To get a feel for the distortion that deconvolution addresses, consider an extended object, which contains many point sources. The image of an extended object is (loosely speaking) a "superposition of PSFs." More precisely, the PSF modifies the object (input) intensity distribution to produce a blurred image (output) intensity distribution. A simple example is shown in Fig. 6.20.

The mathematical expression that rigorously describes the relationship between the object, $O(x,y,z)$, and image, $I(x,y,z)$, intensity distributions is an integral known as a convolution, which is discussed in Box 6.2. In brief, the convolution is an operation (often denoted using the symbol "⊗") that quite generally relates how the output of a system is related to both the input and a blurring function. For fluorescence imaging, the blurring function is the PSF, and the convolution operation reads

$$O \otimes \mathrm{PSF} = I \tag{6.3}$$

Deconvolution is a well-established computational approach for "inverting" the convolution operation and thereby removing blur in fluorescence images. Specifically, deconvolution generates a deblurred estimate of $O(x,y,z)$ (the actual object) based on a knowledge of the PSF and the measured $I(x,y,z)$. To get a basic feel for how deconvolution works, we briefly

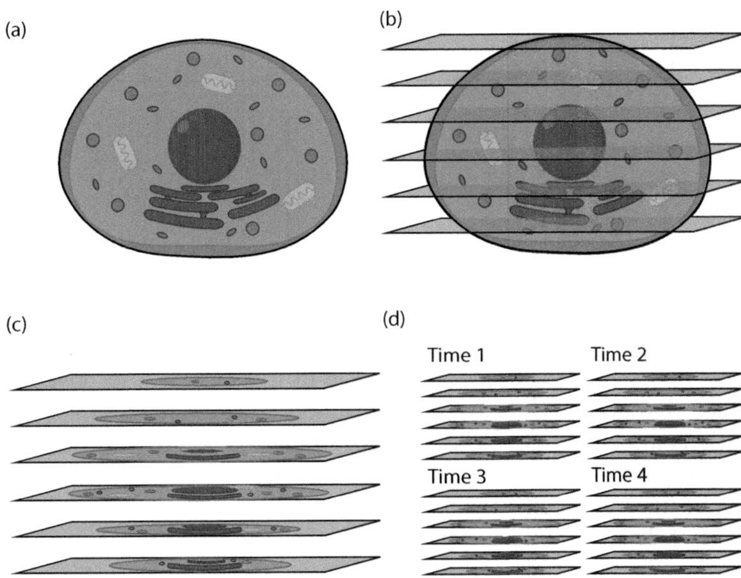

FIGURE 6.19
Schematic of three-dimensional (a–c) and four-dimensional (d) imaging. In Chapter 9, we discuss how to choose the appropriate spacing between optical sections using the Nyquist sampling theorem.

BOX 6.1 COMPLEMENTARY FLUORESCENCE-BASED MICROSCOPY TECHNIQUES

Fluorescence microscopy frequently is used in conjunction with other, complementary fluorescence-based techniques. Here we consider two popular examples – fluorescence recovery after photobleaching (FRAP) and fluorescence correlation spectroscopy (FCS). Both approaches, like confocal microscopy, use a laser to illuminate a small region of the sample and a pinhole to limit the detected fluorescence.

FRAP is a well-established method of quantifying the rate and extent of mobility in a sample. This is accomplished by directing a brief, intense pulse of laser light onto a sub-region of the sample (figure, top panel). This pulse bleaches (renders nonfluorescent) many of the fluorophores in the sub-region. The recovery of fluorescence in the sub-region then is monitored by directing much-attenuated laser light onto the bleached spot and monitoring the temporal dependence of the post-bleach fluorescence. The characteristic timescale of fluorescence recovery is determined by the rates of random (e.g., diffusive) and directed (e.g., driven) motion of the fluorescently tagged molecules. Failure of the fluorescence to recover completely reveals that some fluorescently tagged molecules are immobile. In simple systems, the recovery timescale can be used to measure mobility parameters, such as diffusion coefficients.

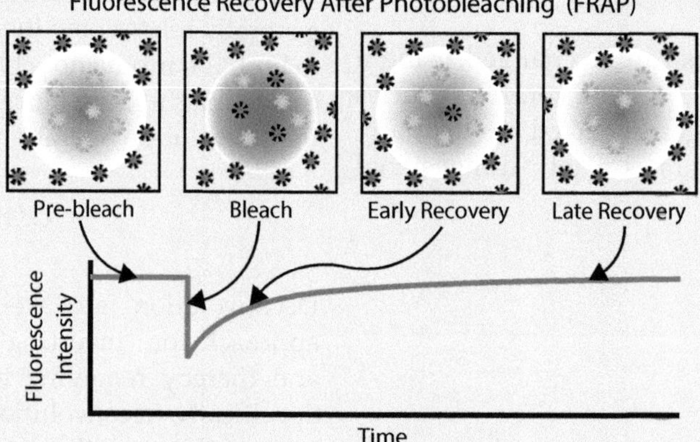

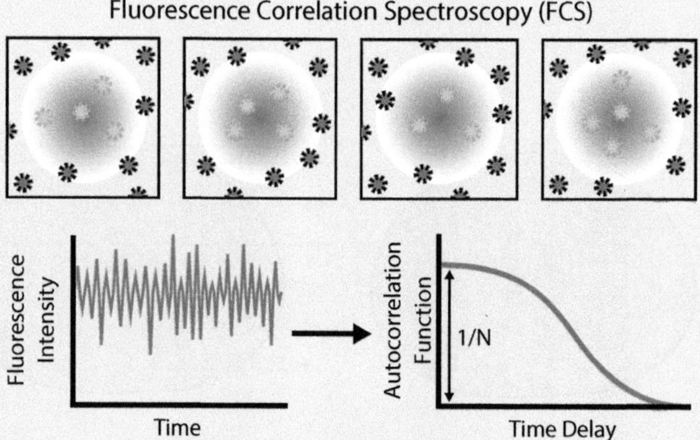

FCS can similarly be used to monitor molecular dynamics. In an FCS experiment, this is accomplished by illuminating an open, sparsely occupied sub-region of the sample and observing fluorescence fluctuations, which reflect changes in the number of fluorescently tagged molecules in the sub-region. Fluorescence fluctuations arise because molecules continually move in and out of the sub-region; thus, the fluctuation timescale reflects the rate of fluorophore motion.

Like FRAP, FCS can be used to quantify mobility parameters in simple systems (e.g., systems containing a single diffusing species). This involves computation of the autocorrelation function (ACF) of the fluorescence intensity. The half maximal point in the ACF gives the diffusion time, τ_d, and the amplitude of the ACF can be used to determine the average number of molecules, N, present in the illuminated area (figure, bottom panel). Interestingly, the diffusion coefficients measured by FRAP and FCS typically are different.

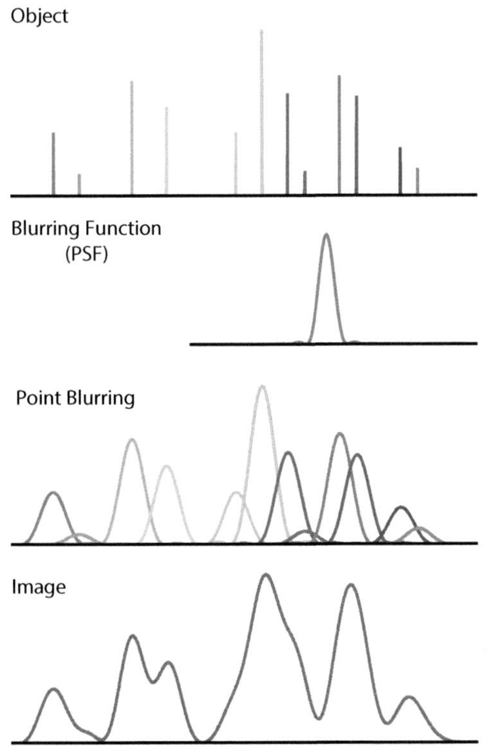

FIGURE 6.20
Example of the distorting effect of the PSF. The effect of the PSF is to generate an image that is a blurred version of the object.

describe a common computational implementation of the method – constrained iterative deconvolution – and show an example of an image deblurred using this algorithm.

Equation (6.3) can, in principle, be solved directly for the object, $O(x,y,z)$. However, in practice, the direct approach is unstable due to noise in the image, sensitivities to low values, etc. These problems can be reduced using constrained iterative deconvolution, which iterates toward an estimate, $E(x,y,z)$, of the object intensity distribution. The quality of the estimates after each iteration is assessed via convolution with the PSF. If the estimate is good, so $E(x,y,z) \cong O(x,y,z)$, it follows from Eq. (6.3) that

$$E \otimes \text{PSF} \cong I \tag{6.4}$$

If Eq. (6.4) does not hold, and $E \otimes \text{PSF} > I$, E is lowered because $E \otimes \text{PSF}$ is too large. For example, in a common iterative approach based on "ratio" updating, E is lowered by scaling it down by the ratio, $I/(E \otimes \text{PSF})$. E is similarly scaled up if $E \otimes \text{PSF} < I$. The convolution yields a value that varies with position (Box 6.2), and updating is implemented voxel by voxel using the position-dependent image and convolution data. E also is subject to a positivity constraint; thus, voxel estimates that become negative during deconvolution are assigned the value zero (reflecting the fact that intensities are never negative). The ratio approach typically converges to a good estimate of $O(x,y,z)$ in 10–15 cycles (Fig. 6.21). Alternative approaches primarily differ in the method used to achieve correction. For example, additive updating also can be used. Constrained iterative methods generally are fast and robust but have the potential to introduce image artifacts, such as rings.

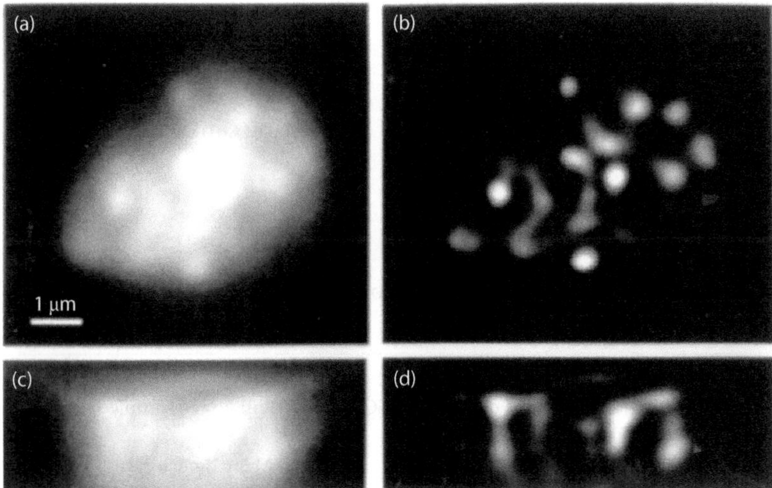

FIGURE 6.21
Comparison of raw (a and c) and computationally deblurred (b and d) images of fluorescently labeled chromosomes. Top panels show lateral views, and bottom panels show axial views. Scale bar = 1 μm. (Adapted with permission from Scalettar, B.A., et al. 1995. Dispersion, aberration and deconvolution in multi-wavelength fluorescence images. *J. Microsc.* 182:50–60. DOI: 10.1046/j.1365-2818.1996.122402.x.)

BOX 6.2 IMAGE FORMATION AND THE CONVOLUTION OPERATION

This box presents the integral form of the convolution operation in Eq. (6.3) and discusses some of its properties. Familiarity with the Dirac delta function, δ, is assumed.

The convolution operation in two dimensions can be expressed as follows (assuming that the magnification is unity):

$$I(x,y) = O \otimes \text{PSF} = \iint_{-\infty}^{+\infty} O(x',y')\text{PSF}(x-x',y-y')dx'dy'$$

Two important conclusions can be demonstrated from this integral form. First, if the object, O, is a point source centered at the origin, which is represented mathematically using the delta function as $O(x',y') = \delta(x')\delta(y')$, the image, I, is the PSF:

$$I(x,y) = \iint_{-\infty}^{+\infty} \delta(x')\delta(y')\text{PSF}(x-x',y-y')dx'dy' = \text{PSF}(x,y)$$

Second, in the (unattainable) limit that the PSF is infinitely narrow, which again is represented using a delta function, the image is a perfect replica of the object:

$$I(x,y) = \iint_{-\infty}^{+\infty} O(x',y')\delta(x-x')\delta(y-y')dx'dy' = O(x,y)$$

More generally, I needs to be rescaled by the appropriate magnification.

6.6 Comparison of Confocal Microscopy and Deconvolution

We close with a discussion of the relative strengths and weaknesses of confocal microscopy and deconvolution (Table 6.1).

Significant differences between confocal microscopy and deconvolution are that the latter does not require scanning and is not limited to the excitation lines provided by a laser. The computational approach also gives stronger signals from dim specimens because there is no signal rejection. However, the computational approach requires that the PSF be accurately determined, usually by experiment, and that a three-dimensional image stack be collected to implement deconvolution correctly. LSCM and computational deblurring can be used together to optimize deblurring.

TABLE 6.1

Comparison of Confocal and Deconvolution Microscopies

	Confocal	Deconvolution
Light source	Laser	Arc lamp/LED
X/Y sampling	Full image sampled one point (LSCM) or ~1000 points (SDCM) at a time	Full image sampled simultaneously
Detector	Point detector (LSCM) Array detector (SDCM)	Array detector
Out-of-focus light	Rejected before reaching detector by pinhole(s) in a conjugate image plane	Detected but reassigned to correct position computationally, via deconvolution
Z sampling	Optional – may be used to obtain three-dimensional information	Required for full deconvolution, even if interested only in a single section

Suggested Reading and Additional Resources

iBiology: Microscopy Series. 2012 – 2015. Fluorescence microscopy. https://www.ibiology.org/online-biology-courses/microscopy-series/fluorescence-microscopy/

Johnson, I. and Spence, M.T.Z. 2010. *The molecular probes handbook (11th Edition)*. Life Technologies, Eugene, OR.

Kubitscheck, U. (Ed.) 2017. *Fluorescence microscopy: from principles to biological applications (2nd Edition)*. Wiley-VCH, Weinheim, Germany.

Molecular Expressions. 2015. Optical microscopy primer: fluorescence microscopy. https://micro.magnet.fsu.edu/primer/techniques/fluorescence/fluorhome.html

Murphy, D.B. and Davidson, M.W. 2013. *Fundamentals of light microscopy and electronic imaging (2nd Edition)*. Wiley-Blackwell, Hoboken, NJ.

Wayne, R. 2019. *Light and video microscopy (3rd Edition)*. Academic Press, San Diego, CA.

HOMEWORK PROBLEMS

6.1 [Vibrational Energies – Estimation] Frequencies associated with transitions between vibrational energy states are on the order of 5×10^{13} Hz. Use this information to estimate the order of magnitude of vibrational energy levels (in eV) and compare your result to typical thermal energies, $k_B T$, where k_B is Boltzmann's constant and T is the temperature. Assume $T = 300$ K. Comment on the significance of your result.

6.2 [Rotational Energies – Estimation] The rotational energies, E_{rot}, of a diatomic molecule with atomic mass m and interatomic spacing r are on the order of $E_{\text{rot}} \sim \hbar^2/mr^2$, where $\hbar = h/2\pi$. Estimate the rotational energy levels for O_2 (in eV) and compare your result to typical thermal energies, $k_B T$ (HW 6.1). Comment on the significance of your result. Hint: Assume the interatomic spacing is 0.1 nm.

6.3 [Transition Energies] Fig. 6.4 shows absorption and emission spectra for anthracene.

a. Estimate the energy differences between the zeroth vibrational state of the ground electronic state (S_0) and the zeroth (red arrow), first (green arrow), and second (blue arrow) vibrational states of the first excited electronic state (S_1).

b. Estimate the energy differences between the zeroth and first and between the first and second vibrational states of the first excited state, using your results from part (a).

c. Compare the energy differences between the vibrational energy levels calculated in part (b) with the difference between the ground and first excited electronic states calculated in part (a).

6.4 [Emission Spectrum – Hydrogen Atom] When hydrogen makes a transition from a higher to a lower energy state, radiation can be emitted. The emission spectrum can be predicted from the hydrogen atom energy levels, which are given by the famous Bohr result $E_n = -13.6(\text{eV})/n^2$, where the quantum number $n = 1, 2, 3, \ldots \infty$.

a. Hydrogen generates four transitions in the visible with wavelengths 656.3 nm, 486.1 nm, 434.0 nm, and 410.2 nm. In all four cases, the final state corresponds to quantum number $n_f = 2$. The initial quantum numbers span the range $n_i = 3 - 6$. Fill in the table below.

Wavelength (nm)	n_i
656.3	
486.1	
434.0	
410.2	

b. Modify Eq. (6.2) appropriately using the Bohr energy levels and confirm your predictions in part (a).

6.5 [Stokes Shift] The Stokes shift plays a key role in fluorescence microscopy because it makes it possible to filter out excitation light and generate an image against a dark background. Stokes discovered this effect in quite a remarkable way using quinine (an ingredient in tonic water and bitter lemon). Find the experiment on the Internet and draw a quick sketch. Hint: Searching "Stokes shift quinine" in Google (check the image section) should bring up a useful picture.

6.6 [Fluorescence Energy Levels – Derivation] Fluorescence emission arises from electronic transitions between quantum mechanical energy levels of fluorescent molecules. The quantitative behavior of these transitions can be modeled using a particle in an impenetrable box, as presented in Section 6.1.2. Derive an expression for the quantized energy levels using a mix of classical and quantum ideas. Specifically, assume that the energy of a particle of mass m is purely kinetic ($E = \frac{1}{2}mv^2 = p^2/2m$), and that the particle's momentum is given by de Broglie's relationship ($p = h/\lambda$). Also assume that the particle's "wavelength" obeys the famous "quantization condition" for a standing wave on a string of length L (see figure). This physics will lead you to Eq. (6.1): $E_j = j^2h^2/(8mL^2)$.

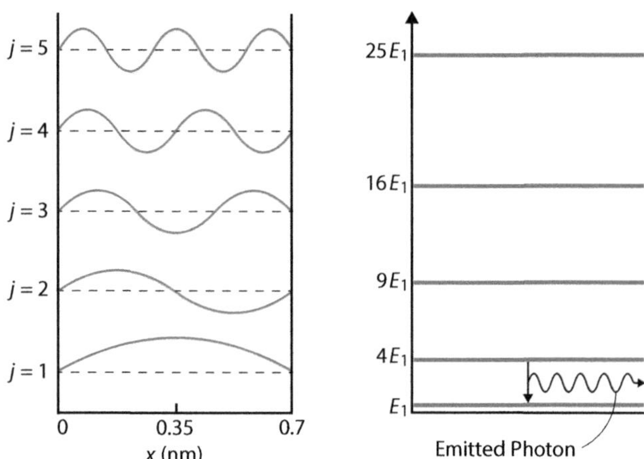

6.7 [Fluorescence Energy Levels – Application] The quantum particle in an impenetrable box provides a reasonable model for fluorescence emission due to electronic transitions (Eq. [6.2]). We can model a typical fluorophore by assuming, for example, that the box has a molecular size ~7 Å. (Most fluorophores include several conjugated rings; the diameter of benzene is 2.8 Å.)

a. Calculate the wavelength of the photon emitted during a transition from the first excited $(j = 2)$ state to the ground (lowest energy) $(j = 1)$ state by assuming that the energy lost during the transition $E_2 - E_1 = E_{photon} = hc/\lambda$. Here, h is Planck's constant, c is the speed of light, and m is the mass of the electron.
b. Repeat the calculation in part (a) for a photon emitted during a transition from the second excited $(j = 3)$ state to the ground state.
c. Are the emitted photons from parts (a) and (b) in the visible part of the spectrum? If so, what is the color of the emitted radiation? If not, in what part of the spectrum are they?

6.8 [Fluorescence Energy Levels – Application] The accompanying figure shows a general structure for a family of cyanine fluorophores. Here, the name and structure of the dye are based on the number of carbons in the polymethine bridge connecting the two ring structures. How would you expect the excitation and emission spectra to change as n (the number of carbons) increases? Explain your answer. Hint: See Eq. (6.2).

Cl^- R_1 N^+ n N R_2

Cy3 ($n = 1$)
Cy5 ($n = 2$)
Cy7 ($n = 3$)

6.9 [Filter Cube] Filter cubes play a central role in fluorescence imaging. Assume that the filter cube shown below works well for fluorophores that are excited by green light and emit red light. This problem probes the roles played by the excitation filter (A), the dichromatic mirror (B), and the emission filter (C). What would happen under the following scenarios, assuming that the arc emits white light and all components behave ideally?

	Color on Sample	Color on Detector
All filters in place	Green	Red
Filter A is omitted		
Dichromatic mirror B is omitted		
Filters A and C are switched		

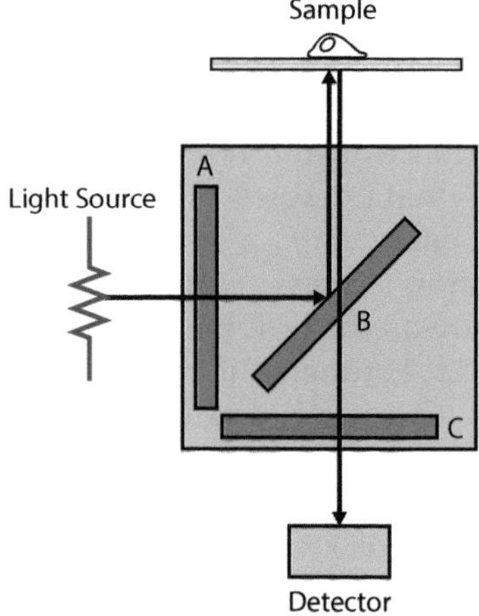

6.10 [Dichromatic Mirror – Theory] The dichromatic mirror plays a key role in fluorescence microscopy. The mirror works on the principle of thin-film interference; this principle also underlies the properties of anti-reflective coatings, and it gives rise to the colors that appear in soap bubbles and in oil films over water. To get a feeling for this important phenomenon, consider the following example.

a. A thin film in air is viewed at nearly normal incidence (see figure). Light reflecting from the top and bottom of the thin film travels a different path length to the eye, where it can interfere. Assume that the standard "path difference" condition for constructive interference, $\Delta x = m\lambda$ (m = 0, 1, 2, …), is valid, where Δx is the path difference, and λ is the wavelength in the medium traversed. Rewrite this expression in terms of film thickness, t, the vacuum wavelength vacuum, λ_0, and the refractive index of the film, n_{film}.

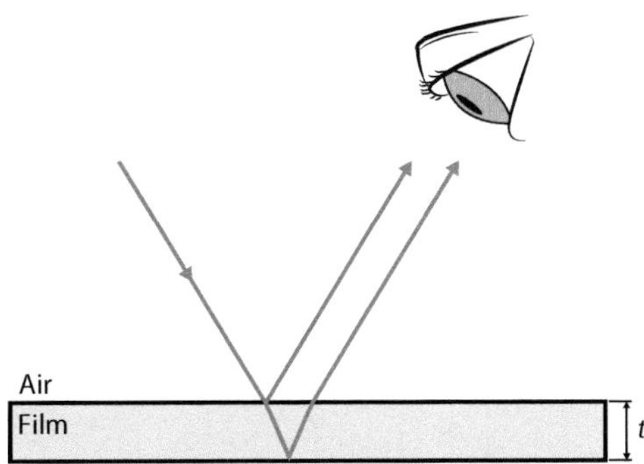

b. The expression in part (A) is slightly incorrect because the waves get out of phase due to the effects of path difference *and* the effects of phase shifts that occur on reflection. When light is incident from a medium of lower refractive index, and it reflects off a medium of higher refractive index, there is a phase shift of π on reflection. Use the fact that a phase shift of π is equivalent to a path difference of $\lambda/2$ to correct your equation for constructive interference in part (A). Assume $n_{film} > n_{air}$.

c. What is the minimum thickness film that will produce constructive interference for light with λ_0 = 600 nm? Assume that n_{film}= 1.25.

6.11 [Dichromatic Mirror – Implementation] Dichromatic mirrors are used to separate excitation and emission light, typically sending the former toward the sample and the latter toward the detector. This is relatively straightforward for single-color imaging and may be accomplished using a longpass mirror. In this exercise, you will design a dichromatic mirror for use with two exemplary fluorophores used in two-color imaging: Alexa Fluor 488 and Texas Red. Their excitation (dotted) and emission (solid) spectra are shown below. Superimpose on the spectra an approximate reflection/transmission profile for a dichromatic mirror that will work for both fluorophores. Briefly explain your logic. (Data for graph exported from ThermoFisher Scientific's Fluorescence SpectraViewer [www.lifetechnologies.com/handbook/spectraviewer].)

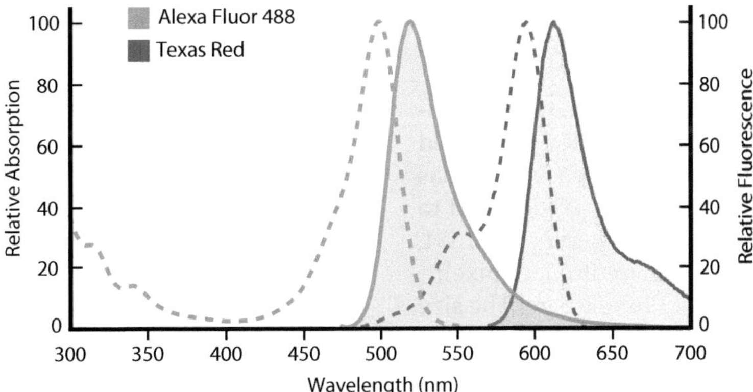

6.12 [Image Brightness – Epi-Fluorescence] In epi-fluorescence microscopy, image brightness, B, varies as $B \propto NA^4/M^2$, where NA and M are the numerical aperture and magnification of the objective.

a. Compare the brightness of 60× and 100× objectives both having NA = 1.4.
b. Compare the brightness of 40×/NA = 1.2 and 60×/NA = 1.4 objectives.
c. Discuss the implications of your results.

6.13 [Light Source Alignment] In both fluorescence and transmitted light microscopy, it is important to align the arc lamp properly. To get some practice, go through the interactive tutorial at the website: http://micro.magnet.fsu.edu/primer/java/kohler/filamentalign/index.html[1]

[1] Permission to use this tutorial is courtesy of "Molecular Expressions.com at Florida State University."

a. Simulate focusing and alignment of the arc using the right default (Bertrand lens out) radio button. Briefly write down the five steps that you went through with the sliders.
b. At the end of the tutorial, you are asked to toggle in the Bertrand lens (using the left radio button) to verify that the filament fills the BFP of the objective. In the BFP, there is an "aperture diaphragm" that you can open and close. Is this what the text calls the condenser aperture diaphragm or the field diaphragm? Justify your answer by looking at the diagram of the Köhler Illumination pathway in Fig. 3.12.

6.14 *[Fluorescence Microscopy – Signal Estimation] It is useful to have a sense for the light flux per pixel in a fluorescence microscopy experiment. To address this issue, answer the following questions:
a. Assume excitation light with $\lambda = 550$ nm. What is the energy per photon?
b. Assume that 2 mW of power is distributed evenly over an area of 1000×1000 pixels^2. What is the photon flux per pixel (in photons/s)?
c. Assume that the fluorescent specimen is dilute and very thin (e.g., a few μm) and thus has a small optical density (OD) ~0.0001. Calculate the number of photons absorbed per pixel per second. Hint: See HW 5.2.
d. Assume that the fluorophore has a quantum yield (ratio of emitted to absorbed photons) of 0.6 and that the detector has a quantum efficiency (ratio of detected to incident photons – see Chapter 9) of 0.4. What is the detected light flux per pixel?

6.15 [Confocal Microscopy – Pinhole Size] The size of the confocal pinhole should be set to match the size of the magnified Airy disk. Use this rule-of-thumb to calculate a reasonable pinhole width for a 100× (NA = 1.4) objective. Assume λ_0 = 500 nm. Hint: Invoke Eq. (4.3), which describes the *radius* of the image of a point object.

6.16 *[Deconvolution Microscopy] The convolution operation (denoted "$\otimes$") describes how the output of a system is related to the input and the system's response function. In microscopes, convolutions describe how an image (the output) is blurred relative to an object (the input). Here, we explore these attributes by computing a discrete (as opposed to continuous) convolution:

$$I(n) = O \otimes \text{PSF} = \sum_{k=-\infty}^{k=+\infty} O(k)\text{PSF}(n-k)$$

of an object $O = (0.5, 2.0, 2.5, 1.0)$ with the point spread function $\text{PSF} = (1.0, 1.0, 1.0, 0.0)$.
a. The convolution measures overlap between a function and a "flipped and shifted" version of a second function, due to the effect of the argument "$n-k$." To evaluate the convolution with the help of a graph, draw a flipped version of PSF(k) similar to that for $O(k)$ in the figure.

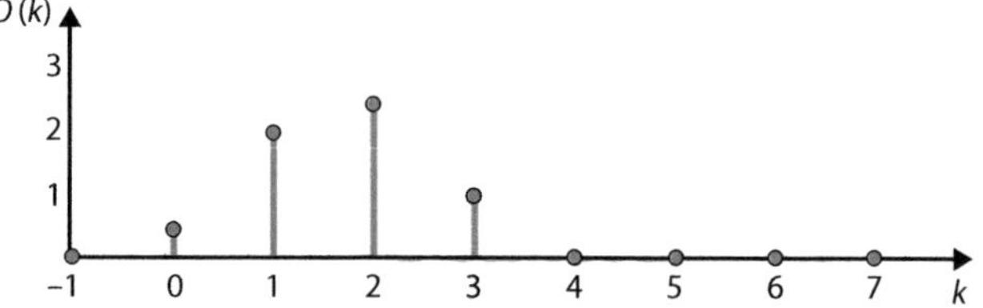

b. Sketch a graph of the flipped version of PSF(k) that also is shifted by $n < 0$ (e.g., $n = -2$).
c. Do the graphs in parts (a) and (b) ever have nonzero overlap? In light of your answer, what is $I(n)$ for $n < 0$?
d. Repeat (b) for n = 3 and use graphical overlap to find $I(3)$. Similarly compute all nonzero values of $I(n)$ and notice the slight smearing effect.

ACTIVITIES[2]

A6.1 Analyzing a PSF

This activity provides an opportunity to visualize and analyze PSF data using the open source Fiji™ (ImageJ™) image processing package. You can download free copies of Fiji and several helpful user guides from the Internet (e.g., at https://imagej.net/Fiji/Downloads and http://imagej.net/docs/guide/user-guide.pdf, respectively). The instructions below should be approximately valid for current versions of Fiji.

A. Open PSF.tif. This is a series of xy images of a sub-resolution-sized fluorescent bead. The images were collected in 0.10-μm focal increments and span 2.00 μm on either side of optimal focus.

[2] Activity files are available for download in the online Support Materials.

B. Go to Process > Math > Gamma and choose 0.5. This manipulation allows dim features in the image to be visualized easily (Chapter 9). Scroll through the images and address the following: What happens to the appearance of the bead as you move away from focus? Comment on changes in width, brightness, and extent of "ringing" in the bead image. Close the file without saving any changes!

C. Reopen the original file and confirm that the integrated intensity in each image is essentially constant, as expected theoretically (Fig. 6.14). To this end, go to Analyze > Set Measurements and check "Integrated Density." Also check "Area, Min & Max, and Mean."

D. Start with the first image and select an analysis area (e.g., using command A on a Mac or using the rectangle tool in the upper left of the Fiji tool bar). Go to Analyze > Measure (often just hitting "M" will work) and collect the data, which will appear in a Results file. Scroll through the images and repeat the measurements for all 40 sections. You can save the Results file and typically open it in Microsoft Excel®. Plot the integrated intensity (density). Is it roughly constant?

E. Is the maximum image intensity also constant? Reconcile the trends of the integrated and maximum intensities.

7

Axially Selective Fluorescence Excitation Techniques

In the previous chapter, we discussed two approaches, confocal microscopy and constrained iterative deconvolution, that produce fluorescence images with significantly reduced blur. In this chapter, we also discuss techniques that yield images with reduced blur, but via a markedly different mechanism – axially selective illumination. This important distinction confers significant advantages over confocal microscopy and deconvolution, including reduced bleaching and phototoxicity. As a consequence, the approaches in this chapter are better suited for noninvasive, extended imaging of live specimens. Two of the approaches also are well suited to imaging deep within tissues and organisms.

7.1 Total Internal Reflection Fluorescence Microscopy (TIRFM)

We start our discussion with total internal reflection fluorescence microscopy (TIRFM). TIRFM also could be included in the next chapter on super-resolution fluorescence techniques because its axial resolution can readily surpass the diffraction barrier. However, we have included TIRFM here due to its reliance on axially selective illumination, as well as its inability to surpass the diffraction barrier in the lateral direction.

TIRFM achieves highly localized axial excitation of fluorophores that are close to an interface by exploiting unique attributes of the electric field associated with the phenomenon of TIR. In biological applications, TIRFM commonly is used to study subcellular processes that occur near an interface generated by a glass coverslip and the surface of adherent cells (Fig. 7.1). Examples include exocytosis and endocytosis, diffusion in the plasma membrane, and molecular binding and unbinding to the membrane. TIRFM is also useful for background reduction in both PALM (Chapter 8) and the molecular genetics technique of CRISPR.

We discussed TIR from the viewpoint of geometrical optics in Chapter 2. There we noted that light incident from a region of higher refractive index that refracts into a region of lower refractive index bends away from the normal (Fig. 2.6). Thus, if the light is incident at sufficiently high angle, it will bend away at an angle of 90°, and there will be no refracted ray in the second medium; the light will be totally internally reflected. The critical (minimum) incidence angle, θ_c, that generates TIR is given by Snell's Law with $\theta_2 = 90°$:

$$\sin\theta_c = \frac{n_2}{n_1} \tag{7.1}$$

(where n_1 is the higher refractive index). All larger incidence angles also produce TIR. For a glass/aqueous interface with $n_1 = 1.52$ and $n_2 = 1.33$, $\theta_c = 61°$.

7.1.1 Mechanism of Axial Discrimination

To understand how TIR achieves axial localization, we need to move beyond the description that emerges from geometrical optics, which is oversimplified and suggests that the EM field vanishes completely in the second, lower index medium. In fact,

(a) Epi-fluorescence

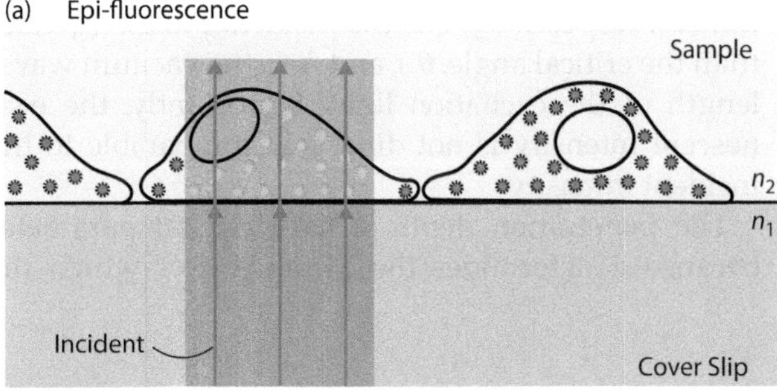

(b) TIRFM

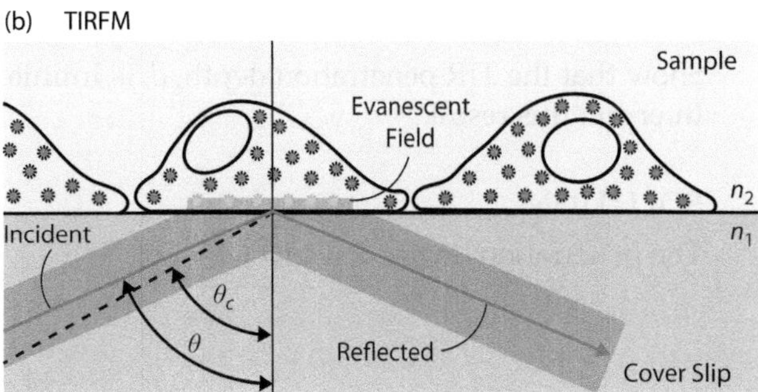

FIGURE 7.1
Comparison of epi-fluorescence (a) and TIRFM (b) illumination (violet) and of the distribution of fluorescing objects (green). Notably, TIRFM illumination and the resulting fluorescence signal are confined to a thin region near the coverslip/cell interface.

DOI: 10.1201/b22076-8

boundary conditions on the waves mandate that a non-propagating "evanescent" electric field penetrate into the lower index medium (Fig. 7.1). However, the field is spatially short-ranged, typically vanishing over a distance from the interface, d, less than a wavelength. Moreover, the field does not, on average, carry energy across the boundary (Box 7.1). The energy of the evanescent field propagates along the surface. The wavefronts are oriented normal to the interface (Box 7.1, Eq. [B7.1.6]).

In TIRFM, the rapid spatial decay of the evanescent field is key because it ensures that more distant fluorophores are not excited, leading to markedly reduced out-of-focus contamination and markedly enhanced axial discrimination (Fig. 7.2). On the other hand, imaging also is restricted to regions near the interface. For biological systems, this typically restricts imaging to the cell membrane and nearby structures.

To gain more insight into TIRFM, we discuss the intensity profile in some detail. The mathematical expression for $I(z)$ is (Box 7.1)

$$I(z) = I_0 e^{-z/d} \tag{7.2}$$

where the penetration depth, d, is given by

$$d = \frac{\lambda_0}{4\pi\sqrt{n_1^2 \sin^2\theta - n_2^2}} \tag{7.3}$$

Here, n_1 and n_2 are the higher and lower refractive indices, respectively, θ is the incidence angle (greater than the critical angle, θ_c), and λ_0 is the vacuum wavelength of the excitation light. Importantly, the evanescent intensity is not dim; I_0 is comparable to the incident intensity.

The penetration depth, d, is a critical parameter because it determines the distance over which the intensity falls to $1/e$ or ~37% of its initial value, I_0, at $z = 0$. To get a feel for the magnitude of d, we insert typical values into Eq. (7.3). For a glass/cell interface, these are $n_1 = 1.52$, $n_2 = 1.38$, $\lambda_0 = 488$ nm, and $\theta = 70°$, which is slightly larger than θ_c. These values yield $d = 105\text{ nm} \ll \lambda$.

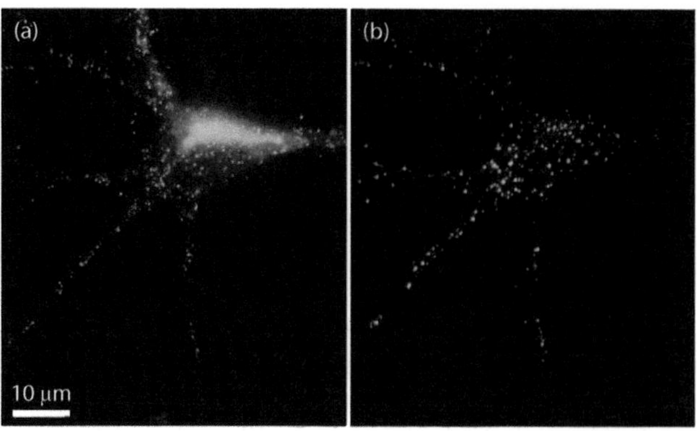

FIGURE 7.2
Comparison epi-fluorescence (a) and TIRFM (b) images of DCVs containing tissue plasminogen activator-EGFP in living hippocampal neurons. The TIRFM image has markedly better contrast, especially in the cell body, due to reduced fluorescence signal from deep within the cell. Bar = 10 μm. (Adapted with permission from Silverman, M. A., et al. 2005. Mechanisms of transport and exocytosis of dense-core granules containing tissue plasminogen activator in developing hippocampal neurons *J. Neurosci.* 25:3095–3106. https://doi.org/10.1523/JNEUROSCI.4694-04.2005. Copyright 2005 Society for Neuroscience.)

7.1.2 Experimental Implementation

TIR will not occur unless the excitation light is incident on an interface at a relatively high angle and from a medium of higher refractive index. This restriction places constraints on the implementation of TIRFM,

SAMPLE PROBLEM 7.1 LIMITS OF TIR

Show that the TIR penetration depth, d, is infinite when the incidence angle equals the critical angle, θ_c. Interpret this result.

SOLUTION

The penetration depth, d, is given by Eq. (7.3):

$$d = \frac{\lambda_0}{4\pi\sqrt{n_1^2 \sin^2\theta - n_2^2}}$$

According to Eq. (7.1), $n_1^2 \sin^2\theta = n_1^2\left(n_2^2/n_1^2\right) = n_2^2$ when $\theta = \theta_c$. Therefore, in this case, the denominator of Eq. (7.3) goes to zero, and d goes to infinity. The interpretation of this result is that, when $\theta = \theta_c$, TIR is transitioning to conventional, unlimited illumination depth.

BOX 7.1 DERIVATION OF THE INTENSITY PROFILE OF THE EVANESCENT FIELD

Here we sketch the derivation of Eqs. (7.2) and (7.3). Our discussion assumes familiarity with representing a plane wave using a complex exponential.

Our analysis begins with the Fresnel equations. These equations specify the *amplitude* transmission coefficients – the ratio of the transmitted amplitude to the incident amplitude – in terms of angles, refractive indices, and polarization (Fig. B7.1.1). The Fresnel equations are derived in Hecht (2017), and we present a similar derivation in Box 10.1 that is applicable to reflection of ultrasound. Our derivation and results are somewhat simplified because sound waves are not vectors.

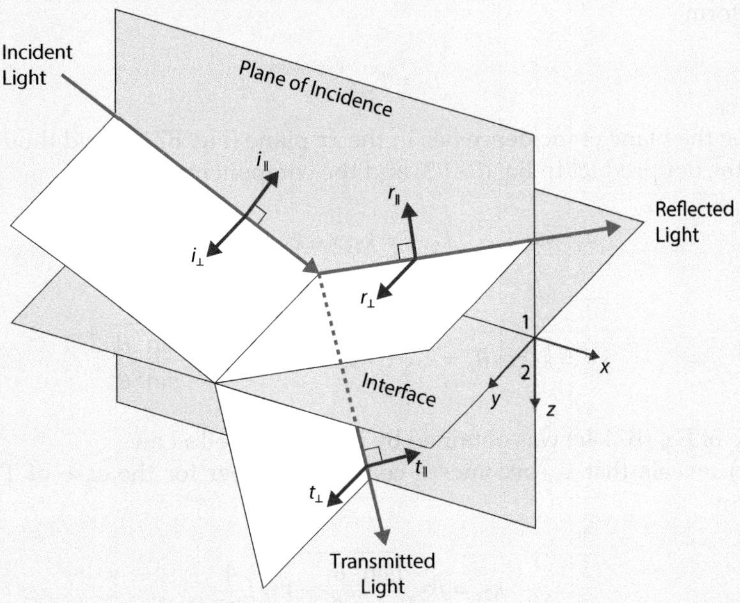

FIGURE B7.1.1
Schematic representation of reflection and transmission of plane waves incident on an interface between homogeneous, isotropic, "dielectric" media. Here, we have inverted the geometry relative to Figs. 7.1 and 7.3 to display the polarizations more clearly.

To set up the notation used in the Fresnel equations, we refer to Fig. B7.1.1, which shows the three-dimensional geometry that describes reflection and refraction at an interface between medium 1, with refractive index n_1, and medium 2, with refractive index n_2. The plane of incidence contains the surface normal and the rays of the incident, reflected, and transmitted fields. The electric field is divided into components that are parallel and perpendicular to the plane of incidence, as are the reflection and transmission coefficients. The Fresnel equations for the amplitude transmission coefficients perpendicular, $t_{\perp}$, and parallel, $t_{\parallel}$, to the plane of incidence are (Hecht 1973)

$$t_{\perp}(\theta_1) = \frac{2\cos\theta_1}{\cos\theta_1 + \left(n_{21}^2 - \sin^2\theta_1\right)^{\frac{1}{2}}} \tag{B7.1.1a}$$

$$t_{\parallel}(\theta_1) = \frac{2n_{21}\cos\theta_1}{n_{21}^2\cos\theta_1 + \left(n_{21}^2 - \sin^2\theta_1\right)^{\frac{1}{2}}} \tag{B7.1.1b}$$

In Eq. (B7.1.1) $n_{21} = n_2/n_1$. When $\theta_1 = \theta_c$, Eqs. (B7.1.1a) and (B7.1.1b) become (HW 7.1)

$$t_{\perp}(\theta_c) = 2 \tag{B7.1.2a}$$

$$t_{\parallel}(\theta_c) = \frac{2}{n_{21}} \tag{B7.1.2b}$$

Equations (B7.1.2a) and (B7.1.2b) reveal that the transmitted field has a nonzero amplitude (the field does not vanish). The presence of a transmitted electric field in medium 2 is a consequence of the requirement for field continuity at the interface. The field cannot abruptly vanish, but it can decay rapidly.

Equations (B7.1.2a) and (B7.1.2b) also reveal that the amplitude transmission coefficients are greater than unity, which means that the amplitude of the transmitted field exceeds that of the incident field. It is also true that the associated *intensity* reflection coefficients are unity. Thus, despite the existence of the evanescent field, all of the intensity is reflected under conditions of TIR; the evanescent wave does not on average carry energy across the interface.

To determine the spatial behavior of the transmitted field, we assume it can be written as a complex exponential of the form

$$\vec{E}_2 = \vec{E}_{02} e^{i(\vec{k}_2 \cdot \vec{r} - \omega t)} \tag{B7.1.3}$$

We also assume that the plane of incidence lies in the xz plane (Fig. B7.1.1), and thus so too does $\vec{k}_2$. Given this, we can write the dot product in Eq. (B7.1.3) and the components of $\vec{k}_2$ as

$$\vec{k}_2 \cdot \vec{r} = k_{2x} x + k_{2z} z \tag{B7.1.4a}$$

$$k_{2x} = k_2 \sin\theta_2 \tag{B7.1.4b}$$

$$k_{2z} = k_2 \cos\theta_2 = k_2 \sqrt{1 - \sin^2\theta_2} = k_2 \sqrt{1 - \frac{\sin^2\theta_1}{\sin^2\theta_c}} \tag{B7.1.4c}$$

The right-hand side of Eq. (B7.1.4c) was obtained by invoking Snell's Law.

Equation (B7.1.4c) reveals that k_{2z} becomes a complex number for the case of TIR ($\theta_1 > \theta_c$). We thus rewrite it in the form

$$k_{2z} = ik_2 \sqrt{\frac{\sin^2\theta_1}{\sin^2\theta_c} - 1} \equiv i\frac{1}{2d} \tag{B7.1.4d}$$

where

$$d = \frac{1}{2k_2\sqrt{\frac{\sin^2\theta_1}{\sin^2\theta_c} - 1}} = \frac{\lambda_0}{4\pi\sqrt{n_1^2 \sin^2\theta_1 - n_2^2}} \tag{B7.1.5}$$

The field in Eq. (B7.1.3) correspondingly becomes

$$\vec{E}_2 = \vec{E}_{02} e^{-\frac{z}{2d}} e^{i(k_{2x} x - \omega t)} \tag{B7.1.6}$$

Equation (B7.1.5) is equivalent to Eq. (7.3) if we switch to the simplified notation in the main text, $\theta_1 = \theta$. To reproduce Eq. (7.2) in the main text, we need to compute the intensity from the (absolute) square of the field (Box 5.3) in Eq. (B7.1.6) to obtain

$$I(z) = I_0 e^{-z/d} \tag{B7.1.7}$$

which are commonly met using a through-objective configuration similar to that shown in Fig. 7.3.

Key components of the setup include a very high NA objective and laser-based excitation, which is directed at a peripheral position in the BFP of the objective (Fig. 7.3). This peripheral position maps into a high incidence angle (Fig. 3.4), ensuring that the condition for TIR is met.

To achieve TIR in cellular applications, the NA of the objective must exceed the refractive index of the sample, as shown in Sample Problem 7.2. Thus, an objective with NA = 1.40, which is a standard high

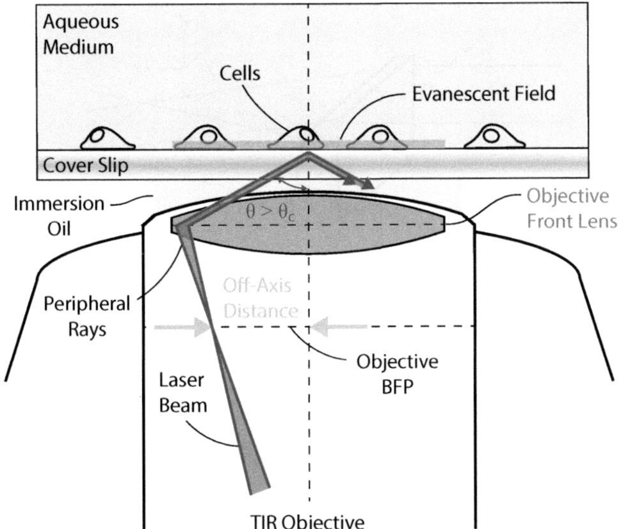

FIGURE 7.3
Objective-based TIRFM system. A laser beam is directed and focused at a peripheral position in the BFP of a very high NA objective. This ensures that the laser light exiting the objective is parallel and thus is fully incident on the glass/cell interface at an angle that exceeds the critical angle required to achieve TIR with a defined and unique penetration depth.

value, is barely adequate, and special TIRF objectives with NA = 1.45 (or higher) typically are used instead. These have a larger range of peripheral positions in the BFP of the objective at which to direct the laser-based excitation. An NA = 1.45 thus makes it easier to aim the laser to achieve TIR, and it allows the excitation to be incident at higher angles that generate more rapidly decaying evanescent fields (HW 7.3 and 7.4).

7.2 Light Sheet Fluorescence Microscopy (LSFM)

LSFM encompasses an array of rapidly evolving techniques that, like TIRFM, rely on illumination that is spatially restricted in the axial direction. However, unlike TIRFM, axially constrained illumination in LSFM can be achieved deep into a sample, and three-dimensional images can be generated. In addition, bleaching and phototoxicity are markedly reduced relative to confocal and widefield microscopy. LSFM thus has emerged as an approach of choice for long-term studies of dynamic (e.g., developmental) processes in relatively large, intact, living specimens, such as *Drosophila* and Japanese rice fish (Fig. 7.4). LSFM also is useful in many other applications, including static, anatomical studies of whole tissues, such as mouse brain, and light-based control of biological function.

Here we will emphasize the key optical principles and implementation of LSFM by focusing on one of the initial, well-known forms of the approach – selective plane illumination microscopy (SPIM). The reader interested in a more exhaustive discussion of variants of SPIM, including scanning-based techniques, is referred to the presentation at Nikon's MicroscopyU website.

7.2.1 Mechanism of Axial Discrimination

In epi-fluorescence microscopy, excitation and emission pathways overlap, and the detection region is elongated along the axial direction (Fig. 7.5).

SAMPLE PROBLEM 7.2 CALCULATION OF MINIMUM NA FOR TIRFM

Show that an NA > 1.38 is required to use objective-based TIRFM to view inside a cell. Assume $n_{cell} = 1.38$.

SOLUTION

To achieve TIR, the laser beam is positioned at the farthest off-axis position in the objective BFP and correspondingly emerges into the immersion oil at the maximum possible angle θ consistent with the NA, which is given by

$$\mathrm{NA} = n_{oil} \sin\theta$$

Snell's Law states that the quantity "$n \sin\theta$" is conserved as the beam traverses various media through to the cell. Given this, and the fact that, at the final interface, the beam bends to an angle of 90° in the cell to undergo TIR, implies that

$$\mathrm{NA} = n_{oil} \sin\theta = n_{cell} \sin 90° = 1.38.$$

Clearly, an NA of 1.40, which is very close to 1.38, is barely adequate, and thus objectives with NA ≥ 1.45 are routinely used instead.

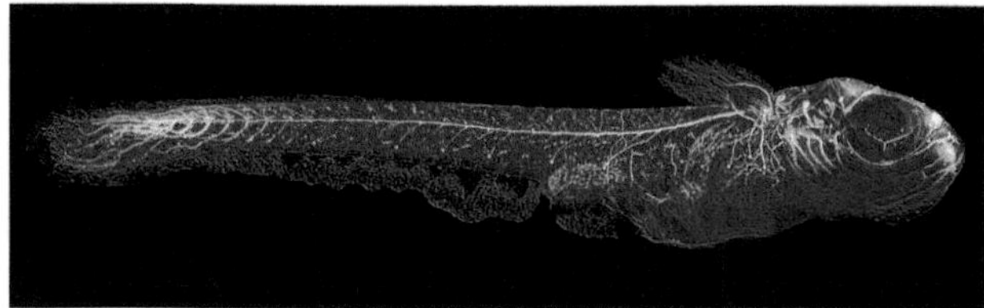

FIGURE 7.4
LSFM image of a juvenile Japanese rice fish. (Image courtesy of Philipp Keller, Stelzer Group, European Molecular Biology Laboratory.)

Thus, epi-fluorescence images tend to be corrupted by out-of-focus signal unless the signal either is rejected (confocal) or computationally corrected (deconvolution). A key distinguishing feature of LSFM is excitation and detection pathways that are orthogonally oriented, such that fluorescence emanates only from a relatively thin sheet (Fig. 7.5). As a consequence, the out-of-focus signal is reduced compared to standard epi-fluorescence microscopy. Equally important, during three-dimensional imaging, each plane is illuminated once, and thus photobleaching and phototoxicity are markedly reduced. The reduction in light exposure can be approximated as the sheet thickness over the sample thickness (HW 7.5).

In SPIM, the sample is illuminated using a light sheet that is created by a cylindrical lens.[1] To illustrate the essential optics, we refer to Fig. 7.6, which emphasizes the absence of curvature along the long axis of a cylindrical lens and the associated failure of light to bend along this axis. As a consequence, parallel rays incident at a given "height" (y coordinate

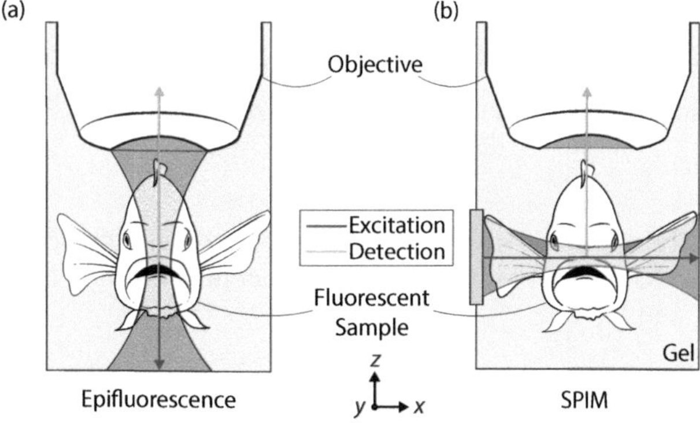

FIGURE 7.5
Comparison of epi-fluorescence and SPIM detection geometries. In epi-fluorescence microscopy (a), the excitation and detection pathways are collinear, whereas in SPIM (b), they are orthogonal.

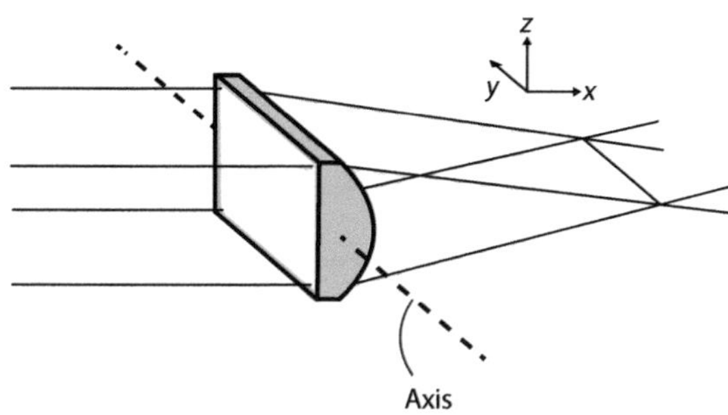

FIGURE 7.6
Creation of a light sheet. Collimated parallel light is focused to a line instead of a point by a converging cylindrical lens. Diverging cylindrical lenses are discussed in HW 7.6, and the use of cylindrical lenses to correct astigmatism is explored in HW 7.7.

in Fig. 7.6) along the long axis focus to a point at the same height, and a set of parallel rays incident at different heights along the axis focus to a line. The light sheet is generated by the entire ray path before and after the line focus. This sheet is viewed by an orthogonally oriented objective in SPIM.

Fig. 7.6 also illustrates some critical attributes of the light sheet. One is that the sheet is not perfectly planar. Instead, it is thinnest at focus and then becomes progressively wider on either side, degrading axial resolution. Moreover, tighter focus leads to more rapid axial expansion, reducing the distance over which the sheet is sufficiently thin to be useful for imaging.

To quantify these attributes, we consider a sheet oriented in the xy plane and spreading axially along z, as shown in Fig. 7.7. For this geometry, the light sheet possesses a hyperbolic profile in the xz plane. The beam waist, w_0, is given approximately by

$$w_0 = \frac{\lambda_0}{\pi \mathrm{NA}} \tag{7.4}$$

In Eq. (7.4), NA is the illumination numerical aperture. The beam expands to a width $w = \sqrt{2}\ w_0$ over a distance along the x-axis known as the Rayleigh

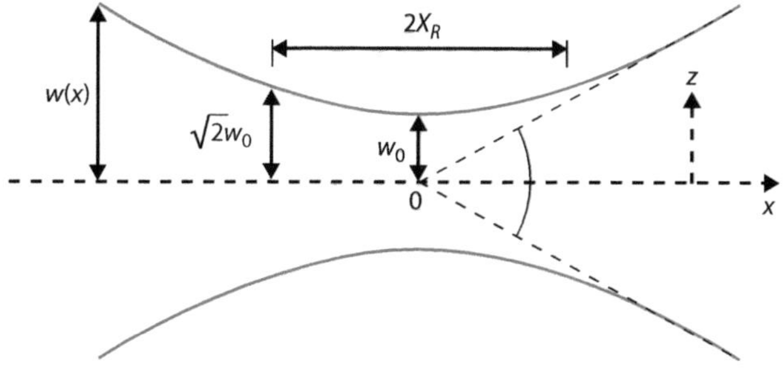

FIGURE 7.7
Hyperbolic beam profile of a light sheet. The beam expands to ~$1.4w_0$ over a distance, X_R. The usable FOV (i.e., an FOV with a sufficiently narrow width) is ~ $2X_R$.

[1] A cylindrical lens alone will suffice, but the NA and degree of correction of cylindrical lenses typically are inadequate. Thus, a cylindrical lens more typically is used in combination with an objective lens.

SAMPLE PROBLEM 7.3 CALCULATION OF RAYLEIGH LENGTH

Use Eq. (7.5) to calculate X_R in a SPIM experiment for beam waists of 3 μm and 10 μm. Assume that $\lambda_0 = 488$ nm. Twice the Rayleigh length is used as a measure of the lateral extent that is usable for imaging (i.e., the FOV).

SOLUTION

Inserting the stated values into Eq. (7.5) yields

$$X_R(3\ \mu\text{m}) = \frac{\pi\ 3^2}{0.488} = 58\ \mu\text{m}$$

$$X_R(10\ \mu\text{m}) = \frac{\pi\ 10^2}{0.488} = 644\ \mu\text{m}$$

These numbers highlight two important aspects of SPIM. The first is the undesirable coupling between axial resolution and FOV. The second is a dramatic increase in FOV as beam waist increases and axial resolution falls.

length, X_R, which is given approximately by (Box 7.2 in Section 7.3.2)

$$X_R = \frac{\pi w_0^2}{\lambda_0} \tag{7.5}$$

Equations (7.4) and (7.5) imply that a thin sheet (small w_0) will have a small FOV (small X_R). This undesirable link between beam waist and FOV is explored in Sample Problem 7.3.

The undesirable link between beam waist and FOV has been eliminated in some scanning-based variants of LSFM. For example, it is possible to generate laterally thin sheets via scanning and then to stitch these together to generate a sheet with a larger FOV. However, this slows data acquisition, and stitching is not trivial to implement.

7.2.2 Experimental Implementation

The key components of a SPIM apparatus are shown in Fig. 7.8. In this figure, sample mounting emphasizes application to developmental biology. Familiar features of the apparatus include the light sheet and the orthogonal orientation of the excitation and detection pathways. Less familiar features include a specimen that is mounted in agarose to hold it in place, a water-immersion objective, and a rotatable sample holder. Long-working-distance objectives are used to avoid collision between the illumination and detection lenses. Water-dipping or water-immersion objectives are used to minimize spherical aberration and facilitate deep imaging. The rotatable holder is used to generate multi-angle views that help to mitigate problems with deep imaging. These include stripe artifacts and shadow artifacts that are due to limited light penetration and detection in thick specimens (Fig. 7.9).

Fig. 7.8 also shows the use of an sCMOS detector. An important strength of SPIM, which reflects an absence of scanning, is speed, and an sCMOS detector is an important component of SPIM's rapid data acquisition (Chapter 9). As a benchmark, SPIM data often are collected at the rate of ~100 images/second, and three-dimensional data stacks are generated in several seconds by incrementing the specimen along z through the sheet.

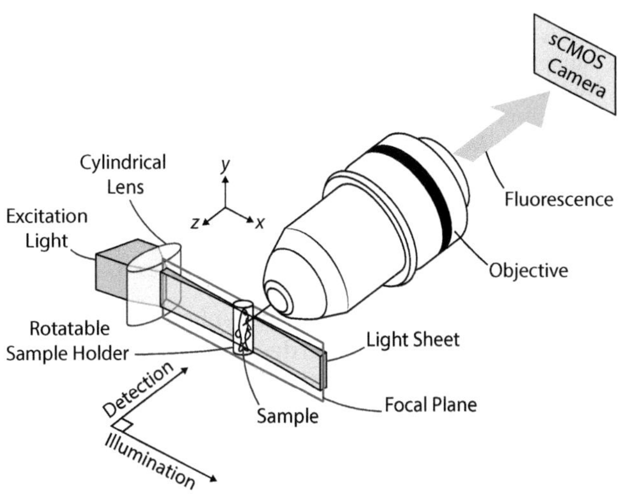

FIGURE 7.8
Schematic of a SPIM apparatus. Laser-based excitation in the form of a sheet impinges on an agarose-embedded specimen, which is mounted in a capillary. The capillary is attached to a motor that rotates the sample. An orthogonally oriented water-immersion objective collects the signal.

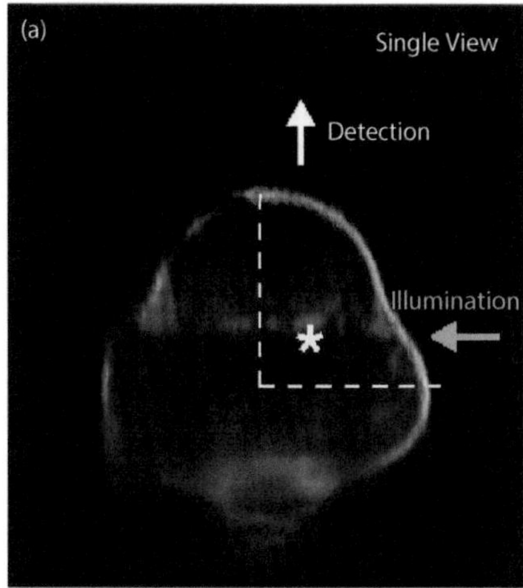

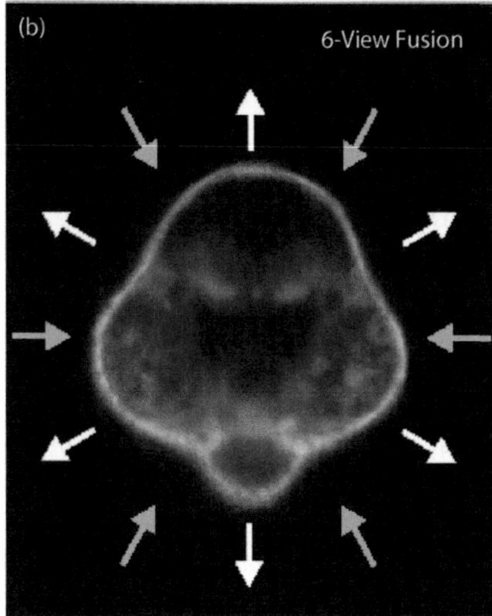

FIGURE 7.9
Elimination of shadow artifacts in SPIM using multiple views. (a) When only one view is collected, the side facing the illumination is much brighter due to limited light penetration. (b) When multiple views are collected and fused, image quality is relatively isotropic. (Adapted with permission from Weber, M., et al. 2011. Light sheet microscopy for real-time developmental biology. *Curr. Opin. Genet. Dev.* 21:566–572. DOI: 10.1016/j.gde.2011.09.009.)

7.3 Two-Photon Fluorescence Microscopy

We close this chapter with two-photon microscopy, which also is well suited to deep imaging. We will discuss this strength in detail in upcoming sections, but, in brief, two-photon microscopy suffers less from deleterious scattering effects that arise in thick samples. In addition, two-photon excitation light is better able to penetrate to deep imaging planes because there is no absorption away from the focal plane. Two-photon microscopy also is uniquely well suited for imaging UV-excitable fluorophores. The main negative relative to LSFM is that conventional two-photon microscopy, which is the emphasis here, is implemented in a point-scanning mode and thus is limited to slower data acquisition.

LSFM and two-photon excitation have been hybridized to capitalize on their combined respective strengths. The resulting state-of-the-art method, two-photon LSFM, can achieve ~2× deeper imaging than one-photon LSFM and ~10× more rapid image acquisition than conventional two-photon microscopy. The reader interested in two-photon LSFM is referred to Truong et al. (2011).

7.3.1 Two-Photon Excitation

To set the stage for a discussion of two-photon excitation, we first review key aspects of the analogous one-photon process.

One-photon (1P) excitation occurs when a photon with energy $E_p = h\nu = hc/\lambda_{1P}$ promotes a molecule in an initial state with energy E_i into a final state with energy E_f. Energy conservation dictates that the photon energy is related to the difference in energy between the states by

$$E_f - E_i = \frac{hc}{\lambda_{1P}} \tag{7.6}$$

The fluorescence emission from a plane, F, induced by one-photon excitation depends linearly on excitation intensity, I, i.e., $F \propto IA$, where A is the area illuminated.

In contrast, two-photon (2P) excitation is achieved via nearly simultaneous absorption of two photons that together have the same energy as the requisite single photon (Fig. 7.10). By analogy with Eq. (7.6), each photon has wavelength λ_{2P}, which is given by

$$E_f - E_i \cong 2\frac{hc}{\lambda_{2P}} \tag{7.7}$$

Equations (7.6) and (7.7) reveal that $\lambda_{2P} \sim 2\lambda_{1P}$. This implies that one- and two-photon excitation spectra generally are markedly different, and that one-photon absorption processes that occur in the UV/visible become two-photon processes in the IR (Fig. 7.10). In contrast, fluorescence emission spectra are identical, because the fluorophore decays via the same pathway whether it is excited by one or two photons.

Two-photon absorption also differs in depending quadratically (nonlinearly) on excitation intensity. The quadratic dependence arises from the requirement that two independent photons arrive together at the same point, where the probability of each event is proportional to intensity. As a consequence of the dependence on I^2, two-photon excitation and emission fall off rapidly with axial distance from focus, and this gives rise to many of the key advantages of

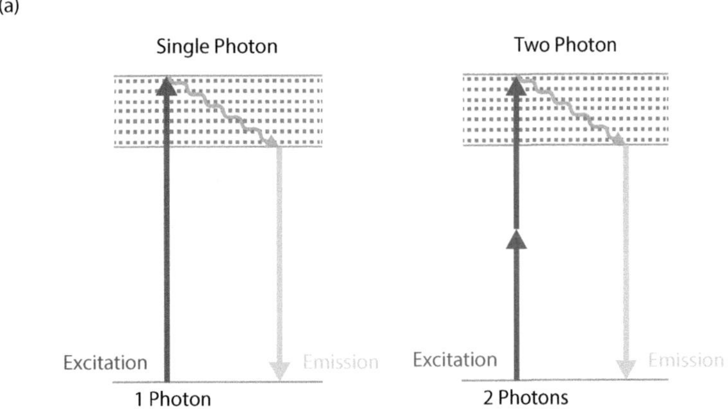

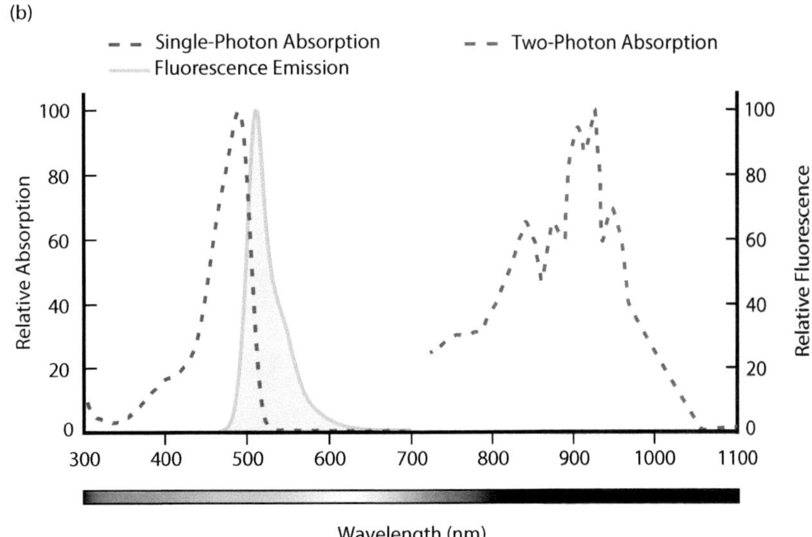

FIGURE 7.10
Comparison of one- and two-photon absorption and emission phenomena. Jablonski diagrams are shown in Panel (a), and excitation and emission spectra for the fluorophore, EGFP, are shown in Panel (b). Unlike the one-photon case, two-photon emission is of *shorter* wavelength than the excitation. (Data for graph exported from University of Arizona Spectra Database [http://spectra.arizona.edu/].)

two-photon fluorescence microscopy for deep tissue imaging. In what follows, we elaborate on this point and focus on the strengths of two-photon microscopy relative to LSCM because both are implemented in a comparable point-scanning mode.

7.3.2 Mechanism of Axial Discrimination

To describe the origins of, and quantify, axial discrimination in two-photon microscopy, we need briefly to address implementation (Section 7.3.4). Two-photon microscopy is implemented by scanning a "Gaussian" laser beam across the sample. Notably, a Gaussian beam propagates to produce an intensity profile that falls off with axial distance, z, as specified by Eq. (B7.2.2) (Box 7.2). The resulting fluorescence signal from a given axial plane, F, is proportional to the integral (HW 7.12)

$$F \propto \int_0^\infty I^2 dA = \int_0^\infty I^2 2\pi r\,dr = \frac{P^2}{\pi w^2} \tag{7.8a}$$

Here P is the total power in the beam, and w is the beam radius, which is given by Eq. (B7.2.1). Invoking the z dependence of w specified in Eq. (B7.2.1) yields

$$F \propto \frac{1}{w_0^2\left[1+\left(\dfrac{\lambda z}{\pi w_0^2}\right)^2\right]} \tag{7.8b}$$

Equation (7.8b) shows that the fluorescence becomes half-maximal when $z = \pi w_0^2/\lambda$ and falls off as $1/z^2$

BOX 7.2 PROPERTIES OF GAUSSIAN LASER BEAMS

The qualitative behavior and the mathematical expressions governing the radius and intensity profile of the Gaussian TEM_{00} mode of a laser beam play critical roles in two-photon and light sheet microscopy. We first consider the behavior of the beam radius. Assume that the beam starts at $z = 0$ as a perfectly flat plane wave with radius w_0. The radius of the beam spreads with axial distance z according to the relationship (see figure)

$$w(z) = w_0\left[1+\left(\frac{\lambda z}{\pi w_0^2}\right)^2\right]^{1/2} \tag{B7.2.1}$$

Moreover, the intensity profile follows the relationship

$$I(r) = \frac{2P}{\pi w^2}e^{-2r^2/w^2} \tag{B7.2.2}$$

Here, P is the total power in the beam, and r is the radial distance in a plane perpendicular to z.

Equation (B7.2.1) reveals that, for large values of z, the beam radius varies as

$$w(z) = \frac{\lambda z}{\pi w_0} \tag{B7.2.3}$$

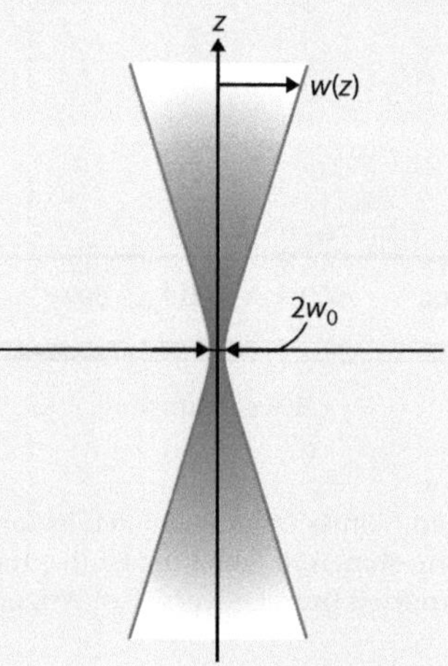

for large z. In contrast, F is z-invariant for one-photon imaging (HW 7.12).

Equation (7.8), and the results that went into its derivation, shows that two-photon excitation and emission are confined to a very thin focal plane. This is an important distinction from confocal imaging, where the illumination has significant axial extent that generates out-of-focus fluorescence signal and causes extensive bleaching (Fig. 7.11). The former problem in confocal microscopy is mitigated with pinhole filtering, but bleaching and toxicity remain.

In contrast, two-photon imaging is "pinhole-free" because out-of-focus fluorescence signal is not generated; there also is minimal bleaching and phototoxicity because excitation is confined to a very thin region (Fig. 7.11). In addition, negative effects of scattering are markedly reduced, facilitating generation of high-contrast fluorescence images from objects deep within thick specimens. This issue is discussed in detail in the next section.

On the other hand, weaknesses of two-photon microscopy include the need for bright, pulsed IR sources (Section 7.3.4), object scanning, and potential problems with heating. The lateral resolution also is somewhat poorer than that of confocal microscopy because the twofold longer excitation wavelength used in two-photon microscopy generates a twofold larger beam waist. The larger waist compromises

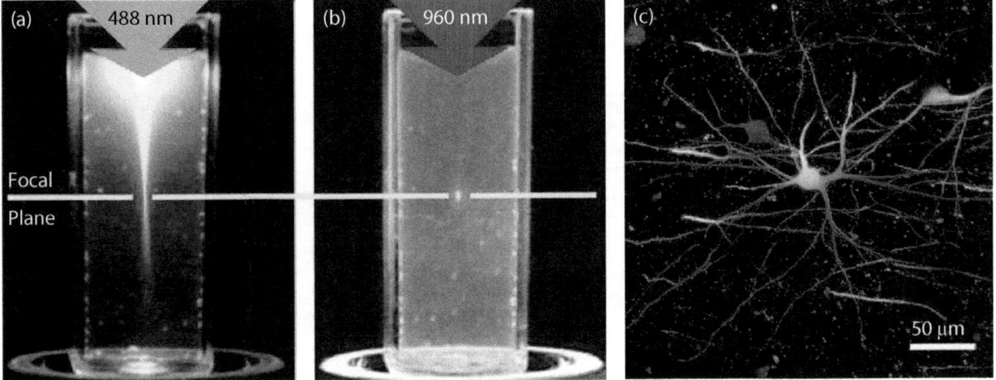

FIGURE 7.11
One-photon (a) and two-photon (b) excitation of a solution of fluorescein showing the localization of signal generated using the latter method. Projection image (c) of mouse neurons 200 μm below the brain surface generated from a three-dimensional two-photon image stack. Scale bar = 50 μm. (Panels [a] and [b] adapted with permission from Zipfel, W. R., et al. 2003. Nonlinear magic: multiphoton microscopy in the biosciences. *Nat. Biotechnol.* 21:1369–1377. https://doi.org/10.1038/nbt899. Panel [c] adapted with permission from Franke, T., et al. 2012. Two-photon microscopy for deep tissue imaging of living specimens. *Micros. Today* 20(4):12–16. DOI: 10.1017/S1551929512000430.)

resolution, but only by a factor of the square root of two. This is because two-photon excitation depends on intensity squared, and this effect restricts the effective beam width relative to one-photon excitation (HW 7.13).

7.3.3 Application to Deep Imaging

The main strength of two-photon microscopy is its ability to generate clear images of samples, especially live animals, with thicknesses that exceed ~50 μm. Several attributes of two-photon microscopy make it much better suited for "deep imaging" than LSCM. These include reduced image degradation from scattering of excitation and emission light and reduced absorption of light beyond focus. We consider these issues in turn.

Emission photons scatter increasingly as imaging depth increases. Fig. 7.12 compares the impact of this increased scattering on confocal and two-photon imaging for the case of thick samples. In the confocal case, emission scattering leads to a reduced signal when in-focus photons miss the pinhole, and it leads to enhanced background when out-of-focus emission photons pass through the pinhole. Both of these problems are avoided with the two-photon method because there is no pinhole (and no out-of-focus emission).

Scattering of excitation light also increases with depth, and this reduces the fluorescence signal generated at focus in both confocal and two-photon microscopy. However, scattering of excitation light does not generate out-of-focus fluorescence signal in the two-photon approach because the probability that two

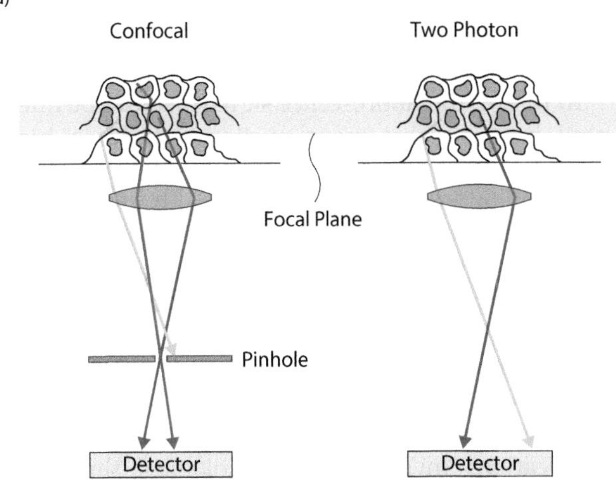

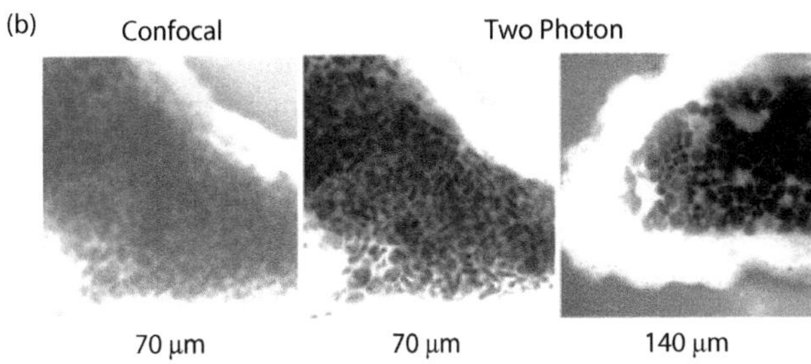

FIGURE 7.12
Comparison of emission scattering effects in (single-photon) confocal and two-photon microscopy. (a) Deleterious effects of scattering that are restricted to the confocal case include in-focus emission missing the pinhole and out-of-focus emission passing through the pinhole. (b) The images show the better penetration of two-photon imaging. Numbers refer to the depth at which the sample was imaged. (Panel [b] adapted from Piston, D. W. 2005. When two is better than one: Elements of intravital microscopy. *PLoS Biol.* 3(6):e207. DOI: 10.1371/journal.pbio.0030207. Open Access.)

DEMONSTRATION 7.1 ABSORPTION OF GREEN LIGHT BY BIOLOGICAL TISSUE

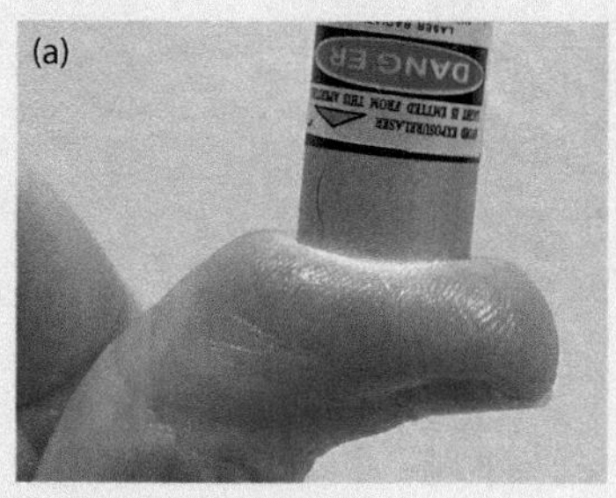

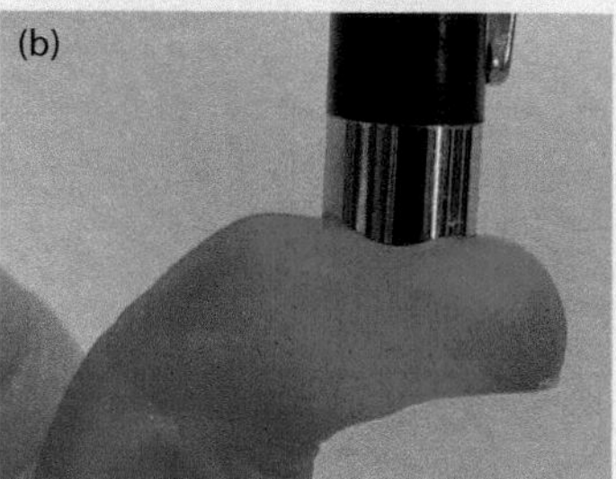

Absorption of green (a), but not red (b), light by biological tissue can easily be demonstrated by placing a finger over green and red laser pointers. This preferential absorption is the reason that light passing through a hand is red when a white flashlight is held against it. (Photographs by the authors.)

scattered excitation photons will be absorbed simultaneously is negligible. In contrast, scattered excitation can generate fluorescence anywhere in the specimen in the confocal approach, and this creates background haze, especially as laser power is increased to aim for deep imaging.

In LSCM, the fluorescence signal also can be attenuated because blue/green excitation light is highly absorbed by endogenous molecules in biological tissue, and excitation light is absorbed by fluorophores outside of focus. In contrast, two-photon microscopy does not suffer from these problems; red/IR light is much less heavily absorbed by endogenous molecules (Demonstration 7.1), and two-photon excitation does not occur outside the focal plane.

In light of these issues, two-photon microscopy is the preferred alternative to confocal microscopy for applications involving thick biological samples. In contrast, the two-photon method does not offer significant advantages for the case of thin samples, except when using UV-excitable fluorophores.

Despite being reduced, scattering and other factors fundamentally limit the utility of two-photon imaging to depths ranging from a few hundred microns to a few millimeters, depending on the level of tissue scattering.

7.3.4 Experimental Implementation

The key components of a typical two-photon microscope are shown in Fig. 7.13. The sample is scanned using tightly focused light from a pulsed laser, such as a titanium:sapphire (Ti:Sa) laser. Use of a pulsed laser is important because this concentrates photons in time and makes it much more likely that two photons will be incident on a fluorophore simultaneously. A PMT is used for detection. Long-working-distance water-immersion objectives are desirable to enable deep imaging.

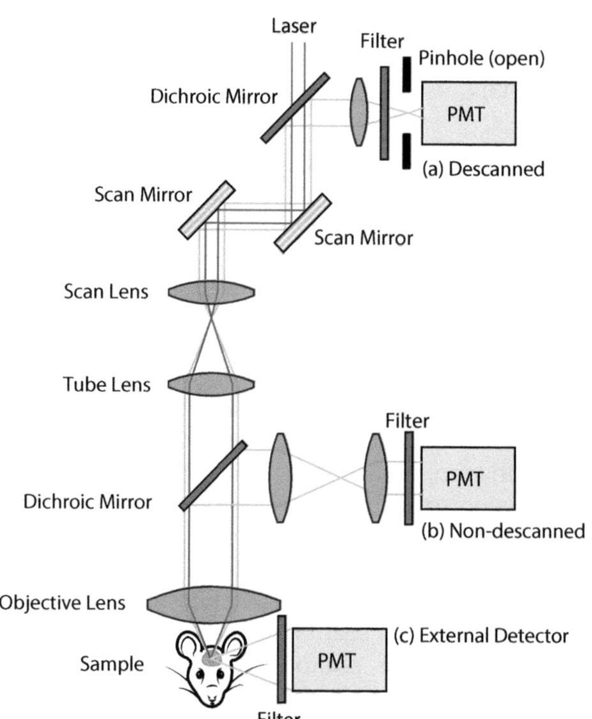

FIGURE 7.13
Schematic of a two-photon microscope showing three collection geometries: (a) descanned, (b) non-descanned, and (c) external. The non-descanned geometry is preferred due to more efficient detection. The non-descanned geometry is an option in two-photon imaging because no pinhole is needed. (Adapted from Benninger, R. K. P., et al. 2013. Two-Photon Excitation Microscopy for the Study of Living Cells and Tissues. *Curr. Protoc. Cell Biol.* 59:4.11.1–4.11.24. https://doi.org/10.1002/0471143030.cb0411s59.)

TABLE 7.1

Comparison of Techniques

Technique	Unique Attributes	Strengths	Weaknesses	Applications
TIRFM	Evanescent fields, axial super-resolution	Fast data acquisition, low bleaching and toxicity	Restricted to surface phenomena/planar imaging, specialized TIRF objectives (for through-objective methods)	Biological processes in living samples near the cell surface
SPIM	Orthogonal detection	Fast data acquisition, three-dimensional imaging, low bleaching and toxicity, relatively good penetration	Shadows and stripes, non-trivial to set up due to marked differences with traditional configurations	Long-term imaging of biological processes in intact living samples, tomographic imaging
Two-Photon	Nonlinear excitation	Three-dimensional imaging, reduced scattering problems, high penetration, low bleaching and toxicity	Slow point scanning, high cost, high instantaneous laser intensity	Long-term imaging of biological processes deep within intact living samples, tomographic imaging

Two-photon microscopy typically uses one of three detection geometries. One option is the relatively familiar descanned geometry, which is analogous to that used for confocal imaging. However, in the two-photon case, there is no pinhole, or the pinhole is fully open. Another option is a relatively unfamiliar, non-descanned geometry, which is preferred because the emission traverses a shorter distance and thus is detected more efficiently (Fig. 7.13). In this alternative, emission is collected by the objective and reflected directly to a PMT; it is even possible to direct the emission to a PMT before it is collected by the objective.

7.4 Comparison of Approaches

In this chapter, we have discussed three fluorescence microscopy techniques that employ axially selective excitation. To facilitate comparison and mastery of the approaches, we end with a concise comparative summary of their important attributes, including applications to the life sciences (Table 7.1).

Suggested Reading and Additional Resources

Allen, J.R. 2020. MicroscopyU. Light sheet fluorescence microscopy. https://www.microscopyu.com/techniques/light-sheet/light-sheet-fluorescence-microscopy

Hecht, E. 1973. Amplitude transmission coefficients for internal reflection. *Am. J. Physics*. 41:1008–1010. https://doi.org/10.1119/1.1987451

Hecht, E. 2017. Optics (5th Edition). Pearson Education Limited, Essex, England.

iBiology: Microscopy Series. 2012 – 2015. Fluorescence microscopy. https://www.ibiology.org/online-biology-courses/microscopy-series/fluorescence-microscopy/

Truong, T.V., et al 2011. Deep and fast live imaging with two-photon scanned light-sheet microscopy. *Nature Methods*. 8:757–760.

HOMEWORK PROBLEMS

7.1 [TIR – Fresnel Equations] Derive Eq. (B7.1.2) starting from Eq. (B7.1.1).

7.2 [TIR – Application] A point source of light emitting in all directions is located 3 m below the surface of water. Calculate the area of the illuminated circle at the surface of the water, assuming that the light from the source emerges from water (n = 1.33) and enters air (n = 1.00).

7.3 [TIR – Dependence on Refractive Index] Calculate the critical angles and penetration depths for coverslips made from silica, glass, and sapphire. Assume that 550-nm light is incident from the coverslip into an aqueous medium (n = 1.33) and that the incidence angle $\theta = 75°$. Which material would best ensure TIR?

Material	Refractive Index	Critical Angle	Penetration Depth
Silica	1.46 at 550 nm		
Glass	1.52 at 550 nm		
Sapphire	1.77 at 550 nm		

7.4 [TIR – Dependence on Wavelength] Calculate the critical angles and penetration depths for glass coverslips at 400 nm, 550 nm, and 700 nm. Assume that the light is incident from the coverslip into an aqueous medium (n = 1.33) and that the incidence angle $\theta = 75°$. Which wavelength has the largest angular range that can be used to create TIR? What are the implications for multi-color imaging?

Wavelength	Refractive Index	Critical Angle	Penetration Depth
400 nm	1.531 in glass		
550 nm	1.519 in glass		
700 nm	1.513 in glass		

7.5 [Estimating Relative Photodamage in LSFM] One major advantage of LSFM is reduced specimen photodamage. The purpose of this problem is to quantify this advantage approximately for the case of LSFM of *Drosophila* embryos.

a. What is the typical thickness of a *Drosophila* embryo? You should be able to find this number on the Internet.
b. Assume that, in an LSFM experiment, a *Drosophila* embryo is illuminated with a light sheet having an average width of 3 μm. Also assume that, in a WF experiment, the embryo is illuminated uniformly throughout its depth with light of similar intensity. Use these assumptions to estimate the relative light exposure in the two types of experiments.

7.6 [Diverging Cylindrical Lenses] Panel (a) in the figure contains a ray diagram that reiterates the focusing attributes of a converging cylindrical lens (Fig. 7.6). Draw an appropriate ray diagram that shows the location of the line focus for the diverging cylindrical lens in Panel (b).

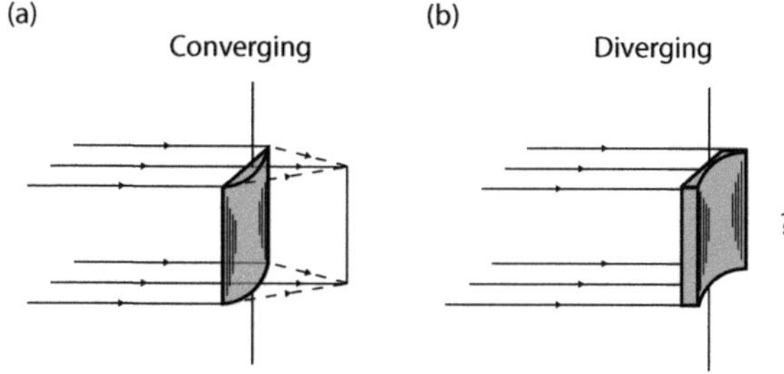

7.7 [Astigmatism – Cylindrical Lenses] Astigmatic corneas are steeper along one axis, so refractive power is different along different directions. The figure at right shows the most common type of astigmatic cornea, which has the greatest curvature in the vertical direction. You plan to correct the problem with a converging cylindrical lens that will equalize power along the vertical and horizontal directions. Draw an appropriately oriented cylinder and explain your logic.

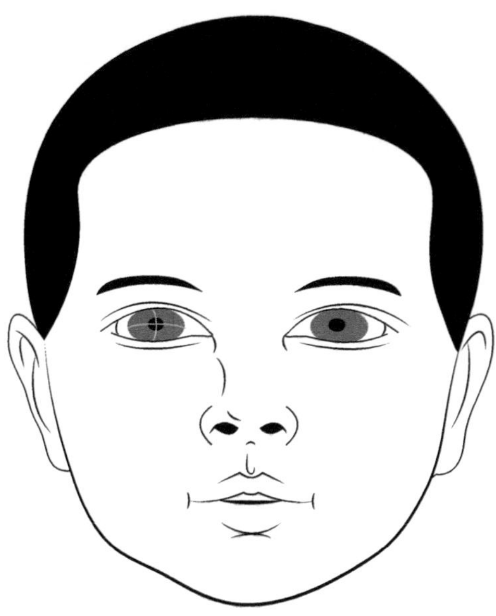

7.8 [Multi-photon Microscopy] Multi-photon microscopy involves simultaneous absorption of two or more lower energy photons to induce the same transition as a single higher energy photon. Assume that a single 500-nm photon will induce an absorption transition in a molecule. If you want to induce the same transition using three identical lower energy photons, what wavelength should you choose for these photons? Justify your choice in a sentence!

7.9 [Focusing and Absorption in Two-Photon Microscopy] Two-photon excitation is confined to the laser focus because the intensity is much higher at focus, and so too is two-photon absorption. To make this statement quantitative, implement the following calculations:

a. Assume that a 1-mm diameter laser beam is focused to a 500-nm diameter spot. What is the increase in intensity at the focal point?
b. Invoke the fact that two-photon absorption is proportional to intensity squared. What is the increase in absorption at focus?
c. Repeat the calculations in parts (a) and (b) to determine the decrease in the two-photon absorption probability (relative to focus) at a point in the sample where the beam diameter is 50 μm.

7.10 [Two-Photon Dichromatic Mirror] Fig. 7.10 shows two-photon absorption and emission spectra for GFP. Sketch a representative reflection/transmission profile for a dichromatic mirror suitable for use in two-photon microscopy with this fluorophore. Assume that the mirror passes excitation light from the light source onto the sample and reflects emitted light from the sample toward the detector, as shown in Fig. 7.13.

7.11 *[Beam Waist] A laser is used to excite fluorescence from a flat two-dimensional surface. The laser is focused through a microscope objective and has a Gaussian beam profile with waist w_1 as described in Box 7.2. The surface is uniformly labeled with fluorophores. The observed fluorescence signal from the illuminated region is F_1 (in photons/s). Suppose that the Gaussian beam profile is changed to waist w_2 without changing the laser power. (a) Use your intuition to guess the new total signal F_2 in terms of the parameters given. Explain your logic. (b) Confirm (or disprove) your guess by integrating Eq. (B7.2.2) over area.

7.12 *[Fluorescence – Axial Variation] (a) For the case of one-photon imaging, show that the total fluorescence signal from an axial plane excited by a Gaussian laser beam does not vary with the plane's axial distance, z, from focus. (b) Confirm the result in Eq. (7.8) via integration using Eq. (B7.2.2).

7.13 *[Two-Photon – Lateral Resolution] Prove that the effective spot size in two-photon microscopy is $\sqrt{2}$ larger than the spot in LSCM. This, in turn, implies a theoretical disparity in lateral resolution of $\sqrt{2}$, with performance in two-photon being worse.

8

Super-Resolution Fluorescence Techniques

In the previous chapter, we discussed one approach, TIRFM, that is capable of significantly surpassing the diffraction limit on resolution in the axial direction but has conventional lateral performance. In this chapter, we follow up with cutting-edge super-resolution fluorescence methods that significantly surpass the diffraction barrier in both the lateral and axial directions.

We start with super-resolution structured illumination (SR-SIM), which can be implemented using most standard fluorescent samples. In contrast, the techniques that follow – stimulated emission depletion (STED) and photoactivated localization microscopy (PALM) – are somewhat less flexible with respect to sample because they rely on switching between optically distinguishable states (Section 8.2). We devote the most space to SR-SIM, in part, because its theoretical foundations are mathematically complex. In addition, the theoretical foundations of SR-SIM are strongly linked to Fourier analysis, which will figure prominently in our upcoming discussions of image processing and analysis, and of medical imaging techniques, especially MRI.

8.1 Super-Resolution Structured Illumination Microscopy (SR-SIM)

SR-SIM uses nonuniform "structured" illumination (SI), usually a grid of lines, to improve resolution. Standard SR-SIM optimally achieves resolution doubling in all three dimensions, while nonlinear SR-SIM approaches can yield further improvements. Because the mathematical foundations of SR-SIM are complex, we focus in the main text on a conceptual and pictorial analysis of the approach. More rigorous mathematical details are presented in Box 8.1.

8.1.1 Mechanism of Resolution Enhancement

Resolution enhancement in SR-SIM stems from the fact that the diffraction limit applies directly to the emitted fluorescence signal, $F(\vec{r})$. However, the fluorophore density, $\rho(\vec{r})$, is the experimental parameter of real structural interest. By exploiting SI, ρ can be determined to higher resolution, which is the goal of SR-SIM. The important parameter that links F and ρ is intensity, I, as given by

$$F(\vec{r}) \propto \rho(\vec{r}) \times I(\vec{r}) \quad (8.1)$$

For Köhler illumination, I = constant, whereas for SR-SIM, I varies with position, typically in a grid pattern (Fig. 8.1).

To cultivate intuition about the utility of grid illumination, consider two tiny objects with equal fluorophore concentration that are more closely spaced than the resolution limit. Under uniform illumination, the image of the two objects will be a single blur. In contrast, under grid illumination (e.g., in the shape of a sine wave), one of the two objects will generally be slightly brighter than the other because it falls at a different position on the sine wave. Moreover, shifting the grid alignment will alter the asymmetry, revealing that the blur arises from a compound object. Comparison of images obtained with different shifts allows reconstruction of finer object detail.

A pictorial and semi-quantitative understanding of SR-SIM can be obtained from two alternative, but equivalent, views of diffraction-limited resolution. The most familiar is that features separated by less than the diffraction limit, d_{min} ~0.2 μm, cannot be resolved. Based on this view, the effect of SI can be understood by appealing to an analogy with moiré patterns.

Consider two closely spaced (high-frequency) patterns, one representing the object density and the other the illumination (Fig. 8.1). Superposition of these will lead to a moiré pattern, representing the fluorescence signal, that contains a wider spacing. The key point is that this coarser pattern might be detectable if its spacing exceeds the diffraction limit, even if the underlying patterns cannot be detected. In this case, the known illumination profile and the observed moiré pattern can be used to detect otherwise unresolvable detail in the specimen. The "pattern free" super-resolution image is reconstructed from the raw patterned images using computers.

To quantify the effect semi-rigorously, we invoke a frequency-domain view of diffraction-limited resolution. This view is founded on Fourier's idea that a signal, such as an image, can be written as a weighted sum of sine and cosine functions with different frequencies (Box 4.3). In the signal's Fourier "decomposition," higher frequency terms encode finer details (Fig. 8.2).

DOI: 10.1201/b22076-9

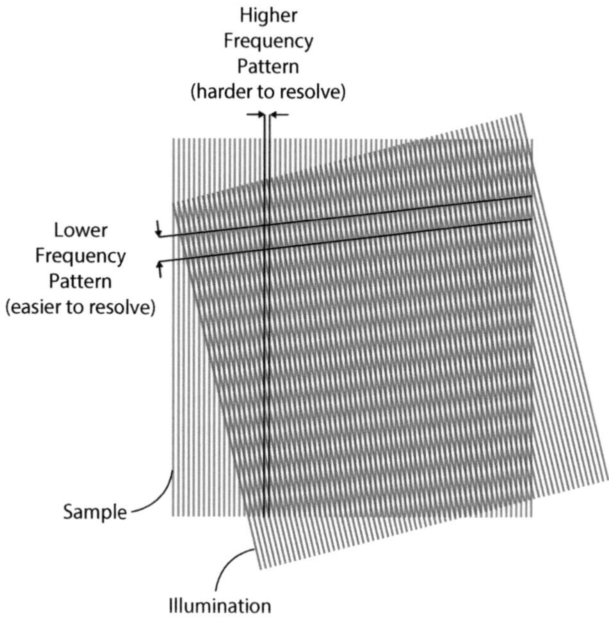

FIGURE 8.1
Moiré effect revealing the production of a more coarsely spaced pattern from superposition of two more finely spaced patterns.

Thus, image resolution will improve if the range of frequencies in the decomposition is extended to higher values (Section 4.1.2).

Armed with the Fourier perspective, consider two sinusoidal functions of time with frequencies ν_1 and ν_2, again representing analogs of the illumination and object. Their superposition will generate a time analog of a moiré pattern – a beat pattern – that similarly contains a lower frequency component (Fig. 8.3). Specifically, the superposition will have a slowly varying envelope that oscillates at a frequency $\nu_{beat} = |\nu_2 - \nu_1|$ (Box 8.2).

To see what a beat analogy implies about resolution enhancement, let ν_{ill} and ν_{obj} represent, respectively, an illumination frequency and an object frequency. Like ν_1 and ν_2 above, these will generate a beat frequency, ν_{beat}, and with $\nu_{obj} > \nu_{ill}$, the object frequency ν_{obj} is given by

$$\nu_{obj} = \nu_{beat} + \nu_{ill} \tag{1.2}$$

We want to quantify the largest accessible ν_{obj} for standard epi-fluorescence microscopy. To this end, we let

BOX 8.1 MATHEMATICAL FOUNDATIONS OF SR-SIM

Our discussion assumes considerable familiarity with convolutions, Fourier transforms, and the Dirac delta function, δ.

The fundamental idea underlying SR-SIM, as mentioned in the main text, is that the diffraction limit on resolution applies to the emitted fluorescence signal, $F(\vec{r})$. In contrast, the fluorophore density, $\rho(\vec{r})$, which is the real structural parameter of interest, can be determined to higher resolution using SI. Alternatively stated, previously inaccessible, high-frequency information can be extracted using SI.

To verify this statement, we start by Fourier transforming Eq. (8.1) in the main text, which expresses $F(\vec{r})$ as a product of $\rho(\vec{r})$ and the intensity, $I(\vec{r})$. Invoking the fact that Fourier transformation converts multiplication into a convolution, we obtain

$$\tilde{F}(\vec{k}) = \tilde{\rho}(\vec{k}) \otimes \tilde{I}(\vec{k}) \tag{B8.1.1}$$

In Eq. (B8.1.1), we have used a tilde to represent the Fourier transform, the symbol ⊗ to represent convolution, and the letter k to denote spatial frequency (Box 4.3). In addition, we have suppressed constants of proportionality.

An appropriate choice for the SI intensity profile is a cosinusoidal pattern given by

$$I(r) = I_0\left[1 + \cos\left(\vec{k}_0 \cdot \vec{r} + \varphi\right)\right]$$

The Fourier transform, $\tilde{I}(\vec{k})$, of this illumination profile can be computed by rewriting the cosine function in terms of complex exponentials. The result is

$$\tilde{I}(\vec{k}) = I_0\left[\delta(\vec{k}) + \frac{1}{2}\delta(\vec{k} + \vec{k}_0)e^{i\varphi} + \frac{1}{2}\delta(\vec{k} - \vec{k}_0)\,e^{-i\varphi}\right] \tag{B8.1.2}$$

Inserting Eq. (B8.1.2) into the integral expression for the convolution (Box 6.2) and using the "sifting" property of the Dirac delta function yields

$$\tilde{F}(\vec{k}) = I_0\left[\tilde{\rho}(\vec{k}) + \frac{1}{2}\tilde{\rho}(\vec{k}+\vec{k}_0)e^{i\varphi} + \frac{1}{2}\tilde{\rho}(\vec{k}-\vec{k}_0)e^{-i\varphi}\right] \tag{B8.1.3}$$

Equation (B8.1.3) shows that the Fourier transform of the fluorescence signal, $\tilde{F}(\vec{k})$, contains information about the fluorophore density at previously inaccessible, high-spatial frequencies, e.g., $\vec{k}+\vec{k}_0$, but the information is encoded in a superposition of three terms. To extract this high-frequency information, three images can be collected at three different phase angles, φ, and the resulting three equations in three unknowns solved algebraically for $\tilde{\rho}(\vec{k})$, $\tilde{\rho}(\vec{k}+\vec{k}_0)$, and $\tilde{\rho}(\vec{k}-\vec{k}_0)$. This procedure will extend frequency information along one direction (Fig. 8.5 in the main text). If data are generated for two additional orientations of $\vec{k}_0$, it is possible to further fill out frequency space (Section 8.1.2).

ν_{beat} be the diffraction-limited, maximum detectable frequency in the fluorescence signal and similarly let ν_{ill} be the diffraction-limited, maximum frequency in the illumination pattern.[1] The largest accessible frequency is then given by

$$\nu_{obj}^{max} = \nu_{beat} + \nu_{ill} = \nu_{max} + \nu_{max} = 2\nu_{max} \tag{8.2}$$

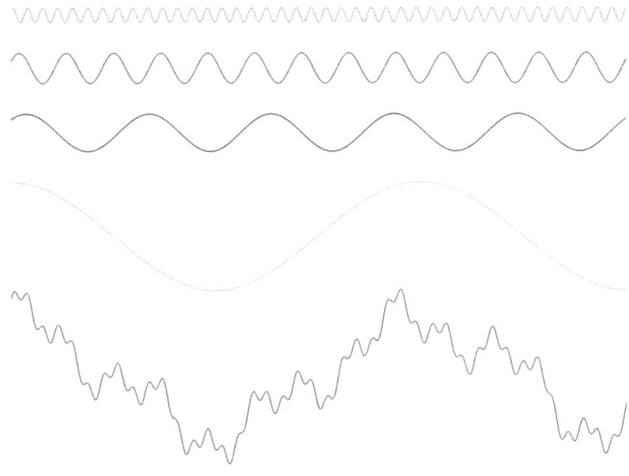

FIGURE 8.2
Fourier decomposition of a complex signal into sine and cosine functions. The bottom function is the sum of the four above it. In general, larger scale details are encoded by lower frequencies, and finer details by higher frequencies. This is visually clear in the figure. It is also consistent with the fact that a constant, which has no variation in detail, is mathematically represented by a zero-frequency cosine term. Similarly, finer details, which are characterized by properties like larger slopes, are described mathematically using higher frequency sine and cosine functions because the largest slope of a sine/cosine function is proportional to its frequency.

[1] Significantly better frequency enhancement can be obtained using much more finely spaced illumination generated using intersecting TIR beams; however, this improvement comes with a loss in imaging depth.

In writing Eq. (8.2), we have neglected the difference between excitation and emission wavelengths, which causes the two maximum frequencies to differ somewhat. Equation (8.2) implies a twofold enhancement in detectable frequencies and thus a twofold enhancement in spatial resolution. The resolution enhancement is actually slightly better due to the neglected wavelength effect. A more rigorous derivation is sketched in Box 8.1.

8.1.2 Experimental Implementation

The implementation of SR-SIM is less complex than the theoretical foundations, although not all of the details will be fully apparent from our simplified treatment in the main text. There is, of course, a need for SI, and one option is a cosinusoidal stripe pattern with the mathematical form (Box 8.1)

$$I(r) = I_0\left[1+\cos\left(\vec{k}_0\cdot\vec{r}+\varphi\right)\right] \tag{8.3}$$

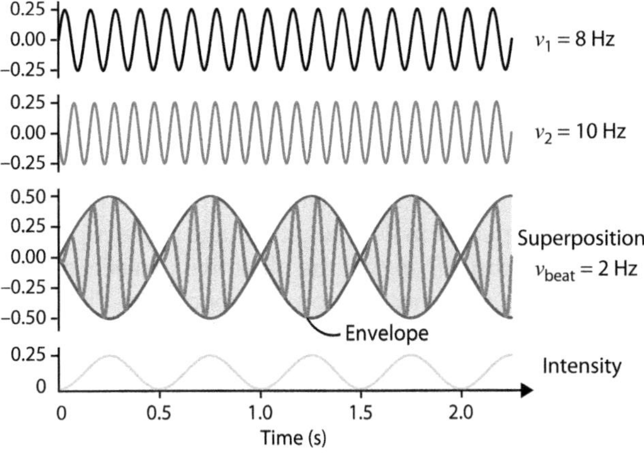

FIGURE 8.3
Beats (lower signal) produced from the superposition of two periodic signals of slightly different frequencies. The beat *intensity* pattern oscillates with a low frequency given by the difference in frequencies of the two underlying signals.

BOX 8.2 BEATS ANALYSIS

Beats are a familiar example of generating a lower frequency pattern from superposition of higher frequency patterns. To demonstrate the effect mathematically, we consider the superposition of two equal-amplitude sinusoidal wave patterns with slightly different angular frequencies, ω_1 and ω_2. At a particular point (e.g., $x = 0$), the individual waves have the functional form (Eq. [2.1])

$$y(x=0,t) = y_0 \sin(\omega t)$$

and their sum is

$$y_{\text{total}}(t) = y_0 \sin(\omega_1 t) + y_0 \sin(\omega_2 t)$$

This superposition can be rewritten in a more illustrative form as

$$y_{\text{total}}(t) = 2y_0 \cos\left[\left(\frac{\omega_1 - \omega_2}{2}\right)t\right] \sin\left[\left(\frac{\omega_1 + \omega_2}{2}\right)t\right] \tag{B8.2.1}$$

using the identity

$$\sin\theta_1 + \sin\theta_2 = 2\cos\left(\frac{\theta_1 - \theta_2}{2}\right)\sin\left(\frac{\theta_1 + \theta_2}{2}\right)$$

The wave in Eq. (B8.2.1) has a slowly oscillating cosine envelope, which is the lower frequency pattern that arises from superposition of the higher frequency patterns. The angular frequency of the envelope is $(\omega_1 - \omega_2)/2$. In music, the envelope gives rise to a beat tone. The beat frequency actually is twice the envelope frequency because the beat tone is determined by intensity, which is related to the square of the wave pattern (Fig. 8.3). We thus have

$$\omega_{\text{beat}} = |\omega_2 - \omega_1|$$

In the main text, we used true frequency, ν, in Hz instead of angular frequency, ω, in rad/s.

To maximize resolution enhancement, the spatial frequency of the striped illumination is set at the diffraction limit.

One method of generating SI is to create an interference pattern. An alternative is shown in Fig. 8.4. A grating is placed in a conjugate image plane, upstream from the sample, so that the objective projects a minified image of the grating on the specimen. This latter approach has several advantages, including compatibility with the use of spatially and temporally incoherent light.

Final super-resolution SR-SIM images are generated computationally from nine (or more) associated raw images. The reason nine images are needed is somewhat subtle; it is outlined schematically in Fig. 8.5 and discussed more rigorously in Box 8.1. In essence, SR-SIM data contain frequency information that is a superposition of three contributions. The three contributions are separated mathematically by taking three pictures corresponding to three phases, $\varphi = 0°, 120°, 240°$, and algebraically solving the resulting "three equations in three unknowns." The phase of the illumination pattern is varied by translating the grating laterally across the optic axis.

The three-phase procedure enhances resolution in one direction (Fig. 8.5). To fill out space further, three grating orientations are used. These can be generated by rotating the grating in Fig. 8.4 about the optic axis.

SR-SIM also can be extended to three-dimensional imaging using an illumination pattern that varies both laterally and axially, yielding resolutions of ~100 nm and ~300 nm in the lateral and axial directions, respectively. A comparison of SR-SIM and confocal images is shown in Fig. 8.6.

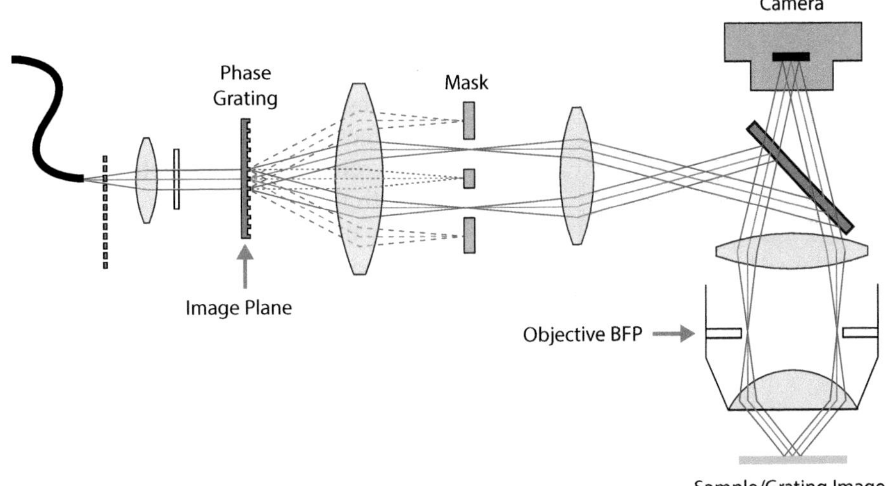

FIGURE 8.4
Schematic of an SR-SIM microscope. Incoherent light from a scrambled laser impinges on a phase grating, which is mounted to allow both translation and rotation. The light is diffracted into multiple orders, but only $m = \pm 1$ are used to generate the illumination pattern. (Adapted with permission from Gustafsson, M.G.L., et al. 2000. Doubling the lateral resolution of wide-field fluorescence microscopy using structured illumination. *Proc. SPIE 3919, Three-Dimensional and Multidimensional Microscopy: Image Acquisition Processing VII.* https://doi.org/10.1117/12.384189.)

8.2 Sequential Readout-Based Approaches

Super-resolution also can be achieved with techniques that sequentially map the positions of very closely spaced fluorophores. The basic idea is that fluorophores that are more closely spaced than the diffraction limit cannot be resolved when visualized simultaneously due to PSF/signal overlap, but they can be resolved when visualized at separate times because overlap is avoided. We will classify approaches that exploit this "loophole" in the diffraction limit as sequential readout approaches.

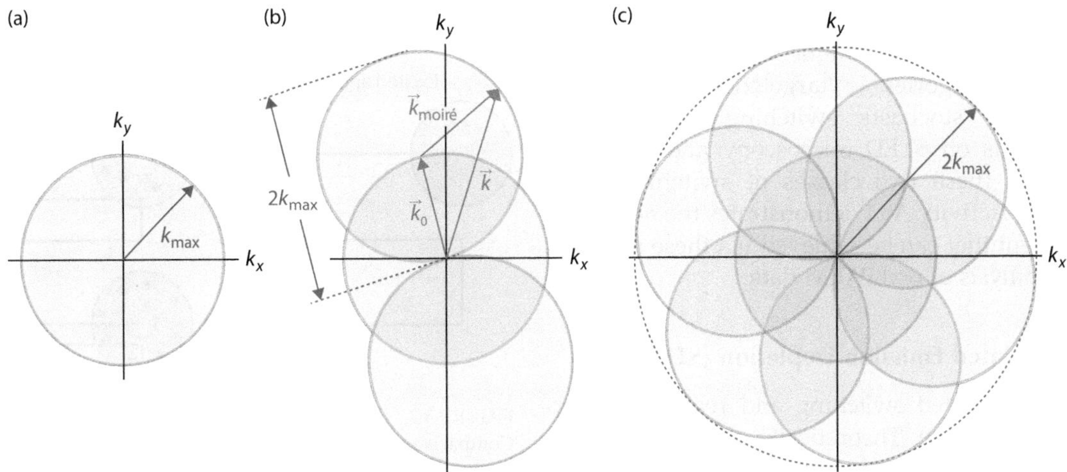

FIGURE 8.5
Accessing frequency information beyond the Abbe limit using SI. (a) The conventional Abbe spatial frequency information is contained in a circle of radius $k_{max} = 2\text{NA}/\lambda$ centered on the origin, as discussed in Chapter 4. When the sample is illuminated with light containing the spatial frequency $\vec{k}_0$, each sample frequency $\vec{k}$ will generate moiré fringes at the difference frequency, $\vec{k}_{moiré} = \vec{k} - \vec{k}_0$, similar to beats. These fringes are detectable if $|\vec{k}_{moiré}| < k_{max}$. (b) This restriction implies that observable $\vec{k}$ must fall within a circle of radius k_{max} centered on the tip of $\vec{k}_0$. Thus, the maximum detectable $\vec{k}$ along the direction of $\vec{k}_0$ has magnitude $k_0 + k_{max}$. With the additional requirement that $k_0 < k_{max}$, the maximum observable frequency has magnitude $2k_{max}$. (c) To more uniformly fill out frequency space, two additional orientations are generated by rotating the grating to change the direction of $\vec{k}_0$. The outcome is a doubling of observable spatial frequencies, with a concomitant doubling of spatial resolution. In practice, achieving isotropic lateral resolution is not trivial and may involve modifications such as collecting more images.

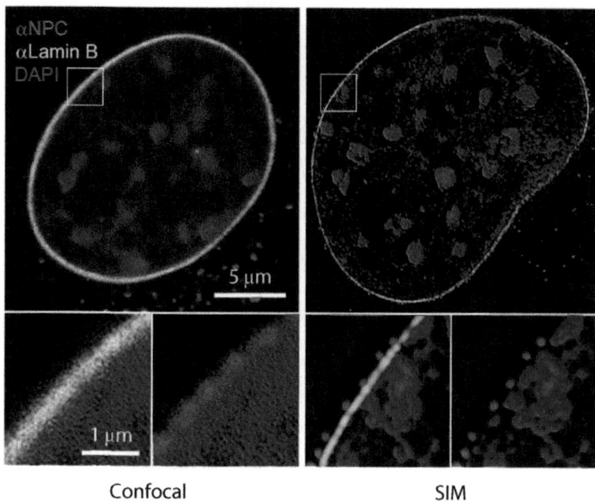

FIGURE 8.6
Comparison renderings of SR-SIM and confocal images of mouse nuclei. Bars = 1 μm and 5 μm. (Adapted with permission from Schermelleh, L., et al. 2008. Subdiffraction multi-color imaging of the nuclear periphery with 3D structured illumination microscopy. *Science* 320:1332–1336. DOI: 10.1126/science.1156947.)

The sequential readout approaches rely on dyes that can be switched between two spectrally distinguishable states (e.g., fluorescing/nonfluorescing). Spectrally distinct states eliminate signal overlap between nearby fluorophores/objects by ensuring that "on" (bright) fluorophores are surrounded by nearby fluorophores that are "off" (dark). Different molecules in the sample are placed in the bright state over time, so super-resolution images of extended specimens can be generated.

Sequential readout techniques fall into two broad mechanistic categories – "targeted switching and readout" and "stochastic switching and readout." We will focus on STED microscopy and PALM as examples of these two classes of switching-based approaches. Activity A8.1 demonstrates the resolution enhancement that can be achieved via these methods through analysis of real PALM data.

8.2.1 Stimulated Emission Depletion (STED)

STED is a "targeted switching and readout" super-resolution technique. Theoretically, there is no lower limit on STED resolution, but currently realizable resolutions are ~20 nm in the lateral direction and ~40 nm in the axial direction. Among super-resolution techniques discussed here, STED is unique in requiring no post-acquisition processing to overcome the diffraction barrier.

8.2.1.1 Mechanism of Resolution Enhancement

STED commonly is implemented as a point-scanning technique; thus, its enhanced resolution is perhaps best exemplified via comparison with LSCM. To this end, we start with a simplistic, but illustrative, analogy. In STED, stimulated emission is used to deplete fluorescence from the periphery of a conventional diffraction-limited scanning spot (Section 8.2.1.2). The remaining undepleted (fluorescing) region has a width that is roughly an order of magnitude smaller than that of a typical scanning spot used to excite fluorescence in LSCM (Fig. 8.7). If we think of the excitation spot as a paint brush, the enhanced resolution obtained with a narrower STED spot can be likened to painting with a narrower brush.

To compare the resolution of STED and LSCM more rigorously, we refer to Fig. 8.7. The figure illustrates the fact that, in both methods, fluorophores are excited by a spot, which impinges on a known position, r_i, in the sample. The important difference is that the excitation spot width, Δr, in STED is engineered to be much smaller than the diffraction-limited spot used in LSCM. The resulting signals are monitored with a point detector (Chapter 9), such as a PMT, and are assigned to the range of positions occupied by the fluorescing molecules, $r_i \pm \Delta r/2$. The spot is scanned over the sample to build up an image, and the sequentially collected signals map features to a resolution, Δr. Notably, the much smaller spot width in STED leads to markedly enhanced resolution (Fig. 8.8 and HW 8.7).

As mentioned briefly above, stimulated emission plays a key role in reducing the dimensions of the spot

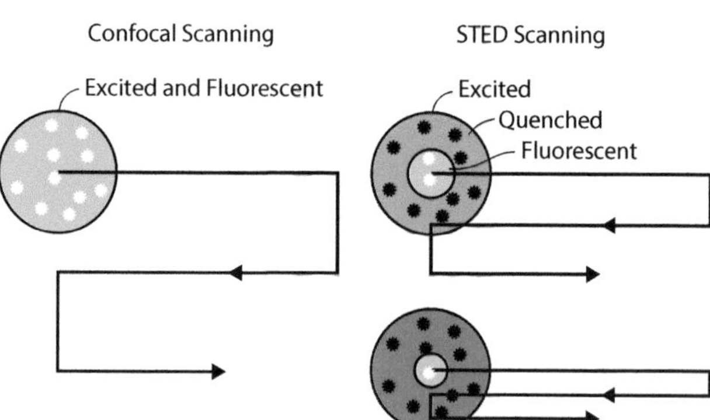

FIGURE 8.7
Comparison of confocal and STED scanning. The left panel shows a hypothetical example of confocal scanning using a diffraction-limited spot with width ~0.5 μm. All of the fluorophores in the confocal spot are excited simultaneously; as a consequence, their signals overlap, and the fluorophores are unresolvable. The right panel shows the analogous case of STED scanning using an unquenched spot with a width that has been significantly reduced by switching peripheral fluorophores off using stimulated emission. Again, fluorophores in the unquenched region are unresolvable. However, the resolution is much better because the unquenched STED spot is much smaller. The two examples show that is possible to tune the size of the unquenched spot. A more intense quenching (red) beam produces a smaller unquenched fluorescing (green) spot and thus higher resolution.

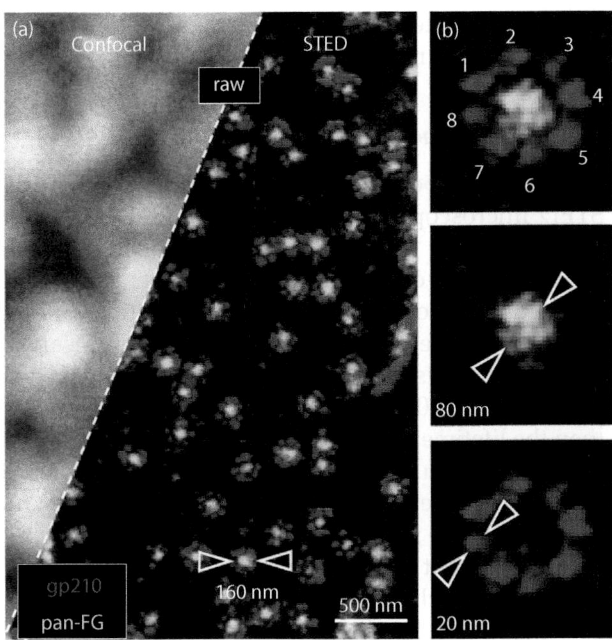

FIGURE 8.8
(a) Comparison of two-color confocal and STED images of nuclear pore complexes (NPCs) in a *Xenopus laevis* epithelial cell line. (b) STED readily reveals the presence, ratio, and arrangement of gp210 (red) and pan-FG (green) subunits within the NPCs. (Adapted from Göttfert, F., et al. 2013. Coaligned dual-channel STED nanoscopy and molecular diffusion analysis at 20 nm resolution. *Biophys. J.* 105:L01–L03. DOI: 10.1016/j.bpj.2013.05.029.)

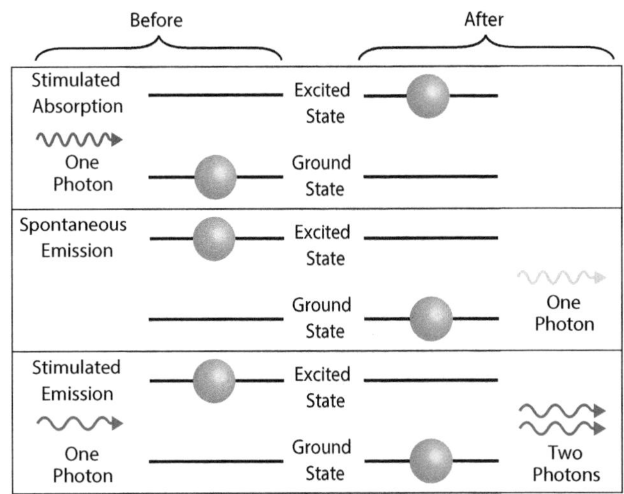

FIGURE 8.9
Comparison of stimulated absorption, and spontaneous and stimulated emission. Stimulated processes require an input photon. Spontaneous and stimulated emission processes output one and two photons, respectively.

that generates fluorescence in STED. We address this key issue next.

8.2.1.2 Spot Engineering via Stimulated Emission

To set the stage for a discussion of stimulated emission, we first review key aspects of two closely related processes – "stimulated" absorption and "spontaneous" emission.

Stimulated absorption is the excitation process that we encountered in Chapters 6 and 7; it occurs when a molecule in a low-energy state is stimulated into a higher energy state by absorption of a photon with energy matched to the difference in energy between the two states (Eq. [7.6] and Fig. 8.9). Spontaneous emission (e.g., fluorescence) similarly is the radiative follow-up that we encountered in Chapters 6 and 7. Spontaneous emission occurs when the excited molecule releases its energy without any outside stimulation. Each type of fluorophore has a characteristic spontaneous emission time, which is known as the fluorescence lifetime. Fluorescence lifetimes are on the order of nanoseconds and represent the average time spent in the excited state before spontaneous release of a photon (HW 8.8).

Stimulated emission is another possible follow-up to stimulated absorption. However, several conditions must be met to produce stimulated emission, so it is much less probable at higher frequencies (HW 8.11). Notably, the molecule must be in an excited state, and it must interact with a photon having energy that is exactly matched to the energy that would be released via spontaneous emission. If stimulated emission occurs, an additional photon is created that is identical to the input photon (Fig. 8.9). The energy of this "output" photon comes from the energy lost by the excited state.

In the context of STED, the key attribute of stimulated emission is that it can be used to deplete (switch off) the excited state and thus quench fluorescence. A beam with high peak intensity generates rapid stimulated emission that knocks molecules from their excited state before they can spontaneously fluoresce. In particular, in STED, depletion by stimulated emission is used to reduce the dimensions of the scanning spot by superimposing a donut-shaped "depletion beam," which is matched to the fluorophore's emission spectrum, on a standard, diffraction-limited excitation beam. Fluorescence from the periphery is quenched by the donut via the process of stimulated emission. In contrast, fluorescence that can be used to map the specimen is preserved in a small central region where the donut intensity is zero (Fig. 8.7).

Another important attribute of stimulated emission is that fluorescence quenching via stimulated emission can be achieved using photons that are tuned to the far-red end of the fluorophore's emission spectrum. These red-shifted photons are not energetic enough to re-excite the quenched fluorophores after vibrational relaxation.

A final key attribute is that the stimulated emission photon travels in the same direction as the initial photon (because the two are identical) and thus also moves away from the detector. Emission filters are needed only to exclude back-scattered depletion and excitation photons, so only useful fluorescence from the center of the spot, which is of different wavelength, is sent on to the detector (Fig. 8.10).

Several additional points are worthy of note. First, quenching by stimulated emission must dominate competing spontaneous decay processes, notably fluorescence emission. This requires a quenching pulse length that is shorter than the fluorescence lifetime and quenching beam intensities on the order of tens to hundreds of MW/cm^2 (HW 8.7 and 8.12). As a consequence, photodamage and sample heating can be a concern. Second, the size of the fluorescing spot can be made progressively smaller by increasing the intensity of the donut beam. Specifically, increasing the donut intensity saturates fluorescence depletion in regions that are ever closer to the donut center (where fluorescence remains unquenched – Fig. 8.7). Third, metastable bright and dark states of reversibly switchable fluorescent proteins can be used to implement STED-like imaging at much lower light intensities.

8.2.1.3 Experimental Implementation

The key components of a typical point-scanning STED microscope are shown in Fig. 8.11. Excitation and depletion of the excited state typically are achieved using overlapping (pulsed) laser beams that are focused onto the sample via the objective lens. An excitation pulse of very short duration (typically a few tens of ps) is followed by a depletion pulse with a duration (e.g., 200–300 ps) that is significantly shorter than the lifetime of the excited state. The time delay, Δt, between the pulses must be short enough to ensure that fluorophores are still in the excited state when the depletion pulse is incident on the sample.

An important attribute of the depletion beam is its donut shape, which allows quenching in the periphery of the excitation beam. A donut shape can be generated using a phase plate.

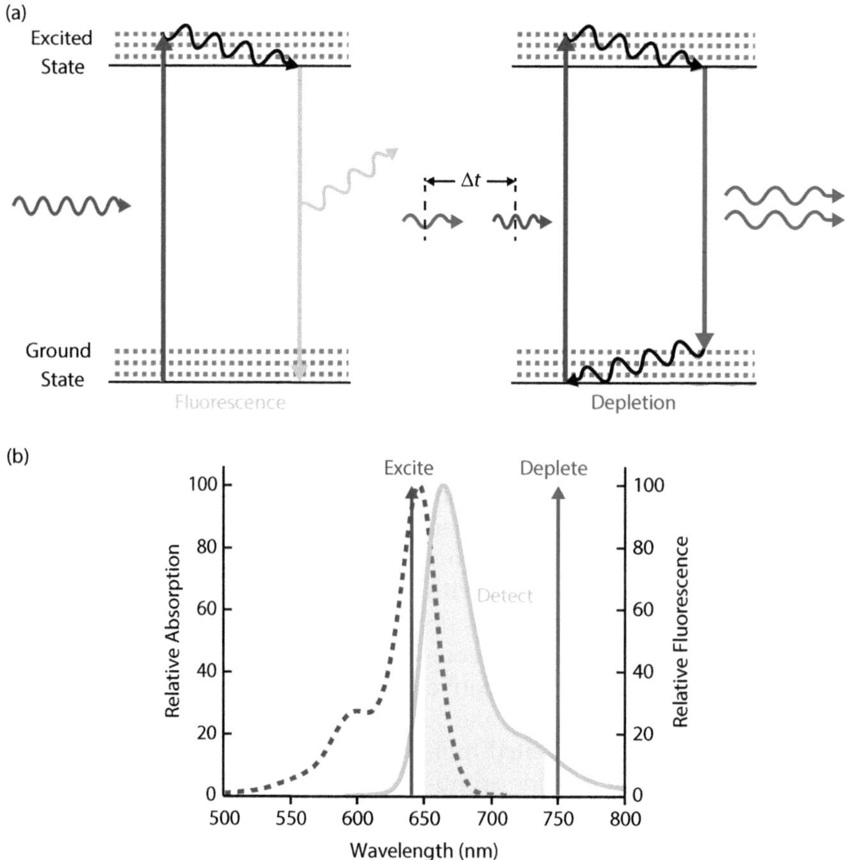

FIGURE 8.10
(a) Jablonski diagrams for spontaneous (left) and stimulated emission (right). (b) Excitation and emission spectra of the popular STED dye, Atto647N. Vertical arrows on the spectra show the optimal wavelengths for excitation (640 nm) and quenching/depletion lasers (750 nm), and the filled region represents the emission regime detected. The depletion wavelength is outside the detection regime and in a regime that minimizes excitation from the ground state. (Data for Panel (b) exported from ATTO-TEC GmbH [www.atto-tec.com].)

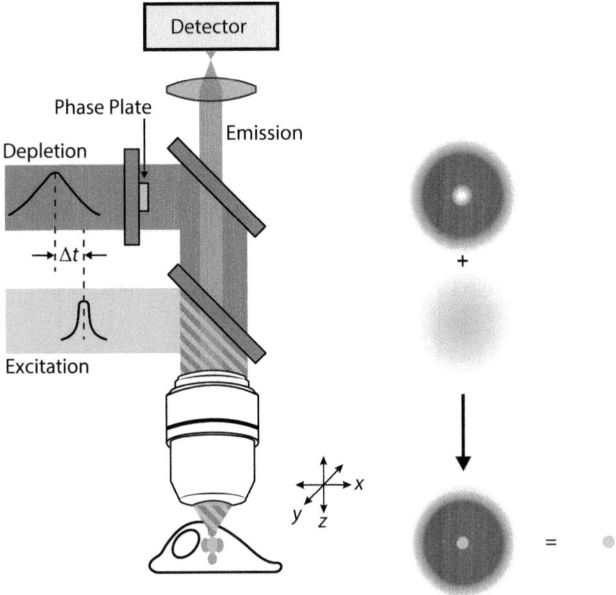

FIGURE 8.11
Schematic of a STED microscope. Synchronized trains of visible (excitation) and quenching (near IR) laser pulses impinge on the sample. Each brief excitation pulse is followed by a quenching pulse that is donut-shaped. The signal can be detected using an avalanche photodiode.

8.2.2 Photoactivated Localization Microscopy (PALM)

PALM is a "stochastic molecular switching and readout" super-resolution technique. Realizable lateral and axial resolutions are similar to those from STED (Table 8.1). Important differences from STED include collecting signals from individual molecules instead of ensembles, an absence of scanning, and mandatory post-acquisition image processing.

8.2.2.1 Mechanism of Resolution Enhancement

The essence of PALM is shown in Fig. 8.12. The sample is labeled with molecules that can be photo-switched between spectrally distinguishable states, which, for the sake of concreteness, we take to be fluorescing and nonfluorescing. We also assume that the sample initially is in the dark state.

To generate a fluorescence signal from the dark sample, a transient switching beam, such as a 405-nm laser line, stochastically turns on a subset of the dye molecules (Fig. 8.12). The activated subset must be sparse enough to ensure that all bright molecules are spatially separated by more than the diffraction limit on resolution. This avoids PSF overlap in the image and ensures that all "on" fluorophores can be resolved. Sparse activation is achieved using a switching beam of appropriately low intensity.

After photo-switching, the sample is illuminated with excitation (also known as readout) light generated using a laser line in the visible, and the activated dye molecules fluoresce until they bleach and turn off. The resulting PALM "sub-image" consists of multiple widely spaced PSFs, which alone typically do not generate an image with suitable structural detail (Fig. 8.12). To create the requisite detail, tens of thousands of sub-images are generated by repeating the activation/excitation/bleaching cycle, with the desired resolution influencing the required number of sub-images (Section 8.2.2.3). The sub-images are processed and superimposed to assemble a final PALM image that is "pointillistic," i.e., created using small dots.

8.2.2.2 Image Processing

In unprocessed PALM sub-images, fluorophores are visible as PSFs that have broad spatial extent, and thus fluorophore coordinates are relatively poorly defined. To circumvent this problem, PALM sub-images are processed, and individual fluorophore coordinates are much more precisely specified (e.g., to within ~10 nm along each lateral direction). Key steps include fitting the PSF to a Gaussian function and determining the best fit fluorophore coordinates (Fig. 8.13). This procedure replaces a diffraction-limited spot by a spot with much smaller width that is set by the uncertainty, $\sigma_{x/y}$, in the determination of the coordinates.

Coordinate uncertainty, $\sigma_{x/y}$, influences resolution and thus is an important quantity in PALM experiments. This uncertainty depends on several

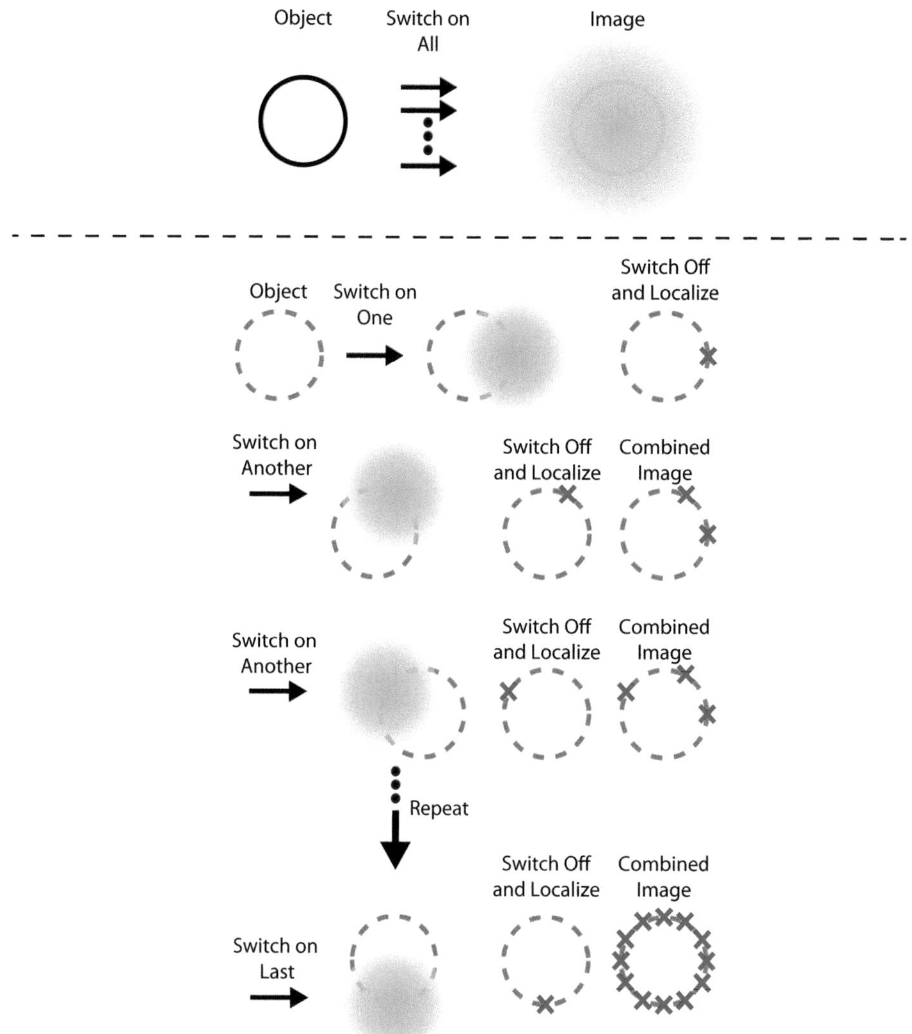

FIGURE 8.12
The essence of single-molecule super-resolution imaging. A small object (shown here as a black ring) is densely labeled with fluorophores that can be switched between "on" and "off" (nonfluorescent) states. In a conventional fluorescence image (represented by the blurry green spot), finer details in the structure of the object are obscured by overlapping diffraction patterns created by closely spaced "on" fluorophores. However, if a subset of fluorophores is stochastically switched "on" at sufficiently low density to avoid overlap, their coordinates can be determined with high precision (indicated by the red × marks). With repeated activation, localization, and bleaching (turning "off") of different fluorophores, sub-diffraction-limited structural features of the object can be resolved. The final result is a map of fluorophore positions, built up like a pointillist painting. (Adapted from Scalettar, B.A., et al. 2015. Biomedical imaging in the undergraduate physics curriculum: module on optical microscopy. *Am. J. Phys.* 83:711–718. DOI: 10.1119/1.4921820. [CC BY 3.0].)

parameters, notably background intensity, pixel size, and signal strength. In the limit of low background and small pixel size, $\sigma_{x/y}$ can be approximated by the expression

$$\sigma_{x/y} = \frac{s}{\sqrt{m}} \tag{8.4}$$

In Eq. (8.4), "s" is a measure of the PSF width, and "m" is the number of detected photons. Equation (8.4) shows, not surprisingly, that a strong fluorophore signal is an important determinant of accurate position assignment. In fact, Eq. (8.4) shows that resolution enhancement requires that each fluorophore generates more than one detected photon. For the case $m \sim 400$, the PALM uncertainty will be 1/20 of the PSF width, or ~10 nm.

It is tempting, and not uncommon, to assume that coordinate uncertainty in PALM is equivalent to "structural" resolution, which is the ability to reproduce (resolve) structural detail; however, this assumption is *incorrect*. This subtle issue is discussed in the next section.

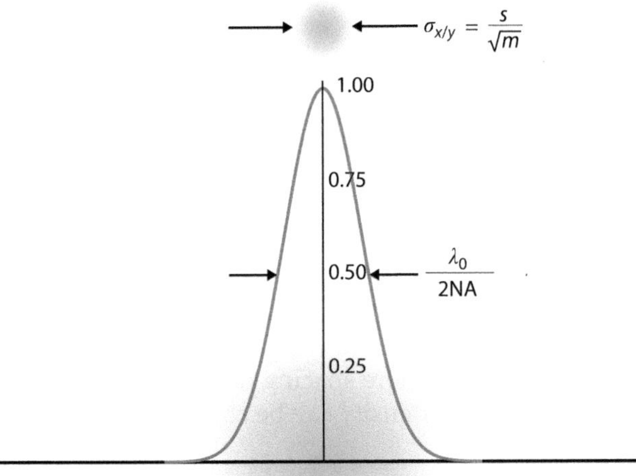

FIGURE 8.13
Schematic of the image processing procedure used in PALM. The raw images of individual fluorophores are diffraction-limited spots with relatively broad spatial extent (large green spot). To refine fluorophore localization, raw spot data are fitted to a Gaussian function (blue curve), and fluorophore coordinates are taken to coincide with the center of the Gaussian. The associated coordinate uncertainty (small green spot) is determined by how precisely the center is localized (Eq. [8.4]).

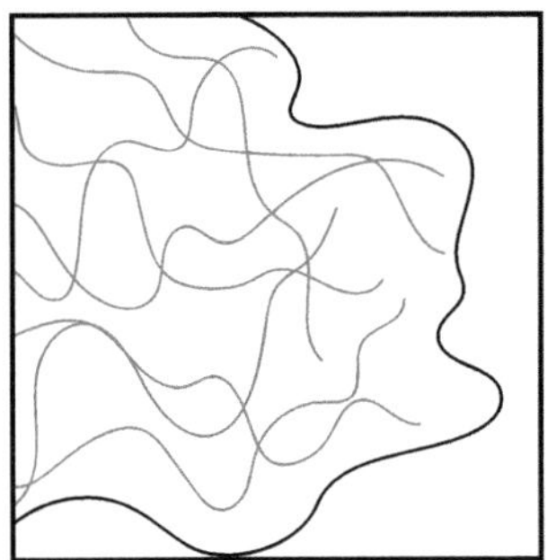
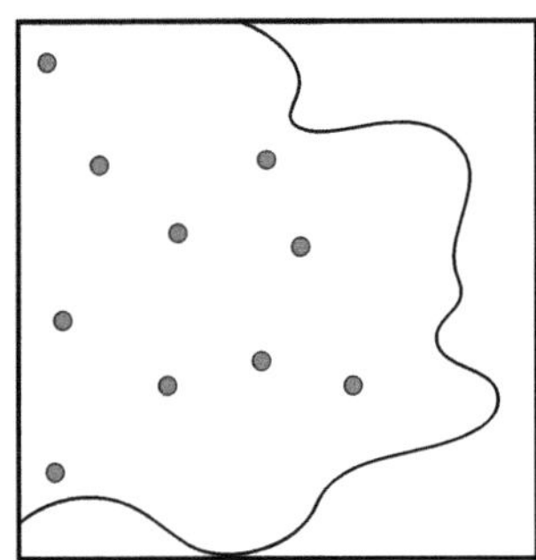

FIGURE 8.14
Sparse, but precise, sampling of filamentous structures yields an image that lacks structural resolution.

8.2.2.3 Image Resolution and the Nyquist Sampling Criterion

The structural resolution of PALM images and the uncertainty in fluorophore coordinates are not identical, as mentioned above. The distinction can be appreciated visually by examining a sparsely, but precisely, sampled image of a complex object (Fig. 8.14). Despite containing well-localized spots, the image lacks structural detail, and thus its structural resolution is poor. This is because sampling is too sparse.

This qualitative argument can be made quantitative by invoking the Nyquist-Shannon sampling theorem, which we will discuss in some detail in Chapter 9. Here we just present the theorem's key message. Assume that the goal is to resolve a smallest feature, or two-point spacing, of size R when reconstructing a signal from spatial "samples" that are separated by a distance, d. The theorem states that the desired resolution will be achieved only if $d < R/2$. In other words, the signal must be sampled more than twice as finely as the desired resolution. Thus, to achieve diffraction-limited resolution of 200 nm, the specimen must be digitized in $d < 100$ nm increments.

The concepts discussed above suggest that resolution in PALM is dictated by two quantities – coordinate uncertainty and sampling frequency. The combined effect of these two parameters has been described approximately by analogy with error propagation, leading to the resolution expression

$$R = \sqrt{\left(\sigma_{x/y}\right)^2 + (2d)^2} \tag{8.5}$$

In Eq. (8.5), "d" is to be interpreted as the *observed* inter-fluorophore spacing because this spacing sets the spatial sampling distance. Equation (8.5) quite reasonably implies that coordinate uncertainty will dominate resolution when d is small (sampling is good), and that sampling will dominate resolution when $\sigma_{x/y}$ is small (coordinate assignment is good). Equation (8.5) also implies that resolution in a PALM image can vary significantly if there are marked differences in inter-fluorophore spacing, with regions of sparse photoactivation having poorer resolution.

SAMPLE PROBLEM 8.1 SPATIAL RESOLUTION IN PALM

Assume that you want to achieve an average resolution of 50 nm in a PALM image and that the coordinate uncertainty, σ, is 20 nm. (a) What is the required average observed inter-fluorophore spacing in the image? (b) What is the required average observed fluorophore density? Take-home message: You must collect enough sub-images to achieve this density.

SOLUTION

a. The relationship between resolution, coordinate uncertainty, and spatial sampling distance is given by Eq. (8.5). This equation can be solved for d, the average observed inter-fluorophore spacing, as follows

$$d = \frac{\sqrt{R^2 - \sigma^2}}{2} = \frac{\sqrt{(50\text{ nm})^2 - (20\text{ nm})^2}}{2} = \frac{\sqrt{2500 - 400}}{2}\text{ nm} = 23\text{ nm}$$

b. The fluorophore density associated with this average inter-fluorophore spacing depends on how the fluorophores are distributed. However, a good estimate of the density can be obtained simply by squaring the separation and taking its inverse:

$$\rho_{\text{est}} = \frac{1}{d^2} = \frac{1}{(23\text{ nm})^2} = 0.00189/\text{nm}^2 = 1890/\mu\text{m}^2$$

This is the fluorophore density required on the labeled structures. The "macroscopic" density may be lower if the labeled structures constitute only a part of the sample (for example, microtubules inside a cell).

8.2.2.4 Experimental Implementation

The key components of a typical PALM microscope are shown in Fig. 8.15. The system contains photoactivation and readout lasers, which are under the control of shutters.

Because PALM involves imaging single fluorophores, signals are intrinsically weak, and detection must be as robust as possible. Background also must be minimized. To this end, a highly sensitive EMCCD or sCMOS camera is used as detector (Chapter 9). In addition, a very high NA objective is used to direct the excitation light onto the sample in TIRF mode, which markedly reduces background in PALM images. A comparison of PALM and widefield images is shown in Fig. 8.16.

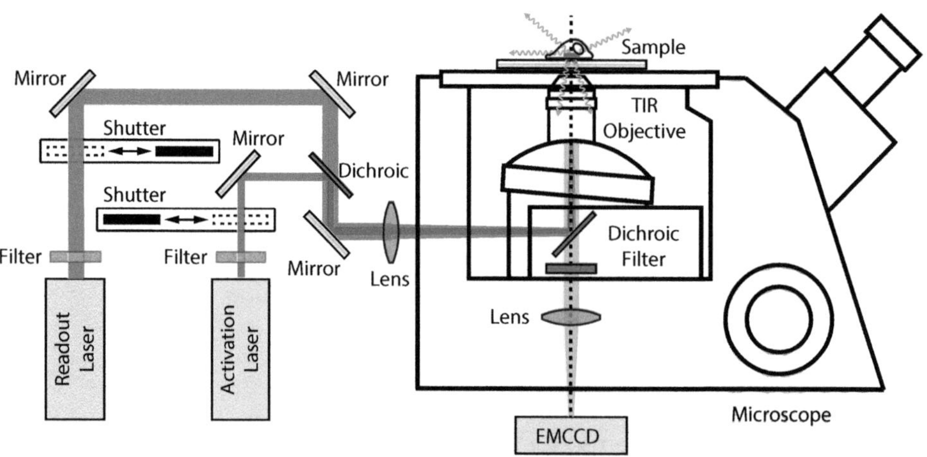

FIGURE 8.15
Schematic of a PALM microscope. Through-objective TIRF-based illumination (readout and activation) typically is used to reduce background signal. In the schematic, the readout (blue) and activation (red) beams are truncated at the objective turret but, in reality, continue on to the sample. The evanescent character of the illumination is depicted in the schematic with a blue bar localized at the glass-cell interface. In contrast, the emission (green) is omni-directional. (Adapted with permission from Gould, T.J., et al. 2009. Imaging biological structures with fluorescence photoactivation localization microscopy. *Nat. Protoc.* (4) DOI: 10.1038/nprot.2008.246.)

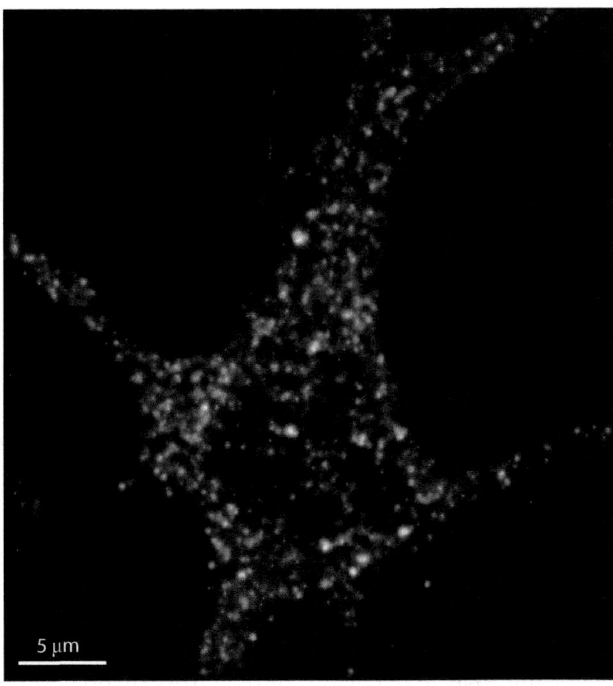

FIGURE 8.16
Comparative PALM and widefield images of DCVs in a hippocampal neuron. In the overlay, DCVs visualized using PALM are encoded red/orange. The PALM image was built up by visualizing individual molecules of tissue plasminogen activator-Dendra2, which localize inside the lumen of the DCV. DCVs visualized using a widefield approach are encoded green. Data related to this image are analyzed in Activity A8.1. Scale bar = 5 μm. (Adapted with permission from Scalettar, B. A., et al. 2014. Super-resolution imaging of neuronal dense-core vesicles. *J. Vis. Exp.* (89), e51394, DOI: 10.3791/51394.)

8.3 Comparison of Approaches

In this chapter, we have discussed three prominent super-resolution techniques. Table 8.1 compares aspects of these techniques. Together with PALM, we have included the closely allied approach, stochastic optical reconstruction microscopy (STORM). When the two approaches were initially introduced, PALM and STORM differed in the nature of the photoswitchable fluorescent tag. PALM utilized photoswitchable fluorescent proteins, and STORM utilized photoswitchable organic dyes introduced via immunostaining. However, these distinctions increasingly are blurred.

TABLE 8.1
Comparison of Techniques

Technique	Unique Attributes	Strengths	Weaknesses	Applications
SR-SIM	Moiré effect used to extract higher frequency information	Uses standard fluorophores, good temporal resolution (acquisition time seconds), straightforward multi-color imaging	Restricted to resolution doubling for linear approaches, mandatory post-acquisition processing that may introduce artifacts	Two- and three-dimensional, multi-color imaging of fixed and living samples
STED	Stimulated emission used to engineer a fluorescent scanning spot to a sub-diffraction-limited size	Theoretically unlimited resolution, no post-acquisition processing, good temporal resolution for small FOVs (acquisition time seconds)	Less amenable to multi-color imaging (due to the need for two lasers to image one fluorophore), trade-off between temporal resolution and FOV for scanning-based implementations, relatively high photodamage and photobleaching	Two- and three-dimensional, multi-color imaging of fixed and living samples

(Continued)

TABLE 8.1 *(Continued)*

Comparison of Techniques

Technique	Unique Attributes	Strengths	Weaknesses	Applications
PALM/STORM	Sparse, stochastic photoactivation used to achieve single-molecule imaging, TIRFM	Very high resolution, allows visualization of single molecules, relatively simple instrumentation (modified widefield microscope)	Requires special probes, relatively poor temporal resolution (acquisition time minutes), mandatory post-acquisition processing, weak signals, sampling considerations heavily impact resolution	Very high-resolution one- and two-color imaging of fixed samples and slow biological processes in living samples, recently extended to three-dimensional imaging

BOX 8.3 COMPLEMENTARY FUNCTIONAL IMAGING TECHNIQUES

The super-resolution techniques presented in this chapter – SR-SIM, STED, and PALM – provide exceptionally detailed information on the distribution and abundance of fluorophores in a sample. However, their current resolution does not allow them directly to observe immediate molecular environments or nearest neighbor interactions. Nevertheless, such information can be obtained using a set of complementary techniques that exploit the influence that local environment, including binding state, has on fluorescence emission. These "functional imaging" techniques differ from those we have studied so far in how they generate contrast. Specifically, instead of basing contrast directly on variations in the intensity of light generated by the sample, these functional techniques base contrast on properties of sample fluorescence that can change during the excited state of the fluorophore (i.e., between absorption and emission). These properties include the ability of the fluorophore to transfer energy to another fluorophore (FRET), the lifetime of the fluorophore (FLIM), and the rotational correlation time of the fluorophore (FAIM), among others. This focus allows the techniques to report on binding interactions, ion concentrations, membrane and cytosol viscosities, and so on. Images can be monochrome but more commonly are pseudo-colored. These techniques can be mixed and matched; for example, both FLIM and FAIM can be used to measure FRET. Finally, although these techniques probe very small, molecular-scale environments, the resolution of the associated images is determined by the technique used to read out the result. These can include widefield, confocal, two-photon, and SIM, among others. The following table describes three common functional imaging techniques.

Technique	Principle	Application
FRET	*Fluorescence resonance energy transfer* Contrast parameter: efficiency of fluorescence energy transfer between a pair of fluorophores This technique produces images by displaying spatial variations in the extent of "energy transfer" between a pair of fluorophores. A "donor" fluorophore, attached to a first molecule of interest, is excited by standard excitation light. However, rather than emitting fluorescence, the excited donor transfers its excited-state energy, without emitting a photon, to a neighboring "acceptor" fluorophore, attached to the same or another molecule of interest. The acceptor can then emit its own fluorescence, typically red-shifted relative to donor fluorescence, or it can lose the energy nonradiatively (quenching the fluorescence). Energy transfer can be assayed by measuring the resulting decrease in donor fluorescence, an increase in acceptor fluorescence, or a decrease in donor lifetime (see FLIM), among others. Significantly, FRET is exquisitely sensitive to the separation, r, between donor and acceptor, falling off as $1/r^6$. For this reason, FRET has been termed a "molecular ruler." Typical distances over which energy transfer can occur are 2–10 nm. Donor and acceptor are chosen such the emission spectrum of the donor overlaps the absorption spectrum of the acceptor. Cyan fluorescent protein (CFP) and yellow fluorescent protein (YFP) are a popular donor and acceptor pair.	FRET is used to determine molecular proximity, including both association and dissociation reactions, in which FRET increases or decreases, respectively. Biological applications include studies of receptor/ligand interactions, the assembly and disassembly of molecular complexes, and conformational changes (which can alter donor/acceptor separation within the same or associated molecules), among others.

Technique	Principle	Application
FLIM	*Fluorescence lifetime imaging microscopy* Contrast parameter: fluorescence lifetime This technique produces images by displaying spatial variations in fluorescence lifetime. The fluorescence lifetime, as described elsewhere in this chapter, is the average time that a fluorophore spends in an excited state before emitting a photon and returning to the ground state. Significantly, the lifetime depends on local environmental parameters, such as pH, ion concentration, binding state, and energy transfer, among others. In contrast, the lifetime does not generally depend on fluorophore concentration (if not too high), the intensity of excitation light, or the thickness of the sample, among others, thereby reducing sensitivity to experimental setup. Lifetime can be measured in the "time domain" or the "frequency domain." In the time domain, fluorophores are excited with a very brief pulse of excitation light (much shorter than the fluorescence lifetime), and the time until first fluorescence emission is measured. This process is repeated many times to build up a decay curve from which the lifetime can be extracted. In the frequency domain, fluorophores are excited continuously with time-varying (e.g., sinusoidal) excitation light. The emission will track the excitation exactly if the lifetime is short. However, if the lifetime is longer, the emission will lag behind the excitation, and its amplitude will flatten out (de-modulate). The lifetime can be extracted from a mathematical analysis of these changes. Fluorescence lifetimes are typically a few nanoseconds but can be microseconds or longer.	FLIM is used to probe fluorophore environment. Biological applications include imaging calcium concentrations, chloride concentrations, and membrane phases, among others. In addition, FLIM is a common method for measuring FRET because energy transfer shortens the fluorescence lifetime of the donor (by providing an additional route for the fluorophore to lose energy).
FAIM	*Fluorescence anisotropy imaging microscopy* Contrast parameter: fluorescence anisotropy This technique produces images by displaying spatial variations in fluorescence anisotropy. Anisotropy is a measure of molecular reorientation during the fluorescence excited state. It is measured by illuminating a fluorescent sample with linearly polarized light and then measuring the anisotropy of the resulting fluorescence. Polarized excitation light preferentially excites fluorophores having a defined orientation. This anisotropic distribution of excited states will lead to an anisotropy in fluorescence emission that depends on how much the fluorophores rotate (or otherwise reorient) during the excited state. Emission from rapidly rotating fluorophores will be isotropic because the fluorophores will be randomly oriented by the time they emit. In contrast, emission from slowly rotating fluorophores will still be anisotropic, because the orientation of the fluorophores will not have changed significantly before emission. Changes in rotation rate are typically brought about by conformational changes and molecular binding. The technique is useful because rates of random diffusive rotation of small molecules are typically on the timescale of the fluorescence lifetime (a few nanoseconds). Thus, binding and unbinding can be measured by an increase or decrease in the anisotropy of the emitted light. Like FLIM, FAIM can be measured using time or frequency-domain techniques. Anisotropy is also similar to FLIM in that it is largely independent of fluorophore concentration, excitation intensity, and sample volume.	FAIM is used to probe fluorophore environment and is particularly suited to studies of molecular orientation and mobility. Biological applications include imaging local microviscosity, receptor/ligand binding, and myosin reorientation, among others. FAIM is also another method of measuring FRET.

Suggested Reading and Additional Resources

iBiology: Microscopy Series. 2012 – 2015. Fluorescence microscopy. https://www.ibiology.org/online-biology-courses/microscopy-series/fluorescence-microscopy/

Kubitscheck, U. (Ed.) 2017. *Fluorescence microscopy: from principles to biological applications (2nd Edition)*. Wiley-VCH, Weinheim, Germany.

Murphy, D.B. and Davidson, M.W. 2013. *Fundamentals of light microscopy and electronic imaging (2nd Edition)*. Wiley-Blackwell, Hoboken, NJ.

HOMEWORK PROBLEMS

8.1 [Beats – Application] You discover a set of tuning forks, including one of unknown frequency. You listen to the beat frequency generated using the unknown fork and a fork of frequency 288 Hz and hear a beat of 32 Hz.

a. What are the possible frequencies of the unknown?
b. When you repeat the experiment with a fork of frequency 341 Hz, you hear a beat of 21 Hz. What is the frequency of the unknown fork?

8.2 [SR-SIM – Beats Analysis (Numerical)] Write a program that adds two harmonic waves of different amplitude and frequency. Study the mechanism underlying beats by adding two waves of different frequency and observing how the frequency of the resultant depends on the difference in frequency.

8.3 *[Beats Theoretical – Sounds of Unequal Amplitude] The superposition of harmonic waves with different frequencies and amplitudes will still produce beats, but the destructive interference is not complete and the contrast is reduced. More specifically, if the two waves have amplitudes y_1 and y_2, the minimum of the envelope of the intensity superposition is given by $(y_2 - y_1)^2$ instead of zero. To verify this, express the two waves in complex exponential form as $y_1 e^{i\omega_1 t}$ and $y_2 e^{i\omega_2 t}$ (Boxes 5.3, 5.4, and 8.1) and prove that the intensity is given by

$$I = \left| y_1 e^{i\omega_1 t} + y_2 e^{i\omega_2 t} \right|^2 = y_1^2 + y_2^2 + 2y_1 y_2 \cos\left[(\omega_1 - \omega_2)t\right]$$

The minimum is achieved when the cosine is minus one.

8.4 [SR-SIM – Sampling Theorem] The lateral and axial resolutions of SR-SIM are stated to be 100 nm and 300 nm, respectively. How finely do you have to sample to achieve these resolutions? Explain your logic.

8.5 [SR-SIM – Resolution Enhancement] Generate quantitative versions of Fig. 8.5a, b (showing circle radii in μm^{-1}) for a 1.4 NA objective assuming λ = 500 nm. For this SR-SIM scenario, what are the largest spatial frequency accessed (in μm^{-1}) and the smallest spatial dimension resolved (in μm)? Hint: $d_{\text{min}} = 1/k_{\text{max}}$.

8.6 [Photon Shot Noise] The emission of light by a fluorophore is a random process that obeys Poisson statistics. This means that, even under identical conditions, the number of photons detected from a fluorescent sample will vary according to the Poisson distribution. This variation is commonly termed "shot noise," or "photon noise," and can never be eliminated. However, it can be reduced by collecting more photons, for example, by increasing the intensity and/or the collection time. The signal-to-noise ratio (SNR) for a Poisson process is given by the following expression: $\text{SNR} = N/\sqrt{N} = \sqrt{N}$, where N is the average number of detected photons. Calculate the number of photons that must be collected to achieve an *SNR* of 10, 100, and 1000.

8.7 [STED Resolution] The resolution in a STED experiment is given by a relatively simple modification to the Abbe formula, $d_{\text{STED}} = \lambda_0 / \left(2\text{NA}\sqrt{1 + I/I_s}\right)$. In this expression, I_s is a measure of the intensity at which fluorescence is reduced by a factor of two.

a. Evaluate d_{STED} in the limit $I/I_s \to \infty$. The result forms the basis for the statement that resolution in STED is theoretically unlimited.
b. Calculate d_{STED} using experimental parameters reported for the fluorophore Citrine, $I_s \sim 81$ MW/cm^2 and $I \sim 810$ MW/cm^2. Assume that $\lambda_0 = 500$ nm and NA = 1.4.

8.8 *[Fluorescence Lifetime] The rate at which fluorescence intensity, $I(t)$, decays is given by $I(t) = I_0 \exp(-t/\tau)$, where I_0 is the initial intensity and τ is the fluorescence lifetime. Substituting $t = \tau$ into these equations shows that τ is the time it takes for the intensity to drop to 1/e = 37% of its initial value. Prove that the fluorescence lifetime also is the average amount of time that a fluorophore spends in the excited state following excitation. Hint: The average value of t, $\langle t \rangle$, can be defined as follows:

$$\langle t \rangle = \frac{\int_0^{\infty} tI(t)dt}{\int_0^{\infty} I(t)dt}$$

8.9 [Rate of Stimulated Emission – Laser] Stimulated emission is at the heart of laser function. Compute the rate of stimulated emission for a 1-mW laser emitting 633-nm photons.

8.10 [STED Microscopy – Equilibrium Ratio of Excited to Ground-State Fluorophores] The ratio of the number of fluorophores in an excited electronic state (e) to the number of fluorophores in the ground state (g) is described, at equilibrium, by the Maxwell-Boltzmann distribution:

$$\text{Ratio}_{e/g} = e^{-\Delta E/k_B T}$$

Here, ΔE is the energy difference between the excited state and ground state, k_B is Boltzmann's constant, and T is the temperature. Calculate this ratio for an energy difference corresponding to a 550-nm photon. Assume $T = 300$ K. Hint: Remember that the energy of a photon is $h\nu$, where h is Planck's constant and ν is the frequency of the photon.

8.11 [STED Microscopy – Equilibrium Ratio of Stimulated to Spontaneous Emission] The ratio of transition rates of stimulated emission to spontaneous emission, at equilibrium, can be calculated from the Maxwell-Boltzmann distribution in HW 8.10 and the fact that (stimulated) absorption must balance stimulated emission and spontaneous emission, yielding the following expression:

$$\text{Ratio}_{\text{stim/spon}} = \frac{1}{e^{h\nu/k_B T} - 1}$$

The variables h, ν, k_B, and T were defined in HW 8.10.

a. Calculate this ratio for an energy difference corresponding to a 550-nm photon. Assume $T = 300$. Discuss the implications of your answer.
b. Repeat part a for a $\nu = 10^{10}$-Hz microwave photon and compare the results.

8.12 [STED Microscopy – Estimating Beam Intensities] In STED, the rate of stimulated emission must outcompete the rate of spontaneous emission. To quantify the implications of this requirement, assume that the rate constant for stimulated emission can be rewritten as the product of an interaction "cross-section," σ and the intensity, i.e., $k_{\text{stim}} = \sigma I$, where the cross-section is an area with dimensions that typically range from 10^{-16} to 10^{-18} cm^2. Also assume that the rate constant for spontaneous emission, k_{spon}, is on the order of the inverse of a typical fluorescence lifetime (Section 8.2.1.2). Use these assumptions, together with the requirement that $k_{\text{stim}} \geq k_{\text{spon}}$, to verify the claim in Section 8.2.1.2 that STED saturation intensities are on the order of tens to hundreds of MW/cm^2. Assume $\lambda = 500$ nm.

8.13 [PALM Microscopy] Calculate the resolution of PALM microscopy under the following conditions: a PSF width of 200 nm, 100 detected photons per fluorophore, and a fluorophore density of $10^4/\mu$m^2. Hint: Review Sample Problem 8.1.

8.14 [FRET] The efficiency, E_{ET}, of fluorescence resonance energy transfer from donor to acceptor, once a donor has been excited, can be described by the following expression:

$$E_{\text{ET}} = \frac{1}{1 + (r/R_0)^6}$$

Here, r is the separation between donor and acceptor, and R_0 is the Förster distance. R_0 is a very complicated quantum mechanical function of donor and acceptor properties, including their relative orientations. However, it also has a simple physical significance.

a. Show that R_0 is the distance at which E_{ET} is 50%.
b. Plot E_{ET} as a function of r for the popular donor/acceptor pair cyan fluorescent protein (CFP)/yellow fluorescent protein (YFP) (R_0 = 4.9 nm). You can construct the plot using a graphing program or by calculating and hand plotting a few key values. Based on the shape of the plot, for what value of r is E_{ET} most sensitive to changes in r? For what values of r is E_{ET} least sensitive?

8.15 [FRET and FLIM] The efficiency, E_{ET}, of fluorescence resonance energy transfer from donor to acceptor can be expressed in terms of their separation, r (HW 8.14). E_{ET} also can be expressed in terms of the fluorescence lifetime of the donor in the presence (τ_{DA}) and absence (τ_D) of acceptor:

$$E_{\text{ET}} = 1 - \frac{\tau_{\text{DA}}}{\tau_D}$$

FLIM is a popular method of measuring FRET for this reason.

a. What is τ_{DA} when E_{ET} = 0% (i.e., when there is no energy transfer)? This result applies when donor-acceptor separations are very large.
b. What is τ_{DA} when E_{ET} = 50% (i.e., when half of the excited fluorophores decay via energy transfer)? This result applies when the donor-acceptor separation equals the Förster distance (HW 8.14).
c. What is τ_{DA} when E_{ET} = 100% (i.e., when all excited fluorophores decay via energy transfer)? This improbable result applies when donor-acceptor separations are very small (among other requirements).

ACTIVITIES[2]

A8.1 Analyzing PALM Data

This activity provides an opportunity to analyze PALM data using the open source Fiji™ (ImageJ™) image processing package. For download instructions, see Activity A6.1. The instructions below should be approximately valid for current versions of Fiji.

A. Open DCV_PALM.tif. This is a PALM image of a fluorescently tagged DCV from a hippocampal neuron. The widefield image is shown in green and the PALM image, which was generated by localizing many individual fluorophores inside the DCV, is shown in red. Your task is to compute the DCV diameter as reported by conventional fluorescence and by PALM. As a benchmark, electron microscopy typically shows that DCVs in hippocampal neurons have a diameter ranging from ~70 nm to 100 nm.

B. Separate the color channels by going to Image > Color > Split Channels.

C. Click on the straight-line tool in the toolbar. Often this is fifth from the left.

D. For one of the channels, draw a straight line through the center of the DCV.

E. Go to Analyze > Plot Profile to obtain a plot of intensity versus distance in pixels. The data can be saved using the "list" option at the bottom of the plot if you want to fit it.

F. Estimate the full width at half maximum (FWHM) in pixels by drawing a line over the half maximum point on the plot (or fit the data to obtain the width). Calculate the number of nm/pixel using the 200-nm scale bar.

G. Repeat for the other channel.

H. What are the two widths of the DCV in nm?

I. Does PALM beat the diffraction limit?

[2] Activity files are available for download in the online Support Materials.

9

Detectors, Sampling, and Image Processing and Analysis

This chapter deals with the capture, enhancement, and analysis of digital images. As a prelude, we briefly describe traditional "analog" images and how digital images are generated from their analog counterparts.

Analog images, such as traditional photographs, and analog signals have smoothly varying properties, such as intensity and color. Analog images therefore are termed "continuous in tone" (Fig. 9.1). In contrast, digital images and digital signals have discretely varying properties. For example, the intensity values in a digital image generated using an "n-bit" (e.g., 12-bit) digital camera are constrained to assume integral values ranging from 0 to $2^n - 1$ (e.g., 0 to $2^{12} - 1 = 4095$) (Box 9.1).

Digital cameras initially encode an image using an analog signal (e.g., a voltage). This then is digitized to yield a (necessarily approximate) digital image. The hallmark of a digital camera is that this conversion is carried out within the camera using an analog-to-digital converter (ADC), or digitizer (see below).

The conversion of an analog image into a digital format involves sampling and quantization. Sampling transforms the image into a finite, two-dimensional array of "samples." Quantization converts the values in the array into integers with values bounded by the bit depth of the detector (Fig. 9.1 and Box 9.1). Sampling is a subtle process that must meet the requirements of the sampling theorem to avoid distortion, as mentioned in Section 8.2.2.3. Additional detail is presented in Section 9.3.

9.1 Detectors

In this section, we discuss light sensors used to detect light and generate digital images in optical microscopy. These can be divided into two types:

- "Point" detectors, such as PMTs and photodiodes, which lack spatial sensitivity and
- "Array" detectors, such as CCD and sCMOS cameras, which are spatially sensitive. We consider point detectors first.

It is helpful to start with a big picture view of detector operation. Typically, these devices convert light into charge and then convert charge into voltage. Ultimately, the (analog) voltage signal is converted into a digital number.

Light → Charge → Voltage → Digital Number

9.1.1 Photomultipliers (PMTs)

We encountered PMTs in our discussion of laser scanning-based forms of optical microscopy, such as LSCM. Attributes of PMTs that are well suited for LSCM include rapid response, which helps offset the slowness of point scanning, and high sensitivity to continually varying light intensity.

PMTs contain several components, which typically are sealed in an evacuated glass tube (Fig. 9.2). These include a conducting surface (e.g., an alkali metal photocathode) that emits "photoelectrons" in response to incident light via the photoelectric (PE) effect and an electrode that accelerates and focuses these photoelectrons onto electron-amplifying dynodes. The photocathode is at a very negative potential, and the dynodes are at successively higher – closer to zero – potentials. Thus, the electric field from one dynode to the next speeds up each electron, such that each electron hits each successive dynode with enough energy to knock out around ten more electrons. This multiplying effect generates up to 10^8 electrons for each electron kicked off the photocathode. Finally, all the electrons are collected at an anode. The anode outputs the resulting current to a circuit. When used to generate an image in LSCM, PMT currents are converted into voltages and then digitized using an ADC.

Advantages of PMTs include a highly linear response to light and a very high electronic gain (current amplification). PMTs also are low noise because their dark current (i.e., their signal in the absence of incident light) is very low (assuming that they are cooled). PMTs thus are well suited to detecting weak signals. Moreover, when light levels are low enough, PMTs can be run in a photon (single-pulse) counting mode that better discriminates against noise by

DOI: 10.1201/b22076-10

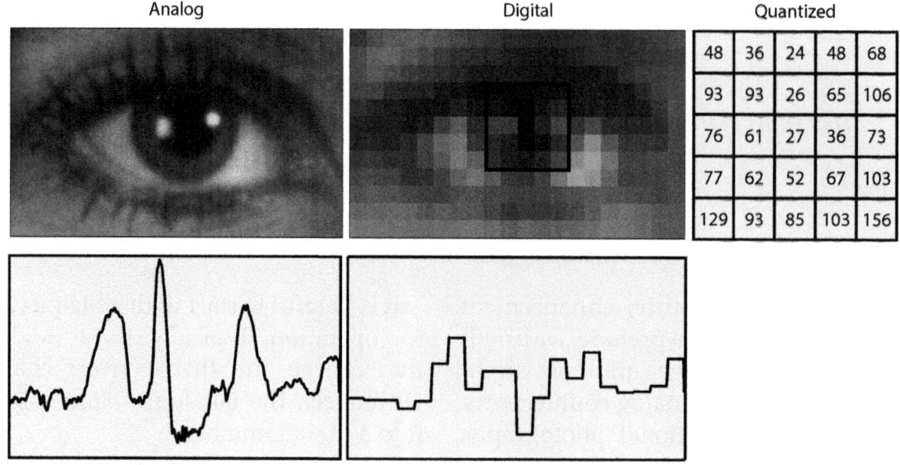

FIGURE 9.1
Comparison of analog and digital images and signals. A digital image is a discrete approximation to its analog counterpart and consists of an array of samples. The samples are quantized to generate an array of integers. The low-resolution digital image shown here consists of a 20 × 12 array. Integers are shown for the 25 pixels in the black box in the center of the eye. The analog and digital signals under the eyes are intensity profiles obtained for lines running through the center of the eye. (Images by authors.)

BOX 9.1 DYNAMIC RANGE AND BIT DEPTH

Dynamic range (DR) and bit depth are important parameters that influence a detector's ability to resolve light intensity. DR is an analog limitation of the detector, and bit depth is a property of the digitizer. DR and bit depth can be loosely distinguished using a simple staircase analogy – DR is similar to the height of the staircase and bit depth to the number of stairs. For a specified DR, greater bit depth will generate smaller steps.

More rigorously, DR describes the ability quantitatively to detect bright and dim regions in an image, which is known as intrascene performance. DR usually is calculated as the highest achievable signal divided by the detector noise. For a charge-coupled device (CCD) camera, the highest achievable signal is determined by the full well capacity of a "pixel" (see below). For a cooled CCD camera, the detector noise is dominated by noise generated by the electronics during signal readout.

To illustrate how to calculate DR, and to develop a feel for typical values, we analyze a cooled CCD camera with relatively small (4 μm) pixels. In this case, detector noise is ~2–5 electrons/pixel. The full well capacity is the product of the electron holding capacity of a pixel, which is ~1000 electrons/μm^2, and the pixel area in μm^2. For the small pixels considered here, the full well capacity is 16,000 electrons. Therefore, assuming an intermediate detector noise of 3 electrons/pixel, the associated DR is ~16,000/3 ~ 5300.

Bit depth describes the range of grayscale values that the digitizer uses to convert an analog signal into discrete values. For example, a 12-bit digitizer is capable of encoding $2^{12} = 4096$ gray levels. Ideally, the digitizer in a camera will utilize a number of gray levels that is matched to the DR to ensure its full use. For our scenario, a 12-bit digitizer would be slightly insufficient because $4096 < 5300$.

Higher bit depths are particularly useful when an image is destined for accurate quantification and/or extensive image processing. Lower bit depths (on the order of 6) will suffice when the goal is just to produce acceptable visual quality because the eye can only distinguish roughly 50 gray levels (on a typical monitor). However, use of very low bit depth will markedly compromise the accuracy of the brightness representation, as exemplified in the accompanying images. (Images by authors.)

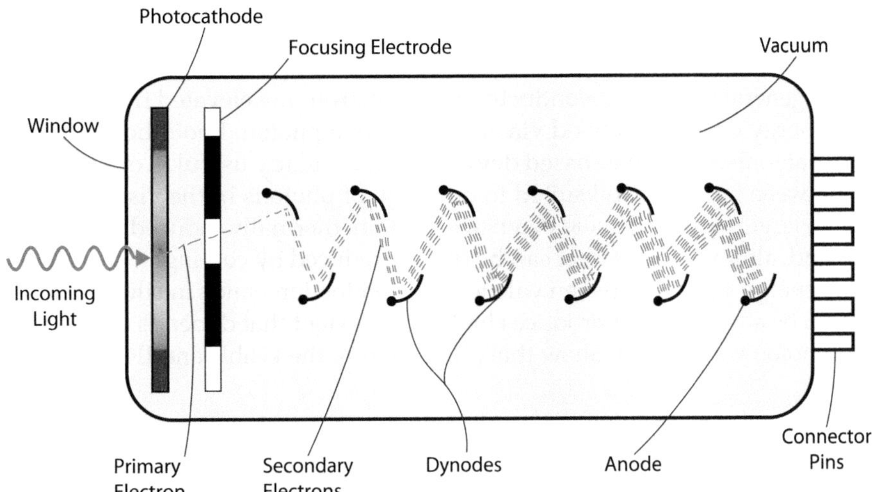

FIGURE 9.2
Schematic of the components and operation of a PMT. An incident photon ejects a primary electron from the photocathode. The primary electron is amplified by a series of dynodes that are held at progressively more positive voltage. As a consequence of the voltage increase, electrons gain kinetic energy and eject progressively more secondary electrons as they move along the dynode chain. Ultimately, the amplified electron signal reaches the anode, where it generates a detectable current.

retaining only pulses that exceed a threshold. PMTs also are capable of detecting very rapid events; their temporal resolution is on the order of a nanosecond. On the other hand, PMTs have a relatively low "quantum efficiency," which means that they detect only a fraction of the incident photons (Fig. 9.3). PMT spectral response also is nonuniform.

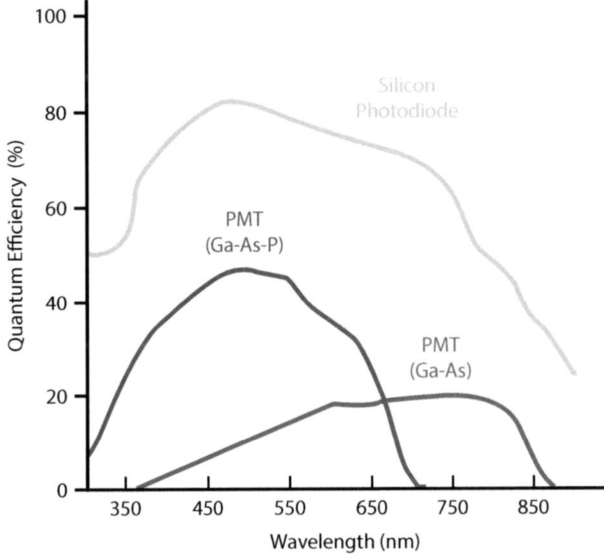

FIGURE 9.3
Comparison of the quantum efficiency of PMTs and photodiodes. Quantum efficiency is the ratio of detected to incident photons. Data are shown for the photocathodes gallium-arsenide (Ga-As) and gallium-arsenide-phosphide (Ga-As-P). (Data for graph exported from http://hamamatsu.magnet.fsu.edu/ articles/ digitalimagingdetectors.html.)

9.1.2 Photodiodes

Like PMTs, photodiodes are devices that convert light into current. Silicon-based photodiodes play a prominent role in the pixels of digital cameras because they are small point detectors that can be assembled into an array that is capable of spatial discrimination. The rationale for the use of silicon-based devices is explored in Sample Problem 9.1.

The current-generating mechanisms of silicon-based photodiodes are founded in relatively complex properties of semiconductors that are somewhat peripheral to our main discussion. In brief, when light of sufficient energy is incident on a photodiode, two types of charge carriers are generated via the PE effect – negatively charged electrons and positively charged "holes." Under the influence of an appropriate applied voltage, these charge carriers create a photocurrent that is a highly linear function of the incident light intensity.

Advantages of photodiodes include their relatively small size, which makes them suited for applications involving small devices. In addition, silicon-based photodiodes have higher quantum efficiencies than PMTs and respond more uniformly to radiation in the UV to IR (Fig. 9.3).

On the other hand, photodiodes lack the gain of PMTs. This limitation is overcome to some extent in avalanche photodiodes. Avalanche photodiodes utilize a high internal electric field to accelerate electrons and thereby produce additional, secondary electrons via impact ionization. The resulting electron avalanche can produce a gain of several hundred-fold, but this comes at the cost of nonlinear response and increased noise. Like PMTs, avalanche photodiodes

SAMPLE PROBLEM 9.1 SOLID-STATE ABSORPTION BANDS

Photoelectrons are generated in a semiconductor when electrons are elevated from a lower energy valence band to a higher energy conduction band via absorption of a photon. Photodiodes can be manufactured from a variety of materials, but silicon-based devices are particularly useful in optical microscopy because the energy gap between bands is well suited to detection of photons in the visible and IR (see figure). In addition, the gap is large enough to avoid extensive random, thermally induced elevation of electrons into the conduction band, although this effect can be further reduced by cooling.

Specifically, the energy gap, E_g, between valence and conduction bands in silicon is 1.1 eV. Photons with larger energies will be absorbed and produce charge to an extent that depends on the quantum efficiency at the associated photon wavelength. Show that photons from the visible and IR are absorbed. Hint: 1 eV = 1.60218×10^{-19} J.

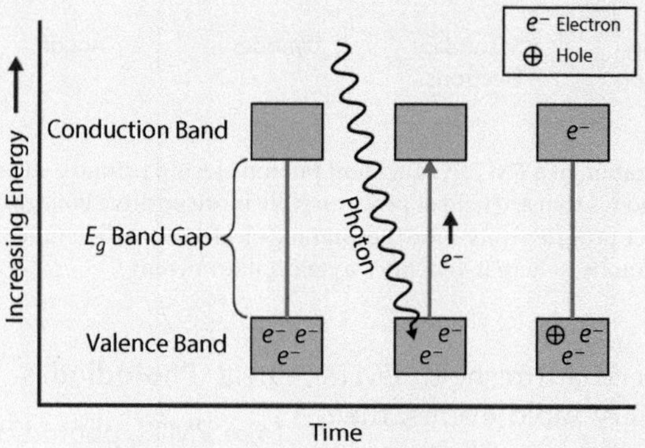

SOLUTION

A photon will be absorbed and promote an electron into the conduction band if the energy of the photon exceeds the energy of the bandgap: 1.1 eV. The relationship between energy and wavelength can be calculated using Eq. (2.4)

$$E = h\nu = \frac{hc}{\lambda}$$

This equation can be solved for wavelength, yielding

$$\lambda = \frac{hc}{E} = \frac{(6.63\times10^{-34}\,\mathrm{J\cdot s})(3.00\times10^{8}\,\mathrm{s})}{(1.1\mathrm{eV})(1.60\times10^{-19}\,\mathrm{J/eV})} = 1.13\times10^{-6}\,\mathrm{m} = 1130\,\mathrm{nm}$$

Thus, a photon with an energy corresponding to the energy gap will have a wavelength in the IR. Higher energy photons ($E > 1.1$ eV) corresponding to shorter wavelengths, including all visible wavelengths, also will be absorbed.

can be run in a photon-counting mode, but they are limited to lower count rates.

9.1.3 Cameras

Digital cameras consist of a two-dimensional array of detectors (pixels). When light from an object is focused onto the camera, a photon-induced charge is generated in each pixel that is proportional to the intensity from the corresponding point in the object (Fig. 9.4). The charge distribution in the detector array thus generates an image of the object.

9.1.3.1 Charge-Coupled Device (CCD) Cameras

For several decades, CCDs were a dominant scientific image sensor. However, CCDs increasingly are a sensor of choice only in niche scientific applications due to marked improvements in complementary metal-oxide semiconductor (CMOS) sensors.

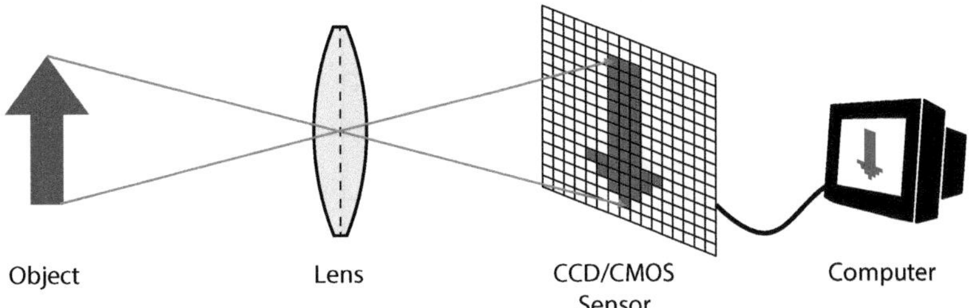

FIGURE 9.4
Schematic of spatial encoding of light distribution using a camera.

Like photodiodes, the operating principles of CCD (and CMOS) sensors are founded in relatively complex semiconductor physics (Fig. 9.5). Similar to a photodiode, photons incident on silicon create electron/hole pairs. Electrons liberated by photons incident on a particular area are collected under a "gate" electrode, which generates a potential well that confines the electrons to a particular pixel. After completion of the exposure, charge is extracted and moved from one storage site to another by manipulating voltages to facilitate readout and charge counting.

The full-frame CCD was the earliest configuration. A schematic of key aspects of image acquisition with this device is shown in Fig. 9.6. Image collection begins when the shutter opens and the array is exposed to photons, leading to photoelectron accumulation in an array of detection pixels known as the "parallel register." Photon detection terminates when the shutter closes. Voltage sequences then are used to move the photoelectrons in the parallel register toward a unique row – the "serial register." During this charge-transfer process, the serial register receives electrons from the adjacent row in the parallel register. In the serial register, charge is transferred pixel-by-pixel to an on-chip circuit, which amplifies the signal and converts it into an analog voltage. Thus, the output is a time sequence of voltages, where each voltage is proportional to the number of photons incident on each pixel. Finally, an ADC assigns a digital value to each voltage. Transfer through the parallel register and into the serial register continues until all rows have been read.

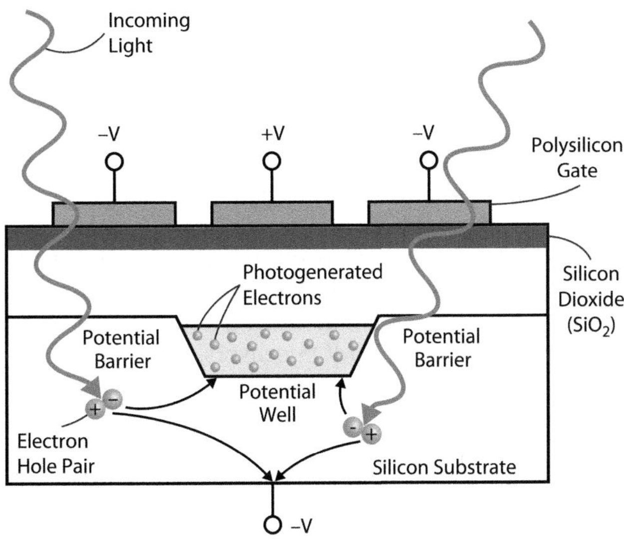

FIGURE 9.5
Key sensing elements in a CCD pixel. A typical pixel includes semi-transparent polysilicon electrodes (gates), a SiO_2 insulator, and a silicon-based substrate. Charge storage is based on applying a positive voltage to a central gate and a negative voltage to surrounding peripheral gates. This creates a well that stores light-generated photoelectrons.

Two variations on the full-frame CCD – the frame-transfer CCD and the interline-transfer CCD – emerged over the years. Both support faster readout than the full-frame device because a new exposure can be initiated without waiting for the serial register to process each pixel from the previous exposure. The frame-transfer CCD accomplishes this by using half of its array for photon detection and the other (masked) half for temporary storage of the previous image while it is processed by the serial register. The frame-transfer design is still utilized in electron multiplying CCDs (EMCCDs) (see below); in contrast, the interline design largely is obsolete.

9.1.3.2 Electron-Multiplying CCD (EMCCD) Cameras

CCD cameras perform very well with relatively bright samples. However, readout noise (Section 9.2) in CCD cameras obscures the signal from dim samples, such as those encountered in single-molecule imaging. Thus, methods like PALM traditionally rely on the use of EMCCD cameras. EMCCDs contain an extended serial register that amplifies the electron signal by ~2% per pixel via impact ionization in silicon. The effect per pixel is small, but it is propagated over ~500 pixels, leading to an overall amplification of $1.02^{500} = 20{,}000$. Relative to the boosted signal, readout noise in EMCCDs becomes negligible. EMCCDs typically are frame-transfer devices that are back-illuminated. The value of a back-illumination design

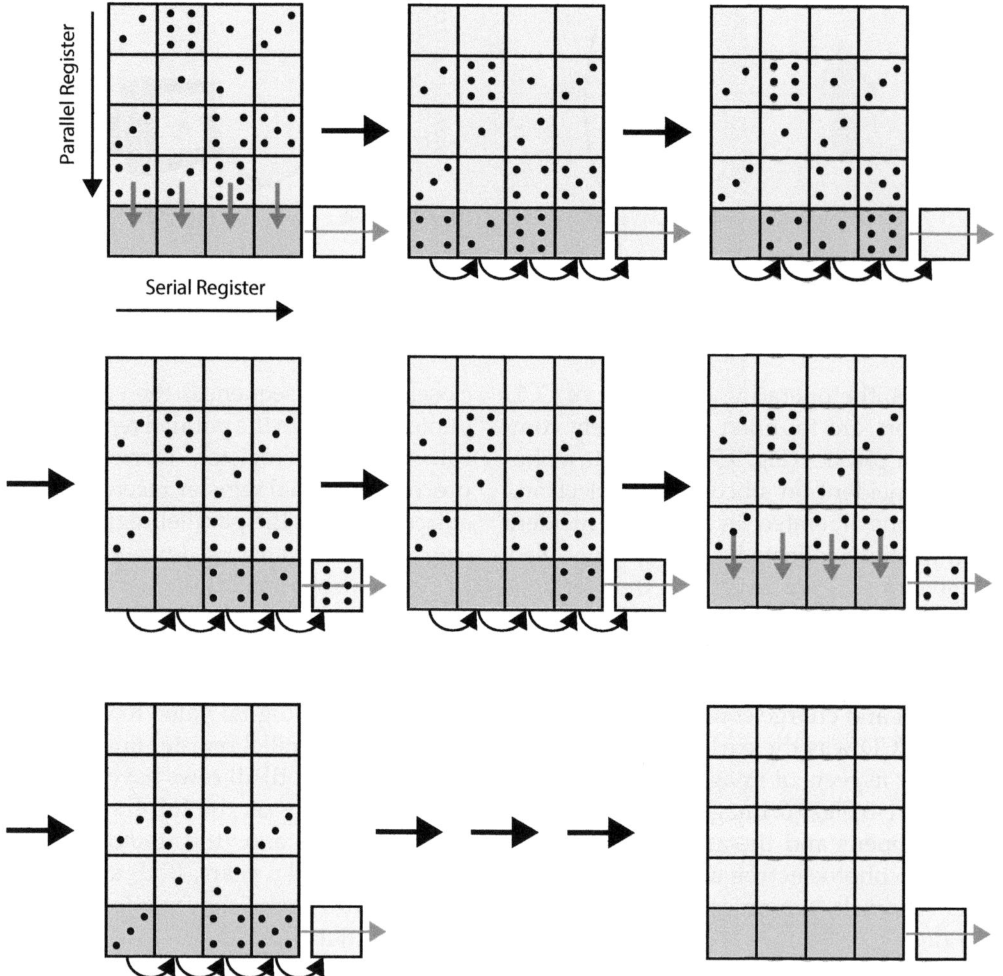

FIGURE 9.6
Schematic of charge accumulation and movement through the parallel (blue) and serial (red-brown) registers in a full-frame CCD. Reading the data from the array during the CCD cycle empties the array. Residual charge is cleared by repeating the CCD cycle but omitting digitization.

is that photons bypass the polysilicon gate electrodes, and this markedly enhances photon detection at shorter wavelengths.

9.1.3.3 Scientific (sCMOS) Cameras

We close our discussion of detectors with a description of CMOS image sensors. CMOS sensors are named for the technology used in their manufacture, which is the same as the technology used to construct integrated circuits. The basic CMOS light-sensing mechanism, the PE effect, is the same as that used by CCD cameras (Sample Problem 9.1).

CMOS sensors have long dominated over CCD sensors in consumer-oriented (e.g., mobile-phone) cameras but have only recently become dominant in high-end science cameras. This is because CMOS sensors did not achieve scientific-level image quality until the relatively recent introduction of sCMOS technology.

A fundamental difference between CMOS and CCD sensors is the site of charge-to-voltage conversion. In the CMOS device, charge-to-voltage conversion occurs at each pixel, whereas in the CCD it occurs at the output of the serial register. This difference reflects the fact that, during manufacture of the more sophisticated CMOS sensors, components like amplifiers and ADCs are added to each pixel (Fig. 9.7). Because charge-to-voltage conversion happens at each pixel in parallel, CMOS devices are potentially much faster than CCDs, which perform the conversion serially (Fig. 9.6).

Another (related) difference is that individual CMOS pixels, or a subset of pixels, can be selected for signal readout. This is possible because CMOS pixels generate a voltage that is transferred directly to the output, and each pixel in the array can be addressed directly with the addition of appropriate electronics. Moreover, in the most sophisticated CMOS devices, pixel-resident

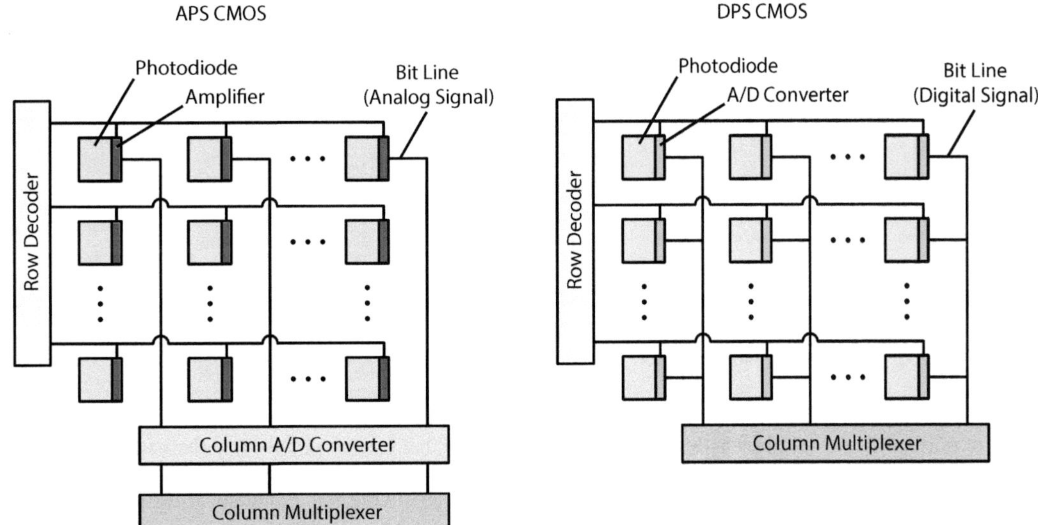

FIGURE 9.7
Schematic representation of two different CMOS sensor configurations. A key feature of the active pixel sensor (APS) CMOS is the presence of an amplifier at each pixel that implements charge-to-voltage conversion. The digital pixel sensor (DPS) CMOS also includes an ADC at each pixel.

ADCs can instantly digitize the analog pixel signals and store them temporarily in memory to facilitate simultaneous, rapid readout of all pixels (Fig. 9.7). In contrast, the charge transfer process used by CCD cameras precludes readout of individual pixels.

One negative of CMOS imagers is the presence of numerous pixel-resident electronic components, which reduces the area that can be used to collect photons. To circumvent this problem, microlenses can be used to focus the light incident on each pixel onto the photon-sensitive area (Fig. 9.8). Another option is back-illumination, which is used by most cutting-edge, high-sensitivity sCMOS cameras (Fig. 9.8).

In summary, CMOS image sensors have improved markedly in recent years and now dominate microscopy-based applications that utilize cameras. Nevertheless, CCDs may still have advantages in special cases. One example is an application, like PALM, that involves very low light levels and thus an EMCCD may provide superior results (electron-multiplying CMOS [EMCMOS] sensors also are under development).

9.2 Noise and the Image Signal-to-Noise Ratio (SNR)

In past chapters, we have focused on the role that spatial resolution and contrast play in dictating image quality, leaving untouched the role of noise. Now,

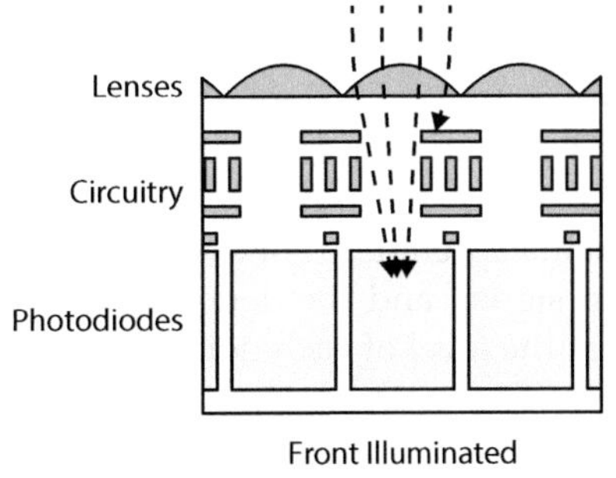

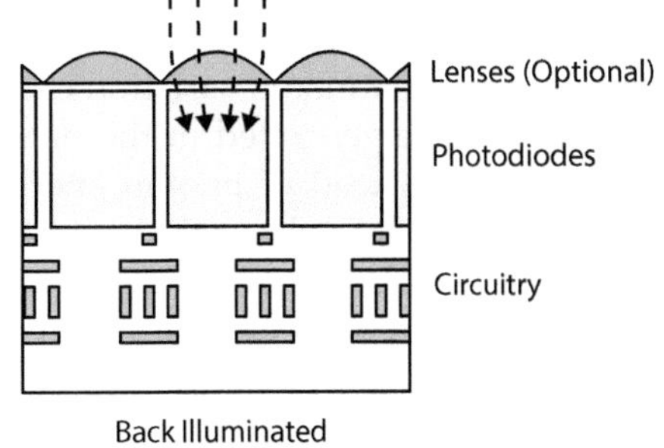

FIGURE 9.8
Schematic comparison of front- and back-illuminated CMOS sensor architecture. In the front-illuminated design, light can be blocked from reaching the charge-generating photodiodes by intervening circuitry. To reduce this effect, microlenses can be used to focus light away from the circuitry. However, microlenses are not particularly effective for light that arrives at non-normal incidence. Therefore, another option is to re-position the circuitry using a back-illuminated configuration that positions the photodiodes before, instead of after, the circuitry.

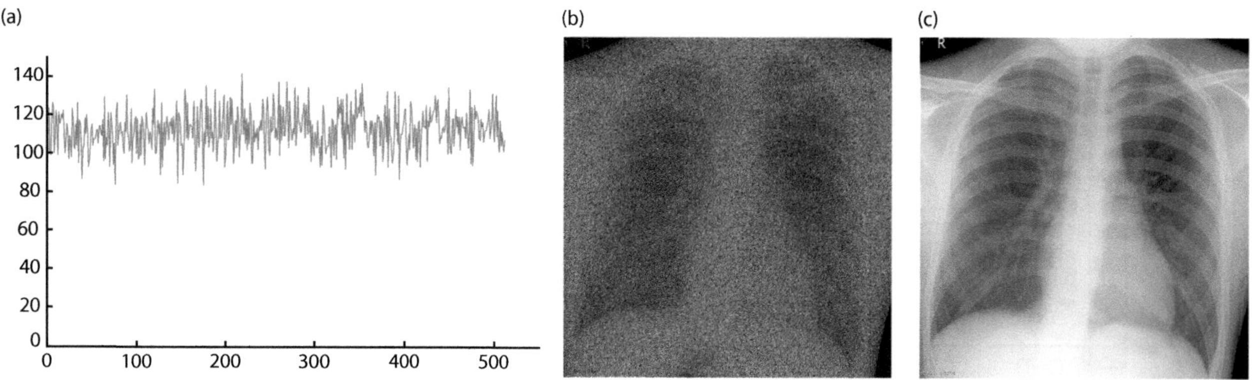

FIGURE 9.9
Visual examples of noise. In Panel (a), noise is depicted as a random signal that is superimposed on the true signal. In Panel (b), noise creates an image that is much grainier than its counterpart in Panel (c), which is relatively noise free. (Images adapted from Lee, S., et al. 2018. Poisson–Gaussian noise analysis and estimation for low-Dose X-ray images in the NSCT domain. *Sensors* 2018, 18:1019. DOI: 10.3390/s18041019.)

armed with the requisite background on detectors, we address prominent sources of image noise, the effects of noise, and how to quantify the signal-to-noise ratio (SNR) for the case of digital cameras.

Noise is the uncertainty that is associated with a signal or measurement (e.g., with pixel brightness – Fig. 9.9). Image noise is a source of random variations in brightness or color that obscure the true signal and that degrade contrast and spatial resolution (Activity A9.1). The visual impact of noise (e.g., a mottled, grainy appearance) is exemplified in Fig. 9.9.

In digital imaging, two important sources of noise are the photon signal and the camera. Photon (shot) noise has its origin in the fact that photon emission from a source is a statistical process (HW 9.2). This gives rise to an intrinsic variation in the number of object-associated photons incident on a region of the detector during successive intervals of time that is described by Poisson statistics (HW 9.2 and 9.3). There similarly is an intrinsic variation in the number of photons incident on equally sized sub-regions (e.g., pixels) of a detector when the specimen is uniform.

Camera-derived noise includes dark and read noises. Dark noise arises from random variation in the number of electrons that are generated in the detector using thermal energy instead of photon energy. Alternatively stated, dark noise is the variation in the dark current. Dark current and its associated noise are a sensitive function of temperature and can be reduced markedly by cooling the detector.

Read noise arises from errors in electronic processes, such as conversion of charge into voltage and subsequent analog-to-digital conversion. Amplifiers are a major source of read noise.

Often it is useful to quantify the level of image noise. For example, it may be important to know if a region of enhanced brightness reflects a real biological effect or is merely a random fluctuation from noise. The SNR can be used to quantify the confidence level for the significance of signal fluctuations, and it can be used to quantify image quality. Given its utility, we turn now to the calculation of the image SNR (HW 9.4).

One well-known expression for the SNR is given by the "CCD Equation," which reads

$$\mathrm{SNR} = \frac{N_s}{\sqrt{\sigma_s^2 + \sigma_B^2 + \sigma_D^2 + \sigma_R^2}} = \frac{N_s}{\sqrt{N_s + n_P\left(N_B + N_D + N_R^2\right)}} \tag{9.1}$$

In Eq. (9.1), n_P is the number of pixels in the specimen being analyzed, N_s is the measured signal summed over the n_P pixels, expressed in electrons, N_B and N_D are the number of electrons per pixel from the background and dark current, respectively, and N_R is the readout noise reported as electrons per pixel. The symbol σ represents the standard deviation (SD) of the signal. The SD is a conventional estimate of noise level because it is a measure of signal variation or uncertainty. For example, N_R is determined from an SD. The mean is an estimate of the true signal level.

The terms in Eq. (9.1) have a relatively straightforward interpretation. For example, the numerator is the mean signal, and the denominator is the total noise, where the level of each noise term is quantified using an SD. The mathematical form for the denominator reflects the fact that the noise terms are uncorrelated. Thus, they add in "quadrature" (i.e., as the square root of the sum of the squares).

In the final version of Eq. (9.1), SDs have been rewritten in terms of average signals. To justify the substitutions, we note that the specimen-, background-, and dark current-derived signals follow Poisson statistics. Thus, for these signals the noise is estimate using

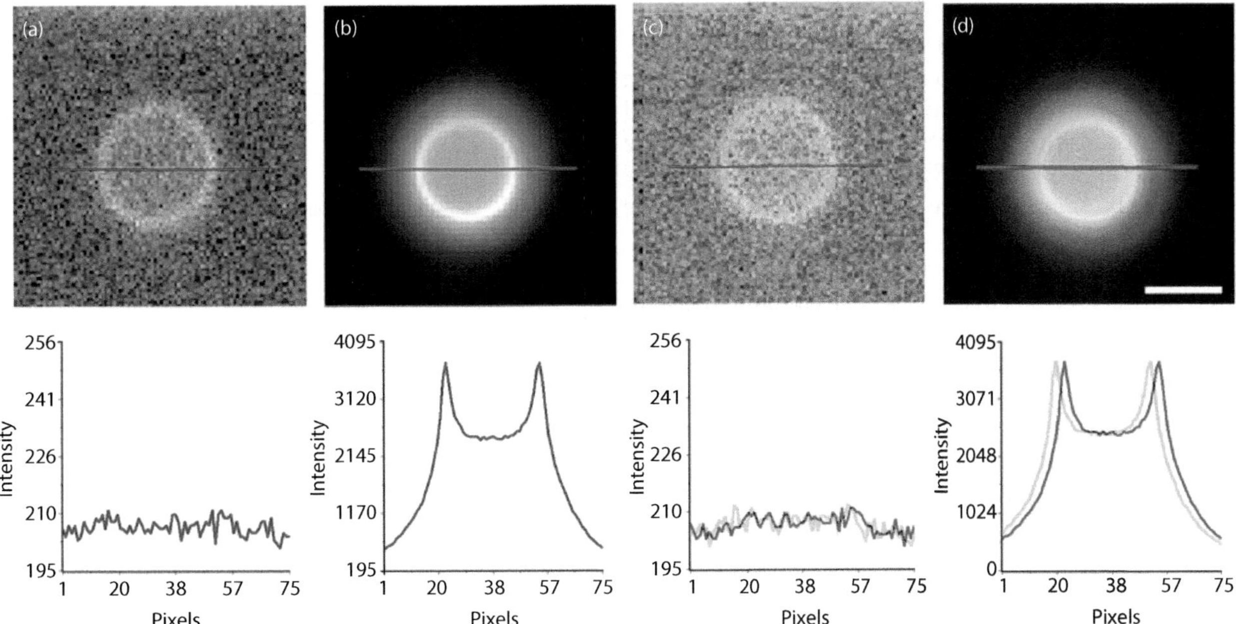

FIGURE 9.10
Visual demonstration of the effect of noise on spatial resolution and contrast. The images were generated using 6-μm beads that were fluorescently labeled on their perimeter. Below each image is a line scan of intensity taken at the position of the pink line. Panels (a) and (b) show successive images of the same bead that differ in exposure time and thus SNR. Only the high SNR image in (b) permits quantitative localization of the bead edges in a line scan. Panels (c) and (d) were generated by taking two successive images of the same bead. The set in (c) was taken with a short exposure time and the set in (d) with a long exposure time. The two images in a set were pseudo-colored, offset by several pixels, and then overlaid. Again, only the high SNR image permits the separation between the beads to be quantified in a line scan. (Adapted with permission from Waters, J.C. 2009. Accuracy and precision in quantitative fluorescence microscopy. *J. Cell. Biol.* 185:1135–1148. DOI: 10.1083/jcb.200903097.) A similar analysis of the impact of noise on resolution and contrast is presented in Stelzer (1998). This earlier article includes some quantitative analysis of the effect of noise on cutoff of high-frequency information (i.e., resolution) in widefield and confocal microscopy.

$\sigma = \sqrt{N}$ (HW 9.2 and 9.3). In contrast, N_R is typically measured and quantified using an SD, and thus this noise term has the form $\sqrt{N_R^2}$. Read noise can be quantified by collecting a very short (zero second) dark image.

In a digital image, pixel values are reported in "analog-to-digital units" (ADUs), e.g., as numbers ranging from 0 to 255. To evaluate Eq. (9.1), signal levels expressed in ADU, N_{ADU}, must be converted into signals in electron counts, N_{e^-}. This can be accomplished using the relationship

$$N_{e^-} = gN_{\text{ADU}} \tag{9.2}$$

In Eq. (9.2), g is the camera gain, which often is set to match the full well capacity of the pixels to the largest number output by the digitizer (HW 9.5). Equation (9.2) reveals that a gain of eight means that each ADU represents eight photoelectrons. The gain may be obtained from the camera specifications, or it can be determined using Janesick's method (HW 9.6).

We close with a brief discussion of the effects of noise on resolution and contrast. Noise often has a large degradative effect on resolution, causing it to be far lower than predicted from numerical aperture and Abbe theory; noise also can markedly degrade contrast. Examples of these effects are depicted in Fig. 9.10. The figure shows that when the SNR is low, the object (in this case, a bead) is indistinguishable from noise in a line scan. In contrast, when the SNR is high, the edges of the bead are well resolved in the line scan. These results are reinforced in color images, where the high SNR image allows the offset between objects to be accurately determined. These effects are explored further in Activity A9.1.

9.3 Sampling

A critical step in creating a digital image is sampling (i.e., recording periodic snapshots) of an associated analog image. Sampling must be done frequently enough to ensure that the digital image reproduces the "essence" of the analog image faithfully. The Nyquist-Shannon sampling theorem provides a rigorous rule for ensuring sufficient sampling to achieve a perfect reconstruction of an analog signal of finite band-width (frequency range). In optical microscopy, the sampling theorem predicts sampling rates that are required to reproduce a certain level of detail in an image and to avoid introducing artifacts (Fig. 9.11).

In Section 8.2.2.3, we mentioned the spatial version of the theorem. In brief, to reproduce a spatial detail of dimension R using spatial "samples" separated by d, it is necessary that $d < R/2$. In other words, the signal must be sampled more than twice as finely as the desired spatial resolution.

The theorem also can be discussed from a frequency-domain perspective. In this case, the sampling theorem states that reproduction of a frequency in the object requires sampling at more than twice that frequency. A simple, illustrative example is shown in Fig. 9.11. Sampling at exactly twice the object frequency can reproduce the desired periodicity, but there is potential for incorrect reconstruction. In contrast, sampling at more than twice the object frequency generates an image with the desired periodicity. Sampling at less than twice the object frequency leads to aliasing, which is the reconstruction of a false, typically lower frequency signal as an outgrowth of insufficient sampling.

The sampling theorem has important implications for choice of image pixel size, and for choice of step size during acquisition of optical sections. For example, pixel size should be <100 nm to achieve diffraction-limited resolution of 200 nm, as mentioned in Section 8.2.2.3. Implications for choice of magnification are explored in Sample Problem 9.2.

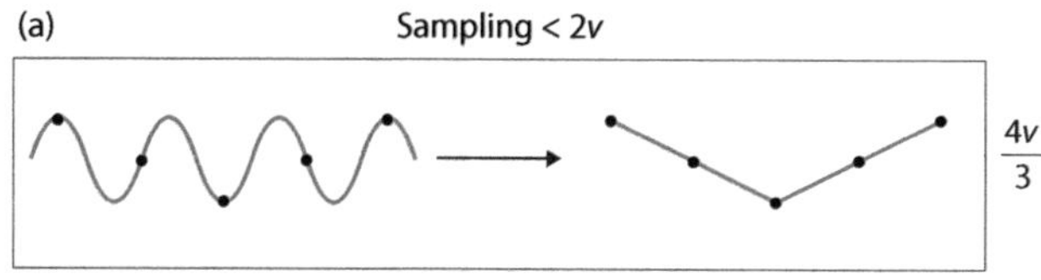

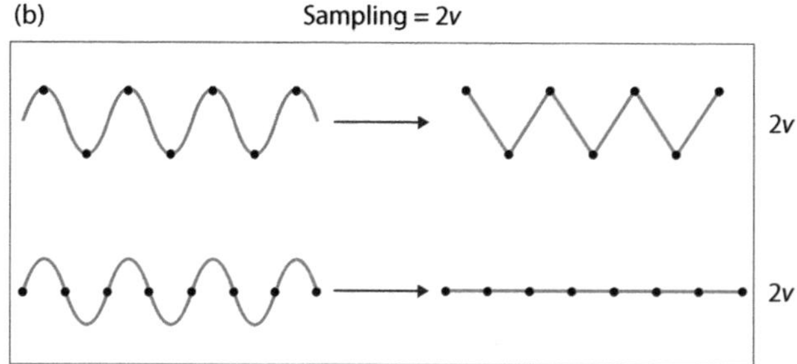

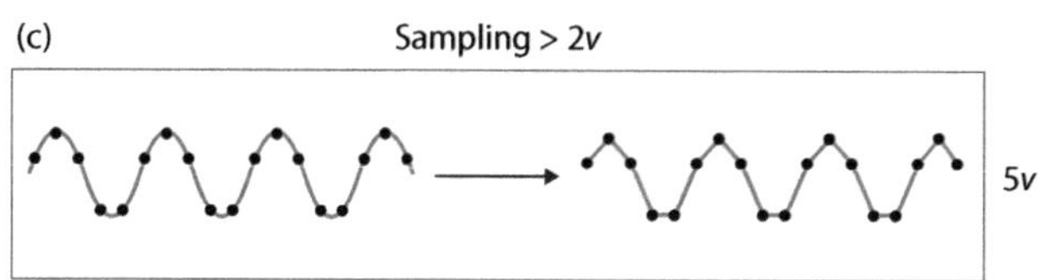

FIGURE 9.11
Comparison of signal reproduction for sampling that does not satisfy the Nyquist requirement (a and b) with sampling that does (c). Insufficient sampling typically leads to a spurious low-frequency signal.

SAMPLE PROBLEM 9.2 SAMPLING THEOREM AND RESOLUTION

The Kodak KAF1401E CCD camera has 6.8-μm pixels. Calculate the magnification that will yield diffraction-limited spatial resolution when detecting 500-nm fluorescence emission using this camera and a 1.3 NA objective.

SOLUTION

The sampling theorem states that a specimen should be sampled at less than half the desired spatial resolution. For a CCD camera, this means that the image should be magnified so that the radius of the Airy disk (i.e., the resolvable element) spans more than two full pixels:

$$\text{Magnification} \times \text{Airy Radius} > 2 \times \text{Pixel Size}$$

$$\text{Magnification} > \frac{2 \times \text{Pixel Size}}{\text{Airy Radius}}$$

The Airy Radius $= 0.61\lambda_0/\text{NA}$. Therefore, the magnification required to obtain the maximum spatial resolution is given by the following expression:

$$\text{Magnification} > \frac{2 \times \text{Pixel Size}}{0.61\lambda_0/\text{NA}} = \frac{2(6.8\ \mu\text{m})(1.3)}{(0.61)(0.500\ \mu\text{m})} = 58$$

This magnification can be provided by a standard 60× microscope objective. Higher magnification objectives also will work, but at the cost of greater photon noise because there will be fewer photons per pixel.

9.4 Introduction to Digital Image Processing and Analysis

One of the most important attributes of digital images is that they can readily be visualized, processed, and analyzed on a computer. Image processing and analysis constitutes a vast array of topics; here, we provide only an introductory discussion aimed at biomedical applications. Readers who are interested in more exhaustive discussions are referred to Russ and Neal (2016) or Gonzalez and Woods (2017).

We begin by distinguishing between image processing and image analysis. Processing involves modification of an image to produce another, generally improved, image, such as one with enhanced detail. Analysis involves obtaining numerical data from images, which can be used on their own or converted into a visual format (Fig. 9.12).

Depending on the goal, one or all of the following general processing and analysis steps can be performed on an image:

- Image Restoration
- Image Enhancement
- Binary Conversion/Image Segmentation
- Measurement

9.4.1 Image Processing

We first tackle processing, which encompasses restoration, enhancement, and binary conversion/image segmentation. Segmentation is directed at simplifying an image, and thereby making it easier to analyze, and thus is linked closely to analysis.

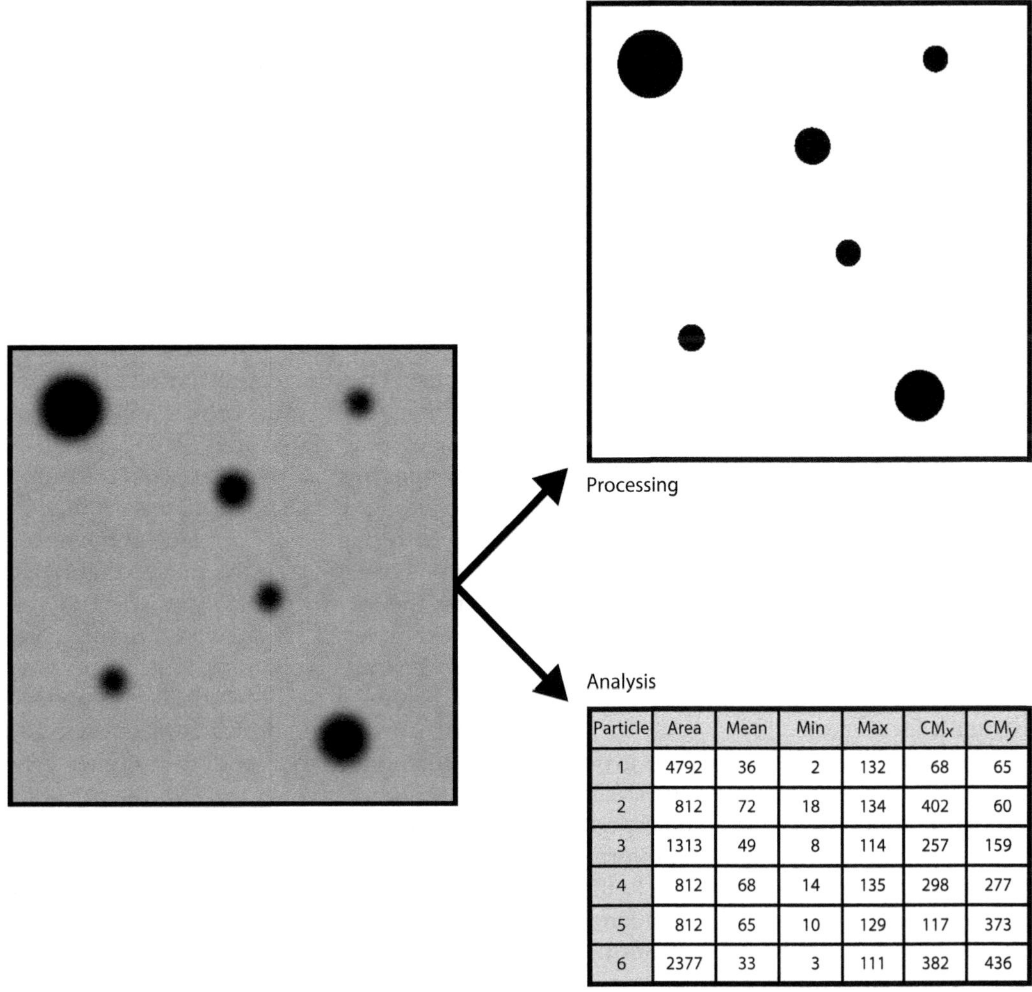

Particle	Area	Mean	Min	Max	CM_x	CM_y
1	4792	36	2	132	68	65
2	812	72	18	134	402	60
3	1313	49	8	114	257	159
4	812	68	14	135	298	277
5	812	65	10	129	117	373
6	2377	33	3	111	382	436

FIGURE 9.12
Schematic of the distinction between image analysis and image processing. In this example, a grayscale image has been processed into a black and white format and has been analyzed to quantify attributes of the spherical particles. CM = center of mass.

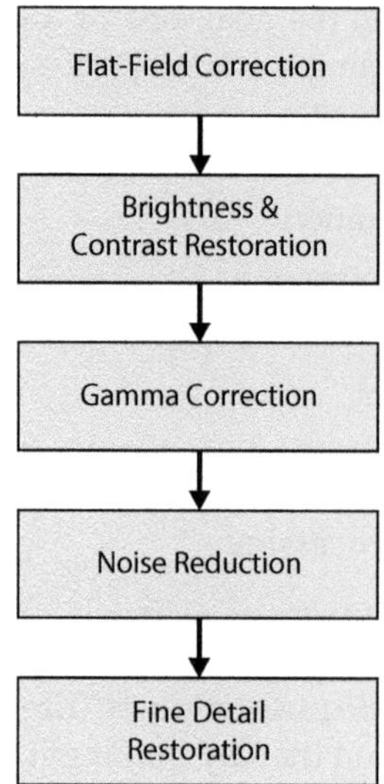

FIGURE 9.13
Flowchart of key steps in an exemplary image processing protocol.

9.4.1.1 Image Restoration

Image restoration involves correcting defects that arise from imperfections in the optics, illumination, detector, etc. There are many restoration tools and possible strategies. In this brief overview, we describe some key tools together with a strategy that is recommended on the Molecular Expressions Website (Fig. 9.13). The focus is on grayscale images. Color image production and processing are addressed in Sections 9.4.1.1.6 and 9.4.1.1.7.

9.4.1.1.1 Flat-Field Correction

The *first recommended restoration step* is flat-field correction, also termed shading correction, which removes degradation from sources like scratches in the optics, detector imperfections, and uneven illumination. The protocol involves generation of flat-field (FF) and dark frames (DF), ideally when the original (raw) image (RI) is collected. These additional images, together with the raw image, are used to implement the correction in Eq. (9.3). Flat-field and dark correction are especially important when the image is destined for quantitative analysis.

A flat-field frame, as the name suggests, is generated using a sample that is uniform (appears flat). A region containing just mounting medium is appropriate for approaches like brightfield microscopy. A uniform solution of dye with spectral attributes matched to those of the sample is appropriate for fluorescence microscopy. The associated image also should be uniform, with deviations from this ideal result arising from imperfections like uneven illumination. The flat-field frame should not be noisy; therefore, it can be produced as an average of multiple frames or from a long exposure.

Similarly, a dark frame is generated without exposing the camera to light, using the sample exposure time, but as an average of multiple frames to reduce noise. Ideally, the image counts in the dark frame would be zero. Instead, there will be nonzero contributions from electronic and thermal noise, and from camera bias offset, which is used to avoid negative numbers in the output. These contributions should be removed.

Image correction using FF and DF can be implemented using the equation

$$\mathrm{CI} = M\frac{\mathrm{RI}-\mathrm{DF}}{\mathrm{FF}-\mathrm{DF}} \tag{9.3}$$

where CI designates the corrected image, and M is a scaling factor calculated as the mean pixel value in FF − DF.

Often, simple background subtraction produces a visual effect that is similar to flat-field correction and thus may be a sufficient substitute if quantification is not the focus. However, the division-based procedure used in flat-field correction is more reliable than subtraction because image data arise from multiplication of flux and exposure time.

9.4.1.1.2 Brightness and Contrast Restoration

The *second step* is to correct brightness and contrast deficiencies in the image. Brightness is a measure of image "lightness/intensity," and contrast is a measure of the extent of grayscale or color variation within an image.

Two tools play an important role in the modification of image brightness and contrast. The first of these is the image histogram, which is a graph showing the allowed image intensities (determined by the number of bits per pixel) on the x axis and the number of pixels in the image corresponding to each intensity value on the y axis (Fig. 9.14). The value of the histogram stems from its use in assessing image quality and in identifying the steps that will aid in image rehabilitation. For example, an image with a tightly clustered histogram has relatively poor contrast because clustering is indicative of lack of variation. In such a case, established corrective approaches can be used to broaden the histogram and produce a higher contrast image (Fig. 9.14).

The second tool is the input-output lookup function or lookup table (LUT), which is a function or rule that transforms an input pixel value into a new output value. A very simple example of a LUT, and its use in contrast enhancement, is shown in Fig. 9.15. One important application of LUTs occurs with image display programs that use a LUT to convert pixel values in a digitized image into a format suitable for display on a computer monitor. For example, a LUT can map a

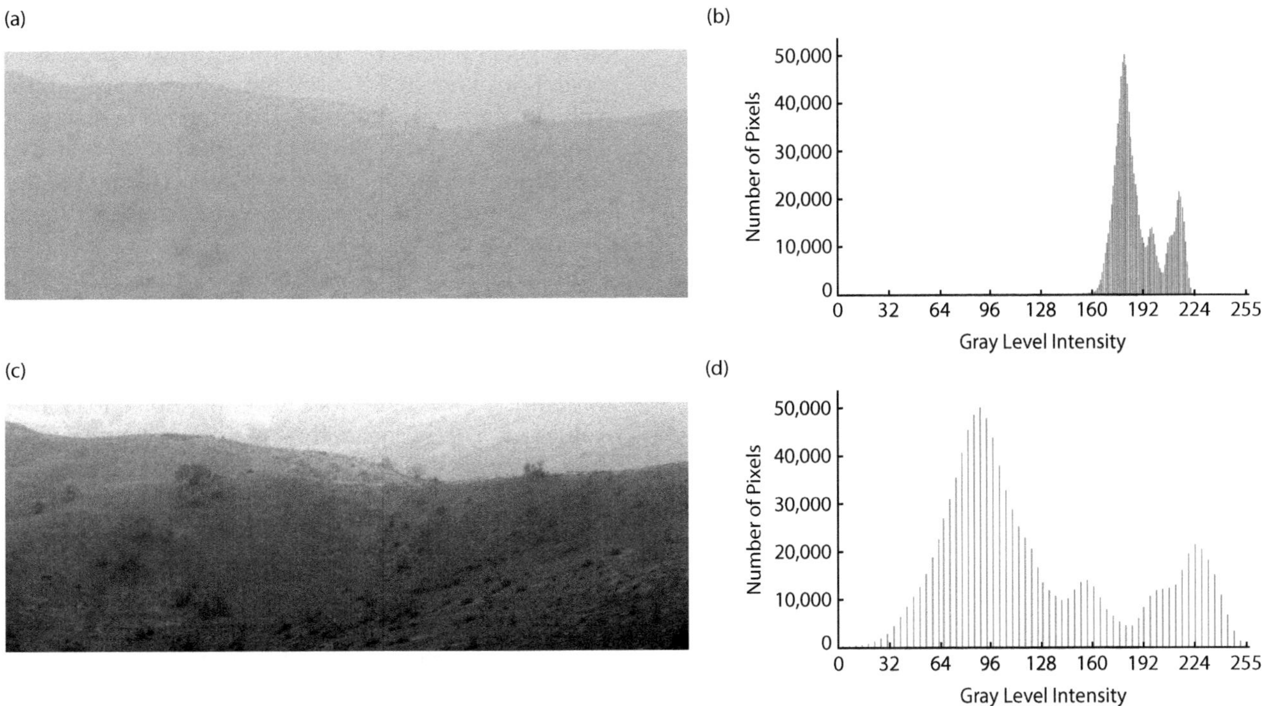

FIGURE 9.14
Comparison of low (a) and normal (c) contrast images and their associated histograms (b) and (d). (Images by authors.)

12-bit image with 4096 gray scales into an 8-bit image with 256 gray scales for monitor display by associating successive groups of 16 values in the 12-bit image with single values in the 8-bit image (e.g., 0–15 map to 0 and 16–31 map to 1).

Histogram stretching is an important example of using histograms and LUTs to implement contrast enhancement. To exemplify the process, we assume that we have a low-contrast 8-bit digital image (gray levels spanning a small subset of values from 0 to 255), and that we want to enhance contrast in the display on an 8-bit computer monitor (Fig. 9.16). The essence of the idea is to use linear interpolation to create a LUT that causes the histogram to span the entire gray level space available. This is accomplished by mapping the image minimum (Min) into zero and the image maximum (Max) into 255 and ramping linearly between the minimum and maximum to connect an Output and Input as follows (HW 9.8):

$$\text{Output} = 255 \times \frac{(\text{Input} - \text{Min})}{(\text{Max} - \text{Min})} \tag{9.4}$$

The visual impact of histogram stretching is shown in Fig. 9.14.

Histogram stretching is one of many contrast-enhancing options. Another is histogram equalization, which is discussed in Section 9.4.1.2.1. Because

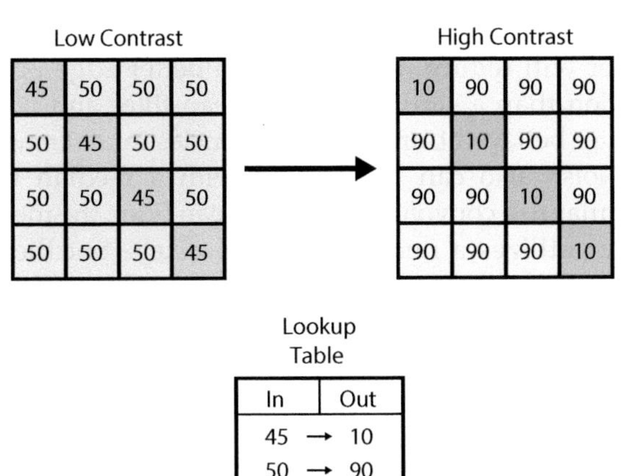

FIGURE 9.15
Simple example of a LUT that enhances image contrast.

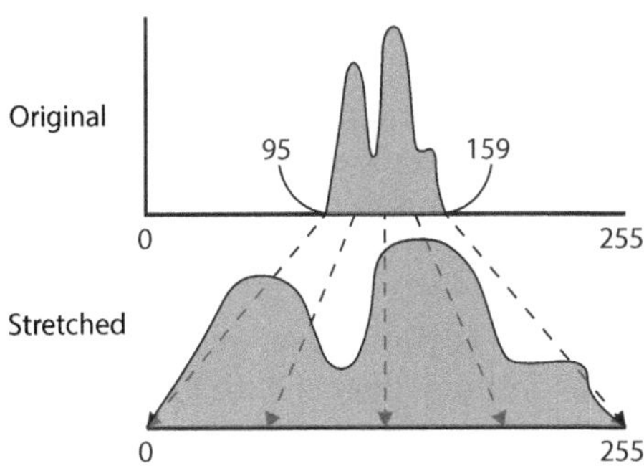

FIGURE 9.16
Schematic representation of the process of histogram stretching.

FIGURE 9.17
Illustration of the effect of gamma adjustment on image brightness and contrast. The original image (middle panel) is too bright at top and too dim at bottom, with both regions showing poor contrast. Gamma values greater than one (upper panel) improve contrast in bright regions. Gamma values less than one (lower panel) improve contrast in dim regions. (Images by authors.)

text space is limited, and experience is the best way to master these tools, methods of manipulating contrast and brightness in a digital image are explored in Activity A9.2.

9.4.1.1.3 Gamma Adjustment

The *third step* is to gamma adjust the image. This is a non-linear (exponential) adjustment of the histogram to facilitate simultaneous visualization of both bright and dim features (Fig. 9.17 and Box 9.2). For an 8-bit image, gamma scaling takes the basic form

$$\text{Output} = 255 \times \left(\frac{\text{Input}}{255} \right)^{\gamma} \tag{9.5}$$

Generally, $\gamma < 1$ enhances differences among dimmer pixels without saturating brighter pixels, and $\gamma > 1$ enhances differences among brighter pixels while deemphasizing dimmer pixels (Fig. 9.17).

9.4.1.1.4 Noise Reduction

The *fourth step* is directed at reducing image noise using a smoothing filter. Of the many filters used for smoothing, we focus on three common choices – mean, Gaussian, and median filters. Fourier filtering, which also can be used to reduce noise and/or periodic artifacts, will be discussed in Section 9.4.1.2.2.

The mean and Gaussian filters are examples of a more general class of "multi-pixel" image processing operations known as spatial convolutions. Spatial convolutions are more complex than the LUTs discussed previously because the new output value that they assign to a pixel contains contributions from *multiple* neighboring pixels. Often this is accomplished using a convolution kernel (Fig. 9.18).

To exemplify the idea, we consider a simple smoothing convolution (a mean filter – HW 9.9) that generates easily verified mathematics. Our example uses a convolution kernel that is a 3 × 3 array of pixels each containing the value one. A better choice for kernel values is 1/9 because this maintains approximate intensity parity between output and input. To address this issue, we will divide output pixel values by 9 at the end of the calculation (Fig. 9.18).

To update a pixel value using the smoothing kernel, the central value in the kernel is placed over the pixel to be modified, and the remaining kernel values are similarly positioned over neighboring pixels (Fig. 9.18). Each pixel in the image is multiplied by the overlying kernel value, and the nine resulting products are summed to yield the new, output pixel value (Fig. 9.18). The process is repeated until all pixels (here neglecting edges) are modified. The output is smoothed by the mean averaging process, as exemplified in the markedly reduced variation in the output in Fig. 9.18. Input pixel values vary by a factor of ~100, whereas output pixel values vary by less than a factor of two.

A positive feature of convolution-based smoothing filters is that they reduce the high-frequency (smaller scale) noise that typically degrades digital images. At the same time, these filters have the undesirable effect of blurring the image and thereby suppressing finer image details (Fig. 9.19). To minimize this problem, pixel values in the kernel can be made to follow a Gaussian function, which gives less weight to more distal neighbors. Like the mean filter, the Gaussian filter will reduce noise, but with better edge preservation than a similarly sized mean filter and with a better behaved transfer of frequencies to the image.

Noise also can be reduced with a median filter. Similar to a convolution-based mean filter, the median filter utilizes a neighborhood of pixels, but there is no kernel. Instead, a box (e.g., 3 × 3) is placed around a central pixel, and the pixel values are sorted. The value of the central pixel is replaced by the median pixel value (Sample Problem 9.3).

Median filters have several advantages. Notably, they deal effectively with more dramatic noise forms that prove troublesome for convolution-based approaches. An example is "salt-and-pepper" (S & P) noise, which arises from data transmission errors.

BOX 9.2 GAMMA SCALING

In the main text, we stated that $\gamma < 1$ will enhance dimmer pixels relative to brighter pixels, and $\gamma > 1$ will suppress dimmer pixels. To verify this assertion, it is useful to generate plots of output and input pixel values for representative values of γ using Eq. (9.5). Three examples are shown in the figure. Notably, the red curve shows that, when $\gamma = 0.5$, the dark pixels utilize more gray levels (the slope initially is large), and output intensities exceed those of the gray curve, which is a straight line with a slope of one. Thus, $\gamma = 0.5$ emphasizes differences among pixels with lower values. In contrast, the blue curve corresponding to $\gamma = 2.0$ has the opposite effect.

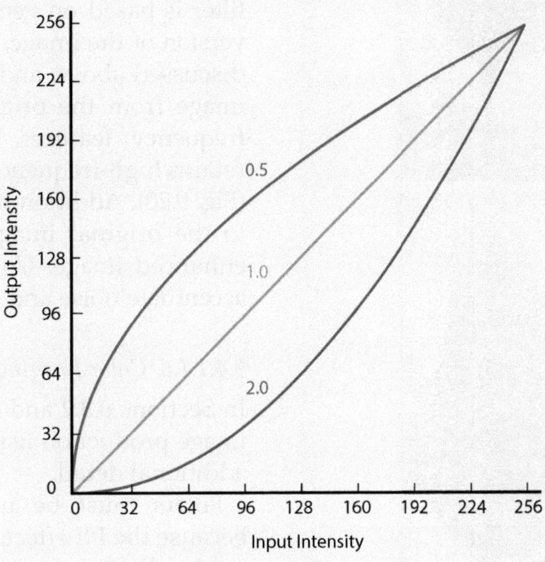

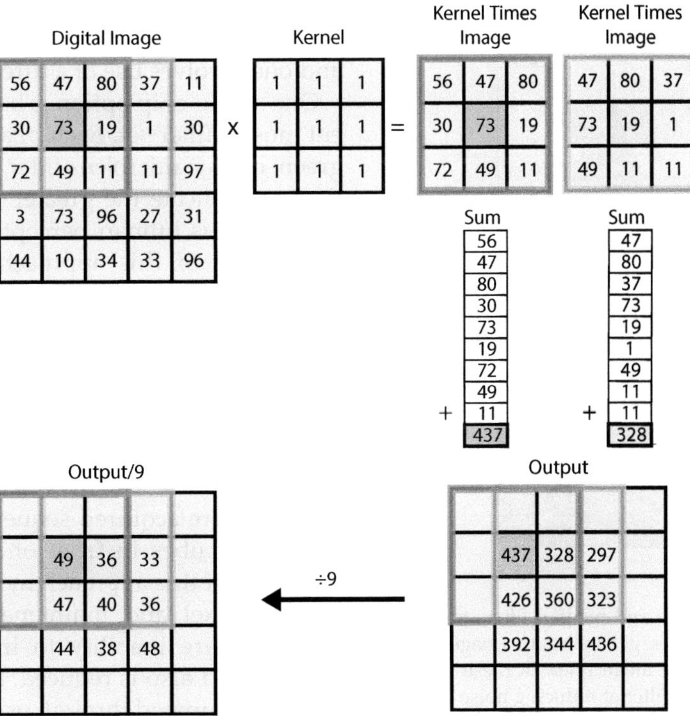

FIGURE 9.18
Schematic of the computations that are involved in mean kernel filtering. Computations are illustrated for the red pixel and its green neighbor to the right.

FIGURE 9.19
Comparison of the ability of mean and median filters to remove dramatic noise. The median filter generates an image that is sharper and less mottled than the mean filter. In these respects, it is more effective than the mean filter at reducing noise and preserving detail. (Images by authors.)

The median filter also is better at preserving fine detail than the kernel-based filters discussed above (Fig. 9.19). A primary negative of the median filter is that it is computationally intensive because extensive sorting is involved.

9.4.1.1.5 Fine Detail Restoration

The *fifth step* is directed at enhancing finer details, such as edges. One popular tool is the unsharp filter. This filter is based on generating an unsharp (smoothed) version of the image, using a tool like the mean filter discussed above, and then subtracting the smoothed image from the original image. This removes low-frequency features and generates an image that retains high-frequency, finer-scale features, like edges (Fig. 9.20). Addition of a portion of the "edge" image to the original image leads to a sharpened, edge-enhanced image. Unfortunately, unsharp filters can accentuate noise and thus must be used judiciously.

9.4.1.1.6 Color Imaging

In Sections 6.3.2 and 6.3.3, we briefly addressed color image production using filter cubes. Here we present additional detail.

Filters must be used to generate color images because the PE effect, which is the basis of light sensing by digital cameras, is not spectrally discriminating (once photon energy is adequate). Three popular options for generating color images are shown in Fig. 9.21. Two involve the use of a single-chip camera, and one involves use of a three-chip camera.

One single-chip option is based on use of an adherent mosaic filter set. Each pixel is covered with a red, green, or blue microlens filter. Green is twice as abundant to replicate the greater green sensitivity of the eye and thus human perception of color (Fig. 9.21a). A strength of this approach is that colors can be imaged simultaneously. On the other hand, each pixel responds only to red, green, or blue, so the actual color associated with a pixel must be reconstructed based on interpolation. Resolution and light sensitivity are also compromised somewhat.

Another single-chip approach involves the use of filter cubes/wheels. This is a method in which different colors are acquired sequentially by rotating filter wheels or cubes in front of the camera (Fig. 9.21b). Strengths of this approach include full color response of each pixel and maintenance of full resolution. Negatives are inability to image colors simultaneously. Speed also is reduced, but this problem can be largely eliminated through use of a liquid-crystal tunable filter, which has no moving components. Instead, color selection is altered rapidly via changing the applied voltage.

SAMPLE PROBLEM 9.3 MEDIAN FILTER

Median filters are used to eliminate uncharacteristic, or outlier, data points. Calculate the value that would be returned by a 3 × 3 median filter acting on the highlighted pixel in the following array.

174	173	177	180	179
188	182	171	189	178
182	185	5	188	173
176	180	190	181	185
186	189	174	177	183

SOLUTION

Median filters replace the value of interest with the median value in the filter window. A 3 × 3 median filter, illustrated below, takes into account nine values: the value of interest (at the center) and its eight nearest neighbors. The filter replaces the center value with the median of the nine values. It does so by sorting the values and then identifying the median as the middlemost value, with equal number of values above and below it. In this case, the median is the fifth highest, or fifth lowest, of the nine sorted values.

Unsorted values: 182, 171, 189, 185, 5, 188, 180, 190, 181

Sorted values: 5, 171, 180, 181, 182, 185, 188, 189, 190

The filter, in this case, replaces "5" with the median value, "182."

174	173	177	180	179
188	182	171	189	178
182	185	**182**	188	173
176	180	190	181	185
186	189	174	177	183

The median filter will always increase low values and decrease high values, making the image more homogenous. However, it eliminates outliers without allowing them to skew the results. In this case, it has eliminated a spuriously low pixel value (an example of "salt and pepper" noise). A mean filter would perform much more poorly (HW 9.9).

The third option involves use of a three-chip camera. A beam-splitting prism sends red, green, and blue to separate chips, as shown in Fig. 9.21c. This method permits simultaneous acquisition of all colors at full resolution, but camera cost is increased significantly.

Once acquired, color images commonly are displayed electronically in RGB format. This format generates a wide array of colors via weighted addition of the three primary colors red, green, and blue (HW 9.12). An alternative display mode – hue, saturation, and intensity (HSI) – that is especially useful for processing is described in the following section.

9.4.1.1.7 Color Image Processing and Color Coordinates

Here we address a few aspects of color image rehabilitation and processing.

One defect that commonly plagues (non-fluorescence-based) color images is inaccurate color balance, which can produce an overall color cast, such as excessive

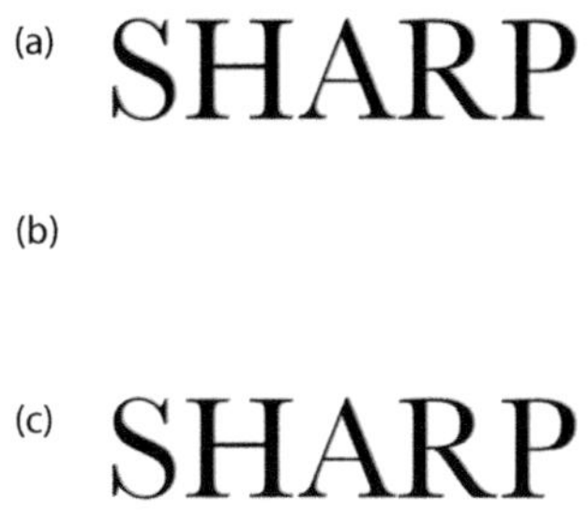

(a) SHARP

(b)

(c) SHARP

FIGURE 9.20
Illustration of the steps involved in implementing unsharp filtering. First, the original image (a) is lowpass filtered. Next, the filtered image is subtracted from the original image to generate a highpass image (b). Finally, a (fractionally) rescaled copy of the highpass image is added to the original image to achieve sharpening (c).

yellow. A common cause of this problem is a mismatch between the color output of the light source and the color balance calibration of the detector. This mismatch can arise because the spectral output of common illumination sources, such as tungsten lamps, depends on temperature (and thus operating voltage). If the operating voltage changes, and the camera's color circuitry is not adjusted accordingly, the detector will not respond correctly and image color will be incorrect. Age also can impact spectral output.

Color balance correction post-acquisition requires adjusting settings for the color channels/coordinates used to display the image. A useful assay is to examine an object that is known to be white or gray. If the object instead is colored, the color balance in the image can be adjusted by altering the RGB color coordinates. The average values of green, red, and blue over a gray object should be equal when color is balanced correctly.

Color image processing extends far beyond color balance adjustment. Color images can be smoothed and sharpened and processed in essentially all of the ways discussed previously. Importantly, despite the familiarity of RGB coordinates, color image processing is better implemented using HSI color coordinates.

In the HSI model, hue describes the dominant wavelength, saturation describes color purity, and intensity describes brightness (Fig. 9.22). Thus, the HSI color model largely decouples the brightness (intensity) component from the color components (hue and saturation). This is not true for the RGB model, and thus if individual RGB coordinates are processed and recombined, the output may suffer from color shifts (Fig. 9.22). In contrast, this problem can be avoided if a color image is converted to HSI coordinates and then only the intensity coordinate is processed. The brightness-altered image then can be recombined

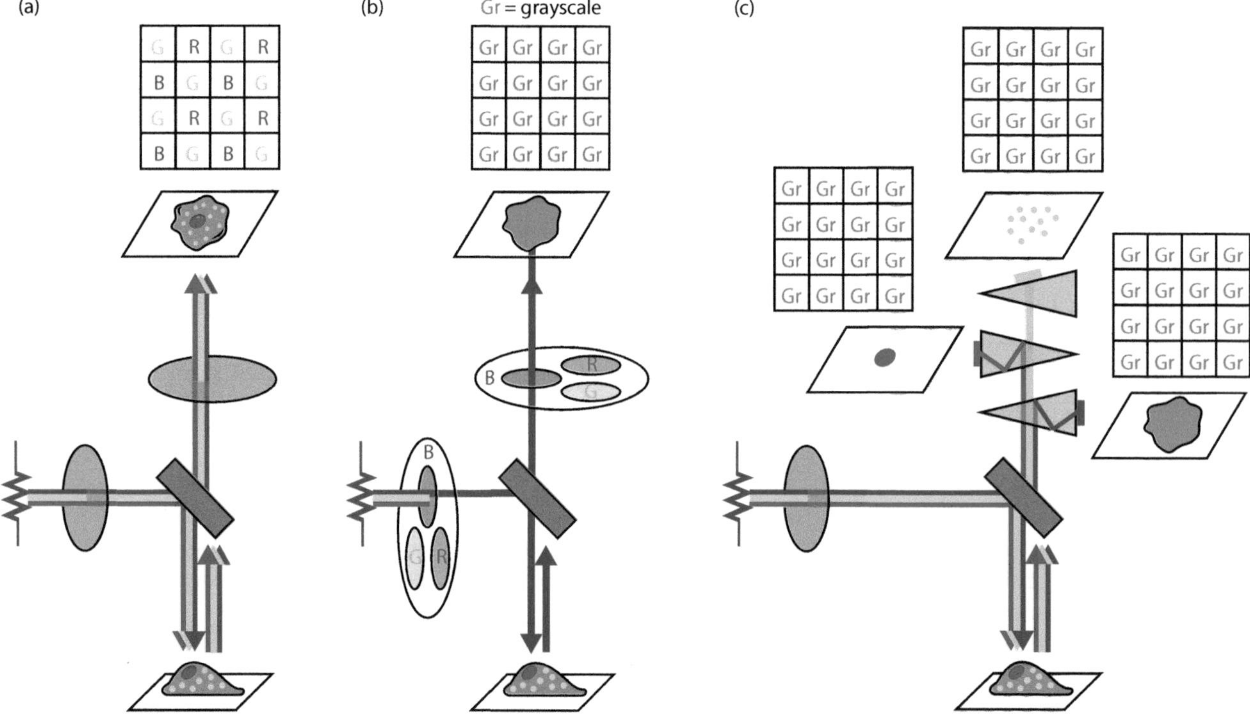

FIGURE 9.21
Schematic representations of options for color imaging. Single-chip options are shown in Panels (a) and (b), and a three-chip option is shown in Panel (c).

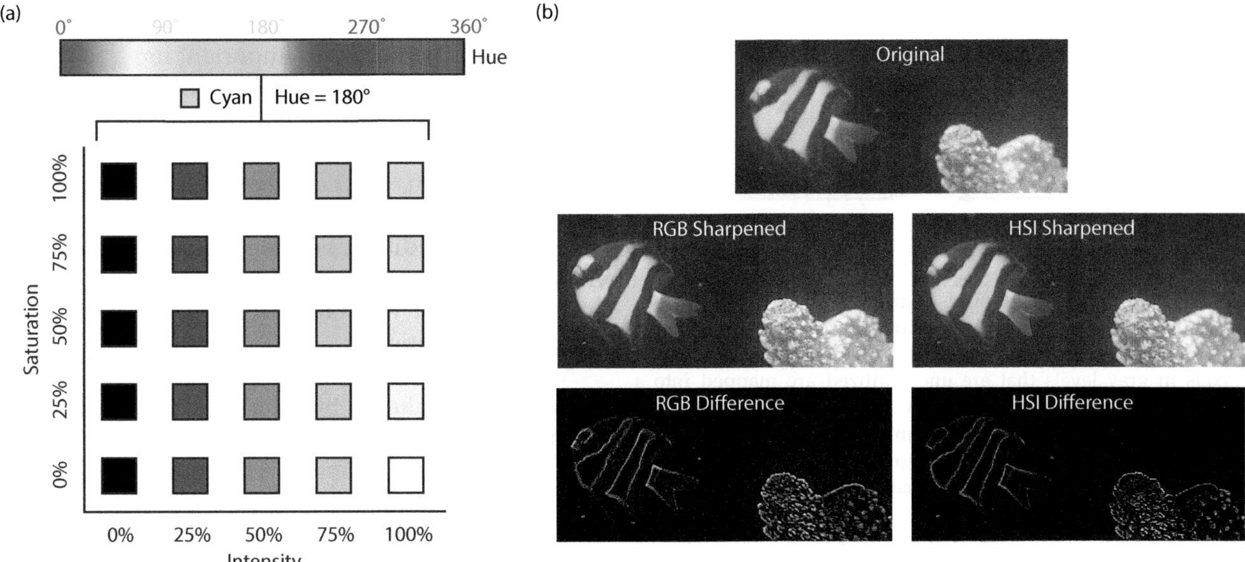

FIGURE 9.22
Illustration (a) of key features of the HSI color model, which is more closely linked to human perception of color than the RGB model. The HSI model also has distinct advantages for color processing. For example, application of a sharpening filter (b) to RGB coordinates can lead to color shifts that are avoided by using HSI coordinates. This is clear in difference (original minus sharpened) images. The RGB difference image shows that red was artifactually introduced into the RGB sharpened image of the fish. (Original image adapted from pdpics.com.)

with the original color components and converted back to RGB format. Color processing using HSI coordinates thus is generally the preferred approach.

9.4.1.2 Image Enhancement

Images can be enhanced, as well as restored. The distinction between restoration and enhancement is slightly murky, but restoration ideally is an objective process directed at correcting image flaws that arise during acquisition. An example is the use of flat-field correction to correct for uneven illumination. In contrast, enhancement is a more subjective process that is directed at altering an image to influence impact on the observer.

Many of the image-restoration tools discussed in the previous section also can be used to achieve image enhancement. Here we describe two additional enhancement tools – local contrast enhancement (LCE) and Fourier filtering.

9.4.1.2.1 Local Contrast Enhancement (LCE)

In Section 9.4.1.1.2, we discussed a global contrast enhancement method, histogram stretching, which maps every pixel in an image with a given brightness into the same new value (Eq. [9.4]). Pixel location, and thus neighboring pixel values, do not influence the output of global histogram stretching. Thus, histogram stretching does not adapt to local brightness in an image, and this can lead to significant contrast loss in small regions. In such cases, it may be useful instead to implement LCE and thereby increase brightness differences locally. For example, pixels that are brighter than their local surroundings can be made somewhat brighter, and pixels that are locally dim can be made still dimmer.

One of the most powerful methods of implementing LCE, used by the open source image processing program, Fiji™, is contrast limited adaptive histogram equalization (CLAHE). One major step in CLAHE is histogram equalization, which is a process that modifies a histogram to utilize gray levels more effectively, similar to histogram stretching. However, histogram equalization, unlike stretching, is aimed at distributing the histogram uniformly over all available gray levels to produce a histogram that is flat (Fig. 9.23). The algorithm used to map gray levels in histogram equalization is described by Fisher et al. (2003).

To achieve LCE, histogram equalization is implemented locally. A small box slides over the image, and the histogram in each small neighborhood is (more or less) flattened. The resulting, locally equalized histogram is used to update the center pixel value. Unfortunately, this process, known as adaptive histogram equalization (AHE), can lead to excessive noise (and contrast) amplification, especially in areas where the intensity in the image is relatively uniform. CLAHE eliminates this problem by using a "clipped-histogram" as explained in the legend to Fig. 9.24. The effect of CLAHE is exemplified in Fig. 9.24.

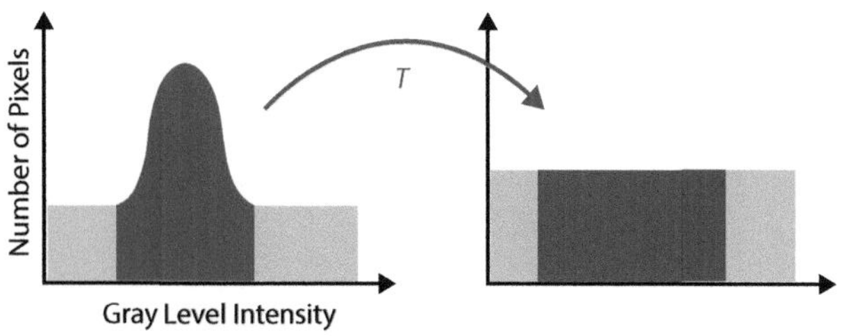

FIGURE 9.23
In histogram equalization, pixels in gray levels that are highly utilized (where the histogram is high) are spread over a larger range of gray levels to lower, and thereby level, the histogram. Similarly, pixels in gray levels that are underutilized are mapped into a smaller area to raise, and thereby level, the histogram. Histogram equalization thus involves a significant change of histogram shape. (Adapted from Wikipedia, "Histogram Equalization." https://en.wikipedia.org/wiki/Histogram_equalization)

9.4.1.2.2 Fourier Filtering

Fourier filtering is another important image enhancement, and restoration, tool. For example, Fourier filtering can be used to remove noise and to smooth images. We refrained from discussing Fourier filtering in Section 9.4.1.1 in part because this method already arose in the context of spatial filtering in Chapter 4. In addition, Fourier approaches are implemented in the "frequency" domain, and the focus in Section 9.4.1.1 was on spatially based methods of image rehabilitation.

As a brief reminder, in Chapter 4 we learned that images can be altered by manipulating (spatially filtering) the object diffraction pattern, which appears in the BFP of the objective lens. In essence, this was

(a)

(b)

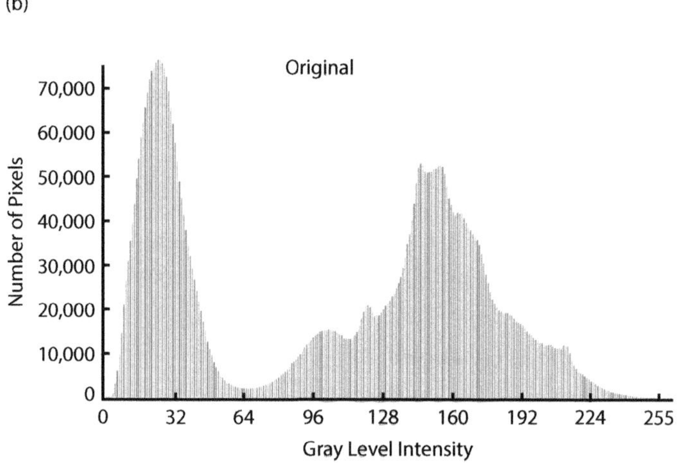

(c)

(d)

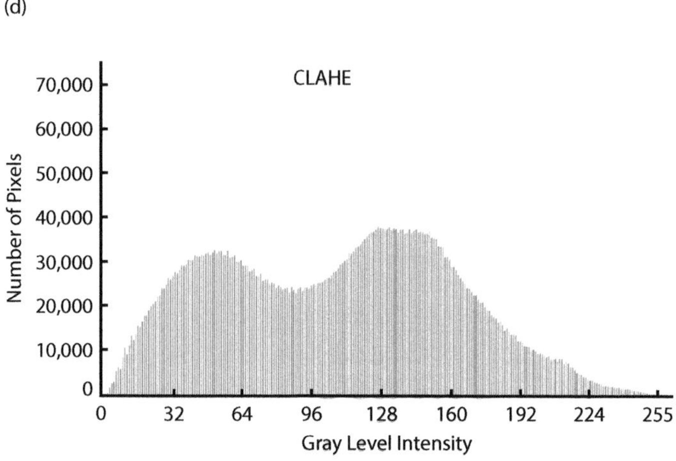

FIGURE 9.24
Illustration of the effects of CLAHE on an image (a and c) and its histogram (b and d). In CLAHE, the histogram is clipped. This limits contrast enhancement because the contrast enhancement at any intensity is proportional to the histogram height at that intensity. (Images by authors.)

an example of Fourier filtering because a diffraction pattern is closely linked to a Fourier transform (FT) (Box 4.3). The primary difference between spatial filtering and digital Fourier filtering is the source of the "transform." In spatial filtering, the objective optically creates an FT of the electric field, which retains all of its phase relationships. In digital FT, a computer generates a transform from the intensities of an image, in which phase relationships have been lost.

Like a lens-generated diffraction pattern, a computer-generated FT contains spots, which identify frequencies that encode details in the image (Fig. 9.25). Coarser details are encoded by more central spots in the FT, and finer details are encoded by more peripheral spots. Moreover, periodic features give rise to spots in the FT that are especially easy to identify, and these spots can be removed by "masking" if the periodicity is an undesirable artifact. For example, Fig. 9.25 shows a NASA image that was degraded by periodic noise, together with the associated image FT (lower left). Also shown is an FT that was "masked" to remove the spots that encode the noise. When the masked transform was back-transformed to the spatial domain, the resulting (enhanced) image was relatively noise free. (In our example, the FT helped eliminate artifacts, but it is important to ensure that it does not introduce artifacts due to limited retention of high frequencies or edge effects.) Additional applications of Fourier filtering are explored in Activity A9.2.

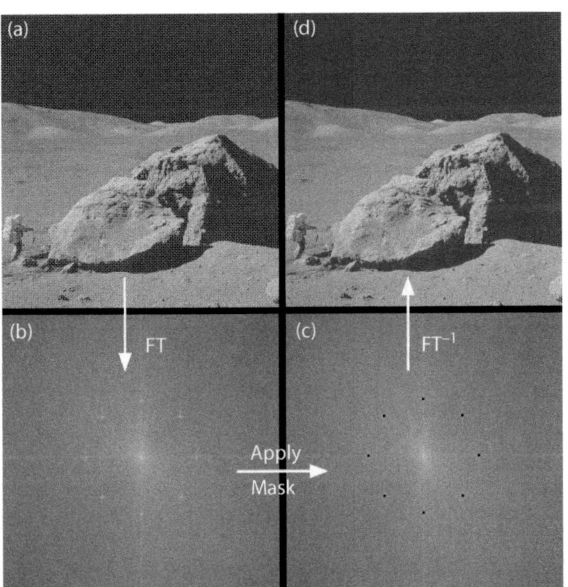

FIGURE 9.25
Example of using Fourier filtering to eliminate periodic noise. First, the original image (a) is Fourier transformed (FT) to generate a frequency-domain image (b) in which the noise is manifest as a set of eight spots located around a ring. Second, the eight spots are removed from the image by applying a mask (c). Finally, the masked image is inverse Fourier transformed (FT^{-1}) to yield a relatively noise-free filtered image (d). (Original image adapted from National Aeronautics and Space Administration [NASA].)

9.4.1.3 Binary Conversion/Image Segmentation

Image processing also encompasses procedures, such as conversion to a binary format, which produce fairly drastic visual changes. Binary images are simple digital images that contain one bit per pixel and thus two gray levels – completely black and completely white. Binary images are of interest in large part because conversion of a grayscale or color image into a high-contrast, black-and-white version often is a useful method of separating interesting features in an image from those that are uninteresting. In particular, binary conversion is a common way to "segment" an image into regions containing pixels with similar properties, usually with a goal of highlighting and identifying objects of interest (Fig. 9.26). Additional processing operations that commonly are applied to binary images are described in Box 9.3.

There are a number of methods that can be used to generate a binary image. Here we discuss histogram-based global thresholding. In the simplest application of this approach, the objects of interest and the (uninteresting) background features differ markedly in intensity, and the histogram thus contains two distinct peaks (Fig. 9.26). In this case, a single threshold in the valley between the peaks can be used to separate objects and background. Pixels with intensity values that are less than the threshold are set to black and those with intensities greater than the threshold are set to white (or vice versa) to produce a binary, segmented image (Fig. 9.26).

In a slightly more sophisticated application, objects of interest might correspond to a central peak in a multi-peak histogram, and then two thresholds can be used to convert the image into a binary format (Fig. 9.26). To this end, pixels in the range bounded by the thresholds could be set to white, and those outside the range set to black.

Unfortunately, not all images are easily segmented using simple global thresholds. Problems with the global approach tend to arise when peaks in the histogram run together, which makes it difficult to separate objects from background (Fig. 9.26). In this case, adaptive thresholding, which is similar in spirit to AHE, may prove useful.

9.4.2 Image Analysis

Image analysis is an immense topic that is directed at extracting numerical data from images. In the interest of brevity, we close this chapter with one example: single particle tracking (SPT). Specifically, we discuss

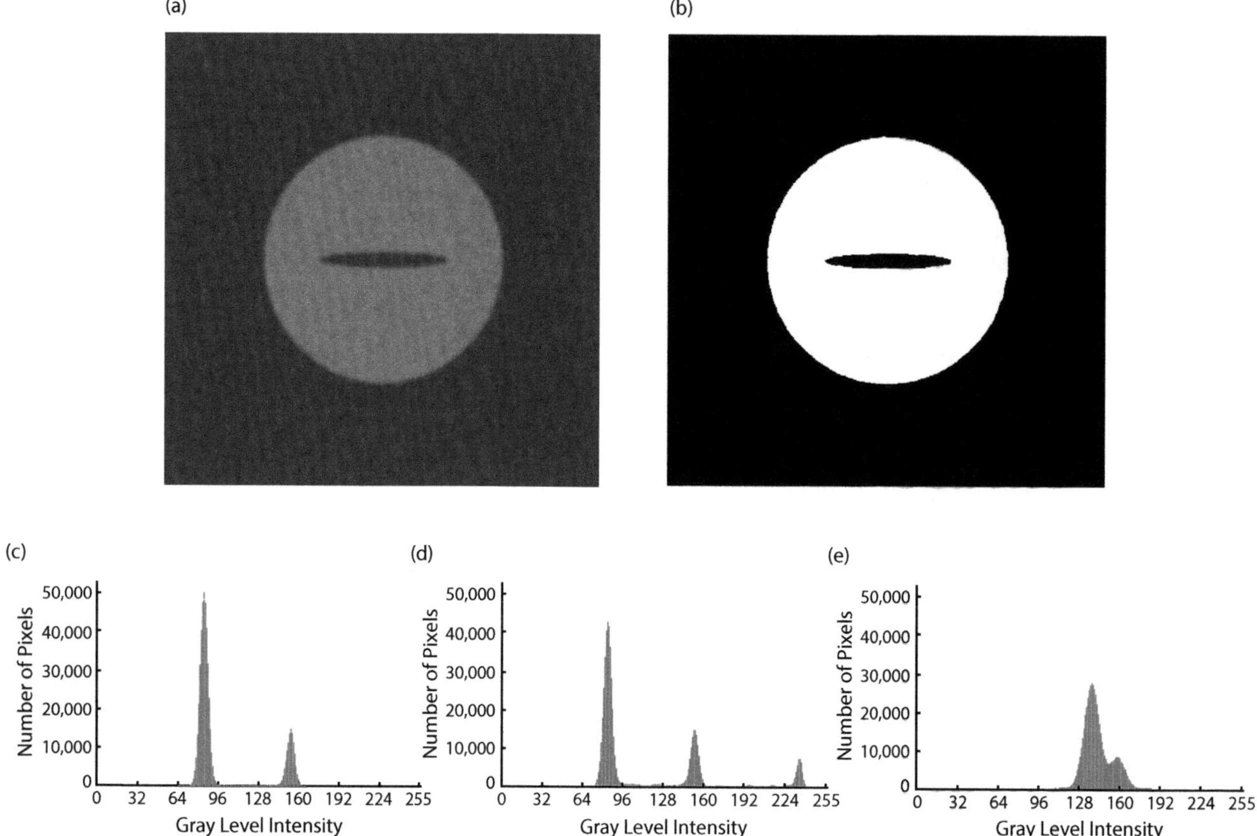

FIGURE 9.26
Schematic example of histogram-based global thresholding and image segmentation. The original image (a) was converted into a binary format and segmented (b) via global thresholding of its associated two-peak histogram (c). Images described by a three-peak histogram (d) also are readily segmented using global thresholding, whereas images with a more blurred histogram (e) are not.

BOX 9.3 ADDITIONAL IMAGE PROCESSING TOOLS

Here we augment our treatment by listing a few additional prominent processing tools and image operations in succinct tabular form. These include the morphological (shape-related) operations of erosion, dilation, opening, and closing, which often are applied to binary images, and the pixel-by-pixel arithmetic operations of addition, subtraction, multiplication, and addition. The reader interested in a comprehensive discussion is referred to the books by Gonzalez and Woods (2017) and Russ and Neal (2016).

Operation Name	Brief Description	Application/Effect
Erosion	Reduces the boundaries of areas of bright (foreground) pixels	Remove small structures and retain large ones
Dilation	Enhances the boundaries of areas of bright (foreground) pixels	Enhance structure visibility and fill small holes
Opening	Erosion followed by dilation	Erode with less "perturbation" to foreground pixels
Closing	Dilation followed by erosion	Dilate with less perturbation to boundary shape
Addition	Add two images or add a constant to an image	Reduce noise via frame averaging
Subtraction	Subtract two images or subtract a constant from an image	Detect motion or determine the differences between two images (Fig. 9.22)
Multiplication	Multiply two images or multiply an image by a constant	Brighten or darken an image via rescaling by a constant
Division	Divide two images or divide an image by a constant	Quantify image differences using a ratio; implement shading correction (Eq. [9.3])

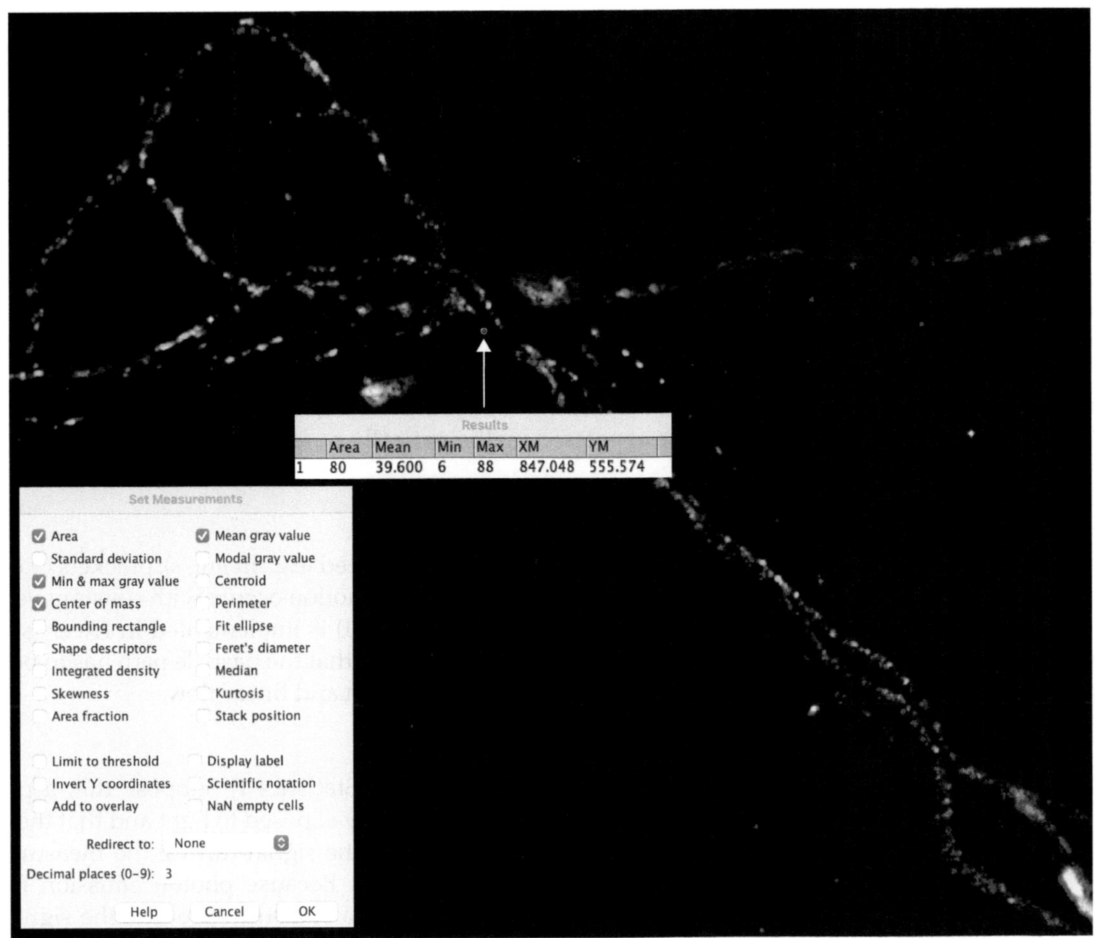

FIGURE 9.27
Example of using Fiji™ to implement SPT of DCVs in a cultured hippocampal neuron. (Image by authors.)

SPT of subcellular organelles in movies generated using time-lapse fluorescence microscopy. We also discuss how to use the SPT data to determine organelle trajectories, speeds, and diffusion coefficients.

The physics and mathematics underlying SPT are discussed in Box 9.4. SPT implementation using Fiji™ is outlined here and explored in Activity A9.3. In brief, implementation involves outlining the organelle in a frame of a movie, and using the Fiji™ measurement option to obtain the x and y coordinates for that frame, which Fiji™ places in a "Results" table (Fig. 9.27). The outlining and measurement processes are repeated in successive frames to produce coordinates as a function of time, t. In the approach adopted in this text, tracking data are entered into Microsoft Excel® for more sophisticated analysis, such as generating plots of mean-squared and root mean-squared displacement as a function of time.

BOX 9.4 SINGLE-PARTICLE TRACKING

In Box 6.1, we discussed extracting diffusion coefficients of cellular constituents using ensemble-based fluorescence techniques (FRAP and FCS). In this box, we discuss monitoring diffusive (and directed) motion by tracking individual particles; this alternative approach is known as SPT. We focus on tracking organelles because these larger structures typically give a fluorescence signal that is robust enough to implement SPT.

To supplement the theory discussed here, we have designed Activity A9.3, which involves manually tracking fluorescently tagged DCVs in a movie using Fiji™. If desired, the tracking data also can be entered into an Excel® spreadsheet that performs calculations required to fit the data to Eq. (B9.4.1). The spreadsheet outputs speeds and diffusion coefficients and the quality of the fit to directed and diffusive motion.

Organelle transport typically can be classified as directed or diffusive. Directed motion is mediated by motor proteins that propel organelles along cytoskeletal tracks. In contrast, diffusive motion is mediated by random molecular forces, such as those arising from solvent bombardment. An organelle might undergo diffusive motion if it detaches from the cytoskeleton.

Directed and diffusive motion produce very different relationships between "displacement" and time. Stated somewhat loosely, displacement depends linearly on time for the case of directed motion (e.g., distance = speed × time) and on the square root of time for the case of diffusive motion (HW 9.13). In SPT, it is traditional to quantify the displacement squared. The signature of directed motion is a parabolic dependence of displacement squared on time, whereas the signature of diffusive motion is an analogous linear dependence (Eq. [B9.4.1]).

More rigorously, in SPT, tracking data are used to compute the mean squared displacement, $\langle r^2 \rangle$, of a vesicle as a function of time, t, and the data are fit to a model that includes the effects of "diffusive" and/or "directed" motion. Such a model can be expressed mathematically as follows:

$$\langle r^2(t) \rangle = 4Dt + \text{v}^2 t^2 \qquad \text{(B9.4.1)}$$

In Eq. (B9.4.1), D is the diffusion coefficient (e.g., in $\mu m^2/s$), and v is speed (e.g., in $\mu m/s$); brackets ⟨ ⟩ denote averaging. The model underlying Eq. (B9.4.1) assumes that directed motion occurs with constant velocity. An established method for computing the time average in Eq. (B9.4.1) is implemented in Excel® spreadsheets that accompany Activity A9.3. One caveat with this analysis is that the particle path has to be quite long, or many paths compared, to reliably quantify a mix of diffusion and linear flow.

Suggested Reading and Additional Resources

Fellers, T.J. and Davidson, M.W. 2015. *Digital image capture and processing. Recommended strategy for processing digital images*. Molecular Expressions. https://micro.magnet.fsu.edu/optics/olympusmicd/imageprocessing/processingsteps.html

Fisher, R., Perkins, S., Walker, A., and Wolfart, E. 2003. *The hypermedia image processing reference*. https://homepages.inf.ed.ac.uk/rbf/HIPR2/welcome.htm

Gonzalez, R. and Woods, R. 2017. *Digital image processing (4th Edition)*. Pearson, Upper Saddle River, NJ.

Russ, J.C. and Neal, F.B. 2016. *The image processing handbook (7th Edition)*. Boca Raton, FL: Taylor and Francis Group.

Stelzer, E.H.K. 1998. Contrast, resolution, pixilation, dynamic range and signal-to-noise ratio: fundamental limits to resolution in fluorescence light microscopy. *J. Microscopy*. 189:15–24.

HOMEWORK PROBLEMS

9.1 [Dynamic Range] A CCD camera includes an array of 13 μm × 13 μm pixels. Hint: See Box 9.1.

a. Calculate the full well capacity of the pixels, assuming that each pixel can hold about 1000 electrons/μm^2.

b. Calculate the dynamic range (DR), assuming that the detector noise is approximately 10 electrons/pixel.

c. Assume that the signal is digitized using an n-bit digitizer. What value of n is well matched to this DR?

9.2 [Poisson Statistics I] Suppose that a pixel is repeatedly exposed to light and that the mean value of the signal during the measurement time is μ. Because photon emission from a source is a statistical process, the signal will vary for the different measurements, where the probability of obtaining the signal result, m, is given by (see figure)

$$P_P(m, \mu) = \frac{\mu^m}{m!} e^{-\mu}$$

This is the famous Poisson distribution. To get a feel for the attributes of the distribution, evaluate and plot $P_P(m, 5)$ for $m = 0$ to 10. What are the most probable values of m? How do these relate to the mean? What is the approximate spread in results? To address spread, estimate the m values at which the probability becomes roughly half maximal and compute their difference. How does this range compare with $2\sigma = 2\sqrt{\mu}$ (HW 9.3)?

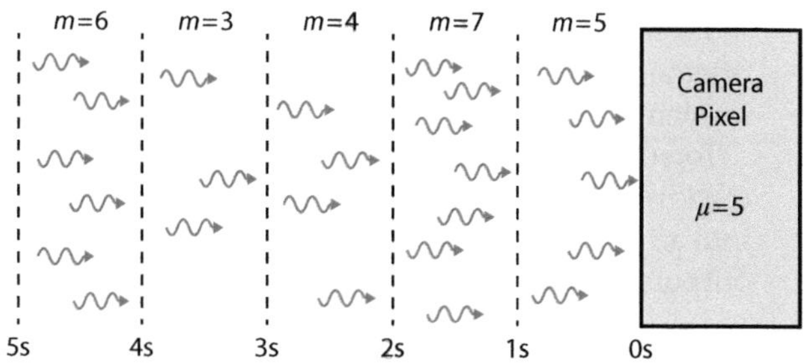

9.3 *[Poisson Statistics II] The Poisson distribution was introduced in HW 9.2. The goal here is to analyze the properties of the distribution rigorously.

a. Verify that the sum of all probabilities is unity by demonstrating that $\sum_{m=0}^{\infty} P_P(m, \mu) = 1$.
b. Verify that the mean is μ by demonstrating that $\sum_{m=0}^{\infty} P_P(m, \mu)m = \mu$.
c. Verify that the standard deviation, $\sigma = \sqrt{\mu}$, by demonstrating that the variance $\sigma^2 = \sum_{m=0}^{\infty} P_P(m, \mu)(m-\mu)^2 = \mu$.

9.4 [Signal-to-Noise Ratio] Calculate the SNR for a digital image with attributes as follows. The specimen signal is 1000 ADU and extends over a square that is three pixels on a side, and the background signal is 30 ADU/pixel. Assume that dark noise is negligible, that read noise is 5 e^- per pixel, and that the camera gain is four. Hint: Use Eq. (9.2) to convert signals in ADU into signals in e^- counts before evaluating Eq. (9.1).

9.5 [Appropriate Camera Gain] Camera gain commonly is chosen to match the full-well capacity of the pixels to the largest number output by the digitizer. Use this rule to determine the appropriate gain for a camera with a full-well capacity of 100,000 electrons that uses a 12-bit digitizer.

9.6 [Camera Gain] The goal of this problem is to verify Janesick's method of calculating camera gain, g. The method is valid when the signal is strong so shot noise dominates, and it relates signal and noise in ADU, which are measurable from the digital image, to the camera gain.

a. To start the derivation, relate the noise (σ_{e^-}) in electron units to the signal expressed in electron units, N_{e^-}, when read noise can be neglected.
b. Use Eq. (9.2) to relate the signal and noise in electron units to their analogs in ADU, N_{ADU} and σ_{ADU}.
c. Use the results from (a) and (b) to show that $g = N_{\text{ADU}} / \sigma_{\text{ADU}}^2$. This is Janesick's method.

Hint: See Section 9.2.

9.7 [Nyquist Sampling] Calculate the largest CCD pixel size that will still allow a microscope system having a 40×/0.7 NA objective to achieve diffraction-limited spatial resolution when viewing 550-nm light. What happens qualitatively to the resolution if the pixels are twice as large as this maximum size? What happens qualitatively if the pixels are half as large? Hint: See Sample Problem 9.2.

9.8 [Histogram Stretching] Derive Eq. (9.4).

9.9 [Mean Filter] The "mean" filter is similar to the median filter described in Sample Problem 9.3, except that it replaces the value of interest with the mean value in the window instead of the median value. The mean value is the straight average of the values in the window and is obtained by summing the values and dividing the total by the number of values. Thus, for a 3 × 3 mean filter, the mean would be the sum of the nine values in the window divided by nine. Calculate the value that would be returned by a 3 × 3 mean filter operating on the center value ("5") in Sample Problem 9.3. How does it compare with the value returned by the median filter? Explain how the difference reflects the different mechanisms used by mean filters and median filters to deal with outliers (uncharacteristic values).

9.10 [Median Filter] Calculate the value that would be returned by a 5 × 5 median filter acting on the value of interest in Sample Problem 9.3.

9.11 [Edge Filter] Smoothing filters, such as the mean and median filters, are used to reduce high-frequency image features, such as those associated with noise. In contrast, edge filters are used to identify abrupt changes in images, like those associated with edges, and to construct new images that emphasize those changes. One popular edge filter, known as the Sobel filter, uses a 3 × 3 kernel. The kernel is asymmetric, unlike the mean and median filter kernels, because it is used to detect systematic differences between pixels to the left and right of, or above and below, the pixel of interest. The following kernel is used to detect vertical edges:

Kernel

1	0	−1
2	0	−2
1	0	−1

a. Apply the filter to the 9 × 9 image in Panel (a). This image contains two vertical edges. The resulting image should be 7 × 7 pixels. The calculations can be performed using a spreadsheet, such as Excel®. Alternatively, the calculations can also be performed manually by recognizing that the simplicity of the image greatly reduces the number of required calculations.

(a) Vertical Edges

10	10	10	45	45	45	90	90	90
10	10	10	45	45	45	90	90	90
10	10	10	45	45	45	90	90	90
10	10	10	45	45	45	90	90	90
10	10	10	45	45	45	90	90	90
10	10	10	45	45	45	90	90	90
10	10	10	45	45	45	90	90	90
10	10	10	45	45	45	90	90	90
10	10	10	45	45	45	90	90	90

(b) Horizontal Edges

10	10	10	10	10	10	10	10	10
10	10	10	10	10	10	10	10	10
10	10	10	10	10	10	10	10	10
45	45	45	45	45	45	45	45	45
45	45	45	45	45	45	45	45	45
45	45	45	45	45	45	45	45	45
90	90	90	90	90	90	90	90	90
90	90	90	90	90	90	90	90	90
90	90	90	90	90	90	90	90	90

b. Now apply the filter to the 9 × 9 image in Panel (b). This image contains two horizontal edges (and corresponds to Panel (a) following a 90° clockwise rotation). The result should again be 7 × 7 pixels. How does the result differ from the result in Panel (a)?

c. How would you change the vertical edge filter so that it would recognize horizontal edges instead?

9.12 [RGB Model] A broad spectrum of colors can be generated from weighted combinations of RGB coordinates. To demonstrate this with specific examples, use Fiji™ (Activity A9.1) to determine the colors (e.g., pink) that are generated with the following RGB weightings:

a. R = 139, G = 83, B = 18.

b. R = 255, G = 3, B = 255. What well-known color is generated when you instead set G = 0?

c. R = 255, G = 201, B = 0.

Note: Fiji™ uses 8-bit color. Therefore, R = 139 means 139/255 × 100% = 54.5% red. G and B have analogous meanings.

Hint: Open the Fiji™ Color Picker (the eyedropper icon that is roughly in the middle of the main window). Then open the foreground or background selection at the bottom of the window. Type in the RGB values specified in the parts above.

9.13 [Estimating Diffusion Times] Assume that a protein has a diffusion coefficient $D = 1\ \mu\text{m}^2/\text{s}$. Estimate the time that it would take the protein to diffuse

a. Across a typical cell with a radius of 10 μm.

b. Down a typical axon of length 1 cm.

c. Comment on the viability of diffusion as a transport mechanism for these two cases.

Hint: Use Eq. (B9.4.1) but set v = 0. This technically is valid for diffusion in two dimensions, but correction for diffusion in a different number of dimensions is a minor issue.

ACTIVITIES[1]

A9.1 Noise and Resolution

The purpose of this activity is to confirm that a poor SNR will degrade resolution. You will use the open source Fiji™ image processing package. Download instructions are provided in Activity A6.1.

A. Open A9.1_yeast.tif. This is a two-image stack of fluorescently labeled yeast vacuoles. The first image has a relatively good SNR, and the second has a poor SNR.

B. Click on the straight-line tool in the toolbar. Often this is fifth from the left.

C. For the bright (first) image, draw a horizontal (zero angle) straight line through the center of the image. Angles, etc. are reported at the bottom of the Fiji™ main window.

D. Go to Analyze > Plot Profile to obtain a plot of intensity versus distance in pixels. Save the plot.

E. Advance to the next image (e.g., using the ">" symbol) and generate a profile plot for an identical line on the dim image. Save the plot.

F. Is it possible to quantify the locations of the vacuole edges from the plot obtained using the bright image?

G. Same question for the dim image.

[1] Activity files are available for download in the online Support Materials.

A9.2 Image Processing

This activity explores several common image processing techniques using the open source Fiji™ image processing package.

1. [Smoothing] This part explores mechanisms for image smoothing, specifically, the ability of median and mean filters to reduce or eliminate salt and pepper noise.
   A. Open Meerkat_noise.tif using Fiji™. This image contains very noticeable salt and pepper noise (i.e., random bright and dark pixels).
   B. Use the median filter (Process > Filters > Median) to reduce the noise. The term "pixel radius" used by this filter specifies the size of the filter kernel: 3 × 3 for radius = 1, 5 × 5 for radius = 2, etc. What is the optimum choice for pixel radius? Why? Save your results.
   C. Repeat the process using the mean filter (Process > Filters > Mean). What do you observe as the radius of the filter is increased?
   D. Compare the results produced by the two filters. Which filter is best for salt and pepper noise?
2. [Enhancing Contrast] This part explores a mechanism for enhancing contrast, specifically, histogram stretching.
   A. Open Meerkat_bright.tif. This image is too bright.
   B. Determine the range of pixel values in the image using Analyze > Histogram. Save the histogram using File > Save. What are the low and high values? What aspect of the histogram predicts that the image will look too bright?
   C. Enhance the image using histogram stretching (Image > Adjust > Brightness/Contrast). Adjust "Min" and "Max" appropriately, and hit "Apply." Save the enhanced image. What values did you choose for "Min" and "Max"? Why?
   D. Determine the range of pixel values in the stretched image using Analyze > Histogram. Save your result. Compare with the histogram of the original image determined in part (B).
3. [Sharpening] This activity explores aspects of image enhancement, specifically, sharpening. Edges and other fine details are first identified by taking a difference between the original image and a blurred version of the image. The image is then sharpened by adding the difference image to the original image:

   $$\text{Sharpened} = \text{Original} + (\text{Original} - \text{Blurred}) \times \text{Scaling Factor}$$

   A. Open Polygon.tif (the "original" image).
   B. Blur the original image using Convolver (Process > Filters > Convolve.) You will need to construct a suitable blurring filter (we recommend a 5 × 5 filter of all 1's) and to choose "normalize." Save the output as Polygon_blur.tif (or similar name) (the "blurred" image).
   C. Subtract Polygon – Polygon_blur to get the edge image Polygon_edge (Process > Image Calculator > Subtract). Note: The edge image is nearly black because differences in the two images are small. However, it can be visualized using histogram stretching. Go to Image > Adjust > Brightness/Contrast and hit "Autoscale" – but don't hit "Apply" and instead hit "Reset" before the next step.
   D. Sharpen the original image by adding Polygon_edge to Polygon. Save your result as Polygon_sharpened. You can also add a fraction of Polygon_edge, corresponding to application of the "Scaling Factor" in the equation above, using "Process > Math > Multiply."
   E. Compare your results above with results produced by the Fiji™ Unsharp filter (Process > Filters > Unsharp Mask).
4. [Fourier Filtering] This activity explores Fourier filtering. Fourier image filtering is a software analog of spatial filtering in the BFP of the objective.
   A. Open Clown.tif.
   B. Compute the fast Fourier transform (FFT) of the image (Process > FFT > FFT).
   C. Scale the background of the FFT image so that it is completely black. To do this, click the "Color Picker" icon on the Fiji™ toolbar. A color palette should appear, with rectangular foreground (F) and background (B) icons at the bottom. Click on the background icon, which will bring up a window for Background Color. Set R = G = B = 0.
   D. Filter the FFT image using a circular mask. First, choose Process > FFT > Make Circular Selection. Next, choose a radius of

100 pixels. Finally, remove the frequency components in the circle using Edit > Cut (the circle should go black).

E. Reconstruct the image from the filtered transform using Process > FFT > Inverse FFT. You should enhance contrast in the image using the technique in part 2 of Activity A9.2. Save both the filtered image and the filter (flatten the filter first using Image > Overlay > Flatten) and explain the change in appearance of the image in terms of the nature of the filter.

A9.3 Image Analysis

This activity explores a common image analysis technique used to track objects in a movie. The technique is SPT, and the objects are DCVs in cultured rat hippocampal neurons.

A. Open Movie_Tracking.tif (using File > Open).

B. Choose "Set Measurements" under the "Analyze" menu. Check the center-of-mass option, which will calculate the x and y coordinates of the center-of-mass of the vesicle to be tracked.

C. Next, identify the vesicle in frame 1 that is marked with the red arrow, and surround it using the ellipse tool in the Main menu. Often this tool is second from the left in the toolbar, and you need to "click" to pull up the ellipse option.

D. Once the vesicle has been surrounded, hit the letter "M" to generate the specified measurements, notably, the coordinates of the vesicle, which Fiji™ places in a "Results" table (Fig. 9.27).

E. Continue the process of surrounding and measuring for *nine* frames in the movie. Run the movie forward using the ">" key and backward using the "<" key.

F. After tracking is complete, save the Results table (File > Save As), which then typically can be opened in Microsoft Excel®. If Excel® protests a bit, try to ignore it.

G. Also track the vesicle marked with the blue arrow in frame 21 for *nineteen* frames.

H. Paste your tracking data (XM, YM) into the Excel® spreadsheet, Tracking Data.xlsx. For the vesicle in frame 1, start pasting in row 4 (columns A and B). For the vesicle in frame 21, start pasting in row 22. Use the resulting data to generate trajectories, speeds, and diffusion coefficients.
The basis of this analysis is described in Box 9.4 (Eq. [B9.4.1]). In brief, in SPT, distance squared is traditionally plotted as a function of time. In directed motion, distance equals speed times time (assuming particles are traveling at constant speed). Therefore, distance squared is proportional to time squared. In contrast, in diffusive motion, distance squared is equal to a constant times the diffusion coefficient times time. In other words, distance squared is directly proportional to time.

I. Construct a scatter plot showing distance squared as a function of time for vesicle 1. What is the shape of the plot? Construct a second scatter plot showing the square root of distance squared as a function of time. What is the shape of the replotted data? What type of motion does vesicle 1 undergo?

J. Construct a scatter plot showing distance squared as a function of time for vesicle 2. What is the shape of the plot? What type of motion does vesicle 2 undergo?

Section II

Medical Imaging

10

Ultrasound

We turn now to medical imaging, which comprises the last four chapters in the text. Unlike our discussion of optical microscopy, which was organized thematically into physical phenomena like resolution and contrast enhancement, the chapters on medical imaging are organized based on technique, and issues like resolution and contrast are addressed separately for each approach. Applications are discussed at the start and the end of each chapter; they also are sprinkled throughout each chapter as appropriate. We begin with ultrasound (US), which is the only approach not founded on the use of EM radiation.

10.1 Essence of the Technique

We present a detailed discussion of US imaging in future sections, but it is useful to start with a brief synopsis of its key features. US is a technique that maps body structures based on echo detection, similar to sound navigation and ranging (SONAR – Fig. 10.1). In brief, a device known as a transducer is used to generate and direct US into the body and to detect US that is reflected or scattered back by tissues and other anatomic structures.

Several attributes of US make it useful for medical imaging. These include a good balance between penetration of the body and interaction with body structures. US transit times within the body also are readily measurable. In particular, echo transit time is used to calculate tissue depth, and echo amplitude is used to generate contrast (Fig. 10.1b). Large amplitude echoes arise from reflections at interfaces with big "mismatches," like soft tissue/bone interfaces, and these appear as very bright areas in US images. Much weaker echoes are generated by reflections at interfaces between "similar" structures, such as different soft tissues, which appear as lighter, variable shades of gray. Because the reflections from soft tissues are relatively weak, some US is transmitted through the interfaces and can be used to generate images of deeper structures (Fig. 10.1b).

US has many strengths and a few weaknesses. It is the technique of choice for fetal imaging during pregnancy because high-resolution images are collected in real time, with no risk to the mother or fetus. US also is a powerful method of imaging the heart and monitoring blood flow and is widely used for screening of soft tissues and organs in the abdomen and pelvis. Significantly, US is cost-effective and portable, and US instruments are widely available. There also is no requirement that a patient remain motionless for a protracted period (like MRI), nor is there use of ionizing radiation (like X-ray or gamma ray-based methods). On the other hand, US is not suited to imaging within bone, including the adult skull, or deep within aerated lung, and has a limited ability to see deeper structures in large patients.

10.2 Attributes of US

US imaging uses sound waves of high frequency, from ~2 to 15 MHz (megahertz). In contrast, audible sounds for humans fall in the range from ~15 Hz to 20 kHz (kilohertz).

In Chapter 2, we mentioned a few properties of sound waves. Here we highlight additional important attributes of sound and briefly reiterate some of the key points mentioned previously. To begin, sound waves differ in several key respects from EM waves. One important difference is that sound is a mechanical wave and thus requires a medium, such as air or water, to support its propagation. Another important difference is that sound is a longitudinal wave, which means that the molecular displacements and pressure disturbances associated with the wave occur along, rather than perpendicular to, the direction of wave propagation (Box 2.1).

Sound waves also have attributes in common with EM waves. For example, the speed of propagation of sound is medium dependent. One important value is the average speed of sound in "soft tissue," which is ~1540 m/s.

The interactions of sound with matter, and associated manifestations, also in many respects mirror those of EM waves. These processes, which include reflection, refraction, transmission, scattering, and absorption (Fig. 10.2), are of paramount importance in

DOI: 10.1201/b22076-12

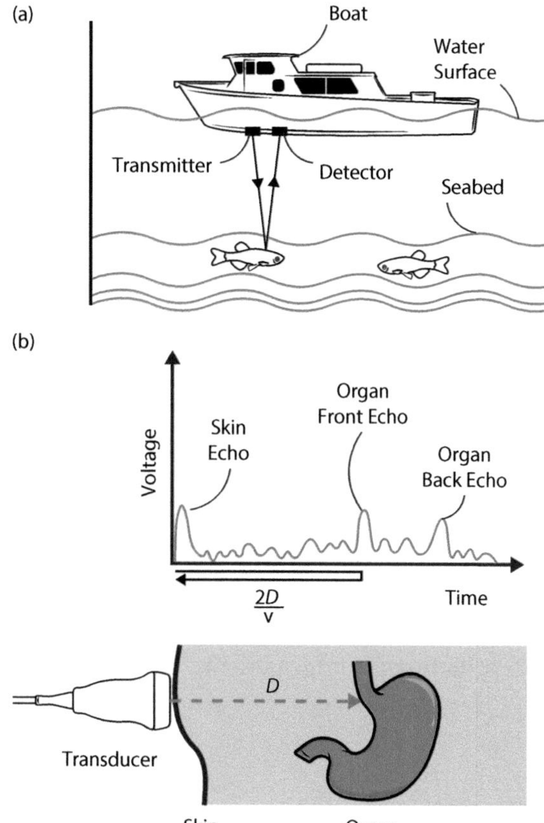

FIGURE 10.1
Cartoon depictions of (a) SONAR and (b) the role that reflection and transmission play in generating an US image. US is both generated, and the echo detected, by a transducer. If an echo is detected at a time, t_{echo}, after the signal was initiated, the tissue depth, D, is determined as $D = v \times t_{echo}/2$, where v = 1540 m/s is the speed of US in the body. The factor of two reflects the fact that the echo time accounts for a traversal to, and from, the tissue.

US-based imaging because they generate the signatures of body structures – depth and brightness – that are used to create an image.

10.3 Generating and Detecting US

US is both generated and detected with a transducer. Transducers are a class of device that convert energy from one form into another. In the case of US transducers, electrical energy is converted into mechanical energy of the outgoing wave using the reverse piezoelectric effect, and mechanical energy of the incoming wave is converted into electrical energy using the direct piezoelectric effect (Fig. 10.3).

The functional component of the transducer is a piezoelectric material. Piezoelectric materials used in medical imaging, which typically are a ceramic crystal (e.g., lead-zirconate-titanate), consist of ordered positive and negative charges with a characteristic equilibrium charge configuration and associated unperturbed element thickness. When a short voltage spike is applied to a resonance transducer element, charge moves, and the element deforms, compressing

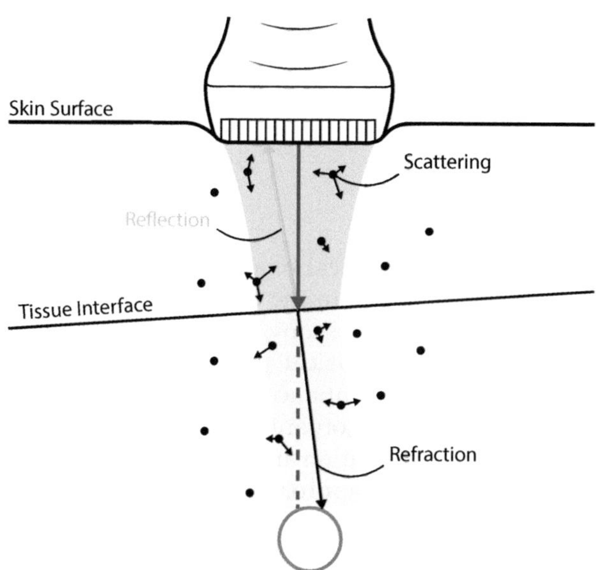

FIGURE 10.2
Schematic representation of several key interactions between US and body structures.

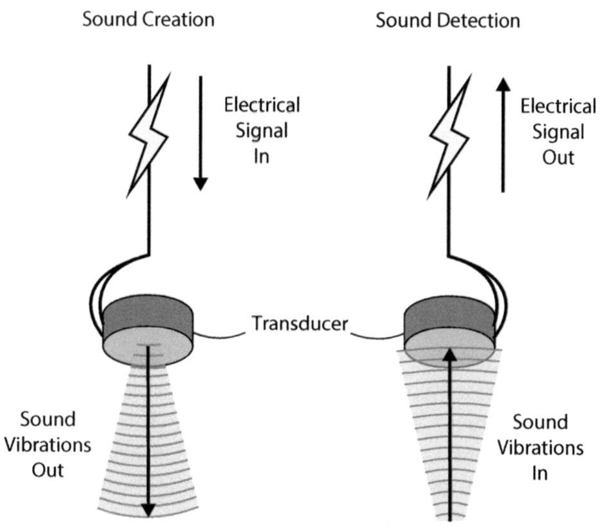

FIGURE 10.3
Essence of how a transducer functions based on the reverse and direct piezoelectric effects. When a voltage is applied, charge redistributes, producing structural change and inducing vibration. Conversely, when an external stress is applied, charge again redistributes, and a voltage is induced. One of the unusual, and exceptionally useful, features of the piezoelectric effect is this "reversibility."

and expanding. The element preferentially oscillates at its resonance frequency and generates an associated US wave. Conversely, when an incident US wave impinges on the element, it compresses and expands. These thickness changes produce an altered charge distribution and an associated voltage with amplitude that is proportional to the incident pressure amplitude; the echo amplitude and time are used to create an image.

Transducers include several other important components, as shown in Fig. 10.4. One is a damping block behind the piezoelectric element, which has several functions. For example, the block is a source of damping; it reduces vibration and facilitates the generation of short US pulses, which are required for good axial resolution (Fig. 10.5 and Section 10.6.2). Another important component is a matching layer in front of the piezoelectric element. This layer reduces the "mismatch" between the transducer and patient, thereby reducing reflection and allowing transmission into the body (Section 10.5). Some additional attributes of transducers are described in Fig. 10.5.

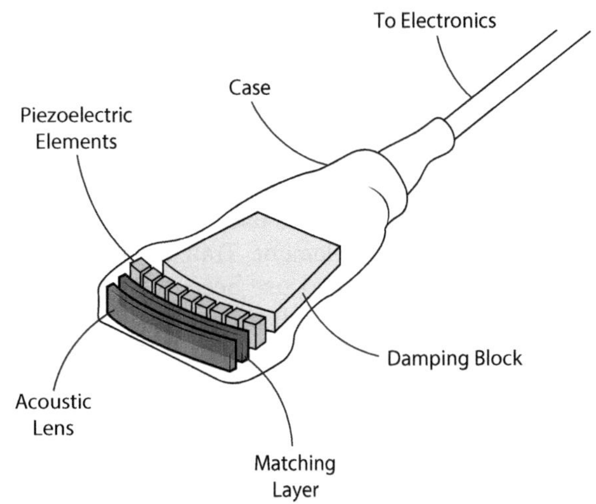

FIGURE 10.4
Key components of a transducer. The damping block absorbs back-directed US and attenuates stray US from the housing. Also shown is a concave acoustic lens, which can be used to focus the wave (e.g., along the elevational direction – Section 10.6).

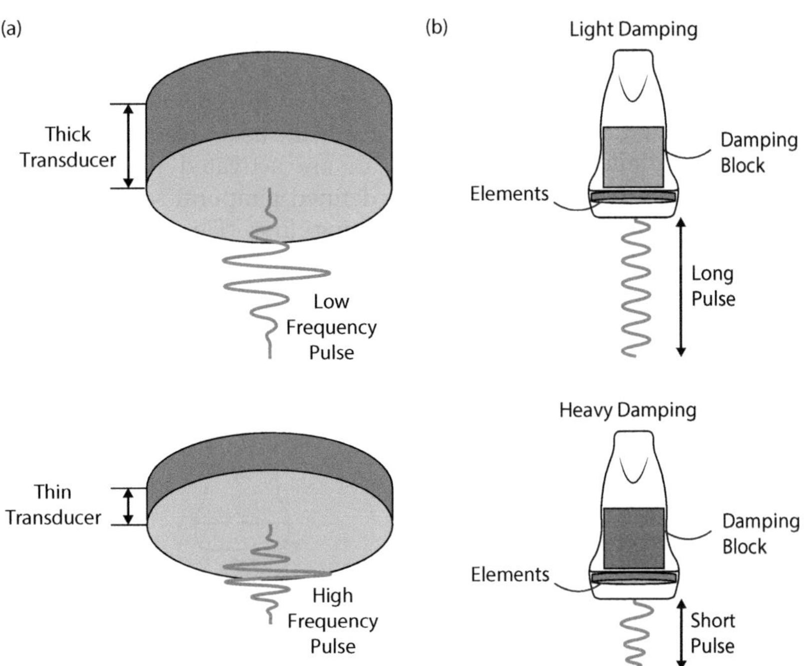

FIGURE 10.5
Pulse attributes of a resonance transducer. (a) Resonance transducers preferentially transmit and receive at the resonance frequency, $\nu = v_{crystal}/2L$, where $v_{crystal}$ is the speed of sound in the crystal element, and L is the thickness of the crystal. Thus, the frequency of a resonance transducer is determined by element thickness, with thicker elements producing lower frequency US. (b) The damping block reduces the purity of the resonance and generates a range of frequencies with width determined by the degree of damping. High damping creates shorter pulses with broader bandwidth (frequency range), and low damping creates longer pulses with narrower bandwidth. The latter are used in Doppler US. Modern multi-frequency transducers can generate and detect several frequency ranges around different "center" frequencies.

10.4 Transducer Design and Beam Attributes

Transducers can consist of one, or multiple, piezoelectric elements. This design attribute affects beam profile, which, in turn, influences FOV and resolution.

Early US imaging systems used transducers having a single piezoelectric element. Transducers with this design are still in limited use because they are cost-effective and, in some applications, such as ophthalmology, an appropriate choice. For a single element, unfocused transducer, the beam width varies with depth, achieving a minimum width that is the approximately the transducer radius, r, at a depth r^2/λ.

Single element, unfocused transducers have several relevant limitations. One is that scanning is implemented mechanically. Another is relatively poor lateral resolution. For example, an unfocused, 1-cm radius transducer will at best yield 1-cm lateral resolution because lateral resolution is dictated by beam width (Section 10.6.1). Much improved (mm) lateral resolution can be achieved by focusing the beam using an acoustic lens or by focusing the beam electronically (Fig. 10.4 and HW 10.2). This latter type of focus is achieved using multi-element transducers, as discussed next.

Most modern US transducers used in medicine contain an array of piezoelectric elements that can implement electronic scanning and also potentially focus and steer the US beam. Transducer attributes, including operating frequency, dimensionality of the array, shape, and the method of sequencing the activation of elements, are tailored to the clinical application. Here we focus on transducers that consist of a one-dimensional array and have a linear shape and distinguish two methods of sequencing element activation to create a two-dimensional, cross-sectional image: linear and phased.

Fig. 10.6 shows schematics of linear- and phased-array transducers. A linear array operates via simultaneous activation of a small subset of elements, whereas a phased array operates via nearly simultaneous activation of all elements. The implications of these differences are clarified below.

Linear arrays are large, consisting of 256–512 elements, and they have a large rectangular FOV that is limited to the region directly in front of the transducer. Linear arrays operate via simultaneous firing of a small set (~20) of adjacent elements, generating a beam that is directed perpendicular to the array (Figs. 10.6a and 10.7a). Additional rounds of firing are generated by shifting the active set by one or more elements. Each echo is detected, synthesized, and assigned to the center of the beam at the appropriate depth. A two-dimensional image is built up from all of the echoes collected in the scan.

Phased-array transducers typically contain 64–128 elements. A two-dimensional image is acquired by sweeping (steering) the beam through multiple angles to scan the ROI (Fig. 10.6b). Phased-array probes produce a sector-shaped FOV that is significantly wider than the transducer footprint; they thus are useful for directing an US beam through a small window that bypasses the ribs and the lungs, permitting imaging of the heart.

In a phased-array transducer, all elements typically are used to generate the beam and collect echoes. To focus the beam electronically, the individual elements all are activated nearly simultaneously in a well-defined temporal sequence, with the outer elements firing first (Fig. 10.7b and HW 10.4). Interference between the slightly delayed waves produces a

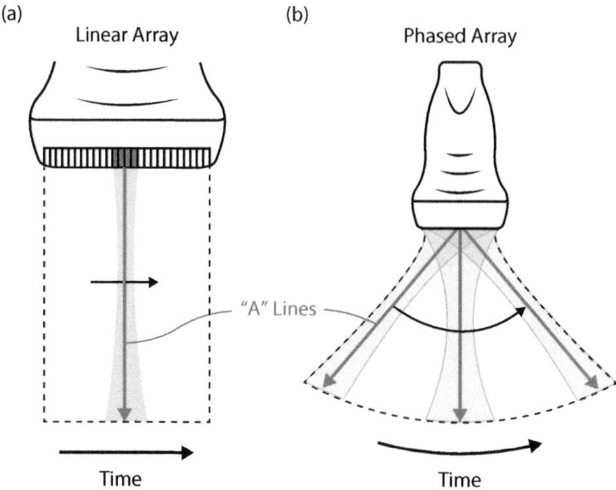

FIGURE 10.6
Creation of a two-dimensional scan using a linear array probe (a) and a phased-array probe (b). An "A" line of data is a set of echoes acquired along one line in the body (Section 10.5.1.1).

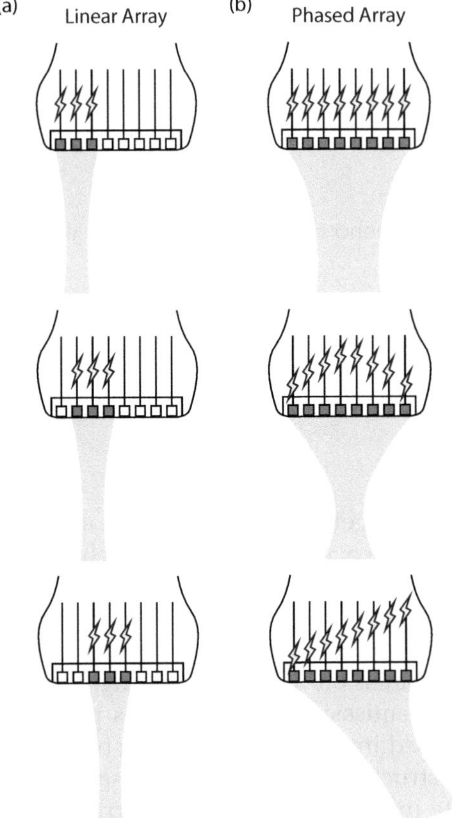

FIGURE 10.7
Element timing patterns and associated beam patterns for linear array (a) and phased-array (b) transducers. Linear arrays are not steered, so the FOV is directly in front, but the activated elements can be fired in a pattern that focuses each line. Steering of the phased array produces a sector-shaped FOV. To implement steering, the beam must be sent along different directions. Thus, the phasing must change for different repetitions (HW 10.3). One phasing pattern, which directs the beam to the left of the sonographer, is illustrated. Yellow lightning bolts denote firing signals, with relative distances between a bolt and its element denoting timing.

converging wave with focal distance determined by the timing delay (phasing); long delays produce a shorter focal distance. Steering, and combined steering and focusing, can similarly be accomplished with slightly different firing patterns (Fig. 10.7).

10.5 Image Formation

In this section, we will discuss image formation, including three echo-display modes – A (amplitude), B (brightness), and M (motion) modes – as well as two- and three-dimensional US image formats.

To start, we need to introduce some key concepts and terminology. First, images are generated using a pulse-echo format (Fig. 10.8). This means that the outgoing US is produced in pulses, where each pulse typically is only a few wavelengths long. These pulses and the associated US energy primarily are confined to a fairly narrow "beam." The transducer generates the outgoing pulses intermittently and spends most of its time detecting echoes (HW 10.5 and 10.6). The time from the beginning of one pulse to the beginning of the next pulse is called the pulse repetition period (PRP). Its inverse, the pulse repetition frequency (PRF = 1/PRP), is the number of pulses/second. A typical PRP is ~500 μs, corresponding to a PRF = 2 kHz. The PRF is not the same as the transducer frequency (i.e., the frequency of the US wave), which is on the order of MHz.

Structure depth in US is encoded in echo time. Several assumptions go into converting echo time into depth and location. First, the propagation speed is assumed to be v = 1540 m/s = 0.154 cm/μs, which is the average speed of US in soft tissue. Second, each pulse and the echoes resulting from that pulse are assumed to travel along the same straight-line transmitted primary beam path between source and reflector (Fig. 10.8). Given this, if the straight-line distance/depth is D, the total distance traveled is $2D$, and the echo time, t_{echo}, can be converted into depth using the expression $2D = v \times t_{echo}$ or

$$D = \frac{0.154\, t_{echo}}{2} \tag{10.1}$$

In Eq. (10.1), the echo time is in μs, and D is in cm. Sample Problem 10.1 uses Eq. (10.1) to verify the "13-μs rule."

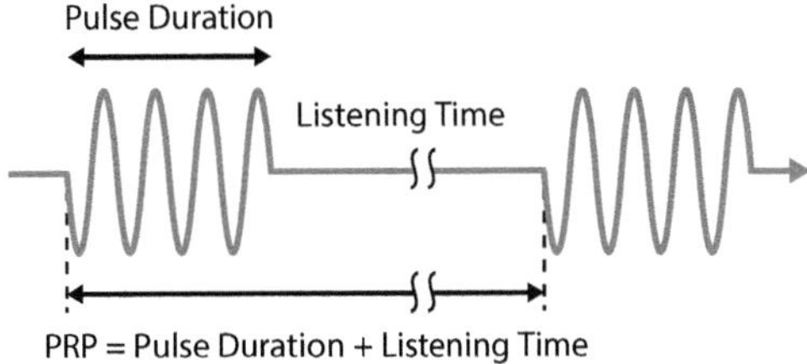

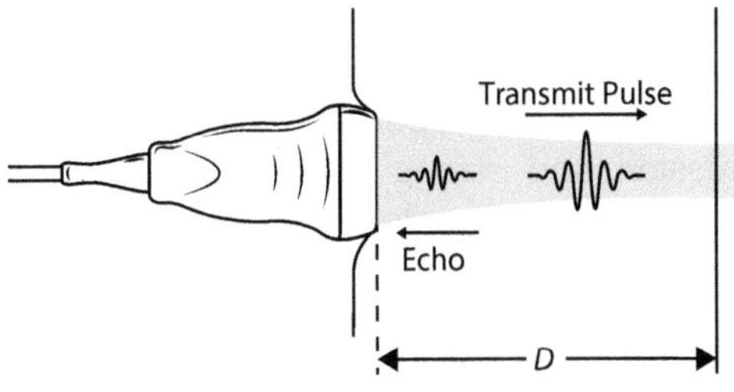

FIGURE 10.8
Examples of the roles of pulses in US-based imaging. The shaded region represents the beam, i.e., the spatial envelope through which most of the US energy travels.

SAMPLE PROBLEM 10.1 VERIFICATION OF 13-μS RULE

Verify the 13 μs rule, which states that each additional 13 μs of "go-return," or "echo," time corresponds to one additional cm of structure depth.

SOLUTION

Equation (10.1) describes the relationship between depth, D, and echo time, t_{echo}. Evaluating this expression for $t_{echo} = 13\ \mu s$,

$$D = \frac{(0.154\ \text{cm}/\mu\text{s})(13\ \mu\text{s})}{2} = 1.0\ \text{cm}$$

shows that 13 μs does correspond to an additional depth of 1.0 cm.

Standard US images display echoes as a white dot on a black background (Section 10.5.1.2). Dot, and thus tissue, brightness is determined by echo amplitude, which is dictated by the strength of any detected specular reflection, diffuse reflection, and/or scattering (see Section 10.7 for more detail). Strong signals appear "whiter" in an image. Relative specular reflection strength can be readily quantified in terms of the reflection coefficient, R, which, for the case of normal incidence, is given by

$$R = \frac{I_r}{I_i} = \frac{(Z_1 - Z_2)^2}{(Z_1 + Z_2)^2} \tag{10.2}$$

The associated transmission coefficient $T = 1 - R$. In Eq. (10.2), I_i and I_r are the incident and reflected intensities, and Z_1 and Z_2 are the characteristic acoustic impedances of the two tissues that form the reflecting interface. As a benchmark, typical intensities at the transducer surface are on the order of 10 mW/cm^2. Z can be calculated from the expression $Z = \rho v$, where ρ is density and v is speed. The standard unit of Z is the "rayl." Equation (10.2) is an exact analog of Eq. (2.8a), which describes reflection of EM waves at normal incidence. A derivation of Eq. (10.2) for the case of arbitrary incidence angle is given in Box 10.1.

Equation (10.2) implies that specular reflections from bone/soft tissue interfaces and air/soft tissue interfaces are very strong (Sample Problem 10.2). Transmission is correspondingly very weak. Weak transmission and strong attenuation (Section 10.7) by bone explain why only superficial images of bone are captured with US. Highly reflecting and attenuating structures like bone also can create image artifacts, termed acoustic shadows, which are echo-free, dark areas behind the obstruction. In a similar spirit, weak transmission at an air/skin interface explains the need for coupling gel between the transducer and skin; the gel is used to eliminate air pockets and thereby reduce impedance mismatch and enhance transmission (Fig. 10.9) (Sample Problem 10.2).

In contrast, US reflections from soft tissue interfaces are relatively weak (a few tenths of a percent), and transmission is correspondingly robust (HW 10.8). Transmission into soft tissues plays a key role in US-based imaging because it can be used to image deeper structures (Fig. 10.1b). Transmission also can generate image artifacts, such as erroneous localization. This can happen if the beam is incident at an oblique angle. The transmitted beam then refracts, and this violates a fundamental assumption in US imaging, which is that the beam follows a straight

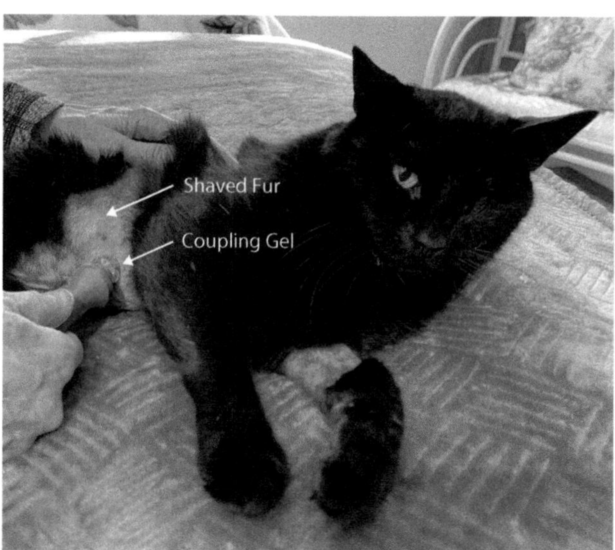

FIGURE 10.9
Picture showing the use of coupling gel (and shaved fur) to facilitate US transmission into and out of the authors' faithful feline friend, Ranger. (Photograph by the authors.)

BOX 10.1 DERIVATION OF EQ. (10.2)

Consider a plane pressure wave that is incident on a boundary and divided into reflected and transmitted waves. Our goal is to derive an expression for the intensity reflection coefficient, R, in terms of characteristic acoustic impedance, Z, starting from the definition:

$$R = \frac{P_r^2}{P_i^2}$$

where P_i and P_r are incident and reflected pressure amplitudes.

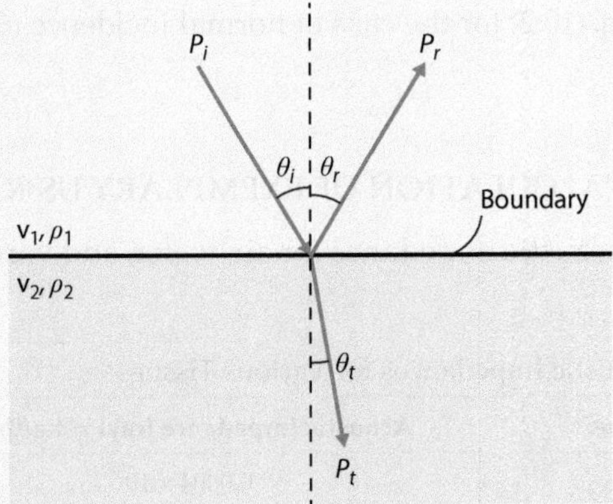

The incident, reflected, and transmitted waves must satisfy two boundary conditions. One is continuity of pressure across the boundary to ensure that there is no net force (otherwise, according to Newton's laws, the boundary would move). The other is continuity of the normal component of particle velocities, u, to ensure that both mediums remain in contact (and thus avoid creating a gap). To implement these conditions mathematically, we use the notation in the figure and write

$$P_i + P_r = P_t \tag{B10.1.1}$$

$$u_i \cos\theta_i + u_r \cos\theta_r = u_t \cos\theta_t \tag{B10.1.2}$$

Dividing Eq. (B10.1.1) by Eq. (B10.1.2), and invoking the Law of Reflection $\theta_i = \theta_r$, we obtain

$$\frac{P_i + P_r}{\cos\theta_i\ (u_i + u_r)} = \frac{P_t}{u_t \cos\theta_t} \tag{B10.1.3}$$

For a plane wave, $Z = P/u = \rho \mathrm{v}$, where ρ is density, and v is wave propagation speed. Thus, to recast Eq. (B10.1.3) in terms of acoustic impedances, we insert the relationships

$$\frac{P_i}{u_i} = \rho_1 \mathrm{v}_1 \quad \frac{P_r}{u_r} = -\rho_1 \mathrm{v}_1 \quad \frac{P_t}{u_t} = \rho_2 \mathrm{v}_2$$

where the minus sign reflects the fact that u_r is in the negative direction. This yields

$$\frac{\rho_1 \mathrm{v}_1 (u_i - u_r)}{\cos\theta_i\ (u_i + u_r)} = \frac{\rho_1 \mathrm{v}_1 \left(1 - \frac{u_r}{u_i}\right)}{\cos\theta_i \left(1 + \frac{u_r}{u_i}\right)} = \frac{\rho_1 \mathrm{v}_1 (1 + \sqrt{R})}{\cos\theta_i\ (1 - \sqrt{R})} = \frac{\rho_2 \mathrm{v}_2}{\cos\theta_t} \tag{B10.1.4}$$

In writing Eq. (B10.1.4), we have invoked

$$R = \frac{P_r^2}{P_i^2} = \frac{u_r^2}{u_i^2} \Rightarrow \sqrt{R} = -\frac{u_r}{u_i}$$

Solving Eq. (B10.1.4) for R yields the result

$$R = \left(\frac{\rho_2 v_2 \cos\theta_i - \rho_1 v_1 \cos\theta_t}{\rho_1 v_1 \cos\theta_t + \rho_2 v_2 \cos\theta_i} \right)^2 = \left(\frac{Z_2 \cos\theta_i - Z_1 \cos\theta_t}{Z_2 \cos\theta_i + Z_1 \cos\theta_t} \right)^2 \quad \text{(B10.1.5)}$$

Equation (B10.1.5) reduces to Eq. (10.2) for the case of normal incidence ($\theta_i = \theta_t = 0$).

SAMPLE PROBLEM 10.2 CALCULATION OF EXEMPLARY US REFLECTION COEFFICIENTS

Table 10.1 shows values of the acoustic impedance for air, water, and various tissues.

TABLE 10.1

Acoustic Impedances for Various Tissues

Tissue	Acoustic Impedance (rayl = kg/(m²s))
Air	0.0004×10^6
Lung	0.18×10^6
Fat	1.34×10^6
Water	1.48×10^6
Skin	1.6×10^6
Kidney	1.63×10^6
Blood	1.65×10^6
Liver	1.65×10^6
Muscle	1.71×10^6
Bone	7.8×10^6

Source: Data from Curry, T. S., III, et al. 1990. *Christensen's Physics of Diagnostic Radiology (4th Edition). Introduction to Radiological Imaging*. Williams & Wilkins. Page 335.

Use Eq. (10.2) to calculate reflection and transmission coefficients for US incident on the following interfaces: (a) air/soft tissue; and (b) soft tissue/bone. (c) Does it matter which direction the US is travelling? Compare this result with reflection and transmission of light at an optical interface (Sample Problem 2.1).

SOLUTION

Equation (10.2) describes the reflection coefficient between two media of acoustic impedances Z_1 and Z_2, respectively. Transmission can be calculated by noting that sound is either reflected or transmitted, so $T = 1 - R$.

a. Table 10.1 lists values of Z for air but not "soft tissue." However, we can approximate Z for soft tissue by taking a median value among the acoustic impedances for kidney, liver, and muscle, namely, 1.65×10^6 rayl. Substituting these values into Eq. (10.2) yields the following expression

$$R = \frac{(Z_1 - Z_2)^2}{(Z_1 + Z_2)^2} = \frac{\left(\left[0.0004 \times 10^6\right] - \left[1.65 \times 10^6\right]\right)^2}{\left(\left[0.0004 \times 10^6\right] + \left[1.65 \times 10^6\right]\right)^2} = 1$$

Significantly, $R = 1$ to within the number of significant figures. Thus, $T = 0$, and essentially no US will enter the body from air (without the help of coupling gel).

b. Repeating the calculation in Part (a) using values of 1.65×10^6 rayl and 7.8×10^6 rayl for the acoustic impedances of soft tissue and bone, respectively, yields $R = 0.42$. Thus, $T = 0.58$.

c. Examination of Eq. (10.2) shows that the proportions of reflected and thus transmitted US are dependent only on the presence of the interface and not on the direction of travel. However, this simple result is valid *only* for US normally incident on an interface. Behavior at other angles is more complicated (Box 10.1). This behavior is consistent with that observed for light incident normally and at an angle on a surface.

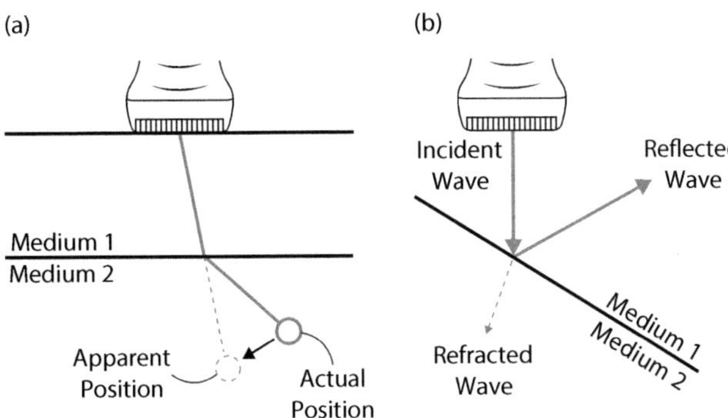

FIGURE 10.10
Imaging problems including (a) mis-localization and (b) signal loss that can arise as a consequence of non-normal incidence of an US beam. Mis-localization arises when the sound beam bends due to refraction at an interface, leading to a displacement in the object's apparent position. An analogous effect is observed with light beams (HW 3.2).

path (Fig. 10.10a). Oblique incidence also can cause the reflected wave to miss the detector (Fig. 10.10b).

10.5.1 Display of Ultrasound Data

US machines can run in different operational states (modes), which display data in different formats. Prominent examples include (1) amplitude or A-mode, (2) brightness or B-mode, and (3) motion or M-mode (Fig. 10.11). Of these, A-mode, which was invented in the 1930s, is the oldest and the most basic. A-mode is used to interrogate a single anatomical direction and generates a graphical output. M-mode, which emerged in the 1950s, similarly interrogates a single direction, but as a function of time, revealing motion. Over time, applications of A- and M-mode have become more specialized. Instead, two-dimensional B-mode, which generates cross-sectional grayscale images, has become the workhorse of modern medical US. Finally, color Doppler mode generates color-coded maps of blood flow in vessels, typically superimposed on a B-mode image (Section 10.8).

10.5.1.1 A-Mode

In A-mode, processed echoes are collected along one "line" within the body, and echo time and a spike, with height proportional to echo amplitude, are displayed (Fig. 10.11). One A-line of data is generated per PRP. The echo time can be used to determine the depth of the reflecting structure in A-mode (and other modalities) using Eq. (10.1). The echo amplitude reflects tissue properties, such as impedance mismatch at an interface. Schematized and real A-mode "images" are shown in Fig. 10.11. Current use of A-mode primarily is restricted to measuring the dimensions of the eye in ophthalmology (Fig. 10.11b) and to therapeutic US where concentration of energy is used for disruptive and/or destructive purposes. Single-element, focused transducers often are used for these applications.

10.5.1.2 B-Mode

A significant limitation of A-mode is the use of one axis to encode echo amplitude in the form of a spike. Thus, A-mode is not well suited to 2D imaging. B-mode overcomes this limitation by encoding echo amplitude as a brightness (i.e., a grayscale) instead of a spike. B-mode display is used in M-mode and to create cross-sectional images (when coupled with scanning). One line of schematic A-mode data, converted to B-mode, is shown in Fig. 10.11.

10.5.1.3 M-Mode

M-mode, like A-mode, does not generate a traditional "picture." Instead, M-mode displays structure motion, as illustrated in Fig. 10.11. In M-mode, rapidly

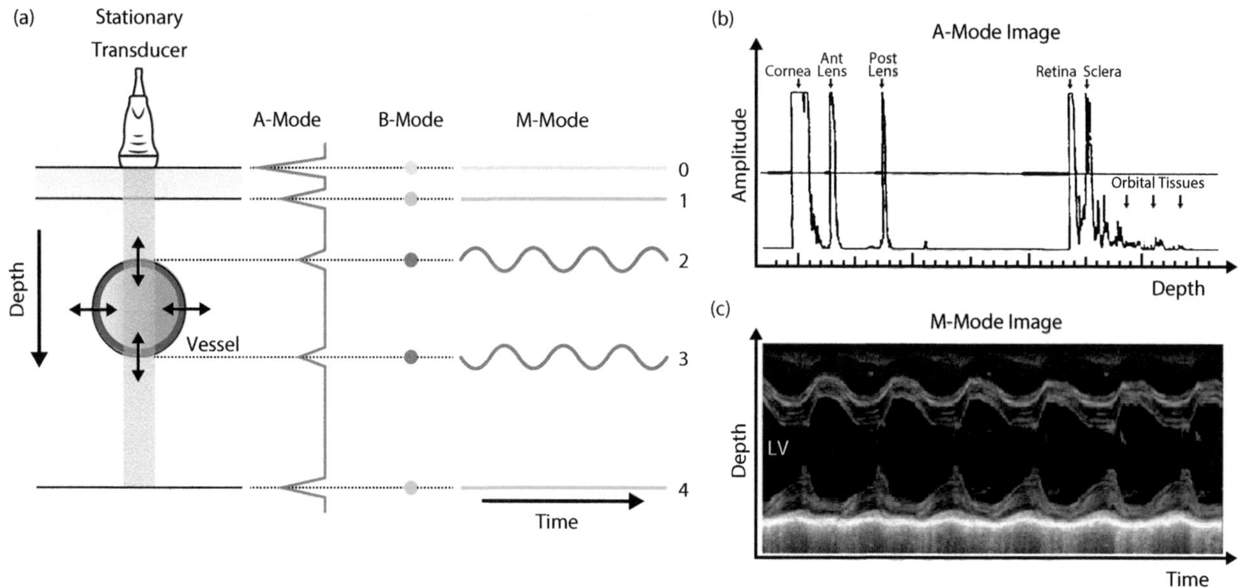

FIGURE 10.11
Ultrasound modes. (a) Schematic representation of US echo-display modes applied to a blood vessel, such as the carotid artery. Here, differences in reflectivity are manifest as differences in spike height or brightness in A-mode and B-mode images, respectively, while changes (↕, ↔) in vessel diameter caused by pressure changes as the heart beats are manifest as time-varying lines in the M-mode image. (b) A-mode image of the human eye showing distances from the transducer to the cornea, lens, and retina, among other features. (Data for graph from Waldron, R.G. 2016. A-scan biometry. http://emedicine.medscape.com/article/1228447- overview.) (c) M-mode image of the left ventricle of a dog heart showing periodic changes in chamber size as the heart beats. (Adapted from Pattanapon, N., et al. 2018. The effects of anesthetic drug choice on heart rate variability in dogs. *J. Adv. Vet. Anim. Res.* 5:485–489. http://doi.org/10.5455/javar.2018.e303. Attribution 4.0 International [CC BY 4.0].)

repeated input pulses are used to generate a succession of A-mode signals. The M-mode image is created by encoding echo amplitude as a brightness, with reflector depth (for a fixed direction in the body) conventionally plotted along the vertical axis and time plotted on the horizontal axis. Thus, changes in dot depth with time reveal motion. Currently, M-mode is used primarily when there is a need for very high temporal and/or axial resolution (Section 10.6.2). For example, M-mode is used in cardiology to study valve motion. In veterinary medicine, M-mode is used to measure chamber size, wall thickness, and heart rate. However, real-time two-dimensional echocardiography has rendered one-dimensional M-mode somewhat less important.

10.5.2 2D and 3D Imaging

A-mode and M-mode were widely used in the early days of ultrasonography. However, as technology has advanced, these uni-directional modes have largely been replaced with two- and three-dimensional imaging, with the former being most common.

10.5.2.1 Two-Dimensional Imaging

Two-dimensional US generates conventional-looking images that are similar to cross-sectional slices (Fig. 10.12). In two-dimensional imaging, echoes are collected along different A-lines in the body using scanning techniques (Fig. 10.6). For each A-line of data, echo times and amplitudes are recorded, and then the depth of the reflecting interface is plotted along the vertical using dots, the brightness of which reflects echo amplitude, i.e., using B-mode display. The two-dimensional image is generated as the beam is swept over the FOV, with each A-line of data assigned to a horizontal distance. Thus, two-dimensional images are plots of tissue depth along the vertical versus transverse location along the horizontal with contrast generated by variations in the amplitudes of echoes generated by reflection and scattering.

Unlike most other medical imaging modalities, two-dimensional B-mode images can be acquired rapidly (HW 10.9). Moreover, the time delay between acquisition and display is short. Thus, two-dimensional B-mode images typically are displayed at rates of ~30 frames per second, permitting visualization of processes in a patient when they occur, i.e., in "real time." Frame averaging, which can be used to reduce the effects of random noise and to smooth the image, reduces frame rate. A single two-dimensional image from a movie of the heart, generated by a student at Lewis & Clark College, is shown in Fig. 10.12.

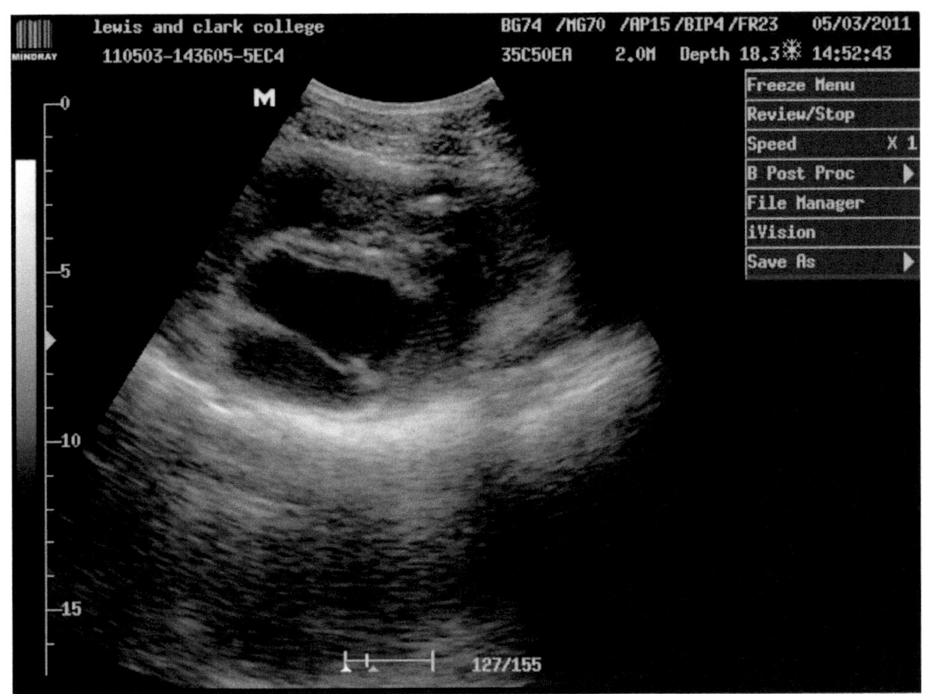

FIGURE 10.12
Frame from an US movie collected by a student in the Biomedical Imaging course at Lewis & Clark College. (Image courtesy of William Steinhardt, Lewis & Clark College.)

10.5.2.2 Three-Dimensional Imaging

Although conceptually straightforward, three-dimensional US has been adopted slowly by clinics. In part, this is because US data are generated using relatively inexpensive, mobile devices, which only recently acquired the necessary real-time data acquisition, reconstruction, and three-dimensional image visualization capabilities. One popular application of three-dimensional imaging is display of the face, hands, and feet of a fetus to the parents.

Three-dimensional US imaging can be implemented by collecting a stack of two-dimensional images, similar to optical sectioning (Fig. 10.13). One method of acquiring the stack is mechanical scanning with a linear transducer. This approach involves translating a transducer across the skin of a patient under the control of a motor/encoder. During this process, a series of evenly spaced two-dimensional images are collected, and these then are assembled into the three-dimensional image.

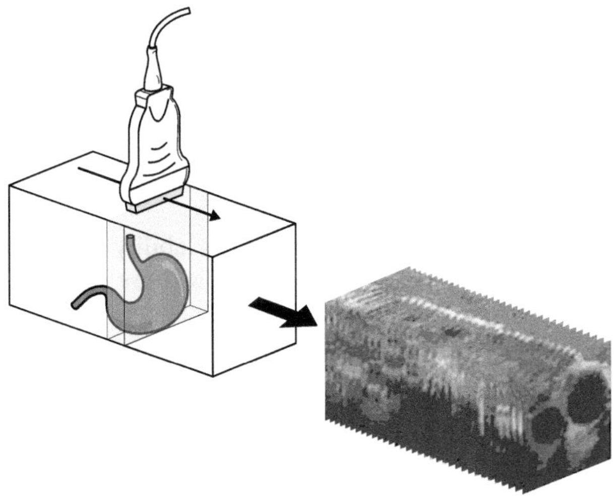

FIGURE 10.13
Generation of a three-dimensional US image from a stack of two-dimensional images. (Stack of tissue slices reprinted with permission from Pollakrit, T., et al. 2011. Reconstruction of 3D ultrasound images based on Cyclic Regularized Savitzky-Golay filters. *Ultrasonics* 51:136–147.)

10.6 Resolution

We turn now to the largely untouched topic of US image quality. The next few sections deal with spatial resolution, followed by contrast.

Spatial resolution in US imaging can be subdivided into resolution along lateral, axial, and elevational directions (Fig. 10.14). Lateral and axial resolution determine ability to distinguish two objects that are closely spaced in directions perpendicular and parallel to the direction of beam propagation, respectively.

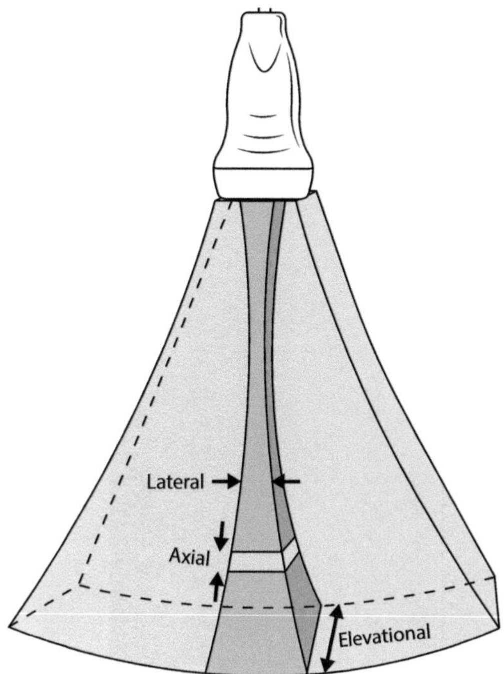

FIGURE 10.14
Diagram showing the three types of resolution that are relevant to US. Axial and lateral resolution are defined within the image plane, and elevational resolution is perpendicular to the image plane. For the one-dimensional arrays discussed here, electronic focusing occurs only in one dimension, and so elevational and lateral resolution differ. Modern two-dimensional arrays also can focus in the elevational direction and thereby achieve similar elevational and lateral resolutions, but they are difficult to fabricate. An acoustic lens also can be used to achieve some focusing in the elevational direction (Fig. 10.4).

Elevational resolution determines ability to distinguish two objects within the thickness of the US beam.

10.6.1 Lateral

For two objects to be resolvable along the lateral direction, their lateral separation must exceed the beam diameter. To rationalize this result, we refer first to Fig. 10.15a, which shows two objects separated by less than the beam diameter. As exemplified in the figure, these closely spaced objects generate overlapping echoes and thus are not resolvable. In contrast, Fig. 10.15b shows objects that are spaced by a distance that exceeds the beam diameter. These generate separate echoes and are resolvable.

Notably, lateral resolution, like beam diameter, varies with depth. Lateral resolution is relatively poor close to, and far from, the transducer and is optimal at a point of beam "focus." Lateral resolution also improves with decreasing wavelength. A good lateral resolution is less than a millimeter (HW 10.2).

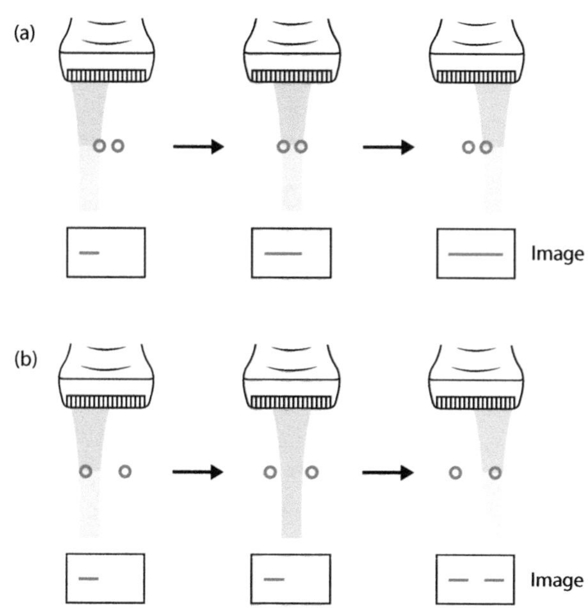

FIGURE 10.15
Cartoon showing that for objects to be laterally resolvable their separation must exceed the beam diameter. (a) The US beam will be continuously reflected by objects that are closer together than the beam diameter. The resulting image will show a single reflection. (b) The US beam will be reflected by the first object, then transmitted, then reflected by the second object when the objects are farther apart than the beam diameter. The resulting image will show two reflections, corresponding to the two objects, separated by a region of zero intensity.

10.6.2 Axial

For two objects to be resolvable along the axial direction, their axial separation, z, must exceed half the spatial pulse length (SPL), where SPL $= n\lambda$ with n typically chosen to be ~3. This result is depth independent.

To derive this result, we refer to the diagram in Fig. 10.16, which shows pulses 1 and 2 reflecting off upper and lower objects, respectively. The structures are axially separated by z. We assume that pulse 2 is generated via transmission at the upper interface when pulse 1 initiates the process of reflection. To avoid overlap, and ensure resolvability, pulse 1 must complete reflection before pulse 2 returns to the upper interface. The time for pulse 1 to finish reflection is

$$t_{\text{upper}} = \text{pulse duration} = nT = \frac{n}{\nu} = \frac{n}{\frac{\text{v}}{\lambda}} = \frac{\text{SPL}}{\text{v}}$$

The time for the front of pulse 2 to arrive back at the upper object is

$$t_{\text{lower}} = \frac{2z}{\text{v}}$$

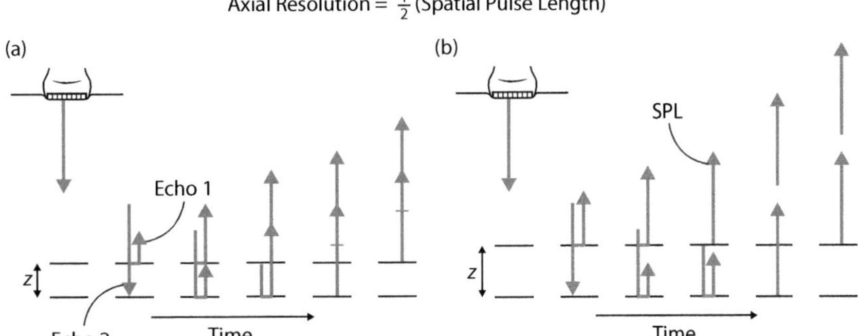

FIGURE 10.16
Cartoon showing that for objects to be axially resolvable their separation must exceed half the SPL. Echoes overlap (a) for objects with an axial separation that is less than SPL/2, whereas reflections remain separate (b) and thus resolvable for more widely spaced objects. In essence, this is because the two pulses remain separated if the upper (initially reflected) pulse is more than half way turned around before the lower (transmitted) pulse arrives at the lower interface.

Resolvability requires $t_{lower} > t_{upper}$, and hence that

$$\frac{2z}{v} > \frac{SPL}{v}$$

Thus, the criterion for axial resolution is

$$z > \frac{SPL}{2} \tag{10.3}$$

In Sample Problem 10.3, we show that axial resolutions are on the order of a few tenths of a millimeter.

Notably, axial resolution improves with increasing transducer frequency. For example, the axial resolution of a 10-MHz transducer is 0.4× that of a 4-MHz transducer (i.e., 0.231 mm), if we again assume that SPL = 3λ. On other hand, penetration degrades with increasing frequency (Section 10.7.3). Thus, the choice of transducer frequency is dictated by application. If the structures of interest are not too deep, so penetration is not an issue, it is best to use higher frequency US and thereby generate an image with improved axial resolution. If penetration is an issue, as in abdominal imaging, lower frequency US is required.

10.6.3 Elevational

Elevational resolution is determined by the "slice thickness," which increases with increasing transducer height and depends on distance from the transducer (Fig. 10.17).

Unlike poor lateral and axial resolution, which smear closely spaced objects, poor elevational resolution leads to image contamination by highly reflective objects from an erroneous plane. To illustrate the origin of this "slice thickness artifact," we refer to Fig. 10.17. For the scenario shown, the elevational profile of the beam near objects A and B causes both objects to contribute to echoes from a scan. Moreover, the echoes from object A will be detected in between those from object B, and thus the resulting image will erroneously imply that A is in the middle of B.

SAMPLE PROBLEM 10.3 US AXIAL RESOLUTION

Calculate the approximate axial resolution of a 4-MHz transducer.

SOLUTION

The axial resolution criterion is $z > SPL/2$. To evaluate this expression, we need a value for SPL. A reasonable SPL is three wavelengths. For a 4-MHz transducer, $\lambda = v/\nu = (1540\ \text{m/s})/\left(4\times10^6\ \text{Hz}\right) = 0.385$ mm. Therefore, a reasonable SPL is 3×0.385 mm $= 1.155$ mm. Finally, using this value, the axial resolution criterion is

$$z > \frac{SPL}{2} = \frac{1.155\ \text{mm}}{2} = 0.578\ \text{mm} \approx 0.6\ \text{mm}.$$

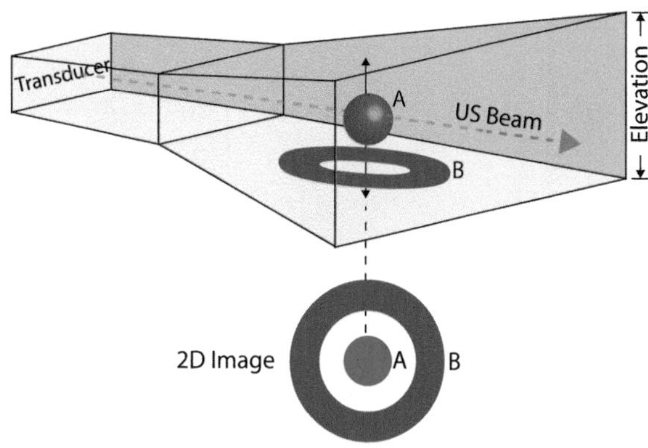

FIGURE 10.17
Role of slice thickness in creating US imaging artifacts. Here, objects A and B lie in different planes, but the elevational profile of the beam causes them both to contribute to echo formation. Consequently, they appear in the same plane in the two-dimensional image.

Elevational resolution generally is poorer than axial and lateral resolution and is on the order of several millimeters.

10.7 Contrast

Contrast in US images reflects variations in object brightness that arise from differences in the absolute size of the echo intensity at the transducer. Echo intensity is influenced by several processes. These include "specular" reflection, which has been our primary focus to this point, as well as diffuse reflection, scattering, attenuation of beam strength due to absorption, and contrast-enhancing agents (if present). Contrast also can be altered via image processing and/or electronic methods.

10.7.1 Specular Reflection

Specular reflection arises when US encounters a "smooth" surface having irregularities with dimensions that are much smaller than λ, such as the diaphragm (Fig. 10.18). Specular reflection from a boundary between tissues or other smooth interfaces is an important determinant of the strength of the US signal that returns to the transducer and thus plays a critical role in generating contrast.

The effect of specular reflection on echo intensity can be quantified by calculating the intensity, I_r, that is reflected from the interface using $I_r = RI_i$ (Eq. [10.2] and Table 10.2). An exemplary calculation is presented in Sample Problem 10.4 in Section 10.7.3

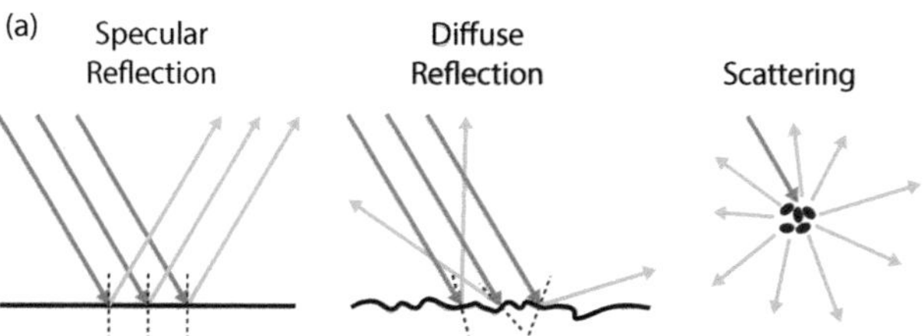

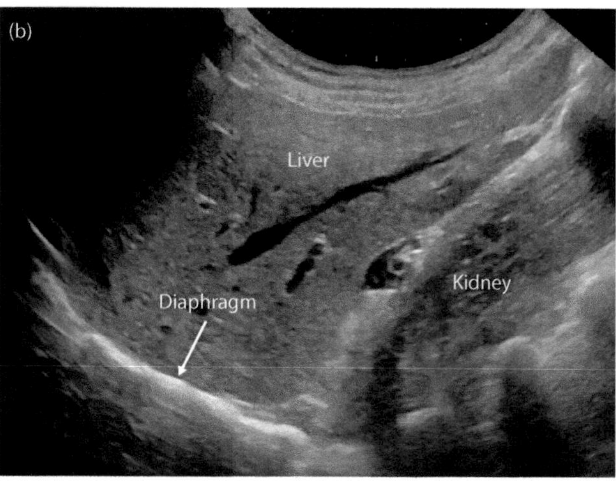

FIGURE 10.18
US signals arise from specular and diffuse reflection and scattering. (a) Schematic comparison of the three processes. The local normal to each interface at the point of ray reflection is shown using a dotted black line. (b) Image of liver showing the characteristic grainy high and low intensity echotexture pattern of organ parenchyma caused by scattering and diffuse reflection. (Image in Panel (b) by Mikael Häggström. Creative Commons CC0 1.0 Universal Public Domain Dedication.)

for a case where the echo intensity also is affected by attenuation.

10.7.2 Nonspecular (Diffuse) Reflection and Scattering

Diffuse reflection arises when US encounters a "rough" surface, with irregularities that are not much smaller than λ. The irregularities cause the normal that appears in the law of reflection to vary in direction over the surface (Figs. 2.5 and 10.18), and thus rays are diffusely reflected (scattered) in all directions. Many soft tissues and the parenchyma of many organs (e.g., the heart, liver, and kidney) act as diffuse reflectors, and their associated scatter signals generate a grainy tissue-specific characteristic appearance in US images, referred to as echotexture (Fig. 10.18). Changes in echotexture suggest disease and thus can provide useful diagnostic information.

Blood also scatters US. Red blood cells (RBCs) are much smaller than the wavelength of US, and they thus generate a scatter signal with intensity that

TABLE 10.2

Reflection Coefficients between Tissues

	Fat	Muscle	Skin	Brain	Liver	Blood	Cranial Bone
Water	0.047	0.02	0.029	0.007	0.035	0.007	0.57
Fat		0.067	0.076	0.054	0.049	0.047	0.61
Muscle			0.009	0.013	0.015	0.02	0.56
Skin				0.022	0.006	0.029	0.56
Brain					0.028	0.00	0.57
Liver						0.028	0.55
Blood							0.57

Source: Data from Erikson, K.R., et al. 1974. Ultrasound in medicine: a review. *IEEE Trans. Sonics Ultrason.* SU-21:144–170.

follows the Rayleigh result, $I \propto \nu^4$. This signal plays an important role in Doppler US.

Diffuse reflection and scattering are multi-directional. Thus, their contribution to echoes is much weaker than the contribution from specular reflections (Fig. 10.18). Nevertheless, this "back-scatter" is detectable with modern US equipment.

10.7.3 Attenuation

Attenuation is the weakening of the US beam as it propagates through the body. The associated loss of beam strength has a major effect on the brightness and quality of US images.

We already have discussed one process that contributes to attenuation – scattering. Another is absorption by the medium and the subsequent conversion of wave energy into thermal energy. Absorption arises from two processes – classical absorption and relaxation absorption. Classical absorption is generated by friction that arises when molecules are displaced by a wave. Relaxation absorption occurs when molecules that are displaced by a compression wave return to equilibrium out-of-sync with the next compression. Energy is dissipated in reversing molecular motion, and the beam weakens. For diagnostic US, absorption is dominated by relaxation effects.

For our purposes, there are two important attributes of attenuation. The first is that the intensity of the remaining, unabsorbed and unscattered part of the beam decays exponentially with distance. Stated mathematically, the intensity varies as

$$I(z) = I_0 e^{-\mu z} \tag{10.4}$$

The second important attribute of attenuation is that it increases with increasing frequency, and this necessitates the use of lower frequency US, with its poorer axial resolution, for probing deeper structures. The frequency effect can be expressed in a mathematically simple form if intensity loss is expressed using a decibel (dB) scale. If I_0 is an initial intensity and I is an attenuated intensity, the dB scale is defined as

$$\text{relative intensity (dB)} = 10 \log\left(\frac{I}{I_0}\right) \tag{10.5}$$

The relative intensity in dB will be positive if $I > I_0$ and negative if $I < I_0$ (as in the case of attenuation). In terms of dB, it is roughly true that US is attenuated by ~0.5 (dB/cm)/MHz in soft tissue (Table 10.3). As a benchmark, a change of –3 dB corresponds to a 50% loss of intensity. Thus, 1- and 5-MHz beams will undergo a 50% intensity loss after travelling 6.0 and 1.2 cm in soft tissue, respectively (Sample Problem 10.4).

The loss of signal that is associated with attenuation can and should be corrected electronically. This process, which is known as time gain compensation (TGC), is depicted schematically in Fig. 10.19. The correction is approximately a reverse exponential function of time, as suggested by Eq. (10.4).

10.7.4 Contrast-Enhanced Ultrasound (CEUS)

Extrinsic contrast-enhancing agents are now routinely used to enhance the quality of US images of areas where intrinsic contrast is low. CEUS is commonly used to enhance the backscatter signal from blood and thereby improve cardiac imaging and evaluation of key parameters, such as blood perfusion of organs. CEUS also improves detection of metastases in the liver based on differences in contrast agent delivery, and washout, between normal liver parenchyma and metastatic lesions.

Intravenously introduced gas-containing microbubbles are a common contrast-enhancing agent (Fig. 10.20). Microbubbles increase contrast in part because the gas has an acoustic impedance that differs markedly from that of surrounding fluids and tissue. In addition, the bubbles are compressible and thus absorb and reradiate more energy than relatively incompressible structures of similar size. Both effects increase echo strength from areas containing bubbles and improve image contrast.

The bubbles also can respond by undergoing nonlinear oscillations at harmonic frequencies that are integral multiples of the incident frequency. Harmonics can readily be distinguished by an appropriately

TABLE 10.3

Attenuation Coefficients for Tissues at 1 MHz

Tissue	Attenuation Coefficient (dB/cm)
Water	0.0002
Blood	0.18
Soft tissues	0.3–0.8
Brain	0.3–0.5
Liver	0.4–0.7
Fat	0.5–1.8
Smooth muscle	0.2–0.6
Tendon	0.9–1.1
Cortical bone	13–26
Lung	40

Source: Data from Bushberg, J.T., et al. 2012. *The essential physics of medical imaging (3rd Edition)*. Wolters Kluwer | Lippincott Williams & Wilkins. Chapter 14.

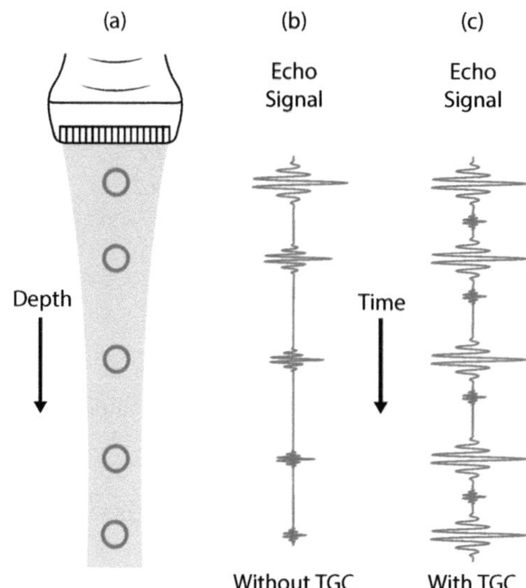

FIGURE 10.19
Time gain compensation corrects US signals for time (depth)-dependent attenuation. In the cartoon, a transducer (a) is aimed at a series of equivalent reflectors from a test object (phantom). The echoes (b) grow progressively weaker with reflector depth due to attenuation. TGC (c) corrects the signal loss and equalizes the echoes.

SAMPLE PROBLEM 10.4 US ATTENUATION

A 4-MHz US device produces a beam with initial intensity I_0 as it leaves the transducer. A portion of the beam reflects off an interface and returns to the transducer, where it is detected. The interface is 1.5 cm from the transducer and has a reflection coefficient $R = 0.05$. Assume that beam attenuation is described by the rule ~0.5 (dB/cm)/MHz. What is the intensity of the echo that returns to the transducer?

SOLUTION

The intensity of the returning echo is the product of the initial intensity of the beam, I_0, the fraction of that initial intensity lost to attenuation, and the fraction reflected at the interface (i.e., the reflection coefficient). The beam intensity decreases ~0.5 (dB/cm)/MHz while traversing tissue. Thus, because the beam travels 2×1.5 cm = 3.0 cm and the frequency of the US beam is 4.0 MHz, the loss during transit is

$$\text{Transit Loss} = 3.0\ \text{cm} \times 4.0\ \text{MHz} \times 0.5 \frac{\text{dB}}{\text{cm-MHz}} = 6\ \text{dB}$$

The intensity drop in dBs can be converted to a percentage drop using the definition of dBs:

$$-6\ \text{dB} = 10 \log\left(\frac{I}{I_0}\right) \quad \Rightarrow \quad I = 10^{-0.6} I_0 = 0.25 I_0$$

Thus, the intensity of the echo that returns to the transducer is the product of the $0.25I_0$ from attenuation and 0.05 from the reflection coefficient:

$$I = 0.05 \times 0.25 I_0 = 0.0125 I_0$$

In other words, the echo intensity is only 1.25% of the initial beam intensity.

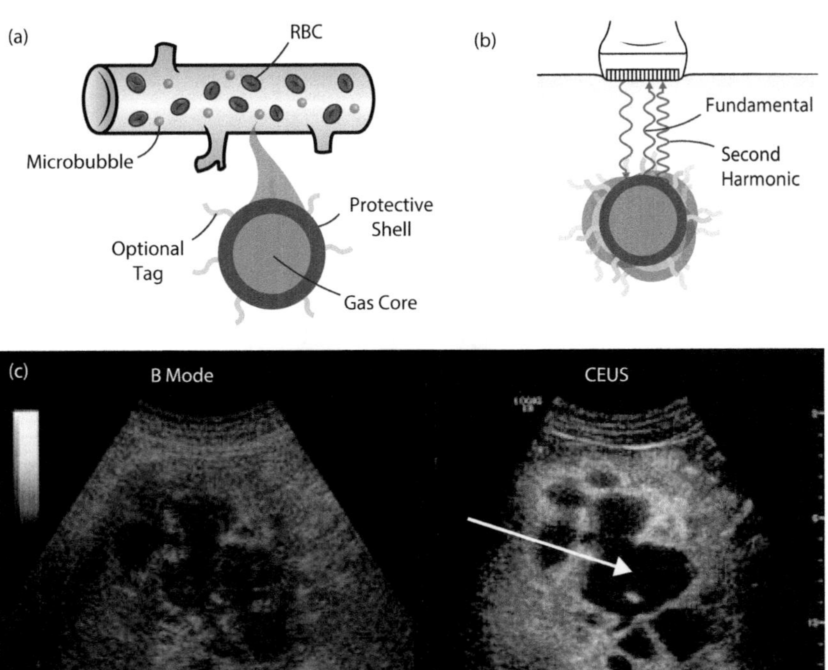

FIGURE 10.20
CEUS using microbubbles. (a) Cartoon depicting the use of microbubble contrast agents. (b) The bubbles can respond nonlinearly and thus generate frequency harmonics. For example, "squaring" $p_0 \sin(\omega t)$ will generate a wave proportional to $p_0^2 \sin^2(\omega t) = p_0^2(1 - \cos(2\omega t))/2$, i.e., a signal at twice the original frequency. (c) Images showing that anechoic hepatic abscesses (arrow) are more visible with a contrast agent. (Panel [c] adapted with permission from Badea, R., et al. 2012. Ultrasound imaging of liver tumors – current clinical applications. *Liver Tumors*, Prof. Alexander Julianov [Ed.]. ISBN: 978-953-51-0036-2. InTech. Available from: http://www.intechopen.com/books/liver-tumors/ultrasound-imaging-of-liver-tumors-current-clinical-applications. Creative Commons Attribution 3.0 License.)

configured transducer and provide better resolution in part because they are at higher frequency. Additional detail on CEUS and harmonic imaging is presented in the legend to Fig. 10.20.

10.8 Doppler US

We turn now to one of the premier applications of US: monitoring blood flow. The essence of the approach is to measure the Doppler effect, which is a change in perceived wave frequency that arises from motion of the wave source and/or the observer (Fig. 10.21). Doppler US uses the frequency change to measure the magnitude and direction of blood flow.

As an important prelude, we review the conceptual underpinnings and essential physics of the one-dimensional Doppler effect. The generalization to two dimensions is straightforward. To start, consider the familiar case of a moving source of sound, like a siren (Fig. 10.21). For this scenario, a change in wavelength gives rise to the change in perceived frequency. To understand why, consider Fig. 10.22, which shows the distance between adjacent crests generated by stationary and moving sources. For the stationary source, the crest-to-crest spacing, λ, is the product of the time between emission of crests, T, and the speed of propagation of the wave, v_w. Specifically, we have

$$\lambda = v_w T = \frac{v_w}{\nu}$$

This is a simple rearrangement of the standard expression relating wave speed, wavelength, and frequency (Section 2.1 and Fig. 10.22).

In contrast, when the source moves, the wavelength also is affected by source speed, v_s. For example, when the source moves toward the observer the wavelength is reduced because the source is "following" the wave (Fig. 10.22). The decrease in wavelength leads to an associated increase in perceived frequency; there is a corresponding increase in wavelength and decrease in frequency on the receding side.

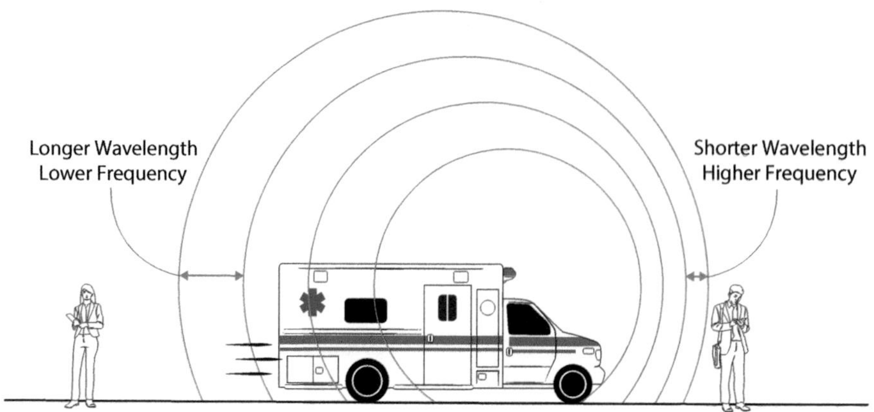

FIGURE 10.21
Cartoon showing a siren creating a signal with a higher and lower perceived frequency based on motion-induced wavelength changes.

The new wavelength, λ', can be computed by referring to the diagram in Fig. 10.22. For the case of a source moving toward an observer, the wavelength is the difference in distance traveled by the wave and the source during one period:

$$\lambda' = (\mathrm{v}_w - \mathrm{v}_s)T = (\mathrm{v}_w - \mathrm{v}_s)\frac{1}{\nu}$$

The perceived frequency, ν', can be calculated from λ' as

$$\nu' = \frac{\mathrm{v}_w}{\lambda'} = \frac{\mathrm{v}_w}{(\mathrm{v}_w - \mathrm{v}_s)}\nu$$

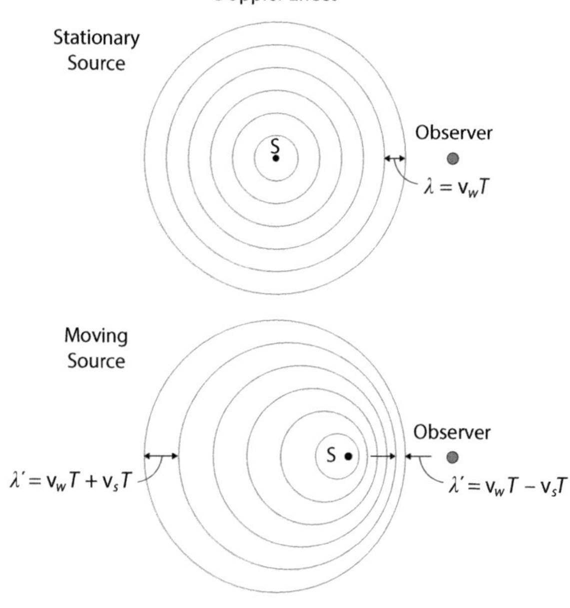

FIGURE 10.22
Cartoon depicting the wavelength changes that arise from a moving source.

where ν is the frequency emitted by the source. For a receding source, the minus sign must be changed to a plus sign. Thus, the general Doppler formula for a moving source is

$$\nu' = \frac{\mathrm{v}_w}{\lambda'} = \frac{\mathrm{v}_w}{(\mathrm{v}_w \mp \mathrm{v}_s)}\nu \tag{10.6}$$

Next, consider the case of an observer moving at speed v_o. In this situation, the frequency (not the wavelength) changes because the moving observer intercepts a relatively larger or smaller number of crests/second, depending on the direction of motion (Fig. 10.23). Deriving the quantitative version of this effect is left for HW 10.15.

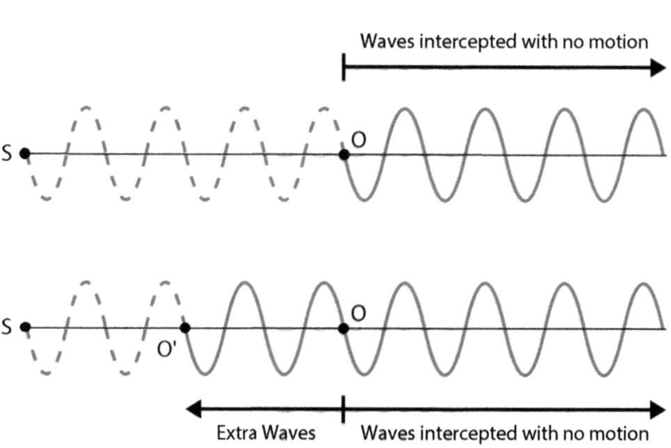

FIGURE 10.23
Origin of the Doppler shift for a moving observer. The frequency increases when an observer moves toward a stationary source because the observer intercepts extra waves per unit time, as shown. The frequency decreases when an observer moves away from a stationary source.

Finally, the general result for the case when both the source and observer are in motion is

$$\nu' = \frac{v_w \pm v_o}{(v_w \mp v_s)} \nu \quad (10.7)$$

In the numerator of Eq. (10.7), the positive sign applies when the observer is moving toward the source, and the negative sign applies when the observer is moving away from the source. In the denominator, the negative sign applies when the source is moving toward the observer, and the positive sign applies when the source is moving away from the observer.

Perhaps surprisingly, the Doppler effect for blood flow involves motion of both the source and the observer. This is because blood acts as a moving observer when it receives US from the transducer, and it acts as a moving source when it backscatters US at the transducer. The mathematics of this more complex case is implemented in Box 10.2. The result for the change in frequency between what is emitted and received by the transducer, $\Delta\nu$, is

$$\Delta\nu = \pm 2 \frac{v_b \cos\theta}{v_w} \nu \quad (10.8)$$

In Eq. (10.8), θ is the angle between the direction of blood flow and the direction of the US beam, v_b is the speed of blood, and ν is the frequency of the wave emitted by the transducer (Fig. 10.24). The change in

BOX 10.2 DERIVATION OF EQ. (10.8)

The derivation of Eq. (10.8) involves a two-step application of Eq. (10.7). The reason is that blood, which we will represent as an RBC, acts first as a moving observer that receives US from the transducer and acts second as a moving source that scatters US back to the transducer.

Consider first the case of the RBC as an observer moving at speed v_b; the transducer is a stationary source emitting at frequency ν. The RBC receives a Doppler-shifted signal at frequency ν_1 given by

$$\nu_1 = \left(\frac{v_w \pm v_b}{v_w}\right)\nu = \left(1 \pm \frac{v_b}{v_w}\right)\nu \quad (B10.2.1)$$

where v_w is the wave speed. Equation (B10.2.1) yields an expression for the first Doppler shift:

$$\nu_1 - \nu = \pm \frac{v_b}{v_w}\nu \quad (B10.2.2)$$

Consider next the case of the RBC as a moving source. The RBC scatters US at frequency ν_1 back to the stationary transducer, which receives a Doppler-shifted signal at frequency ν_2 given by

$$\nu_2 = \left(\frac{v_w}{v_w \mp v_b}\right)\nu_1 = \left(\frac{1}{1 \mp \frac{v_b}{v_w}}\right)\nu_1 \approx \left(1 \pm \frac{v_b}{v_w}\right)\nu_1 \quad (B10.2.3)$$

The approximation in Eq. (B10.2.3) holds in the generally valid limit that $v_b \ll v_w$.

Equation (B10.2.3) yields an expression for the second Doppler shift:

$$\nu_2 - \nu_1 = \pm \frac{v_b}{v_w}\nu_1 \quad (B10.2.4)$$

The total Doppler shift is

$$\Delta\nu = \nu_2 - \nu = (\nu_2 - \nu_1) + (\nu_1 - \nu) = \pm \frac{v_b}{v_w}\nu_1 \pm \frac{v_b}{v_w}\nu \approx \pm 2 \frac{v_b}{v_w}\nu \quad (B10.2.5)$$

In obtaining Eq. (B10.2.5), we have used the fact that the Doppler shift is relatively small, so $\nu_1 \approx \nu$.

For the case where only a component of the velocity, $v_b \cos\theta$, lies along the direction of the beam, Eq. (B10.2.5) is modified to obtain the result in the main text:

$$\Delta\nu = \pm 2 \frac{v_b \cos\theta}{v_w} \nu \quad (B10.2.6)$$

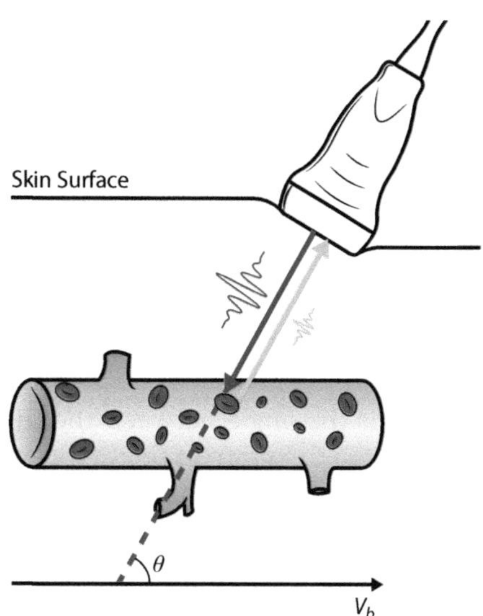

FIGURE 10.24
Geometry of Doppler US measurements.

frequency is positive if the blood is flowing toward the transducer and negative if the blood is flowing away from the transducer. Equation (10.8) holds in the generally valid limit that v_b is much less than the speed of the US wave, v_w = 1540 m/s. Typical frequency shifts are in the audible range (Sample Problem 10.5); thus, the sonographer can listen to the Doppler US signal to facilitate positioning and diagnosis.[1]

An important application of Eq. (10.8) is measurement of v_b to identify abnormal blood flow arising from heart malfunction (Fig. 10.25). The insonation angle, θ (Fig. 10.24), can be determined from an associated B-mode image and is typically adjusted to lie in the range

[1] Although we have emphasized motion-induced frequency shifts, in practice blood velocities are extracted from equivalent, motion-induced phase shifts.

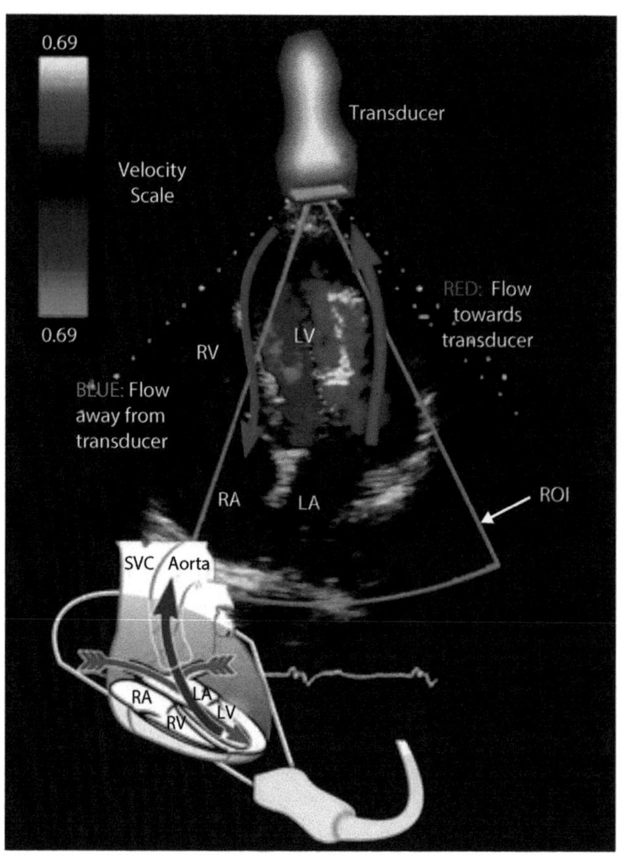

FIGURE 10.25
A color-flow Doppler image. Color-coded blood flow is superimposed on a standard two-dimensional image of the heart. Flow direction is encoded using the format BART (blue away red toward the transducer), and velocity magnitude is encoded with shade (see Velocity Scale). Here L and R are left and right, A and V are atrium and ventricle, and SVC (superior vena cava) and aorta are the blood vessels carrying blood from the body to the heart and from the heart to the body, respectively. (Adapted with permission from Jones & Bartlett Learning.)

SAMPLE PROBLEM 10.5 DOPPLER US

Compute the frequency shift associated with a blood flow velocity of 30 cm/s. Assume that the US is produced by a 3.5-MHz transducer and is incident on the blood vessel at a 35° angle. The US speed is 1540 m/s.

SOLUTION

The frequency shift is produced by the Doppler effect and can be calculated using Eq. (10.8):

$$\Delta\nu = \pm 2\frac{v_b \cos\theta}{v_w}\nu = \pm 2\frac{(0.30\ \text{m/s})\cos 35^\circ}{1540\ \text{m/s}}\left(3.5\cdot 10^6\ \text{Hz}\right) = \pm 1120\ \text{Hz}$$

Thus, the frequency shift is 1120 Hz. Significantly, this value is the audible range and can be used by the sonographer to provide feedback on transducer placement.

between 30° and 60°. Refraction and aliasing can cause problems for angles less than ~30°. Small errors in angle estimation lead to large errors in velocity determination for angles greater than ~60° because the cosine varies rapidly at larger angles (HW 10.19). As the insonation angle approaches 90°, no Doppler shift is detected.

10.9 Endoscopic US

Endoscopic US (EUS) is a hybrid US/endoscopy technique (Box 10.3) that is used when it is important to position the transducer inside the body, very close to

BOX 10.3 STANDARD ENDOSCOPY

Endoscopes are imaging devices that figure prominently in medical procedures that involve "looking" inside the body. One notable example is a colonoscopy, which is an examination of the colon that screens for signs of cancer and its precursors. A colonoscopy often is recommended for people over the age of 50.

To facilitate visualization of structures inside the body, endoscopes must deliver light efficiently over relatively long distances. For this reason, endoscopes contain optical fibers that deliver light with essentially no loss by exploiting the phenomenon of TIR (Chapters 2 and 7). A typical "video endoscope" includes an illumination fiber bundle and a tiny camera at its tip that generates an image of light reflected off the object of interest; see figure. In lieu of a camera at the tip, a fiber endoscope includes an imaging (coherent) bundle that preserves fiber orientation. In the case of an imaging bundle, each fiber collects light from a nearby part of the object, and this spatial distribution of light is maintained during transit out of the body to generate an image. The video endoscope is preferred in many instances, whereas the fiber optic endoscope may be preferred when simplicity, portability, and smaller diameter are important.

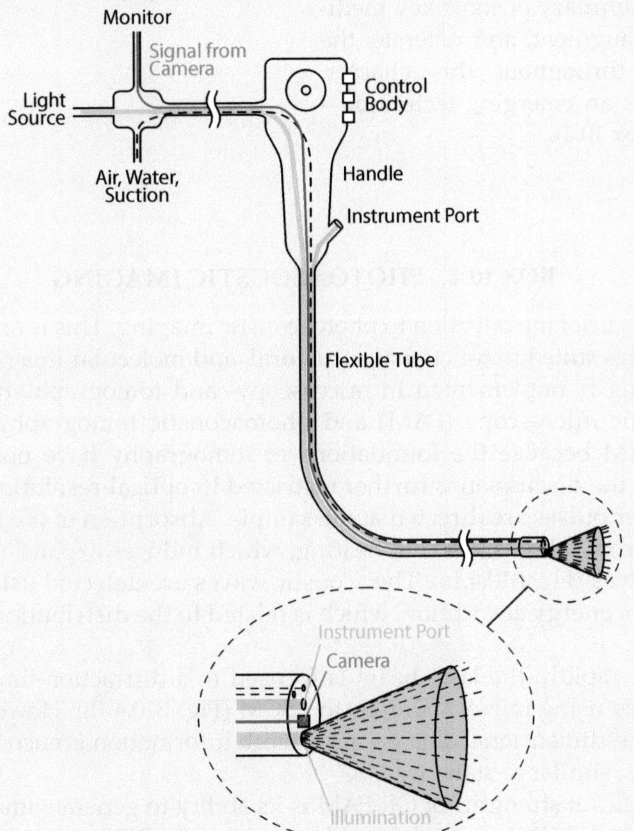

Regular endoscopy yields only a surface view of the digestive tract and other structures, whereas EUS can generate images of all five layers of tissue that make up the gastrointestinal wall.

a structure of interest, to improve image accuracy and image detail. Notable applications include examination of the digestive tract, surrounding organs, such as the pancreas, nearby lymph nodes, and the lungs, and transesophageal echocardiograms (TEEs). A TEE is an alternative to a transthoracic echocardiogram (TTE). TEE is performed by passing a transducer into the esophagus, which is close to the heart. Relative to TTE, the US beam in TEE suffers less from attenuation, and echoes are stronger. As a consequence, TEE images of the heart are clearer, and Doppler data are improved.

In EUS, the transducer is attached to the end of a long flexible tube, and the apparatus is inserted into the body through an appropriate opening (Fig. 10.26). The mouth can be used to image the digestive tract. Images are then generated as in standard US.

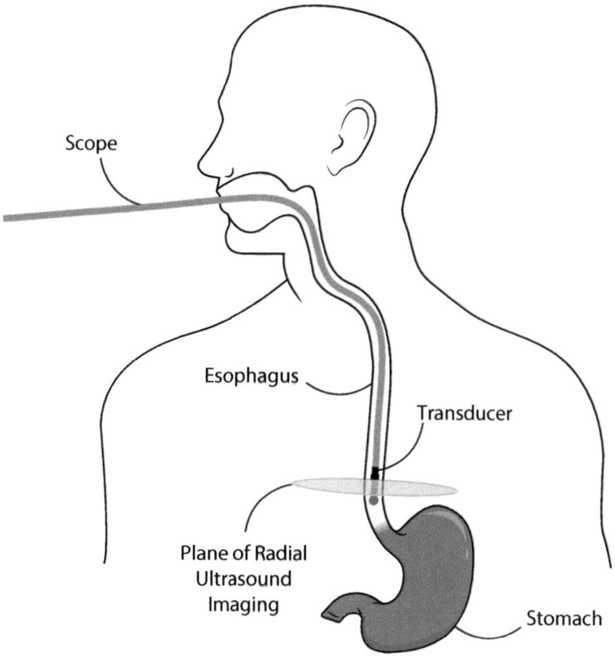

FIGURE 10.26
Cartoon of radial endoscopic US imaging. Endoscopy and US are combined in EUS.

10.10 Applications

We close with a tabular summary of some key medical applications of US to augment, and reiterate, the applications mentioned throughout this chapter. Included in this section is an emerging technique – photoacoustic imaging (Box 10.4).

BOX 10.4 PHOTOACOUSTIC IMAGING

In this box, we provide a brief introduction to photoacoustic imaging. This is an emerging, hybrid (light/acoustic) technique that is suited to functional, structural, and molecular imaging.

Photoacoustic imaging is implemented in microscopy- and tomography-based formats, which are known as photoacoustic microscopy (PAM) and photoacoustic tomography (PAT), respectively. We focus primarily on PAM because the foundations of tomography have not yet been discussed. In the interests of brevity, the discussion is further restricted to optical-resolution PAM (OR-PAM).

In OR-PAM, brief laser pulses are directed at the sample. Absorption of the laser light (e.g., by endogenous or exogenous chromophores) leads to heating, which induces expansion and the associated creation of acoustic US waves (Fig. B10.4.1a). The acoustic waves are detected using transducers. The final image is a spatial map of energy absorption, which is related to the distribution of the absorbing species (Fig. B10.4.2).

To generate an image rapidly, the laser beam is focused to a diffraction-limited spot and then raster scanned over the sample using mirrors, similar to LSCM (Fig. B10.4.1b). However, unlike LSCM, raster scanning generates three-dimensional data because depth information is encoded in the arrival times of the photoacoustic waves, similar to standard US.

One of the most significant strengths of OR-PAM is its ability to generate images of endogenous chromophores, notably DNA and RNA in nuclei and hemoglobin in RBCs, among others. For example, the approach has been used to image unstained nuclei, with the goal of facilitating the study and diagnosis of cancer while bypassing histological staining (Fig. B10.4.2). OR-PAM also has been used to study RBCs in vivo with an aim at elucidating fundamental mechanisms of oxygen metabolism by cells.

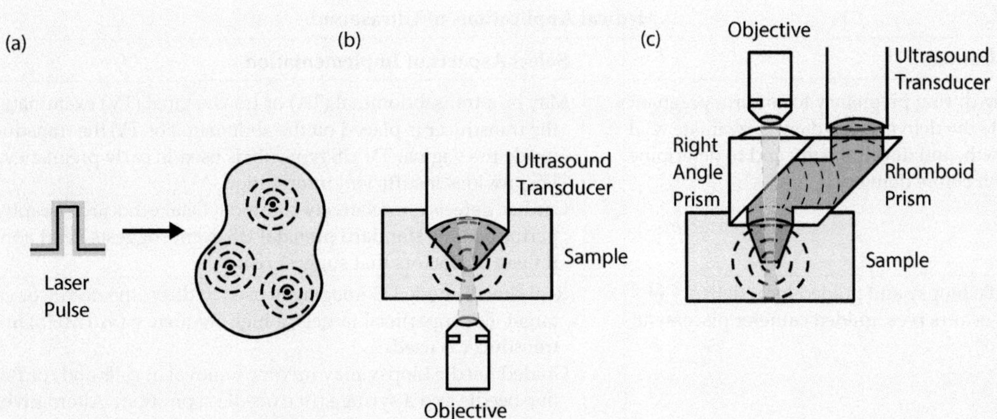

FIGURE B10.4.1
Schematic of the mechanism and implementation of OR-PAM. (a) Light absorbing species (black spots) generate acoustic waves that are detected by a transducer. OR-PAM can be implemented in a transmission (b) or a reflection (c) mode. Three-dimensional images are generated by raster scanning in two dimensions while mapping position in the third dimension using wave arrival times.

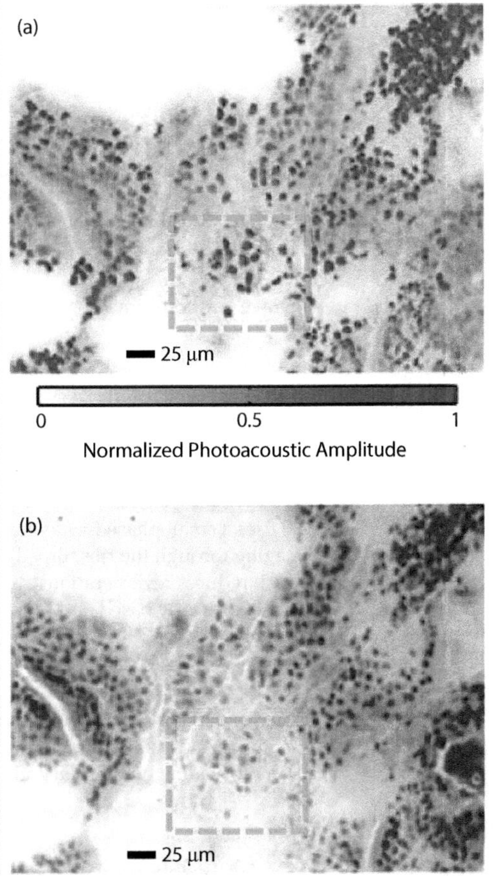

FIGURE B10.4.2
Comparison of OR-PAM and histology. The image in Panel (a) shows a PAM image of tissue from mouse small intestine. The image in Panel (b) shows the tissue stained with the dyes hematoxylin and eosin. The striking similarity underscores the potential utility of OR-PAM. (Adapted with permission from Yao, D-K., et al. 2010. In vivo label-free photoacoustic microscopy of cell nuclei by excitation of DNA and RNA. *Opt. Lett.* 35:4139–4141. © The Optical Society.)

Medical Applications of Ultrasound	
Select Applications	**Select Aspects of Implementation**
Fetal imaging during pregnancy to confirm pregnancy and estimate the delivery due date, to evaluate fetal health, growth, and development, and to determine fetal position before delivery	May be a transabdominal (TA) or transvaginal (TV) examination. For TA, the transducer is placed on the abdomen. For TV, the transducer is placed inside the vagina. TV US typically is used in early pregnancy and when TA US provides insufficient information. Cardiac defects are relatively common. Fetal echocardiography may be performed if a standard prenatal US exam suggests heart abnormalities or if there are factors that suggest risk.
Guided needle biopsy and guided local delivery of anesthetics near nerves, guided catheter placement	Real-time B-mode US imaging is used to direct the needle or catheter to the target. For superficial targets, a high-frequency (>10 MHz) linear array transducer is used. Guided needle biopsy may involve removal of cells and/or fluid using a fine needle and a syringe (fine needle aspiration). Alternatively, "cores" of tissue may be removed using a larger, hollow spring-loaded needle that automatically collects and cuts tissue several times (core needle biopsy). Doppler imaging may be used to identify vessels during delivery of anesthetics.
Organ imaging (e.g., pancreas, liver, kidneys, uterus, bladder)	May involve use of contrast-enhancing, gas-containing microbubbles, which are introduced into the body using an intravenous catheter. May involve use of endoscopic US.
Breast imaging	Uses very high-frequency (e.g., 14 MHz) US. Useful as a follow-up to a standard mammogram that reveals a suspicious area because US can differentiate fluid-filled cysts from solid masses. May involve guiding a biopsy needle to a suspicious area to collect cells to test for malignancy. May be used instead of standard mammography to determine the nature of a breast change when a woman is pregnant (and radiation should be avoided) or when a woman has very dense breasts.
Abnormal cell and tissue destruction (e.g., destruction of tumor cells and uterine fibroids) using high-intensity focal ultrasound (HIFU)	Implements tissue destruction thermally and/or mechanically. Uses a signal generator connected to a focusing transducer to deliver pulses of HIFU (intensity > 1 kW/cm^2) to a millimeter-sized focal region. (Diagnostic imaging typically uses intensities < 100 mW/cm^2.) Destruction is controlled by visualizing the focal spot and target with US or MRI.
Musculoskeletal imaging (e.g., tendons, muscles, nerves, ligaments)	Protocols vary due to the broad spectrum of structures involved. May involve harmonic imaging and/or Doppler US.
Ophthalmologic US (e.g., measurement of eye dimensions and visualization of abnormalities)	Often uses both A- and B-mode US. For A-mode, the transducer is placed directly on the eye with the tear film acting as coupling gel. Eye dimensions may be measured (e.g., to determine the correct power for a lens implant during cataract-correction surgery). For B-mode, the transducer is placed on the eyelid and standard coupling gel is used. Eye structure is visualized to detect abnormalities, such as retinal detachment and cancer of the retina.
Echocardiography (e.g., measurement of heart dimensions and assessment of heart function)	TTE uses a small phased-array transducer with broad bandwidth suited to imaging through the ribs. Broad bandwidth supports harmonic imaging, which reduces reverberation artifacts caused by the ribs. TEE is endoscopy based and may be used if additional image detail is required. Doppler US is used to monitor blood flow, a key indicator of heart function and vessel health.
Therapeutic treatment of sports injuries (e.g., to promote healing and reduce pain)	Implemented in a pulsed or continuous mode, generally by a physical therapist, using maximum intensities that typically are ≤ 3 W/cm^2. May promote healing by thermal and/or non-thermal mechanisms.

Suggested Reading and Additional Resources

Abdulla, S. and Clarke, C. 2011–2022. Radiology Café: FRCR lecture notes: Ultrasound. https://www.radiologycafe.com/radiology-trainees/frcr-physics-notes/ultrasound-imaging

Bushberg, J.T., Seibert, J.A., Leidholdt, E.M., and Boone, J.M. 2012. *The essential physics of medical imaging (3rd Edition)*. Wolters Kluwer | Lippincott Williams & Wilkins, Philadelphia, PA. Chapter 14.

Kane, S.A. and Gelman, B.A. 2020. *Introduction to physics in modern medicine (3rd Edition)*. CRC Press, Boca Raton, FL. Chapter 4.

Støylen, A. 2016. *Basic ultrasound for clinicians*. http://folk.ntnu.no/stoylen/strainrate/Basic_ultrasound. A useful overview of US, from a practitioner's viewpoint, with numerous figures and animations.

HOMEWORK PROBLEMS

10.1 [Transducer – Frequency] The resonance frequency, ν, produced by a typical piezoelectric crystal, such as those shown in Fig. 10.5, is determined by the thickness of the crystal, L, and the speed of sound in the crystal, $v_{crystal}$:

$$\nu = \frac{v_{crystal}}{2L}$$

Assume $v_{crystal} = 4000$ m/s. Calculate the crystal thickness corresponding to the following common transducer frequencies:

Frequency (MHz)	Thickness?	Application
2.0		Deep abdomen, obstetrics, gynecology
3.5		General abdomen, obstetrics, gynecology
5.0		Vascular, breast, pelvis
7.5		Breast, thyroid
10.0		Breast, thyroid, superficial veins, superficial masses, musculoskeletal imaging
15.0		Superficial structures, musculoskeletal imaging

10.2 [Beam Diameter & Resolution – Lateral] At a point of US focus, the beam diameter, BD, is given by a typical diffraction-dictated expression $BD = \lambda f / A$, where f is the focal length and A is the probe active aperture. Calculate a typical beam diameter, and thus lateral resolution, for a 3-MHz transducer with a focal length of 5 cm and an active aperture of 3 cm.

10.3 [Transducer – Steering] Sketch an approximate pulse timing pattern required to steer an US beam to the right of the sonographer. Explain your answer. Hint: See Fig. 10.7.

10.4 *[Transducer – Time Delay for Focusing] Demonstrate that electronic focusing at a distance r from the transducer can be achieved by implementing a time delay for the nth element in an array transducer that is given by $\Delta t_n = y_n / (2 v_w r)$ (see figure). In this expression, v_w is the wave speed, and y_n is the distance of the nth element from the center element. Hint: Use the approximation $y_n / r \ll 1$.

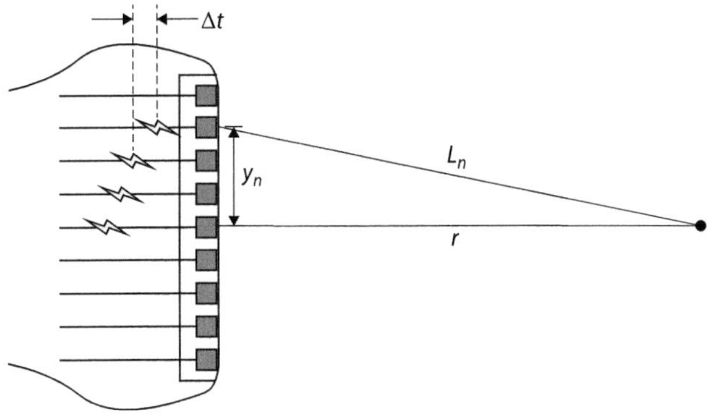

10.5 [Transducer – PRF] US transducers typically generate a short (several wavelength) pulse and then spend a much longer time detecting the ensuing echoes. Transducers repeat this process many times a second to construct an image. In this problem, we estimate the maximum allowable PRF to avoid echo overlap.

a. Estimate the maximum time required to collect all echoes from a given US pulse. Assume that you need to wait long enough to detect the echo off the opposite side of the body, although in practice this echo may be so attenuated that it is undetectable. A maximum (one-way) distance of 0.5 m is probably sufficient. The average speed of US in soft tissue is 1540 m/s.
b. Calculate the largest possible PRF using the return time in part (a).

10.6 [Transducer – Duty Cycle] US transducers typically operate in a pulsed mode: "on" to generate US, "off" to detect echoes. The "duty cycle" of the transducer is defined as the ratio of the time spent in the "on" mode to the total period of the pulse cycle (calculated as the sum of the times spent in the "on" and "off" modes). Calculate the duty cycle of a 4-MHz US instrument, assuming that the pulse length is three wavelengths and that the PRF is 1 kHz. Does the transducer mostly talk or mostly listen?

10.7 *[Reflection Coefficients – Derivation] Assume that an acoustic pressure wave with amplitude $P_i = 1$ is normally incident on an interface between media having different acoustic impedances, Z_1 and Z_2, as shown. At the interface, a reflected wave of amplitude $P_r = r$ and a transmitted wave of amplitude $P_t = t$ are generated. Invoke continuity of pressure amplitude at the boundary (no net force) and conservation of energy to prove Eq. (10.2) describing reflection at the interface. Hint: Recall that the energy, E, of a wave is proportional to amplitude squared. Thus, to implement energy conservation, write $E \propto ZP^2$ for the incident, reflected, and transmitted waves.

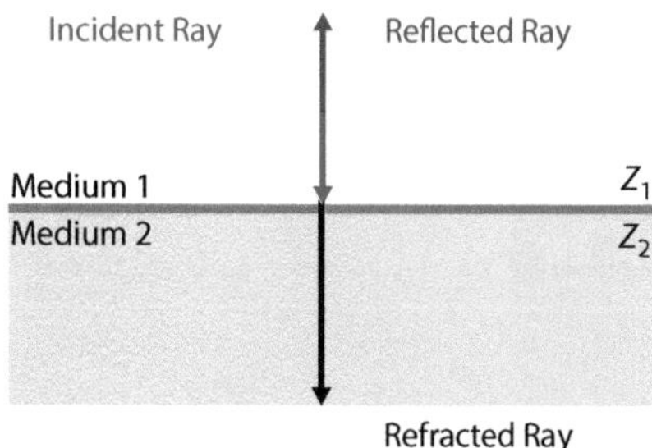

10.8 [Reflection Coefficients – Common Values] Calculate reflection and transmission coefficients for US normally incident on the following medically relevant interfaces, using values of the acoustic impedance in Sample Problem 10.2:

Interface	R	T
Skin:muscle		
Skin:fat		
Muscle:fat		

10.9 [B-mode – Frame Rate] Estimate the B-mode frame rate when the images are to be acquired up to a depth of 5 cm, and the number of lines in each image is 512.

10.10 [Resolution – Axial] Calculate the minimum US frequency that can be used to resolve features 0.5 mm apart in the axial direction, assuming that the pulse is 3 wavelengths long. Hint: See Sample Problem 10.3.

10.11 [Attenuation – Half-Value Thickness (HVT)] The HVT is the thickness of tissue that attenuates an incident US beam by 50%. Show that HVT for soft tissue is given by: $\text{HVT(cm)} = 6/\nu$, where the frequency is expressed in MHz. Hint: Invoke the result that the beam is attenuated by ~0.5 (dB/cm)/MHz in soft tissue.

10.12 [Attenuation – Multiple Tissues] A transducer sends out a 3-MHz US pulse, which then travels 0.7 cm in fat before reflecting off of a fat/muscle interface. Assume that US attenuation in fat follows the rule 0.5 (dB/cm)/MHz and that the intensity emitted by the transducer is 10 mW/cm^2. What is the intensity of the echo (in mW/cm^2) when it returns to the transducer? Hint: See Sample Problem 10.4.

10.13 [Attenuation – Multiple Tissues] An US beam is used to interrogate a liver inclusion. The beam travels through 2 cm of fat, 1 cm of smooth muscle, and 3 cm of liver before reaching the inclusion. A portion of the beam reflects off the inclusion, revealing its location, and returns to the transducer. Calculate the attenuation of 1-MHz and 2-MHz beams, in dB and percent, arising from their transit through the various tissues. Use the smallest attenuation coefficients listed in the ranges given in Table 10.3. *Ignore reflective losses*. Compare the results. Hint: See Sample Problem 10.4.

10.14 [Attenuation and Resolution – Fish Finding] Fish finders are instruments used to locate fish in bodies of water. They operate by emitting ultrasonic pulses and detecting their reflections off fish and other underwater features (Fig. 10.1a). The following entry appeared in 2011 on an Internet boating forum:

> I'm a bit confused about my fish finder. It is a dual transducer 50 kHz/200 kHz device. I am never sure which frequency to use when. I fish shallow water for stripers and mackerel and deeper water for cod and haddock.

a. What frequency would you suggest for the best detail resolution?
b. What frequency do you think that the fisherman should use for shallow water and for deep water? Justify your answers, based on principles discussed in this chapter.

10.15 *[Doppler Shift – Moving Observer] Derive the general Doppler formula for a moving observer (and stationary source):

$$\nu' = \frac{v_w \pm v_o}{v_w} \nu$$

10.16 [Doppler US] The accompanying US image shows a cross-sectional transverse scan of a diseased dog liver. The image consists of a Doppler US scan superimposed on a standard B-mode image. The Doppler scan is color coded according to the BART convention – blue away, red toward – as shown in the vertical velocity scale. The scan was performed to confirm blood flow in the portal and hepatic veins, despite the disease. (Image courtesy of John Feleciano, DVM, DACVR. VDIC Veterinary Diagnostic Imaging & Consultations.)

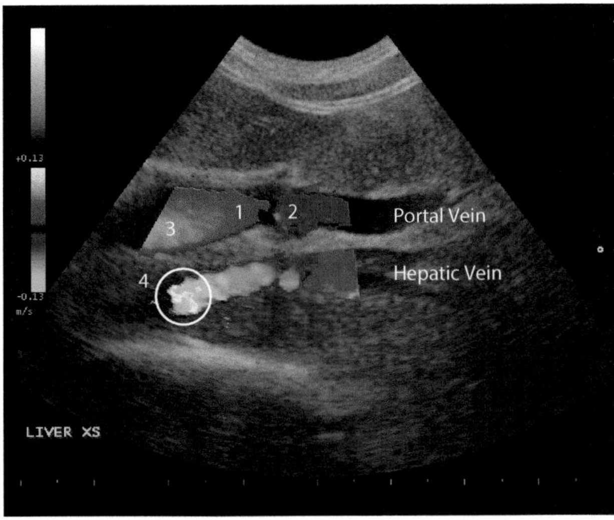

a. Where was the transducer located for this scan, at the top, bottom, or side of the image?
b. The portal and hepatic veins are generally horizontal in the image, and the color coding within them switches (e.g., in the portal vein from red (1) to blue (2), separated by an intervening region that is black). Draw a sketch showing the transducer and the direction of blood flow within the portal vein to illustrate and explain the origin of the red > black > blue color changes.
c. Is the blood flowing faster relative to the transducer at location 1 or 3?
d. The colors in circle 4 most likely reflect aliasing. How could you alter the sampling rate to reduce this artifact? Hint: See Section 9.3.
e. Why do the portal and hepatic veins appear dark in the underlying B-mode image?

Provide explanations for each of your answers.

10.17 [Doppler US] A recent study of blood flow in the carotid artery reported peak resting blood flow velocities of about 75 cm/s. The study used a 7.5-MHz beam directed at a 30° angle to the flow. What is the absolute change in frequency, Δv, between the US emitted and detected by the transducer under these conditions? What is the relative change in frequency, $\Delta v/v$?

10.18 [Doppler US] A handheld Doppler US device has an 8-MHz transducer. Assume that the instrument can detect Doppler frequency shifts as small as 100 Hz. What is the smallest blood flow velocity detectable with this instrument? Hint: Use an insonation angle in the standard 30–60° range to get a real-world answer.

10.19 [Doppler US] Assume that a sonographer uses an angle of 80° to calculate the speed of blood flow when the actual angle is 83°. What is the ratio of the calculated to true blood speeds? What is the percent error? Your result should demonstrate that very significant errors in calculated velocities arise from small errors in angle determination.

10.20 [Doppler Shift – General Application] Imagine that you are riding the Shinkansen – the Japanese "bullet" train. As you approach a sound source, you hear a frequency of 70 Hz. As you recede, you hear a frequency of 50 Hz.
a. Write down the Doppler Shift equations for the approaching and receding phases of the motion in terms of the known shifted frequencies and the unknown speed of the train and the unknown frequency of the source.
b. Solve these two equations for the speed of the train and the frequency of the source.

ACTIVITIES

A10.1 Diagnostic Ultrasound Imaging of the Carotid Artery

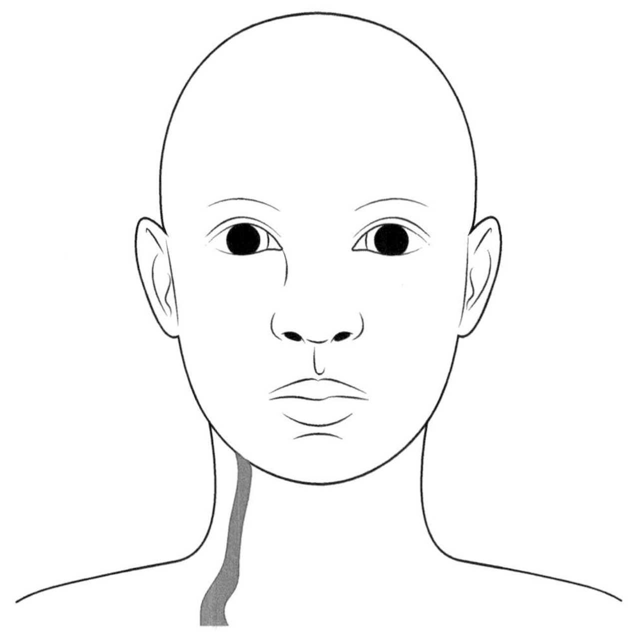

	Male ($n = 325$)	Female ($n = 409$)	You ($n = 1$)
Age, years	35.9 ± 4.5	35.4 ± 4.5	
Carotid lumen diameter, mm	6.09 ± 0.60	5.43 ± 0.48	

Source: Data from Ruan, L., et al. 2009. Correlates of common carotid artery lumen diameter in black and white younger adults: the Bogalusa Heart Study. *Stroke* 40:702–707.

This activity is written under the assumption that there is access to a pocket-sized US device, such as the Wellue EagleView™ dual-head wireless handheld US.

A. Identify the area of your neck for transducer placement by feeling for your carotid pulse. Online videos also give guidance on transducer placement.
B. Apply coupling gel to this region.
C. Place the transducer against your neck and into the gel and try to generate a longitudinal and/or transverse B-mode movie of your carotid. If unsure that you have identified the carotid, online videos show useful nearby landmarks.
D. Measure the diameter of your carotid lumen and compare with the data in the accompanying table.
E. Monitor blood flow through the carotid using color Doppler US.

11

Projection Radiography and Computed Tomography

In this chapter, we return to imaging techniques that utilize EM radiation and discuss projection radiography and X-ray computed tomography (CT). Projection radiography was the first medical imaging technique; it undoubtedly is familiar to every reader from dental visits and airport scans of carry-on luggage. CT was the first medical imaging approach made viable by computers; it impacts our lives somewhat less but is experiencing a rapid increase in use. For example, nearly 90 million medical CT scans were performed in the United States in 2019, and CT recently was introduced at airports to screen luggage more effectively and rapidly.

11.1 Essence of Radiography and CT

In radiography, an X-ray beam traverses a patient positioned between an X-ray source and a detector. The beam is attenuated, along each ray path, to an extent that depends on the anatomical structures encountered because X-ray absorption is material dependent (Fig. 11.1a). As a consequence of differential attenuation, the fraction of the beam that is transmitted through the patient possesses a spatially varying intensity pattern. A radiograph is a picture of this pattern (Fig. 11.1b).

Notably, X-rays and closely related gamma rays are not refracted by common materials; thus, imaging in this (and the next) chapter is "lensless." An important consequence is that images generated by projection radiography are unfocused "shadows" lacking depth information and suffering from structural superposition (Fig. 11.2). Despite this weakness, projection radiography is very widely used because it has excellent spatial resolution, excellent contrast for structures like bone, and a large FOV. In addition, radiography is relatively inexpensive, widely available, and rapid and easy to use. Prominent applications include screening for breast cancer and diagnosis of broken bones.

In contrast, CT yields computer-reconstructed two-dimensional and three-dimensional images that preserve structure depth and that are much better at revealing soft tissues than projection radiographs (Fig. 11.2). Prominent applications of CT include trauma assessment, tumor staging, and angiography to assess blood vessel health and blood flow. On the negative side, CT and, to a lesser degree, projection radiography subject the patient to potentially damaging, ionizing radiation (Box 11.1).

11.2 Properties of X-Rays

X-rays were discovered by Wilhelm Röntgen, who was awarded the first Nobel Prize in Physics in 1901 for this achievement. X-rays are short wavelength, energetic photons that have high penetrating ability, which is essential for generating images of the body from transmitted radiation (Fig. 11.3). In this section, we focus on developing a quantitative appreciation for the energies of X-rays used in medical imaging. This, in turn, will facilitate an understanding of the mechanisms underlying the interactions between "diagnostic" X-rays and matter.

11.2.1 Energies of Diagnostic X-Rays

X-rays often are defined as EM radiation that spans the wavelength regime from ~0.01 nm to ~10 nm. An alternative definition that better describes X-rays and distinguishes them from gamma rays is based on origin: X-rays are generated when electrons interact with matter (or alter their direction), whereas gamma rays are generated by processes that include nuclear decay and particle annihilation (Chapter 12).

To calculate X-ray photon energies, we invoke the famous Einstein relationship between photon energies and frequency/wavelength (Eq. [2.4]):

$$E_{\text{photon}} = h\nu = h\frac{c}{\lambda}$$

where $h = 6.63 \times 10^{-34}$ J·s and $c = 3 \times 10^{8}$ m/s. Inserting the approximate longest (10 nm) and shortest (0.01 nm) X-ray wavelengths into Eq. (2.4) yields bounds on X-ray energies of

$$2.00 \times 10^{-17}\ \text{J} < E_{\text{photon}} < 2.00 \times 10^{-14}\ \text{J}$$

Photon energies, when expressed in joules (J), clearly are very small numbers. Thus, energies are more

DOI: 10.1201/b22076-13

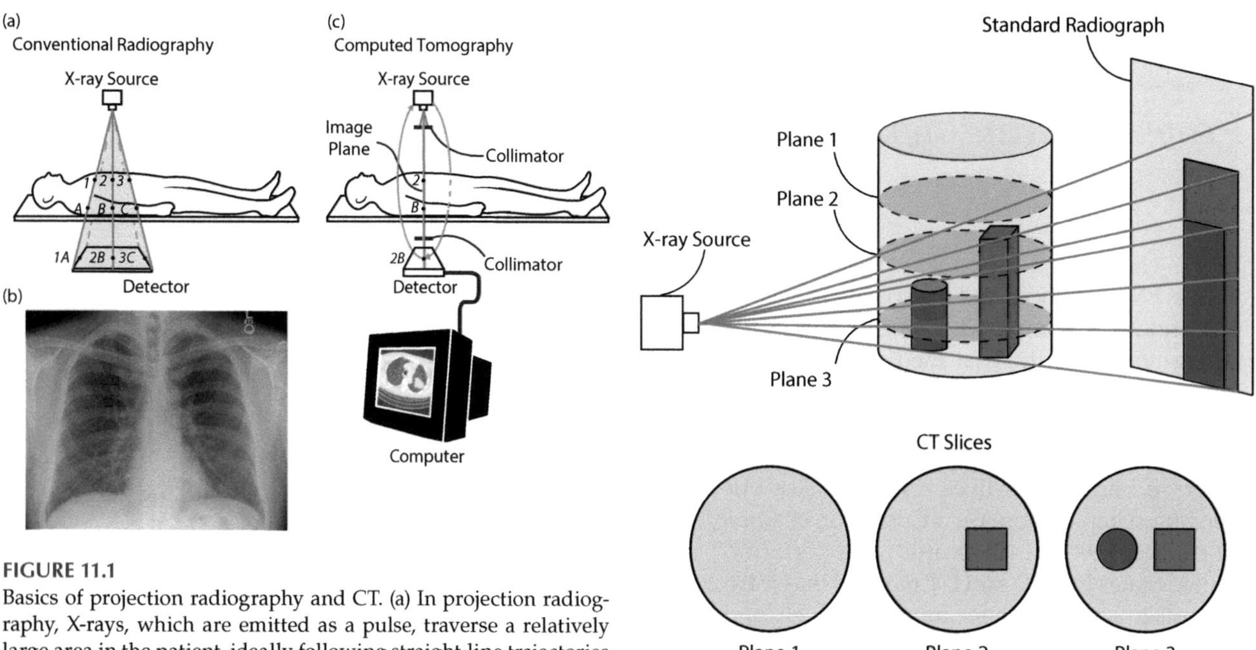

FIGURE 11.1
Basics of projection radiography and CT. (a) In projection radiography, X-rays, which are emitted as a pulse, traverse a relatively large area in the patient, ideally following straight-line trajectories between the source and detector. The transmitted beam generates a signal on a detector that is determined by the summed beam attenuation along the various ray trajectories. The resulting image is a projection of radiopaque (highly attenuating) structures along the beam path. (b) A standard chest X-ray shows that absorbing objects (e.g., bones), which cause fewer photons to hit the detector, are depicted as *brighter* in the image. (c) In CT, the source and detector rotate around the patient, and reconstruction methods convert projections obtained from many angles into cross-sectional views. (Image in Panel (b) by Stillwaterising. Creative Commons CC0 1.0 Universal Public Domain Dedication.)

FIGURE 11.2
Differences between projections and slices (cross-sections). In the case of a projection, each image point reflects summed attenuation information along a straight-line trajectory through the object.

BOX 11.1 HEALTH CONSIDERATIONS FOR MEDICAL IMAGING USING IONIZING RADIATION

X-rays and gamma rays are forms of ionizing radiation that have many valuable uses in the diagnosis and treatment of disease. However, ionizing radiation also can be harmful and must be used judiciously to ensure that its benefits *significantly* outweigh its risks. Thus, there is a constant push in medical imaging to reduce exposure to ionizing radiation while maintaining quality sufficient for diagnosis or intervention. This is summarized succinctly in the "as low as reasonably achievable" (ALARA) principle.

Exposure is a measure of the amount of radiation incident on a patient. Absorption and tissue susceptibility to damage do not influence exposure values. In light of this, risk is better assessed using radiation "dose." Radiation dose can be defined in several ways and quantified using different metrics that provide somewhat distinct measures of risk. Here we will describe two, absorbed dose and effective dose; the latter appears in Table B11.1.

Absorbed dose is a fundamental, *measurable* indicator of risk; it is defined as the energy deposited by a radiation beam in a unit mass of tissue. From a medical perspective, a significant limitation of absorbed dose is failure to account for disparities in damage induced by different types of radiation and disparities in radiation sensitivity of different types of tissue.

Effective dose is a *calculated* quantity that overcomes these limitations and estimates the stochastic, radiation-induced health risk to a patient from an imaging procedure. Specifically, effective dose is a key quantity when irradiation of the body is non-uniform or limited to a specific area. The absorbed dose in irradiated areas is modified to reflect the damaging power of the type of radiation and the sensitivity of the tissue. The results from the irradiated areas are summed to generate an estimate of the uniform

TABLE B11.1

Risk from Imaging Procedures

Procedure		Effective Dose (mSv)	Time for Similar Dose from Natural Background Radiation
Abdomen	CT of abdomen and pelvis	7.7	2.6 years
	Upper GI X-ray study with barium	6	2 years
Bone	Lumbar spine X-ray	1.4	6 months
	Extremity (e.g., hand, foot) X-ray	<0.001	<3 hours
	Bone densitometry (DEXA)	0.001	3 hours
Central nervous system	CT of brain	1.6	7 months
	CT of spine	8.8	3 years
Chest	Chest X-ray	0.1	10 days
	CT of chest	6.1	2 years
Dental	Standard dental X-ray	0.005	1 day
	Panoramic dental X-ray	0.025	3 days
Heart	Coronary CT angiography	8.7	3 years
	Cardiac CT for calcium scoring	1.7	6 months
Nuclear medicine	PET/CT whole body	22.7	3.3 years
Breast	Screening mammography (digital)	0.21	26 days

Source: Data from American College of Radiology: https://www.acr.org/-/media/ACR/Files/Radiology-Safety/Radiation-Safety/Dose-Reference-Card.pdf.

whole-body equivalent dose that carries the same health risk (e.g., for cancer) as the actual procedure. The SI units of absorbed and effective dose are the same (J/kg), but they are distinguished with different names, the gray (Gy) and the sievert (Sv), respectively. An effective dose of one Sv implies a 5.5% chance of developing cancer over a lifetime. Doses commonly are on the order of one-thousandth of a SV (mSv). Occupational exposure for a radiation worker is limited to 50 mSv per year. Medical procedures are guided by the ALARA principle but are not subject to rigid dose limits.

Table B11.1 gives a feel for the health risks associated with some of the medical imaging procedures discussed in this text. Risk is quantified using effective dose in mSv and by reporting timescales that give rise to similar risk from exposure to natural background radiation. In general, the risk in medical procedures is quite low, whereas the benefits are known to be high.

The data in Table B11.1 show that CT entails higher risk than most other X-ray-based imaging procedures. This fact, coupled with some uncertainty about the *very long-term* danger of CT-induced cancer, has led to concern about the rapid increase in CT scanning, especially for children. Children are of particular concern because they are more sensitive to radiation and because they have a longer time window during which to express radiation damage.

The Food and Drug Administration (FDA) has taken concerns surrounding CT, and other higher dose imaging procedures, very seriously and in 2010 announced an initiative to reduce unnecessary radiation exposure from CT, nuclear imaging, and fluoroscopy.

traditionally expressed in electron-volts (eVs), where 1 eV = 1.6×10^{-19} J. Converting to eV yields

$$124 \text{ eV} < E_{\text{photon}} < 124 \text{ keV}$$

Roughly stated, X-ray photon energies lie in the range from ~0.10 keV to ~100 keV (kilo-eV). In contrast, visible photon energies are ~3 eV (Fig. 11.4). Chemical bond energies similarly are on the order of a few eV. Thus, molecules are highly susceptible to X-ray-induced damage. In a cellular setting, this can lead to mutations that cause cancer.

Medical imaging primarily relies on the use of X-rays with energies in the range from ~15 keV to 150 keV. X-rays with energies less than ~15 keV are not useful because they do not pass through the body effectively. These low-energy rays thus contribute to patient dose without contributing to image formation.

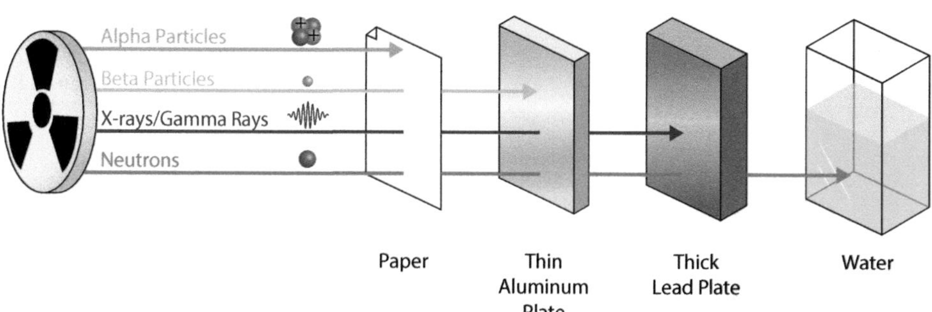

FIGURE 11.3
Cartoon showing the penetrating capacity of various types of particles and EM radiation. Charged (alpha and beta) particles interact strongly with matter and are less penetrating. High-energy photons (X-rays and gamma rays) and uncharged particles (neutrons) penetrate more deeply.

11.2.2 Interactions with Matter

Two primary mechanisms underlie the interaction of diagnostic X-rays with matter: the photoelectric (PE) effect and the Compton effect. We discuss these in turn.

11.2.2.1 The PE Effect

The PE effect refers to the release of charged particles (typically electrons) from a material in response to the absorption of EM radiation. The effect arose briefly in Section 9.1 in the context of radiation detectors. Here we augment our past discussion to address the pivotal role that the PE effect plays in X-ray imaging. We start with a brief review of Albert Einstein's brilliant analysis of the PE effect in metals, which was awarded a Nobel Prize in Physics in 1921.

Einstein's analysis was motivated by PE results that were obtained in the late 1800s by exposing metal surfaces to UV radiation (Fig. 11.4). Several results defied explanation if light is a wave, which was the prevailing view. Instead, Einstein adopted a particle (photon) description of light and was able successfully to explain the puzzles surrounding the PE effect. In the Einstein picture, an EM wave is a stream of photons each possessing a quantum of energy determined by the wave frequency, i.e., $E_{photon} = h\nu$. During irradiation of a metal, a photon can "collide" with an electron and disappear, thereby transferring virtually all of its energy to the electron (Fig. 11.4). An important prediction of this picture, which resolved one of the key PE puzzles, is that electron liberation will occur only if E_{photon} exceeds the minimum electron binding energy. Einstein's theory is described in more detail in Sample Problem 11.1, HW 11.4, and the legend to Fig. 11.4.

The PE effect also occurs when diagnostic X-rays interact with atoms (e.g., carbon or calcium) in the

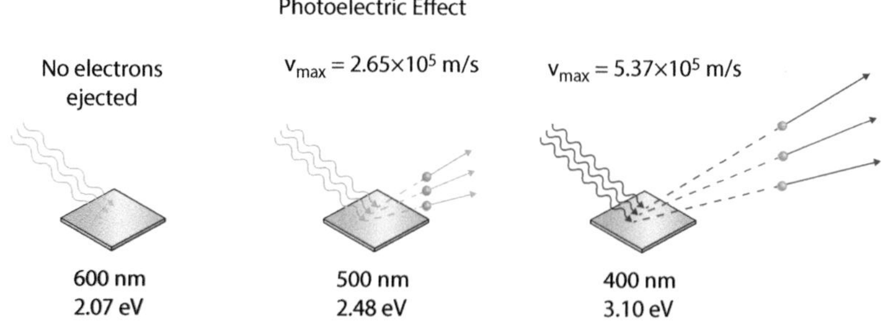

FIGURE 11.4
Attributes of the PE effect in a typical metal (sodium). Energy from incident radiation can eject electrons from sodium as long as the associated photon energy exceeds the surface binding energy, in this case 2.28 eV. Thus, the photon energy required to induce the PE effect in a metal is much lower than X-ray photon energies. 600-nm light (E_{photon} = 2.07 eV) does not induce the PE effect in sodium, no matter what its intensity, whereas 500-nm (E_{photon} = 2.48 eV) and 400-nm (E_{photon} = 3.10 eV) light does. Photon energy in excess of the binding energy appears as kinetic energy of the ejected photoelectron. Thus, the (maximum) speed of photoelectrons increases with increasing photon energy (Sample Problem 11.1).

SAMPLE PROBLEM 11.1 PHOTOELECTRIC EFFECT (ENERGY OF EJECTED ELECTRONS)

Fig. 11.4 shows the PE effect in sodium. Calculate the maximum kinetic energy of ejected electrons for the case of 400-nm (violet) light, and verify the associated velocity value given in the figure.

SOLUTION

The data in the figure can be verified by invoking Albert Einstein's description of the PE effect, which is based on applying energy conservation arguments to a one photon/one electron interaction. In Einstein's description, there is a minimum photon energy, known as the work function, ϕ, that is required to eject the most weakly bound (surface) electrons. Photon energy in excess of ϕ sets the maximum possible kinetic energy of the photoelectron. Stated mathematically,

$$E_{\text{photon}} = h\nu = h\frac{c}{\lambda} = \phi + \left(\frac{1}{2}m_e v_{\max}^2\right)$$

where m_e is the electron mass.

The data in the figure show that $\phi = 2.28$ eV for sodium. This value could be converted into joules and Einstein's equation evaluated using standard "MKS" values for key constants. However, we can solve the problem more rapidly by noting that $hc = 1240$ eV·nm (Sample Problem 6.1). It follows that the maximum kinetic energy in eV is given by

$$\frac{1240}{400} = 2.28 + \left(\frac{1}{2}m_e v_{\max}^2\right)$$

$$0.82 \text{ eV} = \frac{1}{2}m_e v_{\max}^2$$

This is the photoelectron kinetic energy in eV. To get the speed, we now convert to MKS units and insert the electron mass to obtain

$$0.82 \text{ eV} \times 1.6 \times 10^{-19} \text{ J/eV} = \frac{1}{2} \times (9.1 \times 10^{-31} \text{ kg}) \times v_{\max}^2$$

Solving confirms the result in the figure, $v_{\max} = 5.4 \times 10^5$ m/s.

body. Einstein's theory remains valid, but high-energy X-ray photons differ from UV light in dislodging tightly bound inner electrons, which then rapidly lose energy and travel only a short distance in the body. Moreover, quickly after X-ray induced photoemission, an electron in a higher energy state "drops" into the vacated state, and the energy lost by the atom may be released as a secondary, lower energy "characteristic" X-ray (Section 11.3.1 and Fig. 11.5). Secondary X-rays have energies less than a few keV and thus do not escape the body.

The importance of the PE effect in medical X-ray imaging stems from the fact that it generates structure-specific attenuation along the different ray lines through the body. This produces intensity differences at different points on the detector that create contrast (Fig. 11.1).

The extent of PE-based absorption by a target depends on the target's atomic number, Z, and density, ρ, as well as the energy, E, of the beam. Specifically, the probability of absorption via the PE effect is given approximately by the expression

$$\text{Probability}_{\text{PE}} \propto \rho\frac{Z^3}{E^3} \tag{11.1}$$

We will regard Eq. (11.1) as an "empirical" rule. An important implication of Eq. (11.1) is that PE-based absorption is relatively high for targets, like bone, with high atomic number, Z, and high density, ρ (Sample Problem 11.2). Equation (11.1) also implies that absorption is higher for X-rays with energy at the lower end of the diagnostic range because E is in the denominator.

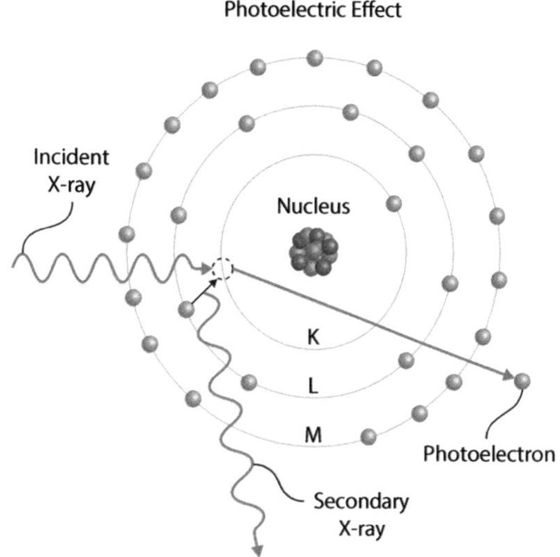

FIGURE 11.5
Mechanism of the PE effect. The PE effect involves photon absorption and ejection of an inner shell electron. In the schematic, an electron is ejected from a K shell. K shells in materials with high atomic number often participate in the PE effect because the associated binding energies are well matched to diagnostic X-ray energies. Quickly thereafter, an electron in a higher energy (L) shell drops into the vacated shell, and the energy lost by the atom is released as a secondary (lower energy) characteristic X-ray.

Key implications of Eq. (11.1) are exemplified in Table 11.1. The data clearly demonstrate that low energy and high effective Z both enhance the PE effect.

The implications for image formation are shown in Fig. 11.6a. Bone/soft tissue contrast is good in a chest image generated with lower energy X-rays due to relatively high PE-based absorption of X-rays by bone. However, contrast degrades at higher energy as the PE-based absorption falls off. These trends also are exemplified graphically in Fig. 11.6b, which shows that attenuation by bone and soft tissue differs markedly at low energy when the PE effect dominates but is very similar at high energy (when instead scattering dominates). This is why bone visibility degrades with increasing energy.

There is an important exception to the general decline in PE interactions with increasing X-ray energy. This "absorption edge" exception arises because PE interactions will only occur if the X-ray energy exceeds the electron binding energy. Moreover, the effect is most probable when the X-ray energy just exceeds the binding energy. As a consequence of this "resonance effect," the probability of a PE interaction increases markedly as X-ray energy transitions from slightly below, to slightly above, the binding energy (Fig. 11.6b).

SAMPLE PROBLEM 11.2 PHOTOELECTRIC EFFECT (RELATIVE ABSORPTION)

The effective atomic number of a compound, or mixture of compounds, is a weighted average of the atomic numbers of the elements comprising the compound(s). In the case of body structures, relevant elements include hydrogen (Z = 1), carbon (Z = 6), nitrogen (Z = 7), oxygen (Z = 8), and calcium (Z = 20). Examples of effective atomic numbers for common body components are shown in the following table:

Material	Effective Atomic Number	Density (g/cm³)
Fat	6.4	0.91
Muscle	7.4	1.06
Air	7.6	0.0012
Bone	13.3	1.75

Source: Effective atomic number data from Allisy-Roberts, P. J. 2008. *Farr's Physics for Medical Imaging (2nd Edition).* Page 13.

Use these values to calculate the relative likelihood of a PE-based absorption event by an average atom in bone and muscle at a particular energy. What is the biggest source of any difference? Hint: See Eq. (11.1).

What would you expect, qualitatively, for bone relative to fat?

SOLUTION

The probability of PE-based absorption by bone relative to muscle can be calculated using Eq. (11.1):

$$\frac{\text{Probability}_{\text{PE,bone}}}{\text{Probability}_{\text{PE,muscle}}} = \frac{\left(\rho \dfrac{Z^3}{E^3}\right)_{\text{bone}}}{\left(\rho \dfrac{Z^3}{E^3}\right)_{\text{muscle}}} = \frac{\left(\rho Z^3\right)_{\text{bone}}}{\left(\rho Z^3\right)_{\text{muscle}}} = \frac{(1.75)(13.3)^3}{(1.06)(7.4)^3} = 9.6$$

Thus, bone is almost ten times as likely to absorb an X-ray as muscle. This difference in probability helps to explain why bone is easy to discriminate from soft tissue using X-rays.

Three terms contribute to the PE absorption probability. However, when computing relative probabilities, the energy terms cancel out. In the present case, differences in effective atomic number contribute about 3.5× more than differences in density, mostly because the term involving atomic number is cubed.

The probability of PE-based absorption by bone relative to fat should be even greater than bone relative to muscle because both the effective atomic number and the density of fat are lower than those of muscle.

From a mechanistic viewpoint, this abrupt increase in interaction probability reflects an abrupt increase in the number of electrons available for interaction.

In the diagnostic X-ray energy regime, absorption-edge effects are particularly significant for high-Z elements like iodine and barium (Fig. 11.6b). This will be important when we discuss the use of iodine and barium as soft tissue contrast-enhancing agents in radiography (Section 11.4.3 and the "Applications Table" in Section 11.6).

11.2.2.2 The Compton Effect

The Compton effect, like the PE effect, demonstrates particle-like attributes of light, and it also garnered a Nobel Prize for its discoverer, Arthur Holly Compton. However, unlike the PE effect, the Compton effect occurs when energy from an incident X-ray is used both to eject a weakly bound outer electron and to generate a scattered X-ray. The scattered ray has lower energy than the incident ray as dictated by energy and momentum conservation (Fig. 11.7).

The physics of the Compton effect is developed in most introductory physics texts. Here we summarize its most medically relevant properties. One important attribute of the Compton effect is illustrated in Fig. 11.8: the effect becomes increasingly important as X-ray energy increases. More specifically,

1. The Compton interaction becomes dominant in soft tissue when energies exceed ~25 keV.
2. The Compton interaction becomes dominant in bone when energies exceed ~60 keV.
3. PE interactions dominate at the lower energies (Fig. 11.8 and Table 11.1).

We derive another important attribute of the Compton effect in Section 11.5.2 – Compton interaction probabilities are roughly proportional to structure density and are largely independent of Z and E. This result, along with its PE analog, is summarized in Table 11.2.

A final important attribute of the Compton effect in the diagnostic regime is the relatively high energy and associated strong tissue penetration of scattered X-rays (HW 11.7). Scattered photons are thus likely to be detected and become a major source of degraded contrast (Fig. 11.9). Additional aspects of the scattering problem, and methods of alleviating it, will be discussed in Section 11.4.4. The PE effect does not contribute to scattering because the incident ray is completely absorbed, and secondary X-rays typically do not escape the body.

11.3 Generating and Detecting X-Rays

US is generated and detected using a single device, the transducer. In contrast, X-rays are generated and detected using distinct devices positioned on opposite sides of the patient (Fig. 11.9).

11.3.1 Generating X-Rays

We consider first the X-ray tube, which is the source. Key components of the tube include a cathode and an anode, which are enclosed in an evacuated structure, and a high voltage source that creates a large potential difference between the cathode and the anode (Fig. 11.10). The big picture view is that electrons generated at

TABLE 11.1

Percentage of PE Interactions

Energy (keV)	Water (Z = 7.4) (%)	Compact Bone (Z = 13.8) (%)	Sodium Iodide (Z = 49.8) (%)
20	65	89	94
60	7	31	95
100	2	9	88

Source: Data from Curry, T. S., III, et al. 1990. *Christensen's Physics of Diagnostic Radiology* (4th Edition). Introduction to Radiological Imaging. Williams & Wilkins. Page 76.

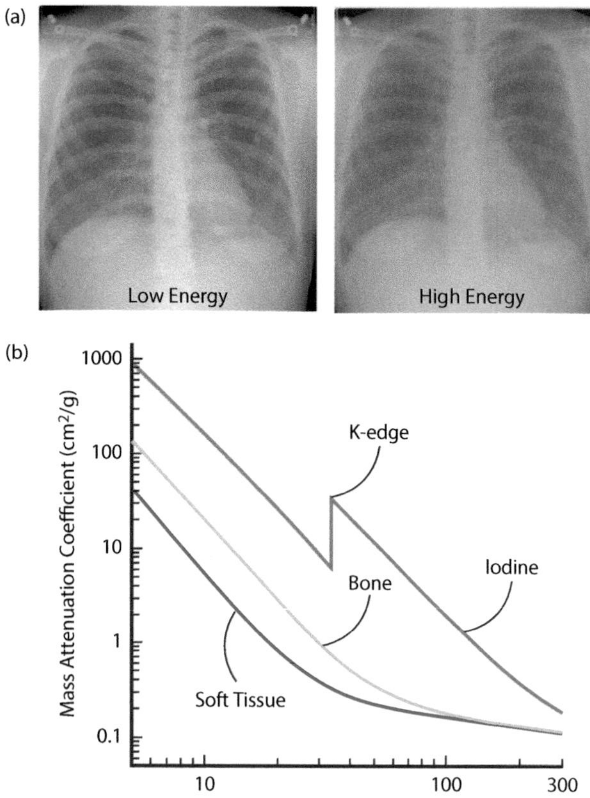

FIGURE 11.6
Energy dependence of contrast. (a) The images show that rib contrast is better at lower energy due to the enhanced contribution by the PE effect. (b) The graphs similarly show that attenuation differences, which are a key determinant of contrast, are enhanced at low energy and near an absorption edge, such as a K-edge. (In this graph, attenuation is quantified along the vertical using the mass attenuation coefficient, which is discussed in Section 11.4.1.) (Panel [a] adapted with permission from https://www.upstate.edu/radiology/education/rsna/. Panel [b] adapted from Seibert, J. A., et al. 2005. X-ray imaging physics for nuclear medicine technologists. Part 2: X-ray interactions and image formation. *J. Nucl. Med. Technol.* 33:3–18.)

the metal cathode are accelerated toward the metal anode. Collisions at the anode cause the electrons to lose kinetic energy. A small fraction of the lost energy produces X-rays, and the rest goes into heat. A more detailed description is given below.

The cathode produces electrons via thermionic emission. Specifically, the cathode assembly consists of a tungsten filament mounted in a focusing cup. The filament is heated by passing current through it, and when a sufficiently high temperature is achieved, some electrons have enough energy to escape from the metal filament.

To generate electron flow between the cathode and anode, a high voltage is applied between them. The cathode is at a negative potential, and the anode at a positive potential, so the anode attracts the electrons.

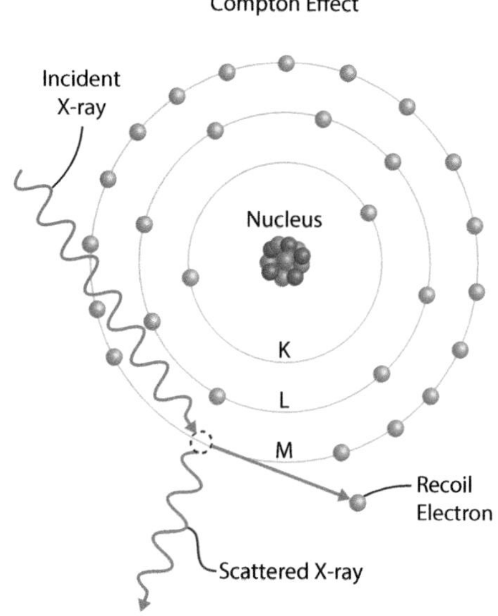

FIGURE 11.7
Mechanism of the Compton effect. The schematic shows that the Compton effect involves photon scattering and ejection of an outer electron.

The resulting electron beam tends to diverge due to charge repulsion, so it is sculpted into a more compact shape by the focusing cup. As a consequence, the beam strikes a small (but finite) area on the anode, known as the "focal spot" (Section 11.4.2). The anode typically rotates so the beam hits different points, thereby minimizing heating problems.

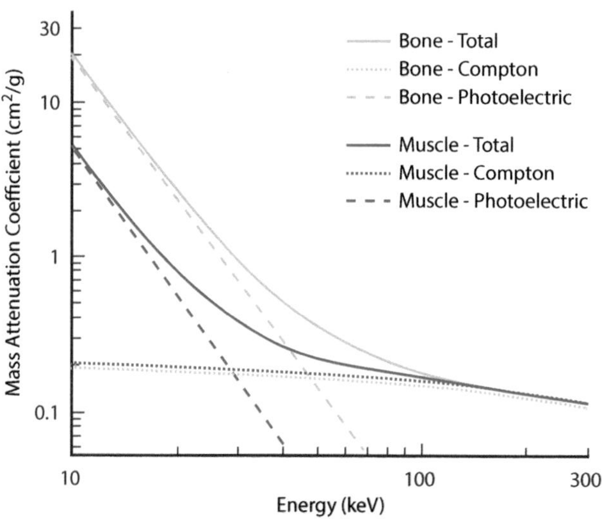

FIGURE 11.8
Energy dependence of the PE and Compton effects. The graph shows the energy regimes in which each effect dominates. Compton interactions are dominant for muscle for most X-ray energies in the diagnostic and therapeutic regimes. (Adapted from Lee, T. C. 2016. Morphology supporting function: attenuation correction for SPECT/CT, PET/CT, and PET/MR imaging. *Q. J. Nucl. Med. Mol. Imaging.* 60:25–39.)

TABLE 11.2
Comparison of Photoelectric and Compton Effects

	Contribution to Interaction Probability			
Mechanism	**Energy (E)**	**Atomic Number (Z)**	**Density (ρ)**	**Energy Range in Soft Tissue**
Photoelectric	$1/E^3$	Z^3	ρ	1–25 keV
Compton	Slow decrease with E	Independent	ρ	25 keV–20 MeV

When electrons collide with the anode, they lose energy and generate bremsstrahlung and characteristic X-rays (Fig. 11.11). Bremsstrahlung (braking) radiation is generated when electrons interact electrically with nuclei and their paths are altered, leading to deceleration. Importantly, a decelerating charge produces radiation. Thus, a sufficiently energetic deflected electron emits X-ray radiation and undergoes an equivalent loss of kinetic energy, which is the origin of the name "braking" radiation. The bremsstrahlung energy spectrum, like loss of kinetic energy, is continuous but is bounded above by the maximum possible kinetic energy of the electron. This is the energy imparted to the electron by the maximum voltage applied between the cathode and anode, which is known as the peak kilovoltage (kVp) (Fig. 11.11).

Characteristic radiation arose briefly in our discussion of the PE effect (Section 11.2.2.1). Characteristic radiation is generated in the anode when an incident electron of high energy ejects an inner electron from an atom. An outer shell electron transitions to fill the vacancy, and a characteristic X-ray may be released with energy given by the difference in energy between the states. The term characteristic is used because the emitted X-rays have energies that are characteristic of the target. Characteristic emission from a tungsten anode is shown in Fig. 11.11.

11.3.2 Detecting X-Rays

Flat panel detectors (FPDs) have become a dominant technology used to convert X-rays (directly or indirectly) into charge. FPDs have achieved prominence because they provide high-quality, efficient, real-time, and large FOV digital detection of X-rays.

A universal component in FPDs is a thin film transistor (TFT) array, which contains millions of detector elements (pixels) arranged in a row-by-column matrix. Each pixel contains a TFT, which functions as a switch, a charge-collection electrode, and a charge-storage capacitor.

During an exposure, the switch is turned off, so charge can collect. Upon completion of the exposure, the switches in the array are activated to implement row-by-row charge readout. Charge flows into data

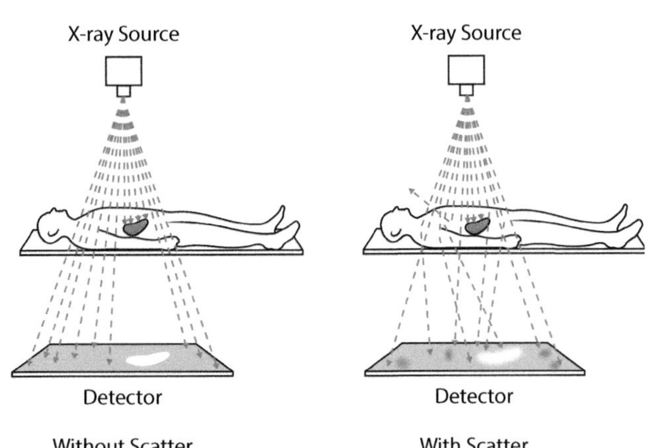

FIGURE 11.9
Effect of Compton scattering on image contrast. On the left is an ideal scenario in which very strong PE interactions in a target (red) generate complete beam attenuation and very strong contrast. On the right is a more realistic scenario in which target contrast is degraded by Compton scattering. The spacing between the patient and the detector has been expanded for clarity.

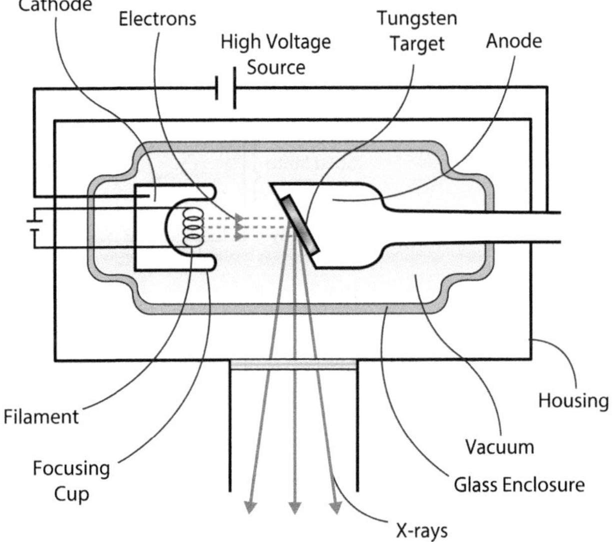

FIGURE 11.10
Schematic of the key components of an X-ray source. Of particular note is the finite (non zero) size of the X-ray producing spot on the anode, which has a major impact on resolution (Section 11.4.2). A collimator (not shown) can reduce the irradiated area and thus reduce patient dose and scatter.

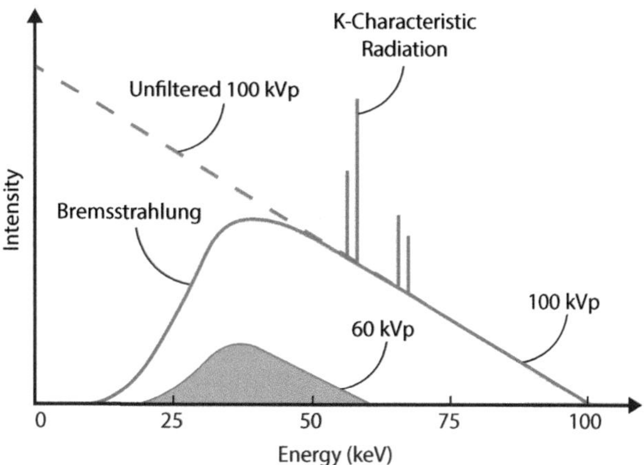

FIGURE 11.11
Comparison of the filtered X-ray energy spectrum generated by a tube operating at 60 kVp and 100 kVp. The maximum associated energies plotted on the horizontal axes are 60 keV and 100 keV because an electron gains 1 eV of energy when it experiences a potential change of 1 V. Also shown is the unfiltered 100 kVp spectrum, which falls off linearly. Filtration of the beam arises in part from components in the tube and in part from metal deliberately placed in the beam to alter its spectrum. Deliberate filtration is used to eliminate low-energy (low penetration) rays that increase patient dose without adding to the detected signal. Characteristic X-rays are not present in the 60 kVp spectrum because the energy is too low. (Adapted from https://en.wikibooks.org/wiki/Basic_Physics_of_Digital_Radiography/The_Source.)

lines, and components outside the active area implement charge-to-voltage conversion and digitize the signal.

FPDs use two different methods of converting X-ray energy into charge: direct and indirect (Fig. 11.12). Direct detection is used when high resolution is of paramount importance (e.g., in mammography). Direct detectors have a layer of amorphous selenium overlying the TFT array. When an X-ray impacts the selenium, a proportional amount of charge is generated that is stored in the local capacitor.

Indirect detection is more common because selenium poorly absorbs the higher energy X-rays used in general radiography. (This is not an issue in mammography, which uses lower energy X-rays.) Indirect detection also is less expensive. Indirect detection is mediated by a scintillator that converts X-rays into more easily detectable visible radiation accompanied by photon amplification (Fig. 11.12). The visible photons then interact with a photodiode on the TFT array to create a charge in the local storage capacitor.

The inorganic crystal, cesium iodide, doped with trace amounts of the impurity thallium, CsI (Tl), is a common scintillator for general radiography. Doping creates luminescent sites with energy levels that are modified, relative to the pure crystal, to generate photons in the visible. Doping also can accelerate

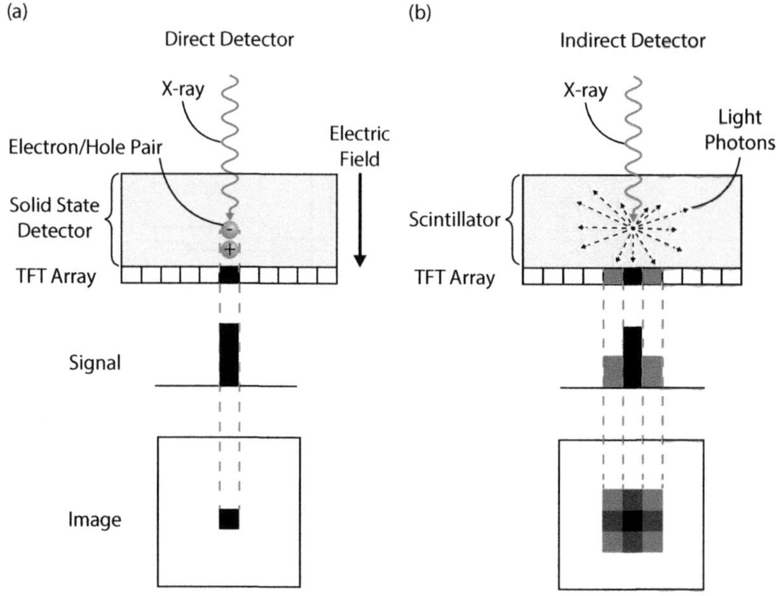

FIGURE 11.12
Comparison of key attributes of direct and indirect X-ray detectors. (a) In a direct detector, the semiconductor amorphous selenium directly produces charge in response to incident X-rays. In the example shown here, the electric field would cause positively charged holes to move toward the pixel array where they collect. (b) In an indirect detector, the scintillator frequently is cesium iodide. Cesium iodide is grown in crystalline columns, which act as pipes that reduce spread of visible radiation and associated blur. Nevertheless, indirect detection generates poorer resolution, as exemplified in the wider signal profile in Panel (b).

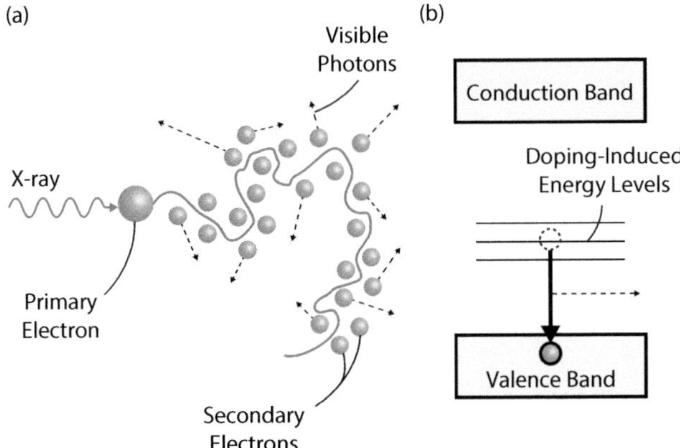

FIGURE 11.13
Schematic depiction (a) of a high-energy primary electron generating thousands of secondary electrons as it traverses the crystal and (b) of doping-induced changes in crystal energy levels that lead to the production of visible photons.

response. An incident X-ray has an energy vastly in excess of that required (~10 eV) for electron excitation and thus can generate a *very energetic* (primary) photoelectron via either the PE or the Compton effect (Fig. 11.13). Typically, the energy from this primary electron excites thousands of secondary electrons, which rapidly de-excite at luminescent sites and generate visible photons. The primary weakness of scintillation-based detectors is compromised resolution because the visible radiation spreads over a spot that is larger than that created by the initiating X-ray (Fig. 11.12).

11.4 Projection Radiographic Imaging

Projection radiographic images are grayscale maps of intensity variations in the transmitted beam produced by attenuating structures in the patient. Our goal now is to quantify the relationship between intensity and attenuation. In subsequent sections, we will apply the results to contrast formation in projection radiography and CT.

11.4.1 Beam Intensity

Individual X-ray photons typically undergo a relatively small number of attenuating interactions (from none to a few) as they traverse matter. These interactions occur stochastically (i.e., randomly). Consequently, the individual rays in a beam penetrate a material to different degrees, even if they have the same energy. Here we derive a mathematical description of the macroscopic beam behavior that results from individual stochastic interactions.

11.4.1.1 Monoenergetic Beams

Consider the scenario in Fig. 11.14, which shows a beam of monoenergetic X-rays traversing a slice in an object. At distance x in the object, the beam intensity is $I(x)$. As a consequence of absorption and scattering, X-rays are removed from the beam, and the intensity drops. We describe this mathematically by saying that as the beam traverses an infinitesimal depth dx, the intensity drops by an amount dI. We expect dI to be proportional to I and to dx and thus write

$$dI = -\mu I dx \tag{11.2}$$

The negative sign ensures that intensity falls. The constant of proportionality, μ, is known as the linear attenuation coefficient and has units of inverse distance (e.g., cm^{-1}).

In diagnostic radiography, the linear attenuation coefficient primarily contains contributions from the PE and Compton effects (Section 11.2.2). Expressed mathematically

$$\mu = \mu_{PE} + \mu_{Compton} \tag{11.3}$$

The energy, atomic number, and density dependence of μ_{PE} and $\mu_{Compton}$ are given in Table 11.2.

Equation (11.2) can be integrated to yield (Sample Problem 11.3)

$$I = I_0 e^{-\mu x} \tag{11.4}$$

where I_0 is the initial intensity, I is the final transmitted intensity, and x is the thickness traversed. Equation (11.4) is known as Beer's Law. The reciprocal of μ, $1/\mu$, is the average photon range (HW 11.8). In HW 11.9, Eq. (11.4) is used to estimate X-ray energies that will lead to acceptable transmission through thicker body structures.

A material's attenuating capacity also can be characterized using the mass attenuation coefficient, which is defined to be μ/ρ, where ρ is density. The key distinction between linear and mass attenuation coefficients is illustrated in Table 11.3. Specifically, in a case in which the chemical makeup is unaltered,

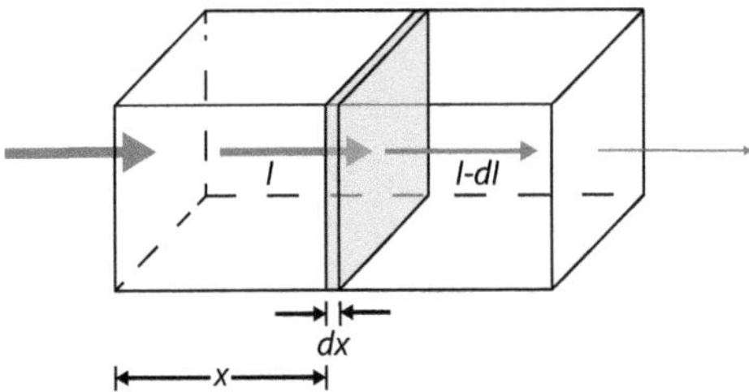

FIGURE 11.14
Schematic of the mechanism and mathematics of X-ray attenuation by an absorbing material.

SAMPLE PROBLEM 11.3 DERIVATION OF BEER'S LAW

Derive Eq. (11.4) starting with Eq. (11.2).

SOLUTION

To obtain Eq. (11.4), we first rearrange Eq. (11.2) as follows:

$$\frac{dI}{I} = -\mu dx$$

This expression is integrated over a slice of thickness x, where the limits of integration are determined by setting the initial intensity at $x = 0$ equal to I_0 and the final transmitted intensity at x equal to I. Mathematically we have

$$\int_{I_0}^{I} \frac{dI}{I} = \int_{0}^{x} -\mu dx$$

The integral on the left is a natural logarithm, and thus we have

$$\ln(I)\Big|_{I_0}^{I} = -\mu x\Big|_{0}^{x}$$

Subtracting the upper and lower limits yields

$$\ln(I) - \ln(I_0) = \ln\left(\frac{I}{I_0}\right) = -\mu x$$

Finally, we exponentiate both sides to obtain Eq. (11.4):

$$I = I_0 e^{-\mu x}$$

TABLE 11.3

Comparison of Linear and Mass Attenuation Coefficients of H_2O.

50 keV			
Linear Attenuation Coefficient (cm^{-1})	**Mass Attenuation Coefficient (cm^2/g)**	**Density (g/cm^3)**	**Relative Distances for Equivalent Attenuation**
0.214	0.214	1	Water 1 cm
0.196	0.214	0.917	Ice 1.09 cm
0.000128	0.214	0.000598	Water Vapor 1670 cm

but density changes, the linear attenuation coefficient can change markedly, whereas the mass attenuation coefficient will remain constant. In other words, changes in the mass attenuation coefficient better mirror changes in chemical makeup because the effect of density has been removed. (Another alternative descriptor of attenuation, the half-value layer (HVL), is discussed in Box 11.2.)

In general, the X-ray beam will successively traverse multiple attenuating structures with different thicknesses, x_i, and attenuation coefficients, μ_i (Fig. 11.15). In this case, successive application of Beer's Law yields the transmitted intensity as

$$\begin{aligned} I_{\text{trans}} &= I_0 e^{-\mu_1 x_1} e^{-\mu_2 x_2} e^{-\mu_3 x_3} \ldots \\ &= I_0 e^{-(\mu_1 x_1 + \mu_2 x_2 + \mu_3 x_3 + \ldots)} = I_0 e^{-\sum \mu_i x_i} \end{aligned} \tag{11.5}$$

In writing down Eq. (11.5), we have used the fact that the intensity exiting structure "i" is the input intensity for structure "$i+1$" (Fig. 11.15). Equation (11.5) will be used in our upcoming discussions of contrast and CT.

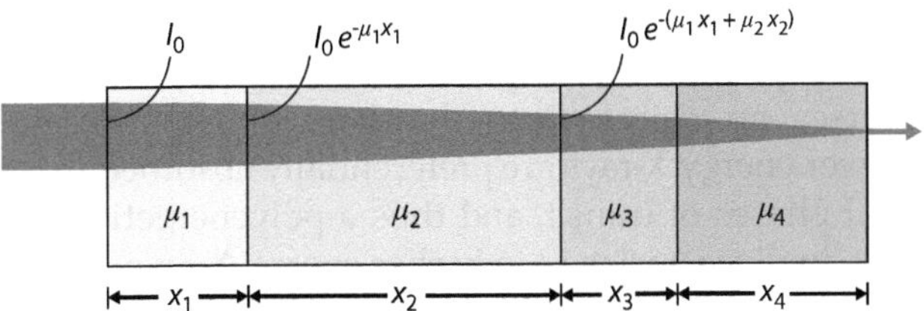

FIGURE 11.15
Schematic depiction of repetitive attenuation by discrete structures. The input for each structure is the output from the structure at its immediate left.

11.4.1.2 Polyenergetic Beams and Beam Hardening

In the previous section, we assumed that the image-producing beam was monoenergetic. In reality, the beam includes a wide continuum of energies because Bremsstrahlung dominates characteristic radiation (Fig. 11.11). For simplicity, we generally will continue to assume that the X-ray beam is monoenergetic, and thus that the simple exponential decay law given in Eq. (11.4) is valid. However, here we discuss one important consequence of using a polyenergetic beam, which is beam hardening (Box 11.2).

BOX 11.2 THE HALF-VALUE LAYER AND EFFECTIVE ENERGY OF POLYENERGETIC BEAMS

The HVL is defined to be the material thickness that is required to reduce the intensity of the transmitted beam to one half of its initial value (see figure). The HVL and μ are inversely related, as follows (HW 11.10):

$$\mu = \frac{\ln(2)}{\text{HVL}} = \frac{0.693}{\text{HVL}}$$

Typical values for the HVL for tissue in the diagnostic regime are a few cm; additional values are shown in the following table.

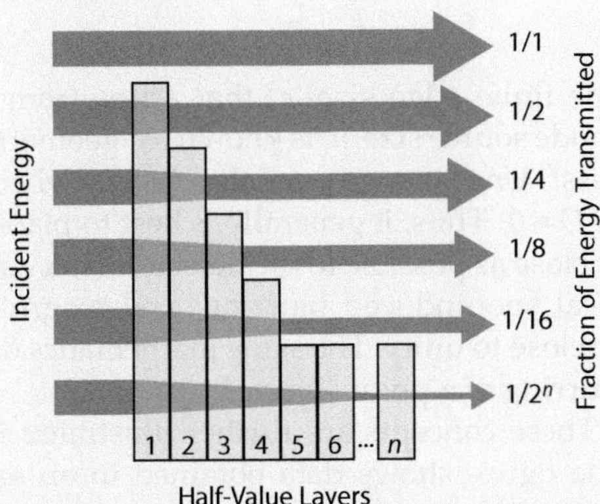

	HVL Thickness (mm)	
Peak Voltage	Lead	Concrete
50	0.06	4.32
100	0.27	15.10
150	0.30	22.32
200	0.52	25.0

Source: Data from Iowa State University Center for Nondestructive Evaluation. https://www.nde-ed.org/Physics/X-Ray/HalfValueLayer.xhtml

The HVL is useful for several reasons. One is that the HVL is more intuitive and somewhat better suited to attenuation estimation. For example, after a beam traverses three HVLs, its intensity is $(1/2)^3 = 1/8$ of its initial value.

Another use of the HVL is as a tool for quantifying the penetrating ability of realistic, polyenergetic beams in terms of an "equivalent" monoenergetic beam. To this end, the HVL is measured using filters and a meter. The polyenergetic beam then is assigned an "effective energy," which is the energy of the monoenergetic beam that would generate the same HVL. Effective energies associated with a polyenergetic beam from an X-ray tube typically are 30%–50% of the maximum tube energy.

Beam hardening refers to an increase in average X-ray photon energy as a polyenergetic beam traverses an object. Beam hardening arises because lower energy X-rays are preferentially absorbed in initial phases of transit, and thus a polyenergetic beam gradually is enriched in higher energy X-rays.

Beam hardening can be beneficial. For example, soft rays in a polyenergetic beam are poorly transmitted through the body. Thus, soft X-rays contribute to patient dose without generating significant signal, as mentioned previously. To reduce this effect, soft X-rays are removed using a metal filter that prehardens the beam before sending it toward that patient. On the other hand, beam hardening can be a significant source of artifacts in CT. We will address this issue in Section 11.5.2.

11.4.2 Resolution and Magnification

Spatial resolution in projection radiography is dictated by multiple factors. The overall spatial resolution, R_{total}, arising from N independent contributions, R_i, can be calculated using the analog of an error propagation formula for a sum or difference:

$$R_{total} = \sqrt{\sum_{i=1}^{N} R_i^2} \tag{11.6}$$

Some of the factors that dictate resolution already have arisen in this chapter, including the extent of light spread when X-rays are indirectly detected using a scintillator. Pixel size also is important because it sets sampling rates (Section 9.3). However, we have not yet addressed a primary determinant of resolution in projection radiography, namely, the extended nature of the radiation source at the anode. This leads to magnification-dependent loss of resolution.

To understand this effect, we first need to discuss magnification in projection radiography, which arises because the beam diverges as it propagates from the source. To verify this, refer to Fig. 11.16, which shows a diverging beam that creates two triangles extending between the source (ideally assumed to be a point) and the object and between the source and the image. The two triangles are similar. Thus, the bases and altitudes of these triangles are proportional, and we can write the magnification, $M = i/o$, as

$$M = \frac{i}{o} = \frac{C+D}{C} = 1 + \frac{D}{C} \tag{11.7}$$

Clearly, M is always bigger than unity, but it becomes approximately unity when the object is very close to the detector ($D = 0$).

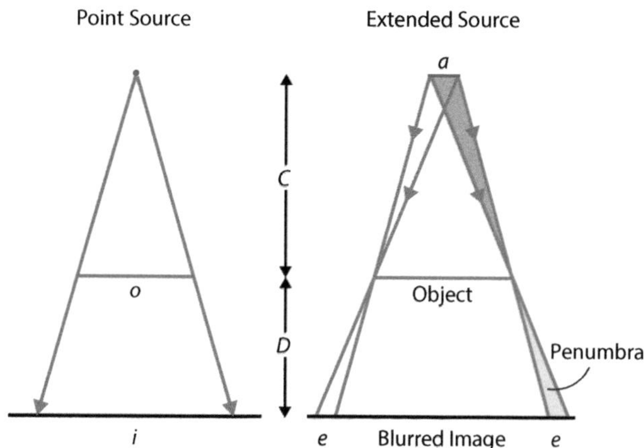

FIGURE 11.16
Schematic representation of magnification and magnification-dependent unsharpness effects in radiography. The beam is relatively narrow when it exits the source, but it becomes wider and correspondingly less intense as it propagates toward the patient and onto the detector. This spread generates depth-dependent magnification. In addition, distinct point sources in the extended focal spot create their own image of the edge. This generates blur.

The effects of magnification and finite anode source size on resolution also are illustrated in Fig. 11.16, which compares the image of the edge of an object for the case of point and extended sources. The point source generates an edge image that is highly localized, whereas the extended source generates an edge image that is blurred. The size of the edge image "e" can be related to the size of the source "a" by noting that the blue-shaded and magenta-shaded triangles are similar, yielding

$$\frac{e}{a} = \frac{D}{C} \Rightarrow e = a\frac{D}{C} \tag{11.8}$$

The finite edge size, e, that arises from a non zero anode source size, a, is known as "geometric unsharpness." Unsharpness vanishes ($e = 0$) when $a = 0$ and/or $D = 0$. Thus, it generally is best to place the patient as close as possible to the detector. This will minimize focal spot-induced blurring, and magnification will be close to unity. The same mathematics describes the blurring of a point object.

These concepts are further illustrated in Fig. 11.17. The figure shows data obtained in an experimental study of the finest line pair spacing that can be resolved under specified imaging conditions. Consistent with the qualitative prediction of Eq. (11.8), large and small focal spots yield essentially the same limiting resolution (3 [line pairs]/mm), when the object and detector are in "contact." In contrast, resolution is compromised (by a factor of two in this case) by use of a large focal spot when the object and detector are spatially separated.

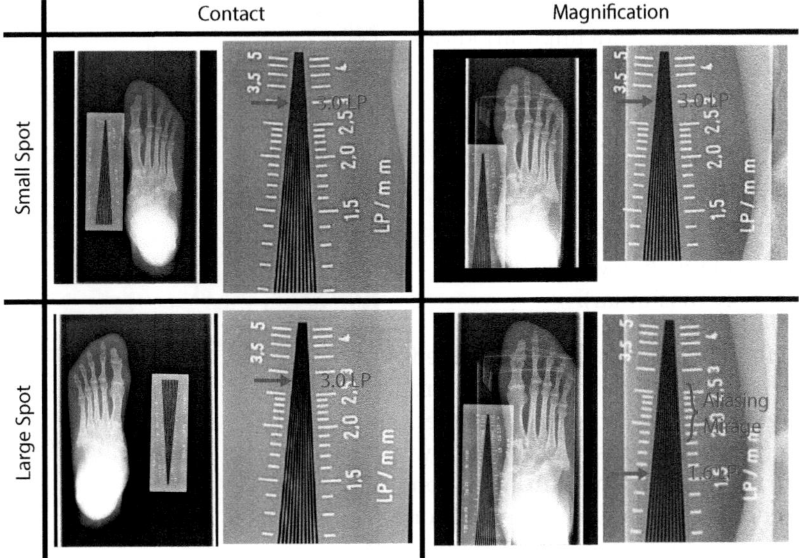

FIGURE 11.17
Experimental data showing the effect of focal spot size, a, on resolution as a function of object-to-detector separation, D. The panels on the top were obtained with a relatively small focal spot size, 0.6 mm. The panels on the bottom were obtained with a larger spot, 1.2 mm. Interestingly, the resolution obtained with the large spot appears to improve between about 2.5 and 3.0 LP/mm; however, this is a mirage caused by aliasing (Section 9.3). (Images adapted with permission from www.upstate.edu/radiology/education/rsna/.)

The data in Fig. 11.17 also provide quantitative insight into the spatial resolution of projection radiography. Resolutions of a few tenths of a mm are routinely obtained with digital detectors.

11.4.3 Contrast

Like resolution, contrast in radiography is dictated by multiple factors. In keeping with the spirit of our microscopy discussion, we will focus on object-generated contrast. Other important factors that influence contrast include detector attributes and post-acquisition processing, as discussed in Chapter 9.

We start with a general definition of contrast of an object "O" relative to background "B," which is

$$C = \frac{\Delta I}{I} = \frac{I_B - I_o}{I_B} \tag{11.9}$$

where I_o and I_B are the intensities that impact the detector along paths that traverse the object and nearby background. In Eq. (11.9), $I_B > I_o$ (i.e., the object is attenuating), so the definition yields a positive contrast.

To relate this general definition to specific object and background attributes, consider the scenario shown in Fig. 11.18. The object and background are assumed to have linear attenuation coefficients μ_o and μ_B and thicknesses x_o and x_B, respectively. In light of Eqs. (11.5) and (11.9), we have

$$C = \frac{I_0 e^{-\mu_B x_B} - I_0 e^{-\mu_B(x_B - x_o) - \mu_o x_o}}{I_0 e^{-\mu_B x_B}}$$

$$C = 1 - e^{-(\mu_o - \mu_B)x_o} \tag{11.10}$$

The contrast, C, in Eq. (11.10) will be high (close to unity) when the exponential term is close to zero and

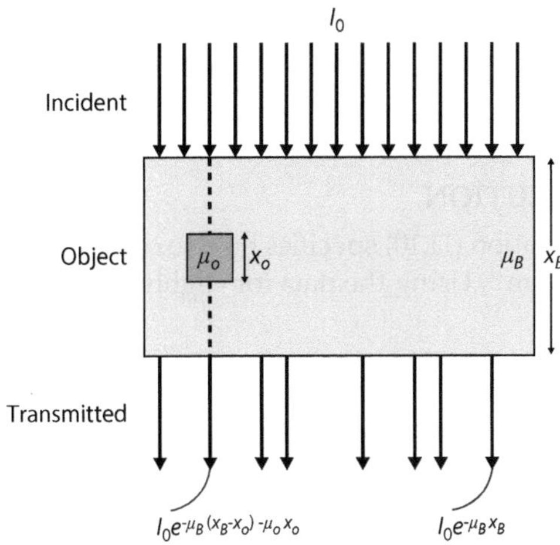

FIGURE 11.18
Geometry and mathematical results used in the derivation of Eq. (11.10).

will be low (close to zero) when the exponential is close to one. This has a number of important implications for object contrast. Specifically, contrast will be high if

1. Object and nearby background differ significantly in mass attenuation coefficient or density (i.e., $\mu_o - \mu_B$ is large).
2. The object is relatively thick (i.e., x_o is large).
3. X-ray energies are relatively low (i.e., the values of μ are relatively dissimilar).

Thus, objects like ribs can be readily visualized against nearby soft tissue when beam energies are relatively low because all three requirements are met. On the other hand, soft tissue contrast tends to be low due to similarities in attenuation. The third criterion also shows that low energy generally is preferred for thinner body parts where penetration is not a problem because contrast will be better.

The low intrinsic contrast of soft tissues frequently can be overcome by using "contrast media." Contrast media are nontoxic liquids containing radiopaque (high μ) elements. When introduced into a specific soft tissue in the body, a contrast medium increases the disparity in the attenuation coefficient between the labeled region and nearby surrounding tissue, and contrast correspondingly increases (Fig. 11.19). The magnitude of the effect can be quantified using Eq. (11.10) (Sample Problem 11.4).

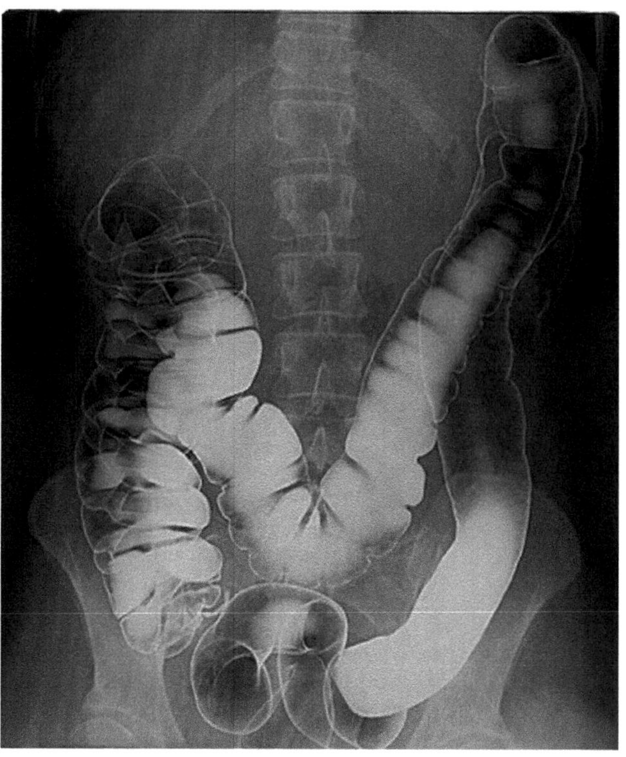

FIGURE 11.19
Example of the use of a barium contrast-enhancing agent to image the digestive tract, which normally would exhibit poor contrast. (Image from "Double-contrast barium enema." 2019. In Wikipedia. https://en.wikipedia.org/wiki/Double-contrast_barium_enema.)

SAMPLE PROBLEM 11.4 X-RAY CONTRAST AGENTS

Compare the contrast between soft tissue and a 1-mm-diameter artery that has been contrast medium-stained with an Omnipaque™ (iodine contrast) stain. Compare the contrast for photon energies slightly above, and slightly below, the iodine K absorption edge, which lies at 33 keV, based on the following data:

$\mu = 0.6\ \text{cm}^{-1}$ for Omnipaque™ for energies slightly below the edge
$\mu = 2.0\ \text{cm}^{-1}$ for Omnipaque™ for energies slightly above the edge
$\mu = 0.4\ \text{cm}^{-1}$ for surrounding soft tissue at both energies

SOLUTION

Equation (11.10) specifies how to compute the contrast. Here, $x_o = 0.1$ cm, $\mu_B = 0.4\ \text{cm}^{-1}$, and $\mu_o = 0.6\ \text{cm}^{-1}$ or $2.0\ \text{cm}^{-1}$. Using the data for slightly below the K-edge, we have

$$C_{\text{below}} = 1 - e^{-(0.6-0.4)(0.1)} = 0.02$$

Slightly above the K-edge, we similarly have

$$C_{\text{above}} = 1 - e^{-(2.0-0.4)(0.1)} = 0.15$$

Thus, there is a 7.5-fold increase in contrast generated by using X-rays with energies just above the absorption edge.

Iodine and barium are elements that are commonly included in contrast media because they strongly attenuate medical X-rays. This is because iodine and barium have relatively high atomic numbers, Z, as well as absorption edges in the diagnostic energy regime. They thus undergo strong PE-based absorption of medical X-rays (Section 11.2.2.1).

The visibility of structures also can be enhanced using dual-energy radiography. For example, dual-energy chest radiography is a method of generating a "subtraction" radiograph that emphasizes bone or soft tissue and suppresses the other component (Fig. 11.20). The approach is founded on the different energy dependencies of the attenuation coefficients of bone and soft tissue. Figs. 11.6 and 11.8 show that, at low energy, the mass attenuation coefficient of bone markedly exceeds that of tissue, whereas at high energy, bone mass attenuation is nearly indistinguishable from that of tissue. These different energy trends make it possible to construct a bone-enhanced or tissue-enhanced radiograph using an appropriately weighted linear combination of a standard high- and low-energy radiograph (HW 11.15). In conventional chest radiography, the dual images typically are generated using ~60 kVp- and 120 kVp-beams. In the analogous CT approach, ~80 kVp- and 140 kVp-beams are commonly used.

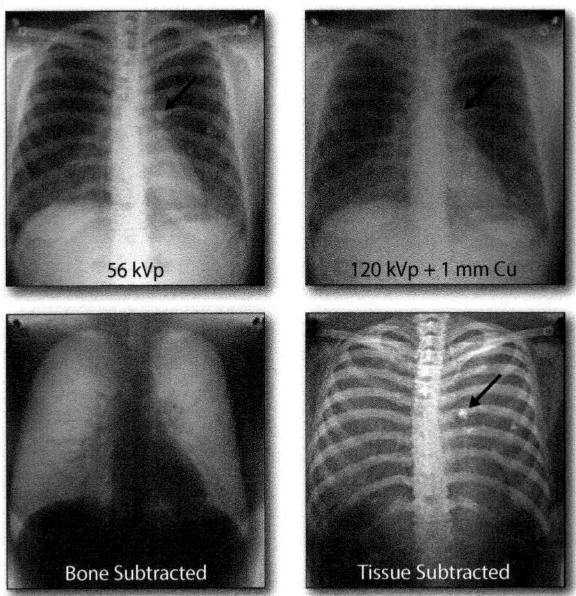

FIGURE 11.20
Dual-energy radiography can generate bone- and tissue-subtracted images that emphasize soft tissue and bone, respectively. Dual-energy imaging also is the basis of bone density scans and is an exciting new addition to CT. For example, dual-energy CT holds great promise as a method of improving the characterization of lesions and renal stones and of improving vascular imaging. Here, low- and high-energy chest X-ray radiographs were generated using 56 kVp and 120 kVp sources, respectively. Difference images were constructed to emphasize soft tissue and bone. Significantly, the difference images show that the nodule visible in the original low-energy image is associated with bone and not with soft tissue. (Images adapted from "Basic Physics of Nuclear Medicine/ Dual-Energy Absorptiometry." 2021. In Wikibooks. https://en.wikibooks.org/wiki/Basic_Physics_of_Nuclear_Medicine/Dual-Energy_Absorptiometry.)

11.4.4 Scattering

To this point, we have focused on the effects that "primary" X-rays have on the signal that impacts a small region of the detector. Primary rays escape the patient without interaction. However, the patient also is a source of "secondary" scattered rays that do not align with the primary beam. Many of these also reach the detector and can markedly degrade contrast (Figs. 11.9 and 11.21).

To quantify the effects of scattering, we assume that scattering arises from the entire background volume, whereas scattering from the small object is negligible. We thus add a background-proportional scattered intensity, $I_s = RI_B$, to each primary intensity in Eq. (11.9). This yields a reduced contrast in the presence of scattering, which is given by

$$C_{\text{scatter}} = \frac{I_B + RI_B - (I_o + RI_B)}{I_B + RI_B} = \frac{I_B - I_o}{I_B(1+R)} = \frac{C}{1+R} \quad (11.11)$$

Equation (11.11) shows explicitly that scattering reduces contrast by the factor $1/(1+R)$. The most common method of reducing this effect is to place an anti-scatter grid between the patient and detector (Fig. 11.21). The grid consists of alternating X-ray-absorbing strips and non-absorbing material; the strips typically are tilted to accommodate primary beam divergence from the focal point. This ensures that most primary rays pass through the non-absorbing regions, whereas scattered rays largely impact the absorbing regions (Fig. 11.21).

11.5 Computed Tomography

We change the focus now to CT, which has several advantages over projection radiography. These include generating cross-sectional and three-dimensional images of the body that preserve depth and lack the structural superposition that plagues projections (Fig. 11.22). CT also suffers less from scatter-induced image degradation because the beam is much narrower (e.g., ~1 cm versus up to ~43 cm). One key consequence is that contrast sensitivity in CT is significantly

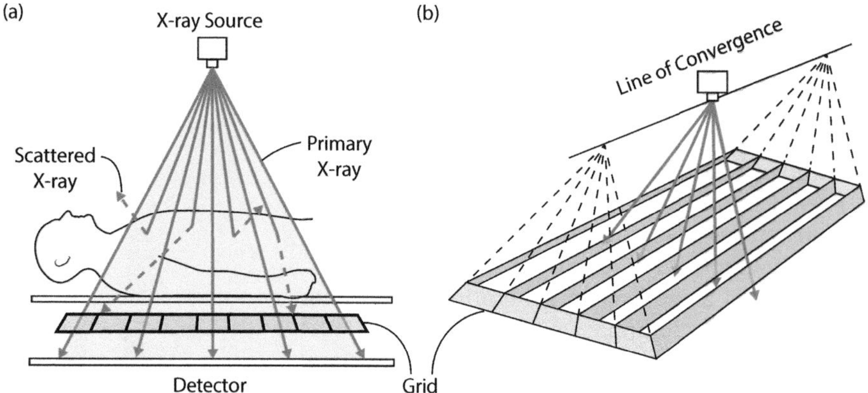

FIGURE 11.21
Image-degrading effects of scattered rays can be mitigated by using an anti-scatter grid (a). Most grids have a linear structure (thin lead sheets that reduce scatter in one dimension), as shown in Panel (b). Occasionally, grids have a cross-hatched structure (not shown) and reduce scatter in two dimensions. Contrast degradation due to scatter also can be reduced using digital contrast enhancement approaches and by confining the beam to the ROI. A typical grid density is ~150 strips/in.

improved over projection radiography. CT can resolve attenuation differences of ~0.5%, whereas projection radiography can resolve differences of ~2%. On the other hand, CT has somewhat poorer spatial resolution than projection radiography (~0.5 mm) and subjects the patient to a much higher radiation dose.

It is useful to start with a big picture view of image production in CT. The essential ideas are nicely illustrated by discussing the first-generation CT scanner, which was introduced in the early 1970s by Godfrey Hounsfield. In recognition of his contributions to CT, Hounsfield was awarded the 1979 Nobel Prize in Medicine in conjunction with Allan Cormack, who made key contributions to the theoretical underpinnings of the technique.

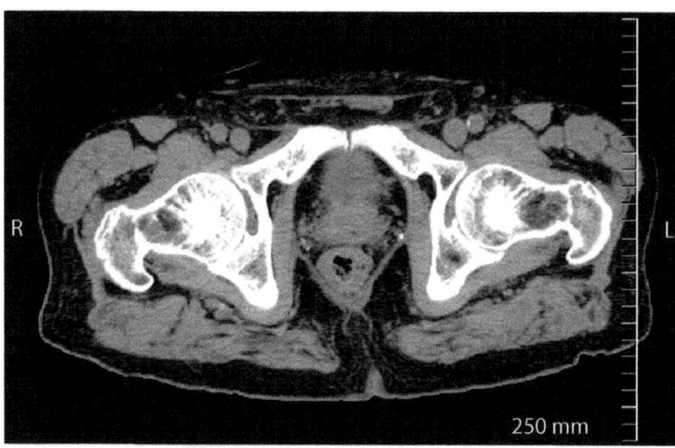

FIGURE 11.22
CT image of a human pelvic region showing right and left hip joints. An inward-facing ball (the femoral head) at the top of the thigh bone fits inside an outward-facing socket (the acetabulum) in the pelvis, creating a ball-in-socket joint that allows each leg to move in a variety of directions. (Image courtesy of HRA Irrevocable Trust.)

For Hounsfield's early scanner, X-ray transmission measurements corresponding to many (~160) ray paths through an axial (z) slice were collected by scanning a thin "pencil" beam across the patient in synchrony with a detector on the opposite side (Fig. 11.23). The set of transmitted intensity data generated by the scan constituted a "view." Additional views (totaling 180) were collected by rotating the detector-tube apparatus around the patient in 1° increments. Reconstruction algorithms described below were then used to reconstruct a matrix of attenuation values in the patient from the 28,800 = 180 × 160 measurements collected from the views. Reconstructed attenuation values were encoded as gray levels to create a cross-sectional (x/y) image.

11.5.1 Data Acquisition

Image acquisition in CT has advanced considerably since the 1970s, with associated reductions in scan times from several minutes to roughly a second. Here we review essential aspects of data acquisition for state-of-the-art, multi-slice CT (MSCT)[1] scanners. Key attributes of images and image reconstruction are discussed in the next sections. Readers interested in additional details are referred to Bushberg et al. (2012).

One key innovation in modern CT scanners is replacement of translation, which is inherently slow, with rotation. Modern CT instruments bypass source/detector translation by using a "fan" or "cone" beam that is wide enough to span the patient width together with an array of detectors that captures the entire transmitted beam (Fig. 11.24). In addition, the source

[1] MSCT scanners often also are called MDCT (multi-detector CT) scanners.

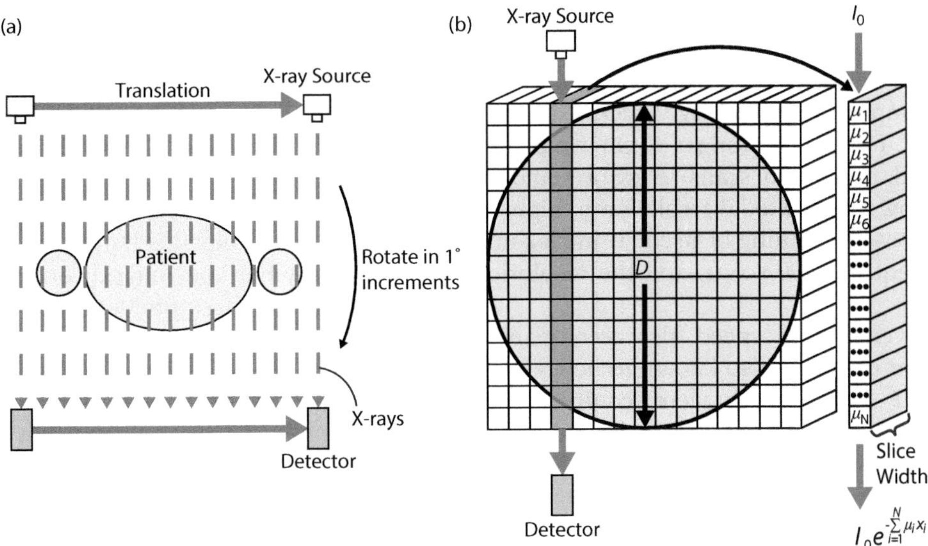

FIGURE 11.23
CT data acquisition and image reconstruction illustrated in the context of a first-generation CT scanner. Data acquisition (a) involved irradiation of a thin slice in the patient and generation of multiple one-dimensional views of transmitted intensity at angles differing successively by 1°. The method of creating one view using a set of vertically downward ray paths (dashed arrows) is shown schematically in Panel (a). Image reconstruction (b) from the view data was directed at computation of an attenuation map for the FOV (Section 11.5.4). To this end, Hounsfield created an $N \times N$ reconstruction matrix by dividing the FOV, and its associated scan circle (blue), into a set of small voxels. For a circle of diameter, D, the pixel dimensions along x and y were D/N; the pixel dimension along z was determined by the slice thickness. Transmitted intensity data for a ray path through the dark green column in the matrix, together with the desired attenuation map, are shown in Panel (b).

and detector are mounted on one platform and rotate through 360° around the patient, in synchrony, to generate ~1000 views at different angles.

Another key innovation is the design of the detector. Older CT scanners utilized detector arrays that were one dimensional. The entire transmitted beam width along z was intercepted by single detectors, and thus the array lacked spatial discrimination along z (Fig. 11.24). In contrast, the beam fan in the (x/y) plane was intercepted by several hundred detectors. Because the beam was undivided along z, each scan generated a single slice along the length of the patient, and these instruments thus are referred to as "single-slice" CT (SSCT) scanners.

One major disadvantage of SSCT systems is that many scans are required to image a modest area in the patient. For example, 100 scans of 1 mm in width are required to cover 10 cm. Associated scan times are relatively long, and thus imaging usually is implemented by collecting wider, lower resolution slices. The beam width and resolution are altered by adjusting a collimator (Fig. 11.24).

In contrast, MSCT systems use a two-dimensional detector array (Fig. 11.24). For example, along the z direction, a scanner might have 64 detectors of width 0.625 mm. Because two-dimensional arrays are capable of spatial discrimination along z during a single scan, MSCT systems can acquire multiple slices

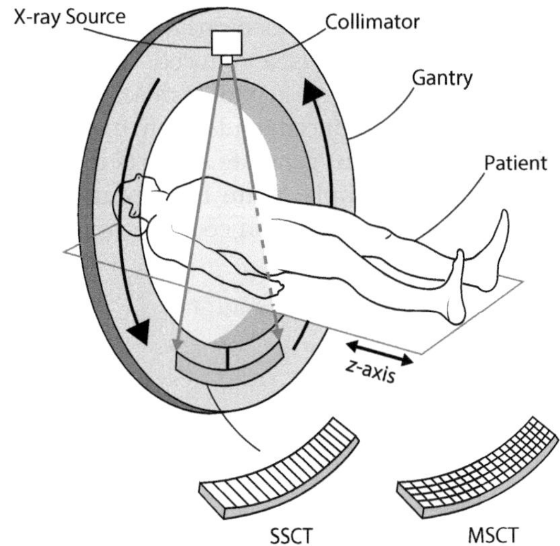

FIGURE 11.24
Third generation (rotate-rotate) platform. Key components include a gantry, which is the rotating part of the scanner that houses the X-ray source and detectors, and the patient table. The table can tilt to alter slice orientation. The patient translates along z (i.e., perpendicular to the plane of the gantry) to collect multiple slices. Also shown are SSCT and MSCT detectors. An SSCT detector has elements that are undivided along z. Thus, one slice is collected at a time. Slice thickness is determined by a collimator that restricts beam width to ~1–10 mm. In contrast, an MSCT detector has a divided array along z that can collect multiple slices simultaneously. Slice thickness is dictated by array element width.

simultaneously. For example, 64 high-resolution (0.625 mm) slices can simultaneously be acquired using a beam of width 40 mm = 64 × 0.625 mm. The associated temporal improvements relative to SSCT are very substantial. MSCT-based scanning of 40 cm in a patient is accomplished in just 10 scans using a standard 64-detector array, and the axial resolution is high. In contrast, SSCT-based scanning of 40 cm, using a 2-mm beam width, requires 200 scans, and the resolution is lower.

Major improvements in speed also accompanied the development of spiral (also known as helical) CT. Before spiral CT, data typically were acquired using a "sequential" CT scan. In the sequential mode, the patient/platform are stationary during axial data collection, and then the two are moved while the beam is off (Fig. 11.25). In contrast, in a spiral scan, the patient/platform move at a constant speed while the gantry rotates around that patient. This significantly improves data acquisition speed relative to the sequential start/stop approach.

11.5.2 Image Attributes

CT images are generated using high-energy (hard) X-ray beams. Hard beams are created by using a high tube voltage (e.g., 120 kVp) in conjunction with tin filters, which absorb the lower energy, soft X-rays.

We already have mentioned one advantage of high-energy beams, which is a reduced content of low energy X-rays that increase patient dose. High-energy beams also have a reduced tendency to undergo beam hardening, which can generate artifacts in CT reconstructions. For example, beam hardening can cause the reconstructed attenuation coefficients of identical tissue to vary (fall) with depth. One common manifestation is an artificially dark region (corresponding to lower attenuation coefficients) in the center (the thickest part) of a uniform cylindrical material. This is known as the "cupping" artifact. Beam hardening also can give rise to spurious detail, including dark and light streaks near highly attenuating structures, such as bone and metallic hardware.

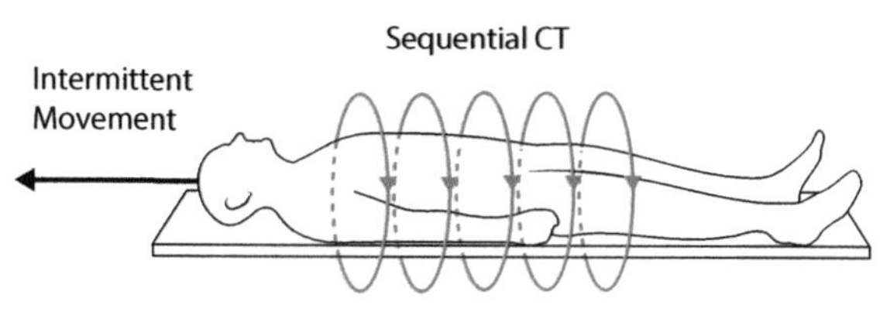

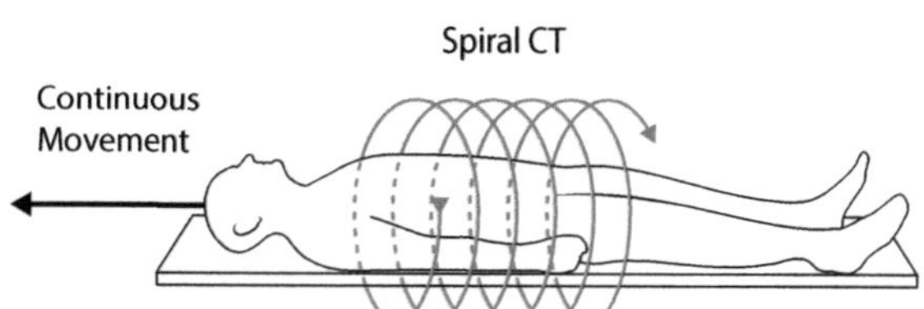

FIGURE 11.25
Comparison of sequential and spiral CT. In sequential CT, scanning and patient translation alternate (multiple times). In spiral CT, scanning and patient translation occur simultaneously. Sequential scans still are used for dose-saving applications.

An initially hard beam (and correction algorithms) helps to reduce these problems. Nevertheless, beam hardening, especially due to metal implants, remains a common source of artifact and is a major focus of efforts to improve the quality of CT images.

Use of a hard beam has important implications for the interactions and tissue properties that contribute to a CT image. With soft (low energy) beams, PE effects are dominant, and thus density, atomic number, and energy all influence contrast, with the latter two being particularly important. This follows from the fact that (Table 11.2)

$$\mu_{\mathrm{PE}} \propto \rho \frac{Z^3}{E^3}$$

In contrast, with hard beams, Compton scattering is dominant, and thus density alone is the major determinant of soft-tissue contrast (Table 11.2 and Fig. 11.8). To justify the relatively minor effect of Z, we note that Compton scattering results from a photon interaction with a loosely bound electron. This is true of all electrons in atoms with low Z and of most electrons in atoms with high Z. Thus, the Compton interaction probability and the associated attenuation coefficient depend on the number of electrons per unit volume. Stated quantitatively (Sample Problem 11.5)

$$\mu_{\mathrm{Compton}} \propto \left(\mathrm{electrons/cm}^3\right) \propto \rho N_A \frac{Z}{A} \qquad (11.12)$$

In Eq. (11.12), N_A is Avogadro's number, ρ is the density in g/cm^3, and A is the molar mass in g/mole.

Significantly, $Z/A \sim 1/2$ for common atoms in soft tissue (except hydrogen) because these lower molecular weight atoms have the same number of protons and neutrons. Thus, the soft tissue attenuation coefficient in CT becomes Z independent, varying as

$$\mu_{\mathrm{Compton}} \propto \frac{\rho N_A}{2} \qquad (11.13)$$

Equation (11.13) demonstrates that μ_{Compton} is largely independent of Z (for soft tissue). Fig. 11.8 similarly shows that μ_{Compton} is largely independent of E. Thus, for soft tissue, the attenuation map that comprises the CT image essentially is a map of tissue density.

SAMPLE PROBLEM 11.5 PROOF OF EQ. (11.12)

Prove Eq. (11.12).

SOLUTION

Our derivation is based on the fact that the Compton attenuation coefficient is determined by the density of electrons available for interaction with incident X-rays. We thus have

$$\mu_{\text{Compton}} \propto \left(\text{electrons/cm}^3\right) = \left(\text{electrons/g}\right) \times \left(\text{g/cm}^3\right) = \frac{(\text{electrons/mole})}{(\text{g/mole})} \times \rho = \frac{ZN_A}{A} \times \rho = \rho N_A \frac{Z}{A}$$

11.5.3 The CT Grayscale

In CT, reconstruction generates a set of average linear attenuation coefficients, $\mu(x,y,z)$, in voxels at points (x,y,z) in the patient, and these are converted into a grayscale to create an image. For historical reasons, it is standard to use CT numbers expressed in Hounsfield units (HU), in honor of the CT pioneer Sir Godfrey Hounsfield, to generate the grayscale map. The CT number is defined as

$$\text{CT}_{\text{number}}(x,y,z) = 1000 \times \frac{\left(\mu(x,y,z) - \mu_w\right)}{\mu_w} \tag{11.14}$$

where μ_w is the linear attenuation coefficient of water for the X-ray spectrum used to generate the image. Typical values of CT numbers for different tissue types are given in Table 11.4. Brightness increases with increasing CT number. Notably, the CT grayscale, unlike most grayscales, is physically meaningful because it is directly linked to an anatomically relevant parameter, μ.

TABLE 11.4

CT Numbers for Common Tissues

Substance	CT Numbers (HU)
Air	–1000
Lung parenchyma	–910 to –850
Fat	–100 to –50
Water	0
White matter	20–30
Kidney	20–40
Spleen	35–55
Gray matter	37–45
Blood	45–65
Liver	45–65
Hematoma	40–90
Bone	700–3000

Source: Data from Hounsfield units – scale of HU, CT numbers. http://radclass.mudr.org/content/hounsfield-units-scale-hu-ct-numbers. Accessed September 12, 2021.

11.5.4 Image Reconstruction

A key first step in image reconstruction is conversion of intensity data, which depend exponentially on attenuation along ray paths, into logarithmic data. Logarithmic conversion has an important linearizing effect. To demonstrate this, we start with Eq. (11.5), which states that the final transmitted intensity falling on the kth detector, after the beam traverses multiple attenuating structures (indexed by i) is given by (Fig. 11.26)

$$I_k = I_0 e^{-\sum \mu_i x_i}$$

Taking the natural logarithm (ln) of both sides yields logarithmic ray-attenuation data, R_k, at the kth detector given by

$$R_k = \ln\left(\frac{I_0}{I_k}\right) = \sum \mu_i x_i = \sum a_i \tag{11.15}$$

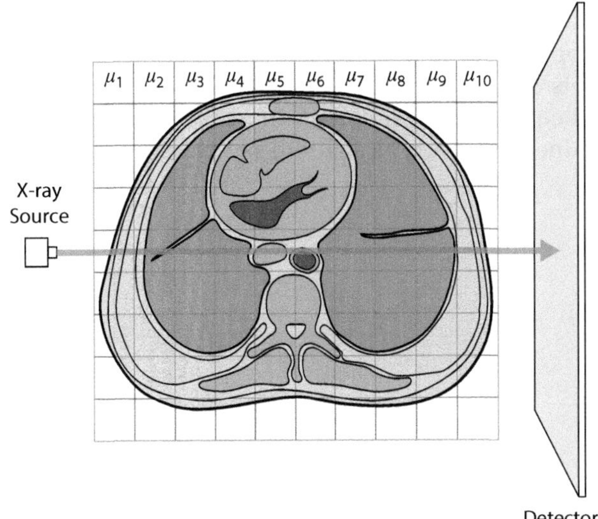

FIGURE 11.26

Schematic representation of the passage of an X-ray beam through multiple attenuating structures in a thin slice through the body. (Thoracic cross-section adapted with permission from Patrick J. Lynch.)

Equation (11.15) shows that logarithmic ray data are sums of attenuations, a_i, in the voxels along the ray path. The reconstruction algorithms described next use this property to reconstruct a_i values from measured R_k values where a set of R_k values obtained at a given angle is a projection. Ultimately, each a_i is converted into an associated μ_i from the known voxel width x_i for the ray path.

There are several possible methods of reconstruction. One conceptually straightforward method is a brute force algebraic approach. The basic idea is to use the projection data to write down a set of simultaneous linear equations for the unknown attenuation coefficients and then to solve the equations algebraically. The equations often are overdetermined (more equations than unknowns) and generally are corrupted by noise.

To illustrate the approach, consider the object and the six ray sums shown in Fig. 11.27. In this case, we have the following set of equations

$$a_1 + a_3 = R_1$$

$$a_2 + a_4 = R_2$$

$$a_1 + a_2 = R_3$$

$$a_3 + a_4 = R_4$$

$$a_1 + a_4 = R_5$$

$$a_2 + a_3 = R_6$$

For specified R_k, these equations can, in principle, be solved for the a_i (HW 11.16). Unfortunately, this approach becomes impractical for the very large data sets acquired in CT because hundreds of thousands of equations are involved. In addition, an overdetermined system typically will not have a solution that satisfies all equations simultaneously (i.e., the equations are inconsistent). For simplicity, we assume consistency here.

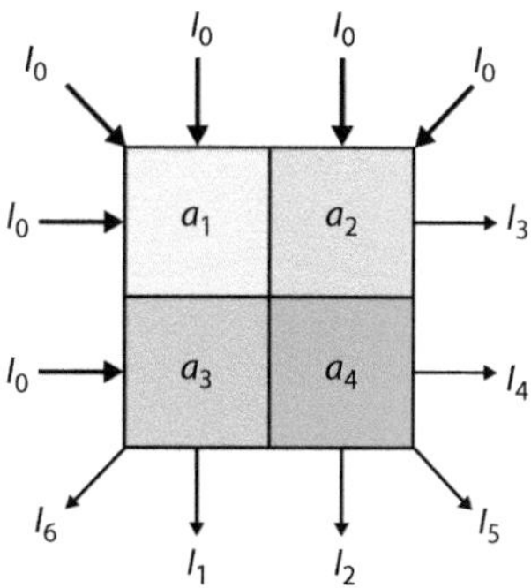

FIGURE 11.27
Simplified example of generating intensity projection data from attenuating structures.

Iterative algorithms are an alternative to a direct, algebraic solution. They are more efficient, but still are computationally demanding, and thus are most commonly used for reconstruction in nuclear emission computed tomography, which generates smaller data sets. However, with improvements in computer speed and implementation, iterative algorithms are increasingly being used in CT. Iterative approaches will be discussed in Chapter 12.

Filtered back projection (FBP) is another alternative. FBP, which is based on a formal mathematical solution to the reconstruction problem, has long dominated CT because it is efficient and produces high-quality reconstructions. In the main text, we focus on a conceptual treatment of FBP; the rigorous theoretical foundations, valid for the case of non-diffracting sources and a parallel beam geometry, are developed in Box 11.3.

FBP is an improvement on an older, common-sense reconstruction approach known as simple back projection (SBP) (Fig. 11.28). In SBP, each view is back projected by dividing the measured data evenly over every image pixel along the path of acquisition. The final reconstruction is the sum of all of the back-projected views (Sample Problem 11.6). When many views are used, SBP generates a reconstruction that reflects object structure but is contaminated by blur (Fig. 11.28).

The presence of blur in SBP reconstructions suggests a need to enhance higher frequency and suppress lower frequency information in the image. In Box 11.3, we demonstrate that a ramp filter given by

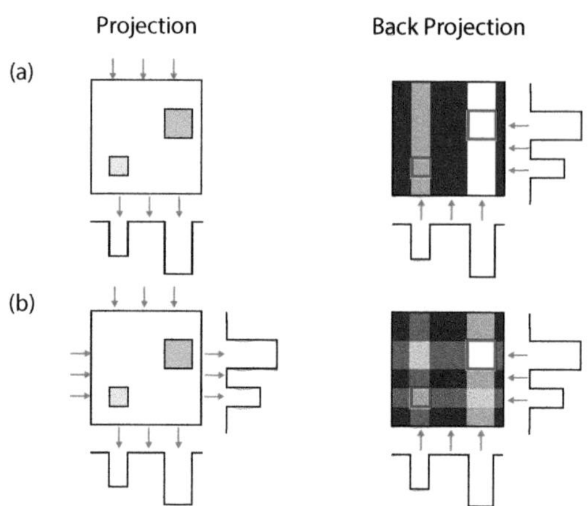

FIGURE 11.28
The essence of SBP for a simple two-dimensional object. Panel (a) shows a single view (projection) of the object and the outcome from SBP. Panel (b) shows analogous results obtained using two views.

BOX 11.3 THEORETICAL FOUNDATIONS OF FBP

The content of this box is very mathematical. The box is included primarily to alert the reader to the fact that FBP has a rigorous mathematical foundation. In particular, the big take-home message from this box is that it is rigorously correct to reconstruct $\mu(x,y)$ by (1) filtering forward projections using a ramp, (2) smearing the filtered projections equally along all points on the ray line to the object, and finally (3) summing the back-projected results. Examination of Eqs. (B11.3.2) and (B11.3.3), which show projections, *P*, being filtered by the ramp to reconstruct μ, will serve to illustrate the mathematical foundations of FBP.

Our treatment is succinct and assumes considerable familiarity with the Dirac delta function and Fourier transforms, as was the case for the material in several past boxes. The reader is referred to Kak and Slaney (2001) for more detail.

For objects composed of discrete subparts, Eq. (11.15) shows that each logarithmic intensity measurement obtained in CT is a sum of attenuation values along a ray path. More generally, for objects that are continuous, the measurement is a line integral of attenuation values along a ray path of the form

$$R_\theta(t) = \int_{\text{Line}(\theta,t)} \mu(x,y)ds$$

Here Line(θ,t) is a line specifying a ray path corresponding to a distance parameter t and a projection at specified angle θ, and ds is a differential element along the line, as shown in Fig. B11.2.1.

The points (x,y) on Line(θ,t) satisfy the equation

$$x\cos\theta + y\sin\theta = t \tag{B11.3.1}$$

Given this, the line integral that generates the ray measurement can be rewritten in terms of the Dirac delta function, δ, as

$$R_\theta(t) = \int\int_{-\infty}^{+\infty} \mu(x,y)\delta(x\cos\theta + y\sin\theta - t)dxdy$$

$R_\theta(t)$ is known as the radon transform of $\mu(x,y)$ over the line Line(θ,t). A projection is a set of line integrals (a set of ray measurements). The analysis that follows is valid for projections generated by collecting a set of parallel ray measurements for different t at fixed angle θ (Fig. B11.2.1).

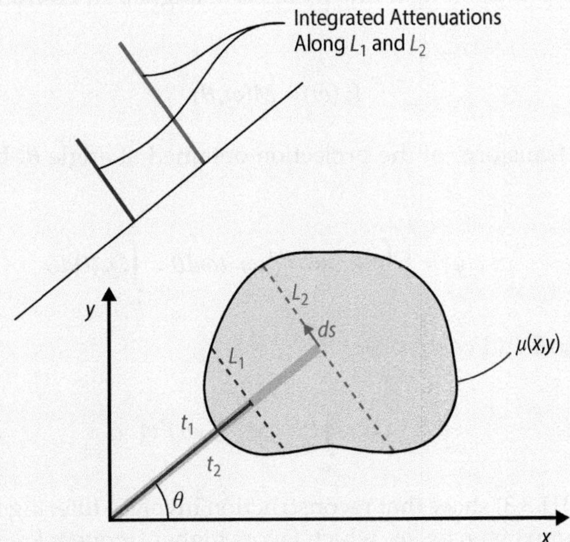

FIGURE B11.2.1
Coordinate system used in filtered back projection.

Image reconstruction in CT is directed at inverting the radon transform and reconstructing $\mu(x,y)$ from projections measured at different angles. Here we derive an inversion algorithm based on FBP. To this end, $\mu(x,y)$ is written in terms of its Fourier transform $M(u,\mathrm{v})$ as

$$\mu(x,y)=\iint_{-\infty}^{+\infty} M(u,\mathrm{v})e^{2\pi i(ux+\mathrm{v}y)}\,du\,d\mathrm{v}$$

It is useful to switch to polar coordinates (ω,θ) via the transformations

$$u=\omega\cos\theta;\quad \mathrm{v}=\omega\sin\theta;\quad du\,d\mathrm{v}=\omega\,d\omega\,d\theta$$

to obtain

$$\mu(x,y)=\int_0^{2\pi}\int_0^{\infty} M(\omega,\theta)e^{2\pi i(\omega\cos\theta x+\omega\sin\theta y)}\omega\,d\omega\,d\theta$$

This integral can be broken into two parts as follows:

$$\mu(x,y)=\int_0^{\pi}\int_0^{\infty} M(\omega,\theta)e^{2\pi i\omega(\cos\theta x+\sin\theta y)}\omega\,d\omega\,d\theta+\int_0^{\pi}\int_0^{\infty} M(\omega,\theta+\pi)e^{2\pi i\omega(\cos(\theta+\pi)x+\sin(\theta+\pi)y)}\omega\,d\omega\,d\theta$$

Invoking the relationships

$$M(\omega,\theta+\pi)=M(-\omega,\theta);\quad \cos(\theta+\pi)=-\cos\theta;\quad \sin(\theta+\pi)=-\sin\theta$$

yields

$$\mu(x,y)=\int_0^{\pi}\int_{-\infty}^{+\infty} M(\omega,\theta)e^{2\pi i\omega(\cos\theta x+\sin\theta y)}|\omega|\,d\omega\,d\theta=\int_0^{\pi}\int_{-\infty}^{+\infty} M(\omega,\theta)e^{2\pi i\omega t}|\omega|\,d\omega\,d\theta$$

where we have set $x\cos\theta+y\sin\theta=t$.

To proceed further, we need to invoke the Fourier Slice Theorem. This famous theorem, which is proved in Kak and Slaney, relates the Fourier Transform of a projection of $\mu(x,y)$, obtained at angle θ, to a "slice" of the Fourier Transform of $\mu(x,y)$ along a line inclined at angle θ in Fourier space. Mathematically, the theorem states

$$P_\theta(\omega)=M(\omega,\theta)$$

where $P_\theta(\omega)$ is the Fourier transform of the projection obtained at angle θ. In light of the slice theorem, we have

$$\mu(x,y)=\int_0^{\pi}\int_{-\infty}^{+\infty} P_\theta(\omega)e^{2\pi i\omega t}|\omega|\,d\omega\,d\theta=\int_0^{\pi} Q_\theta(t)\,d\theta \quad \text{(B11.3.2)}$$

where again $t=x\cos\theta+y\sin\theta$ and we have set

$$Q_\theta(t)=\int_{-\infty}^{+\infty} P_\theta(\omega)e^{2\pi i\omega t}|\omega|\,d\omega \quad \text{(B11.3.3)}$$

Equations (B11.3.2) and (B11.3.3) show that reconstruction involves filtering the Fourier transform of the projection, $P_\theta(\omega)$, using a ramp given by $|\omega|$, which favors higher frequencies, to generate a filtered projection $Q_\theta(t)$. The filtered projections contribute equally to all reconstruction points on a line specified by (θ,t) and are summed over all angles to reconstruct $\mu(x,y)$. This is the essence of FBP – filtered projections are smeared equally and thus contribute equally to all points on the ray line to the object.

SAMPLE PROBLEM 11.6 SBP RECONSTRUCTION

Confirm that the process of SBP generates a blurred, but qualitatively correct, reconstruction of the object below. To this end, use the attenuation characteristics of the object to generate two views in the specified (bottom and right) directions. Then use the two views to generate an SBP (Fig. 11.28).

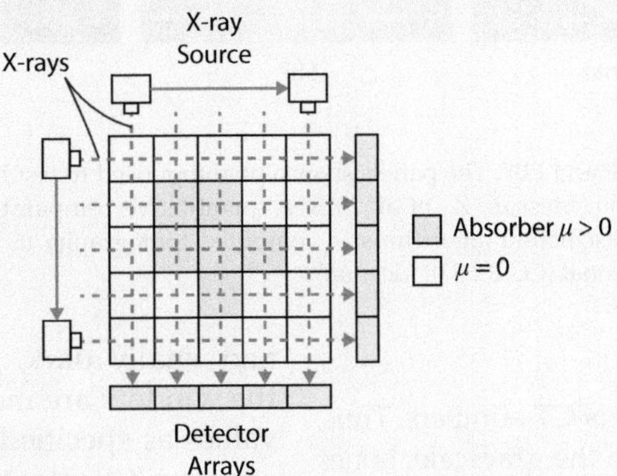

SOLUTION

The solution is presented pictorially below. The two views (i.e., the raw projection data) are obtained by summing the number of attenuating structures along each ray line and thus are labeled "forward projections" in the solution figure; the values are listed near the detector arrays. Results for each ray line are then smeared uniformly back along the line of acquisition (here without normalizing) and summed. The results of back projection are tabulated in the middle image matrix below. The final result is a blurred reconstruction, as shown on the right side of the figure.

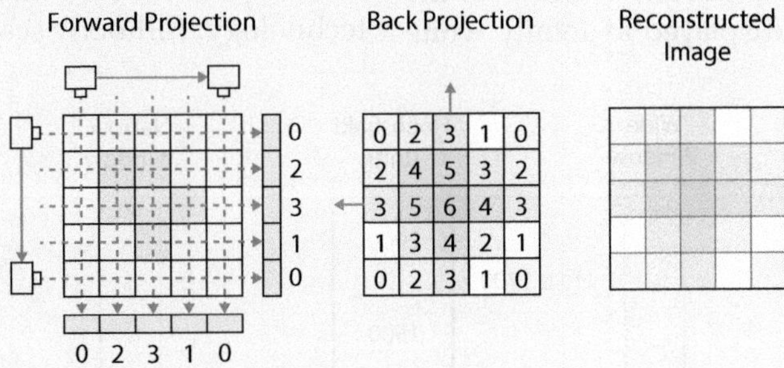

the absolute value of the frequency (i.e., $|\nu|$) compensates correctly for frequency problems that arise in SBP (Fig. 11.29). Even in the absence of proof, it is clear that a ramp filter has the correct qualitative behavior. It gives low weight to low frequencies and successively higher weight to higher frequencies. For the ramp, the weighting between the minimum and the maximum is linear.

Filtering can be implemented in the frequency domain or the spatial domain. The key conceptual point is that filtering before back projection yields a marked improvement in reconstruction quality relative to unfiltered back projection (Fig. 11.30).

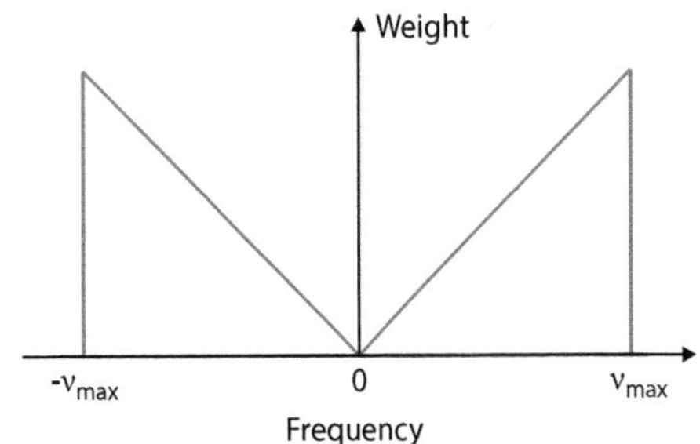

FIGURE 11.29
Graph of ramp filter. The ramp filter has a tendency to amplify noise in projection data, and so other filters frequently are used instead to minimize this problem.

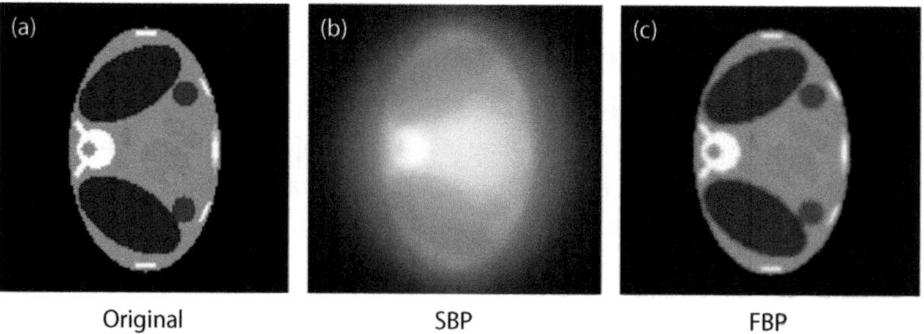

FIGURE 11.30
Comparison of results obtained with SBP and FBP. The panels show a phantom used to test image quality (a) and images reconstructed using SBP (b) and FBP (c). (Adapted from Messali, Z., et al. 2015. A quantitative comparative study of back projection, filtered back projection, gradient and Bayesian reconstruction algorithms in computed tomography (CT). *Int. J. Prob. Stat.* 4:12–31. DOI:10.5923/j.ijps.20150401.02. Attribution 4.0 International [CC BY 4.0] License.)

11.5.5 Image Display

CT images often span a wide range of CT numbers. Thus, if the entire range is mapped onto the grayscale, bones may appear as a saturated white, and soft tissue may appear essentially black. To circumvent this problem, windowing is used to display a subset of the CT numbers optimally and thereby permit visualization of subtle differences in tissue density (Fig. 11.31). Windowing plays a pivotal role in the high contrast sensitivity of CT.

The basic idea is to identify CT numbers corresponding to the tissue densities of interest and to set the window width in an image-processing program to encompass this range of CT numbers. The effect is that CT numbers that are outside the window at the high end and low end are displayed as totally white and totally dark, respectively. CT numbers within the window are mapped onto intermediate grayscale values as specified by the LUT, which often is a linear map (Chapter 9). The effects of different window widths on image contrast are shown in Fig. 11.32.

11.6 Applications

We close with a tabular summary of some key medical applications of X-ray-based imaging to augment, and reiterate, the applications mentioned throughout this chapter. We also discuss briefly discuss a related technology: full-body scanners (Box 11.4).

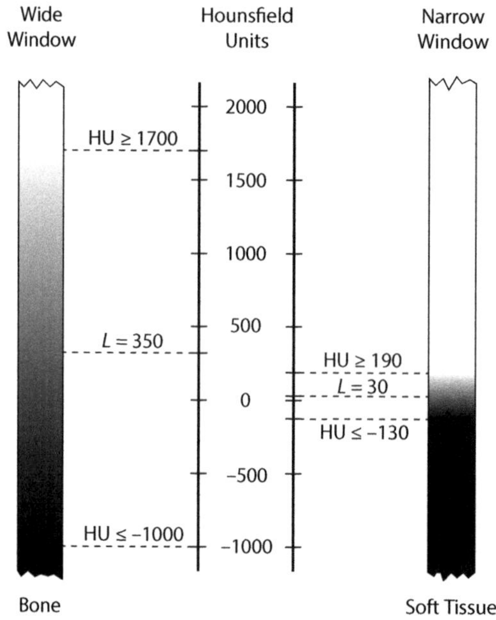

FIGURE 11.31
Essence of windowing. The grayscale range corresponding to structures of interest is identified, and then two parameters are used to set values below and above this range to black and white, respectively. The parameters are the level, which defines the center value of the grayscale range of interest, and the width of the grayscale range. The examples depict representative windows for bone and soft tissue. As the window width is reduced, image contrast in the ROI increases and fewer CT numbers are displayed.

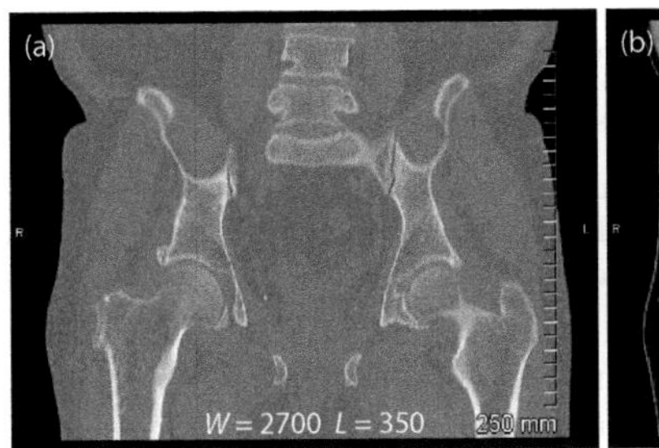

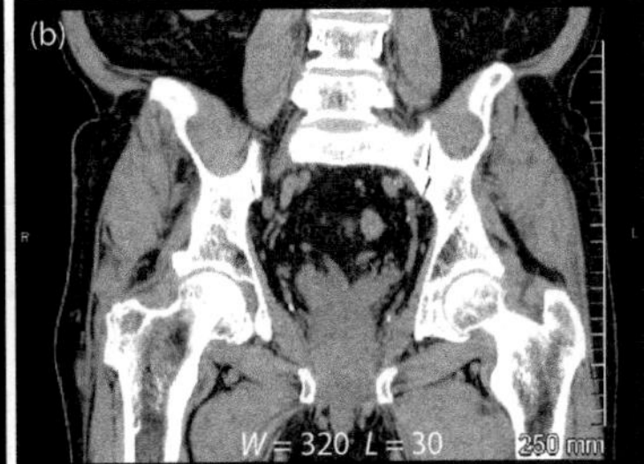

Bone Window Soft Tissue Window

FIGURE 11.32
Effects of windowing on CT image contrast. (a) The "bone window" shows bone more distinctly. (b) In contrast, the "soft tissue window" shows soft tissue more distinctly. These views are taken from the same study shown in Fig. 11.22. L = level, W = window width. (Images courtesy of HRA Irrevocable Trust.)

BOX 11.4 FULL-BODY SCANNERS

This text focuses on imaging techniques used to view internal features of cells, tissues, and whole organisms. Here, we take a brief detour to explore techniques used to image *external* features, namely, those employed by full-body scanners. These now ubiquitous devices are used at airports, courthouses, and other facilities to screen people for weapons and other contraband hidden under clothing.

Most full-body scanners utilize one of two primary types of radiation, X-rays or millimeter waves. Significantly, both techniques monitor backscatter, not transmission, creating images of surfaces. In this respect, both approaches resemble ultrasound.

X-ray backscatter scanners use low-energy X-rays. The X-ray source and X-ray detector are located on the same side of the body, in contrast to conventional radiography and CT scans. The low-energy X-rays penetrate clothing but are mostly blocked by the body. Some X-rays are backscattered, via Compton scattering, and are detected by the detector. The backscattered X-rays are used to create a two-dimensional image of the body, revealing contraband. Differences in X-rays scattered by the body and contraband can be used to identify the contraband.

Millimeter-wave scanners use lower energy, millimeter waves. These typically are sent toward the body and detected from two sides simultaneously using two rotating transmitters, which also function as receivers. The reflected waves are used to construct a three-dimensional image. Millimeter-wave scanners have, by act of Congress, replaced X-ray backscatter scanners at U.S. airports due to concerns about the health effects of X-rays.

The images formed by both of these techniques are too good in the sense that they reveal anatomical information that many find disturbing. Therefore, they are often blurred or replaced with cartoon images* for presentation, both to protect privacy and accentuate areas of concern. In some cases, a colored box or other symbol is superimposed over a silhouette of the body to indicate areas of concern. These areas may be patted down by hand

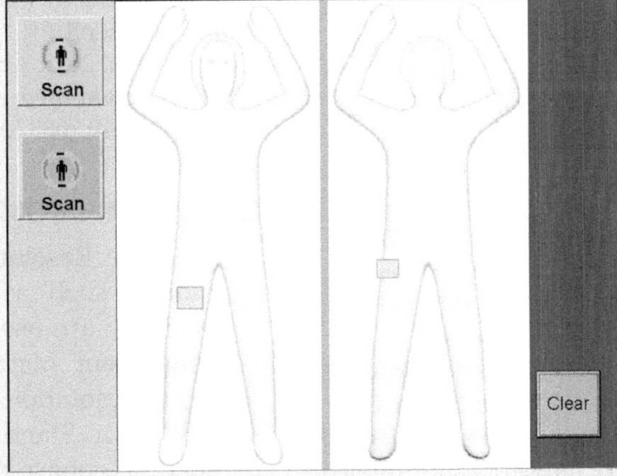

* Image courtesy of U.S. Transportation and Security Administration (TSA).

Medical Applications of Projection Radiography and CT	
Select Applications	**Select Aspects of Implementation**
Fluoroscopy – real-time viewing for placement of medical devices (e.g., catheters, stents, bone plates, and prostheses) and for diagnosing diseases of the GI tract ….	Involves production of real-time images at rates of 3–30 frames per second. Uses markedly reduced X-ray dose per frame to allow production of ~10,000-frame movies. Requires compensatory ultra-sensitive, low-noise detection.
Mammography – detection of breast cancer and other abnormalities	Involves breast compression to reduce tissue thickness, structure overlap, patient movement, and scatter. Uses specialized mammography systems that generate a small focal spot and low-energy X-rays and that have high-resolution detectors.
Chest radiographs – detection of broken bones, lung cancer, pneumonia, pulmonary edema ….	May involve dual-energy radiography to "subtract out" bone or soft tissue and thereby emphasize the remaining component.
Radiography of the GI tract – detection of ulcers, polyps, and tumors ….	Involves administration of a barium contrast medium to coat structures (e.g., the wall of the large intestine) and enhance contrast. Often uses fluoroscopy.
Urography – X-ray examination of the urinary tract for detection of kidney stones, infection, cancer ….	Involves injection of iodine contrast medium to facilitate visualization of the structure of the urinary system and assessment of function (e.g., as manifest in reduced rates of iodine excretion).
Emergency assessment – rapid evaluation of trauma to the chest, abdomen, and the musculoskeletal system, and of suspected stroke, pulmonary embolism ….	Uses spiral CT with or without contrast agents.
Dental X-rays – detection of cavities, impacted teeth, jaw issues ….	Uses projection radiography and/or "cone beam" CT.
Planning radiotherapy for cancer treatment	Uses CT to localize the tumor accurately and define its three-dimensional extent, localize nearby normal tissue, and provide electron densities for dosimetry. Uses external skin landmarks (skin tattoos) and immobilization devices to ensure reproducible patient positioning.
Coronary angiography (CA) – screening for blockage of the coronary arteries	***Conventional CA (invasive).*** Involves cardiac catheterization to introduce iodine contrast medium into arteries of the heart. Uses fluoroscopy for catheter guidance and subsequent imaging. ***Computed tomography CA (noninvasive).*** Involves intravenous introduction of iodine contrast medium and may include medication to slow heart rate to make the images somewhat clearer. In addition to coronary arteries, head and neck arteries also commonly are studied.

Suggested Reading and Additional Resources

Abdulla, S. and Clarke, C. 2011–2022. Radiology Café: FRCR lecture notes: X-ray imaging. https://www.radiologycafe.com/radiology-trainees/frcr-physics-notes/x-ray-imaging

Bushberg, J.T., Seibert, J.A., Leidholdt, E.M., and Boone, J.M. 2012. *The essential physics of medical imaging* (3rd Edition). Wolters Kluwer | Lippincott Williams & Wilkins, Philadelphia, PA. Chapters 7 and 10.

Curry, T.S., III, et al. 1990. *Christensen's physics of diagnostic radiology* (4th Edition). Williams & Wilkins, Media, PA.

Kak, A.C. and Slaney, M. 2001. *Principles of computerized tomographic imaging*. Society of Industrial and Applied Mathematics, New York, NY. Chapter 3.

Kane, S.A. and Gelman, B.A. 2020. *Introduction to physics in modern medicine* (3rd Edition). CRC Press, Boca Raton, FL. Chapter 5.

Mylott, E., et al. 2011. An easily assembled laboratory exercise in computed tomography. *Eur. J. Phys.* 32:1227–1235. doi:10.1088/0143-0807/32/5/010

HOMEWORK PROBLEMS

11.1 [Relative Resolution] Medical imaging techniques, such as X-ray imaging and MRI, typically are used to study the body and its constituent bones and organs. The lateral (x/y) resolution of these techniques is about 0.1–1 mm. Standard microscopy techniques, such as phase-contrast, DIC, and fluorescence, typically are used to study cells and their constituent organelles. The optimum lateral resolution of these techniques is about 0.2 μm.

a. Estimate the dimensions of the area viewed in a typical chest X-ray. Compute a relative resolution by taking a ratio of the absolute resolution to the lateral dimension.
b. Estimate the dimensions of a typical cell. Compute a relative resolution by taking a ratio of the absolute resolution to the cell dimension.
c. Compare the relative resolutions of medical imaging and optical microscopy.

11.2 [Projection Radiography – Overview] Explain, using words and a drawing, how a projection radiograph is generated.

11.3 [Properties of X-rays] Describe the two primary mechanisms underlying the interaction of diagnostic X-rays with matter. Explain how each mechanism is affected by physical properties of matter and the implications for X-ray imaging.

11.4 [Photoelectric Effect] A major experimental puzzle surrounding the PE effect in the late 1800s was the observation that each metal has a threshold frequency ν_T for ejection of electrons. Specifically, radiation with frequency below ν_T will not generate photoelectrons no matter how intense, and this fact is at odds with the wave theory of light. In contrast, ν_T is predicted and easily calculated using Einstein's model. Calculate the threshold frequency and threshold wavelength for sodium. Are your results consistent with Fig. 11.4?

11.5 [Photoelectric Effect] Calculate the relative likelihood of a PE-based absorption event by an average atom in muscle and fat using the values for effective atomic number and density given in Sample Problem 11.2. Compare the result with that calculated for bone and muscle in the Sample Problem. What are the implications of any differences?

11.6 [Photoelectric Effect – Shielding] The health hazards associated with exposure to X-rays can be reduced by minimizing exposure times and by shielding portions of the body that are not being imaged. Lead is a common shielding material due to its low cost and excellent shielding properties. It is used in partitions that separate the X-ray technician from the X-ray imager and in protective garments, such as aprons and blankets, that protect unimaged parts of the patient. Calculate the relative likelihood of a PE-based absorption event by a lead atom ($Z = 82$; $\rho = 11.3$ g/cm^3) and generic soft tissue ($Z = 7.42$; $\rho = 1.0$ g/cm^3).

11.7 [Compton Effect – Photon Energy] The energy of a Compton-scattered photon, E_s, is given by

$$E_s = \frac{E_0}{1 + \frac{E_0}{511\ \text{keV}}(1 - \cos\theta)}$$

where E_0 is the incident energy and θ is the photon scattering angle relative to the direction of incidence. Use this result to show that a Compton interaction of an incident 85-keV photon will produce a scattered photon with an energy of at least 64 keV. This result underscores the relatively high energy of Compton-scattered photons.

11.8 [Beer's Law – Average Photon Range] Use Beer's Law, Eq. (11.4), to demonstrate that $1/\mu$ is the average range (i.e., penetration depth) of a photon in a medium with attenuation coefficient μ.

11.9. [Beer's Law – Energy Dependence of Transmitted Intensity] The linear attenuation coefficient of fat is 0.50 cm^{-1} and 0.17 cm^{-1} at 20 keV and 60 keV, respectively. Calculate the fraction of transmitted X-rays at each of these energies assuming the rays traverse 20 cm of fat. What does this sample calculation imply about the utility of low-energy X-rays for imaging thicker body structures?

11.10. [Half-value Layer] Show that the half-value layer, HVL, and attenuation coefficient, μ, are inversely related as specified in the following expression:

$$\mu = \frac{\ln(2)}{\text{HVL}} = \frac{0.693}{\text{HVL}}$$

Hint: See Box 11.2.

11.11. [Attenuation – Shielding] Lead aprons are used to shield parts of the body from unnecessary X-ray exposure, as noted in HW 11.6.

a. Assume that a lead apron contains a 0.75-mm-thick layer of lead and compute the fraction of incident 140 keV X-rays that are transmitted by the apron. The mass attenuation coefficient of lead for 140 keV X-rays is $\mu/\rho = 2.0$ cm^2/g, and the density of lead is $\rho = 11.3$ g/cm^3.
b. What is the required thickness for a lead wall that is designed to transmit only 1% of the incident intensity of a 10.0 keV X-ray beam? The mass attenuation coefficient of lead for 10.0 keV X-rays is $\mu/\rho = 127$ cm^2/g.

11.12 [Projection Radiography – Mammography] Equation (11.10) shows how contrast depends on attenuation coefficients and object thickness.

a. Calculate the contrast between an object consisting of a 0.20 mm microcalcification and the surrounding breast tissue for 20 keV and 60 keV X-rays. Use values in the table.

X-Ray Energy (keV)	Breast (cm^{-1})	Calcification (cm^{-1})	Water (cm^{-1})
20	0.50	4.8	0.76
60	0.17	0.55	0.20

b. Repeat the calculation for a 0.20-cm lump of water in breast tissue for 20 keV and 60 keV X-rays.

c. Which cases would be detectable if the required contrast is at least 1%?

11.13 [Blur and Magnification] Compute the blur and magnification using values for source-patient distance, patient-detector distance, and focal spot size as follows: C = 40 cm, D = 20 cm, and a = 0.1 mm (Fig. 11.16). Are these numbers consistent with the need to resolve objects with size $\leq$ 0.1 mm in a high-resolution mammogram?

11.14 [Computed Tomography – Overview] Explain, using words and a drawing, how a CT scan is generated.

11.15 *[Dual-energy Imaging] This problem outlines the mathematical foundations of dual-energy imaging as applied to conventional chest radiography (Fig. 11.20). The method is implemented by acquiring two radiographs – one at a low energy, E_1, and the other at a high energy, E_2. Each encodes attenuation data that reflect traversal of bone and soft tissue. Thus, Eq. (11.5) implies that

$$I_j = I_0 e^{-(\mu_{bj} x_b + \mu_{tj} x_t)}$$

In the above equation, μ and x denote linear attenuation coefficients and structure thicknesses for bone (b) and soft tissue (t). The two energies are distinguished using the subscript j. As with CT, the exponential data are linearized to yield:

$$P_1 = \ln\left(\frac{I_0}{I_1}\right) = \mu_{b1} x_b + \mu_{t1} x_t$$

$$P_2 = \ln\left(\frac{I_0}{I_2}\right) = \mu_{b2} x_b + \mu_{t2} x_t$$

The goal is to construct a linear combination image, LC, of P_1 and P_2 that contains only the bone (or tissue) component. To this end, we write LC in terms of unknown weighting factors (W) as follows:

$$LC = W_1(\mu_{b1} x_b + \mu_{t1} x_t) + W_2(\mu_{b2} x_b + \mu_{t2} x_t)$$

Show that LC can be converted into a tissue or bone image using the following weighting ratios:

$$\frac{W_1}{W_2} = -\frac{\mu_{b2}}{\mu_{b1}} \quad \text{(yields tissue image)}$$

$$\frac{W_1}{W_2} = -\frac{\mu_{t2}}{\mu_{t1}} \quad \text{(yields bone image)}$$

11.16 [Computed Tomography – Image Reconstruction] Fig. 11.27 shows a simple 2 × 2 object and six associated ray sums (two vertical, two horizontal, and two diagonal). Write down four algebraic equations describing the ray sums indicated below in terms of the object "attenuation" coefficients ($a_1 \cdots a_4$) and solve for the four attenuation coefficients given the following numerical values:

$$R_2 = 10 \quad R_3 = 14 \quad R_4 = 6 \quad R_5 = 12$$

11.17 [Full Body Scanners] The millimeter-wave region of the EM spectrum spans wavelengths ranging from about 1 mm to 10 mm. Calculate the associated photon energy at these two wavelengths in eV. How do millimeter-wave energies compare to diagnostic X-ray energies?

ACTIVITIES

A11.1 Transmission Imaging (Projection and Tomographic)

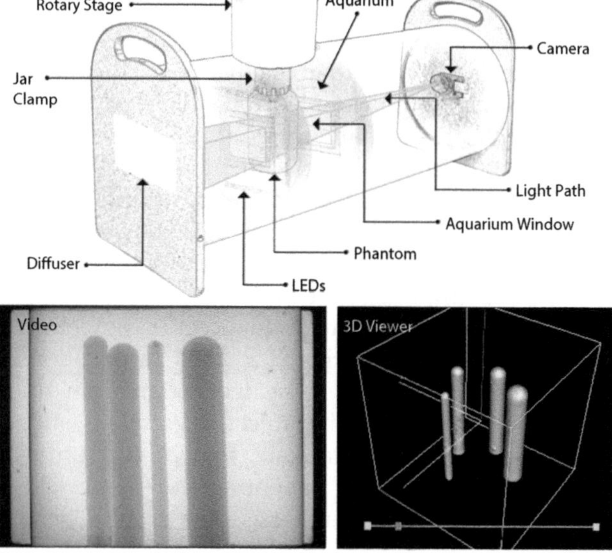

The purpose of this "activity" primarily is to alert readers to

1. Educationally oriented devices that are capable of reconstructing two- and three-dimensional images from projections collected at many angles.
2. Possible experiments and activities. Detail is omitted because it is very device specific.

A cost-effective commercial option, which is pictured here and used at Lewis & Clark College, is the DeskCAT™ multi-slice optical CT scanner developed by Modus QA.[2] There also are non-commercial options, such as the tomography device described by Mylott et al. (2011). Both options avoid the danger associated with ionizing radiation by using visible/IR radiation and translucent samples.

The Modus and Mylott devices allow students to collect projections and generate reconstructions, as well as analyze and process images. The DeskCAT system, in particular, comes with ~15 laboratories that allow students to explore

1. The differences between projection and tomographic images.
2. Reconstruction algorithms.
3. Resolution and contrast.
4. Fan and cone beam CT.
5. Dual-energy CT.

[2] Images courtesy of Modus Medical Devices: www.modusqa.com/products/deskcat-optical-ctscanner-for-biophysics-education/.

12

Planar Scintigraphy and Emission Tomography

In this chapter, we continue to discuss medical imaging using ionizing radiation, but our focus shifts from X-ray transmission imaging to gamma-ray emission imaging. A key message in what follows is that transmission and emission imaging are largely complementary and often are used together to improve the efficacy of diagnosis and treatment. Specifically, X-ray transmission techniques primarily assay for structural changes induced by trauma or disease. Emission techniques, in contrast, assay for disease-associated changes in function, which may or may not be accompanied by structural change. Typically, the functional changes are visible as an altered distribution and/or rate of delivery of a radioactive tracer (Fig. 12.1).

Emission imaging is experiencing a renaissance and currently is the most rapidly growing medical imaging modality. Growth has been spurred by technological developments and the thrust toward personalized medicine informed by molecular imaging.[1] Emission approaches are especially well suited to molecular imaging because of their exquisite sensitivity and associated freedom from pharmacological side effects.

12.1 Essence of Emission Imaging

In emission imaging, the patient is exposed to a radiopharmaceutical (radiotracer), which consists of a radioactive material associated with a pharmaceutical that localizes to a particular site, e.g., a tumor, in the body. Once labeled, the site becomes a source of high-energy, gamma-ray photons that are generated by nuclear decay events. The gamma rays escape the body, and the spatially varying "count profile" that they produce on a detector is used to generate static and/or time-lapse images of the distribution of the radiopharmaceutical at the site of accumulation (Figs. 12.1 and 12.2). Nuclear imaging modalities include both projection and tomographic approaches, similar to X-ray imaging.

Projection nuclear imaging (planar scintigraphy), like projection radiography, generates images that lack depth information and that suffer from structural superposition. Planar scintigrams are, nevertheless, a common diagnostic tool. Prominent applications include whole-body bone scans to evaluate fractures, arthritis, and cancer, as well as thyroid, renal, and pulmonary imaging to study organ function.

In contrast, nuclear emission computed tomography (ECT), like CT, yields computer-reconstructed cross-sectional and three-dimensional images of radionuclide distribution. Tomographic images provide better localization of radioactive tracers and thus facilitate more accurate assessment of function. Prominent applications of ECT include evaluation of the functional status of the heart (e.g., adequacy of blood flow) via myocardial perfusion imaging and identification of tumors (i.e., cells with a high metabolic demand) based on enhanced accumulation of radioactively tagged glucose analogs. Emission imaging also is used to diagnose cognitive impairment and is a powerful method of studying brain activity in normal individuals and mapping sites involved in key tasks. Cutting-edge nuclear/CT and nuclear/MRI imaging systems can be used to evaluate function and anatomy simultaneously.

12.2 Physics of Radioactive Materials

We begin with a discussion of the essential physics of radioactivity. This background material will serve as a foundation for an understanding of nuclear imaging techniques.

12.2.1 Atomic and Nuclear Structure

Atoms consist of a nucleus surrounded by a cloud of electrons. The nucleus contains two types of "nucleons" – positively charged protons and uncharged neutrons. The number of protons is specified by the atomic number, Z. The number of protons plus neutrons is specified by the mass number, A. (The number of neutrons, N, is given by $N = A - Z$.) The atomic

[1] Molecular imaging has been defined as "the visualization, characterization, and measurement of biological processes at the molecular and cellular levels in humans and other living systems." Pomper, M., et al. 2008. What is molecular imaging? *J. Nucl. Med.* 49:163P.

DOI: 10.1201/b22076-14

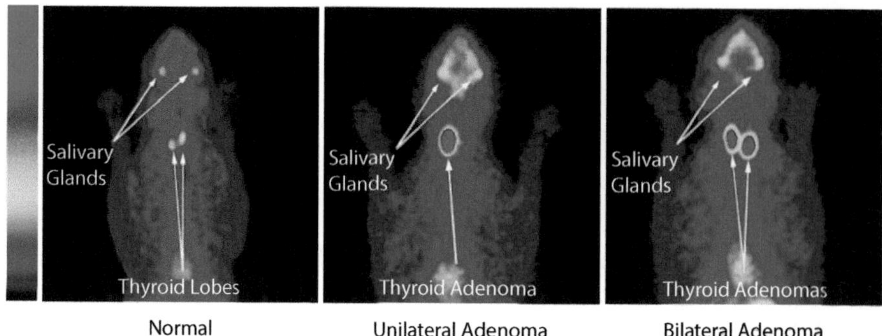

FIGURE 12.1
Emission imaging of a normal cat and cats with unilateral and bilateral adenomas. Uptake of a tracer that localizes preferentially to the thyroid is enhanced in the cats with thyroid adenomas. Warmer (redder) colors denote higher activities. (Adapted with permission from Michael Broome, Advanced Veterinary Medical Imaging.)

number identifies an element "X," which is denoted using the symbol ${}^{A}_{Z}X$. In a neutral atom, the nucleus is surrounded by a cloud of Z negatively charged electrons.

Most chemical elements exist in several "isotopic" forms, which differ in the number of neutrons present in the nucleus. Frequently, some of the nuclear configurations are unstable, and the associated isotopes are radioactive. Radioactive forms of different elements are known as radionuclides.

Nuclear stability is determined by the interplay between two forces. In a typical nucleus, there are many protons packed into a tiny volume, and they exert strong repelling, electrostatic forces on one another. This repulsion is counteracted by an attractive strong nuclear force between nearby neighboring nucleons (Fig. 12.3). The strong nuclear force is approximately the same between two protons, two neutrons, and a proton and a neutron; it completely stabilizes many smaller nuclei.

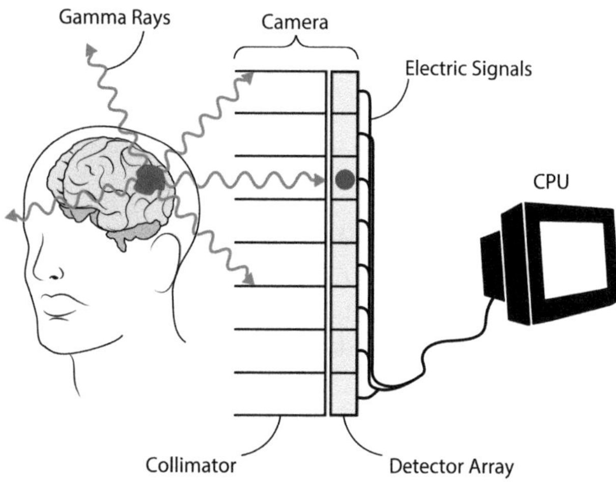

FIGURE 12.2
Cartoon depicting the basics of planar nuclear imaging. Gamma-ray photons are preferentially emitted from a site (such as the brain). The camera and associated electronics detect and count the escaping photons and localize the originating nuclear decay events. Photon "counts" are mapped to pixels, and, over time, the count profile becomes a meaningful representation of distribution that can be used to create an image.

12.2.2 Stable and Unstable Nuclei

There are a few thousand known nuclei, and a few hundred of these are stable. For the lightest nuclei, stability is greatest when $N = Z$ (Fig. 12.4). In contrast, for heavier nuclei, stability is greatest when $N > Z$. A simple qualitative rationale for the latter trend is that electrostatic repulsion increases as proton number increases, and this enhanced repulsion can best be offset by adding neutrons, which augment the attractive strong nuclear force without also augmenting repulsion. However, the strong nuclear force saturates, and thus when $Z > 83$ nuclei inevitably are unstable.

Unstable nuclei play a key role in nuclear imaging because they decay into a more stable configuration and, in the process, release energy that can be used to generate an image. There are many decay schemes whereby an unstable "parent" nucleus will undergo a spontaneous transition to generate a more stable "daughter" nucleus (Fig. 12.5). Here we are interested in decay processes that lead directly or indirectly to the generation of high-energy, monoenergetic, gamma-ray photons, which are able to escape the body and create an image.

12.2.2.1 Isomeric Transitions

Many radioactive decay processes yield daughter nuclei that are in an excited state. Nuclei of this type can further decay to a lower energy state via a photon-emitting transition that is the nuclear analog of spontaneous emission (e.g., fluorescence) from atoms and molecules.

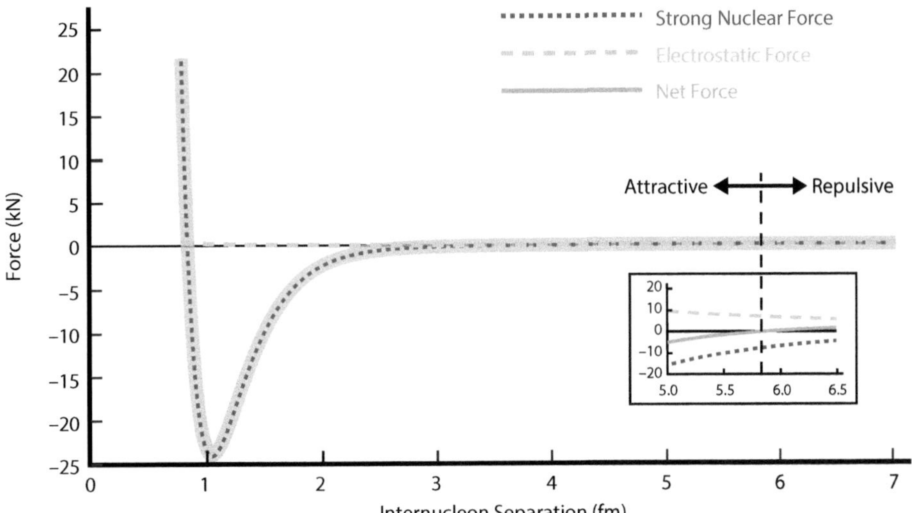

FIGURE 12.3
Schematic plot of relevant forces that act between protons in the nucleus. The repulsive (positive) Coulombic force is relatively long-ranged. The attractive (negative) strong nuclear force is very short-ranged. It is negligible when the protons are more than a few femtometers (10^{-15} m) apart. (As a benchmark, an atom is roughly 10^{-10} m in size.) However, over short distances the strong nuclear force is much stronger than the Coulombic force. Over *very* short distances, the nuclear force is repulsive. Curves were generated from standard theoretical expressions.

Transitions in this class, which do not involve a change in A or Z, can be described by the equation

$$ {}^{A}_{Z}X^{*} = {}^{A}_{Z}X + \text{energy} $$

where "*" denotes the excited state.

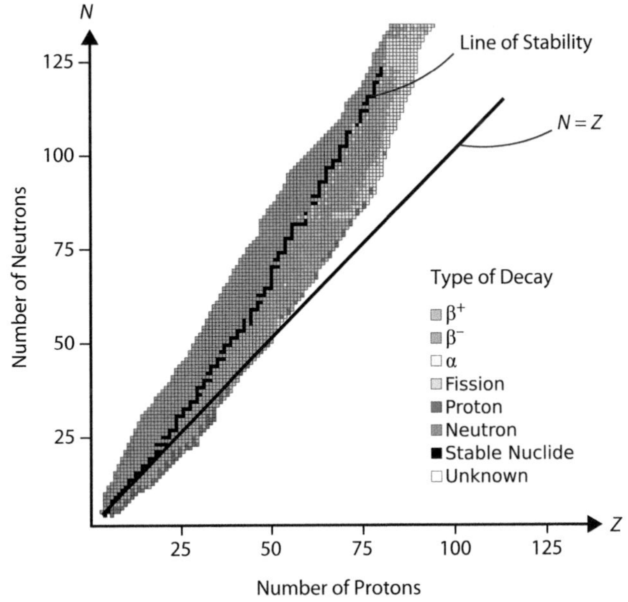

FIGURE 12.4
Stability curve for nuclei. Smaller stable nuclei have an equal number of protons and neutrons, whereas larger stable nuclei have more neutrons than protons. (Data for graph exported from a work by Steven Legg. https://en.wikipedia.org/wiki/Radioactive_decay#/media/File:Table_isotopes_en.svg.)

Excited nuclear states generally are very short-lived. However, some can persist for relatively long times (up to several years). These are known as metastable states, and thus the letter "*m*" is appended after the mass number. An important example in nuclear medicine is technetium-99*m* (Tc-99*m*), which has a characteristic decay time (a half-life [Section 12.2.3]) of ~6 hours.

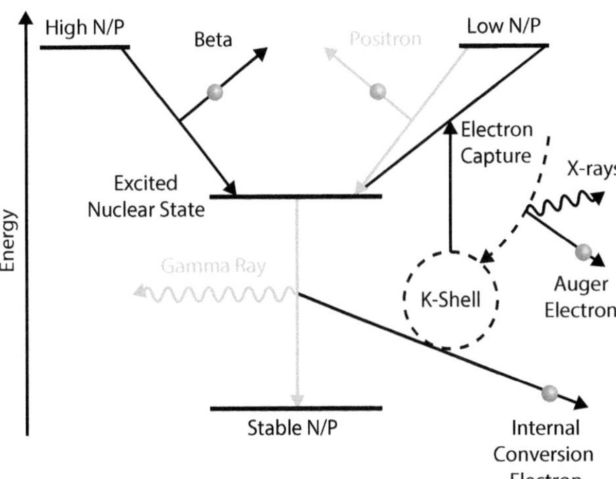

FIGURE 12.5
Schematic diagram of nuclear decay mechanisms. The two mechanisms of greatest interest here are gamma-ray emission from an excited nuclear state and positron emission from a neutron-deficient (low N/P) nucleus. These are highlighted in green. (Adapted with permission from the Sprawls Educational Foundation.)

SAMPLE PROBLEM 12.1 NUCLEAR PARTICLE IN A BOX

The quantum "particle in a box" model was introduced in Chapter 6 to estimate energies associated with electronic transitions in fluorescent molecules. The model showed how those energies depend on the size of the box and the mass of the particle.

Invoke the model to estimate energy changes associated with nuclear transitions. How do these energies compare with the energies of gamma rays used in medicine?

SOLUTION

The particle in a box model provides the following expression for energy levels (Eq. [6.1])

$$E_j = \frac{j^2 h^2}{8mL^2} \quad j = 1, 2, 3\ldots$$

where E_j is the energy level associated with quantum state j, h is Planck's constant, m is the particle mass, and L is a characteristic box dimension. This expression can be evaluated using the mass of a proton ($m = 1.67 \times 10^{-27}$ kg) and a typical nuclear dimension ($L = 1 \times 10^{-14}$ m) to calculate the energy associated with a representative transition:

$$E_2 - E_1 = \left(2^2 - 1^2\right)\frac{h^2}{8mL^2} = 3\left[\frac{\left(6.63 \times 10^{-34}\ \text{m}^2\text{kg/s}\right)^2}{8\left(1.67 \times 10^{-27}\ \text{kg}\right)\left(1 \times 10^{-14}\ \text{m}\right)^2}\right] = 9.9 \times 10^{-13}\ \text{J} = 6.2\ \text{MeV}$$

This energy difference is somewhat larger than, but in the ballpark of, the energy of medical gamma rays.

Nuclear transitions yield a photon of much higher energy than atomic and molecular transitions. Specifically, typical photon energies are on the order of hundreds of keV to MeV (Sample Problem 12.1). For example, Tc-99*m* emits photons with an energy of ~140 keV. This energy is similar to that of a high-energy diagnostic X-ray.

12.2.2.2 Positron Emission

Another important decay process involves release of a positron (β^+), which is the positively charged antimatter counterpart of the electron. Positron emission typically occurs when instability reflects a shortage of neutrons and helps to reduce the shortage by converting a proton into a neutron (Fig. 12.5). The equation describing positron emission is

$${}^{A}_{Z}\text{X} \rightarrow {}^{A}_{Z-1}Y + \beta^+ + \nu + \text{energy}$$

where ν is a neutrino. The neutrino is a chargeless, essentially massless, very weakly interacting particle. Positron emission initially generates a daughter atom with a net negative charge. However, the daughter quickly loses an electron to achieve electrical neutrality.

In medical applications, a key feature of positron emission is downstream production of two, 511-keV photons, which travel in anti-parallel directions (Fig. 12.6). The photons are generated after the positron

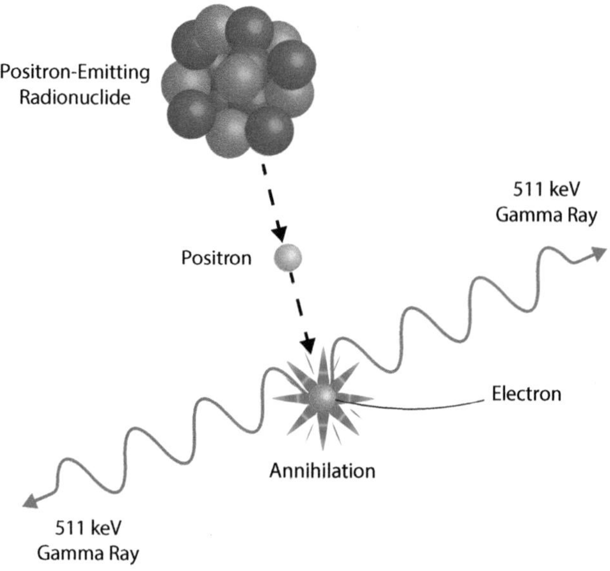

FIGURE 12.6
Annihilation photons are created after positron emission. The photons have anti-parallel trajectories as dictated by conservation of momentum.

has lost most of its kinetic energy via interaction with its surroundings. When essentially at rest, the positron interacts with an electron, and the two particles (each with a rest-mass energy of 511 keV) are annihilated. Their energy is converted into two 511-keV "annihilation" photons, which generate the detected signal if they escape the body. An important example of a positron emitter is fluorine-18, ^{18}F, which decays according to the equation

$$^{18}_{9}\mathrm{F} \rightarrow ^{18}_{8}\mathrm{O} + \beta^{+} + \nu + \text{energy}$$

12.2.3 Decay Kinetics

Production of gamma-ray photons is one important attribute of radionuclides used in medical imaging. Another is a characteristic decay time/lifetime that falls in the range from a few hours to a few days. Timescales in this range are long enough to permit production of an image but short enough to generate adequate signal quickly using small doses, which minimizes harm to the patient. Given the central importance of lifetimes and decay kinetics, we discuss these issues in some detail.

Our discussion will parallel that in Section 11.4.1 because nuclear decay is another example of a random process. Our goal is to develop a mathematical description of the macroscopic, bulk loss of radioactive nuclei resulting from individual stochastic (decay) events. We assume a constant probability of decay for a nucleus per unit time, λ, and expect that the change in the number of radioactive nuclei, dN, in a time interval, dt, will be proportional to N and dt. The equation is

$$dN = -\lambda N dt \tag{12.1}$$

where the negative sign ensures that N decreases with time. As in Section 11.4.1, Eq. (12.1) integrates to yield an exponential decay law:

$$N = N_0 e^{-\lambda t} \tag{12.2}$$

where N_0 is the number of radioactive nuclei at $t = 0$.

In many cases, radioactive decay is quantified in terms of activity, A, which is the number of decays per second. In light of Eq. (12.2), we have

$$A = -\frac{dN}{dt} = \lambda N_0 e^{-\lambda t} = A_0 e^{-\lambda t} \tag{12.3}$$

where $A_0 = \lambda N_0$ (HW 12.1). Common units of activity are the becquerel (Bq) and the curie (Ci). One Bq is equal to one decay per second. One Ci is equal to 3.7×10^{10} Bq.

The characteristic timescale associated with radioactive decay frequently is described using the half-life, $\tau_{1/2}$, instead of λ. The half-life is the time required for half of the nuclei that are present at a given time to decay and is related to λ by (HW 12.4)

$$\tau_{1/2} = \frac{\ln(2)}{\lambda} = \frac{0.693}{\lambda} \tag{12.4}$$

Table 12.1 lists some commonly used radionuclides, together with key properties and applications. The first

TABLE 12.1

Key Attributes of Prominent, Medically Relevant Radionuclides

Nuclide	Decay Mechanism[a]	Half-Life	Photon Energy	Prominent Application
Technetium-99*m*	Gamma emitter	6.0 hours	140 keV	Bone and brain scans
Iodine-123	Electron capture	13.2 hours	159 keV	Diagnosis of thyroid disease
Indium-111	Electron capture	2.8 days	171 keV and 245 keV	Detection of infection
Iodine-131	Beta minus	8.0 days	364 keV	Treatment of hyperthyroidism and thyroid cancer
Fluorine-18	Positron emitter	109.8 minutes	511 keV (after annihilation)	Tumor detection, brain imaging
Carbon-11	Positron emitter	20.4 minutes	511 keV (after annihilation)	Early cancer diagnosis
Nitrogen-13	Positron emitter	10.0 minutes	511 keV (after annihilation)	Myocardial perfusion imaging
Oxygen-15	Positron emitter	2.0 minutes	511 keV (after annihilation)	Oxygen metabolism, blood flow

Source: Data from Flower, M.A. (Editor), *Webb's Physics of Medical Imaging* (2nd Edition). CRC Press, Boca Raton, FL. Pages 204–206, 209–213.

[a] Electron capture, like positron emission, is a decay mode that helps to stabilize neutron-deficient isotopes. The nucleus captures an inner electron, converting a proton into a neutron and causing the emission of a neutrino. In contrast, beta minus (electron) decay helps to stabilize proton-deficient isotopes. The neutron decays into a proton and an electron, and the latter is emitted together with an antineutrino. These decay modes and the others in Table 12.1 are illustrated in Fig. 12.5.

four radionuclides are used in planar imaging and its tomographic counterpart (SPECT). The second four are used in positron emission tomography (PET).

12.3 Generation and Detection of Radioactivity

Many radioactive nuclei occur naturally. However, medicine relies primarily on the use of radioactive nuclei that are generated artificially in one of three ways – using a cyclotron, a nuclear reactor, or a radionuclide generator. Here we summarize the cyclotron and generator approaches because these are the sources of two of the most prominent radionuclides used in medical imaging, ^{18}F and Tc-99*m*.

12.3.1 Generating Radioactivity

Cyclotrons are a source of positron-emitting radionuclides used in PET. The basic operating principle is to accelerate charged particles, e.g., protons, to high energy, so they can penetrate the Coulombic barrier generated by target nuclei and induce reactions that generate radionuclides (Fig. 12.3). For example, a proton incident on oxygen-18 (in ^{18}O-enriched water) can lead to capture of the proton, release of a neutron (n), and the creation of ^{18}F, which is unstable. The stable isotope of fluorine is ^{19}F.

A schematic of a cyclotron is shown in Fig. 12.7. Charged particles are generated intermittently from a source that is centrally located in a gap between two "Dees." The particles can be either positive ions (e.g., a proton or a deuteron) or negative ions (e.g., H^-). "Negative ion cyclotrons" have several advantages, but radionuclides ultimately are produced by nuclear reactions generated by bombardment of the target by a positive ion. Thus, electrons must be stripped from negative ions before the actual reaction occurs (Fig. 12.7).

To increase their energy, particles are injected into the gap and accelerated toward one of the Dees by a high voltage. Once inside the Dees, the particles (moving with velocity $\vec{v}$) are exposed to a magnetic field, which exerts a force given by the Lorentz force law, $\vec{F} = q\vec{v} \times \vec{B}$. The Lorentz force is directed radially inward and supplies the centripetal force that causes the particles to execute circular motion (HW 12.8 and 12.9). After traversing a semicircle, the voltage across the Dees is reversed, and the particles pick up additional energy and move into the second Dee at higher speed. As a consequence, the particles move in a circular orbit with a larger radius. This process is repeated many times, causing the particles to gain energy and execute a spiral trajectory, eventually exiting the Dees and impinging on a target consisting of stable nuclei. A nuclear reaction between the particle and target generates a radioactive product. For example, bombardment of a ^{68}Zn target by a proton produces ^{67}Ga, which is used to study inflammation and infection.

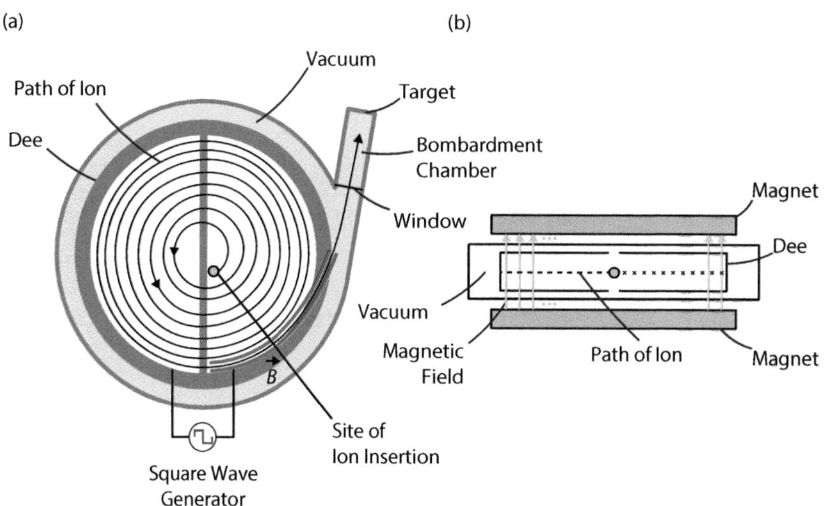

FIGURE 12.7
Schematic representation of a cyclotron. Panel (a) shows a top view, and Panel (b) shows a cross-sectional view and the magnetic field lines. The Dees are white. The mechanism of removing the charged particles from the Dees and directing them onto the target depends on their charge. Positive ions are removed using a negatively charged deflector plate. Negative ions are exposed to an electron stripping foil to generate positive ions that participate in the actual reaction. The magnetic field exerts an oppositely directed force on the positive ions that deflects them toward the target.

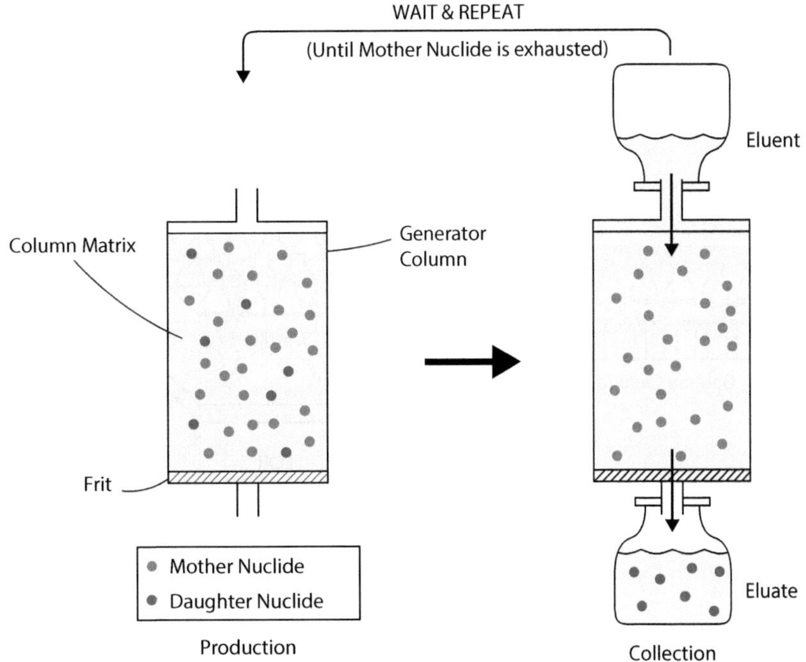

FIGURE 12.8
Schematic of a radionuclide generator. For the technetium generator, the parent is Mo-99 in the form of ammonium molybdenate, $(NH_4)^+(MoO_4)^-$, which is bound to an alumina resin. The daughter, Tc-99*m*, is readily removed from the column using normal saline because the chloride ions in the salt displace $(TcO_4)^-$ but not $(MoO_4)^-$.

Generators are a useful and relatively inexpensive source of selected radionuclides, like Tc-99*m*. Generators house a longer lived parent that is linked to a column and that has chemical properties differing from those of the daughter (Fig. 12.8). The parent population continually produces daughters via radioactive decay until the parent is depleted. The daughters are much less tightly bound to the column and thus can be eluted by passing a solution over the column.

As an example, we consider beta minus decay of the parent molybdenum-99 (Mo-99). This reaction produces the important daughter Tc-99*m*, and an antineutrino $(\bar{v})$, as follows

$$^{99}_{42}\text{Mo} \rightarrow \,^{99m}_{43}\text{Tc} + \beta^- + \bar{v} + \text{energy}$$

12.3.2 Detecting Radioactivity

The most frequently used detector in planar nuclear imaging is the Anger gamma scintillation camera; its key components are a collimator, a large continuous scintillation crystal spanning a chest-sized area, PMTs, and processing electronics (Fig. 12.9). Together these components detect and localize decay events and build up a tally of decay events as a function of location to create an image.

We start with the role of the collimator. To create a meaningful projection, each image point should reflect signal that maps to a well-defined line from the patient. In transmission radiography, this criterion is met by unscattered X-rays because X-rays emanate from a small and distant (approximately point) source (Fig. 11.1). In contrast, in radionuclide imaging, many widely distributed point sources emit isotropically, and the image of each point would span the entire detector, in the absence of collimation (Fig. 12.9). Thus, collimators are mandatory in radionuclide imaging (PET is an exception). Collimators act in essence like a straw, passing only rays along a well-defined line of response (LOR) containing the source. Typically, the LOR is normal to the detector (Fig. 12.9).

State-of-the-art collimators consist of hexagonal holes separated by lead walls called septa, which are designed to absorb most of the photons that do not follow trajectories that are aligned with the holes (Fig. 12.9). Thus, septal thickness is tailored to the application. For example, septa are fabricated to be thicker when designed for use with photons of higher energy. Longer wall lengths and smaller hole sizes are used when a collimator is designed for high resolution, but this negatively affects sensitivity because more gamma rays are excluded.

The collimator conventionally is followed by a thallium-activated sodium iodide, NaI(Tl), scintillation

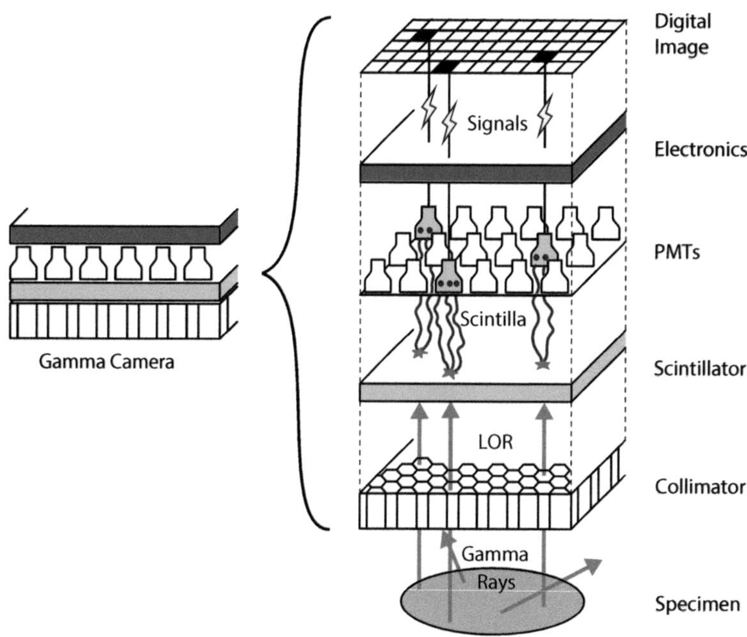

FIGURE 12.9
Schematic illustration of the components of a gamma camera and digital image encoding. Alternatives, such as CCD- or CMOS-based detectors, also can be used. However, the gamma camera is particularly well suited for aspects of nuclear imaging, such as detecting individual events and discriminating events based on energy.

crystal (Section 11.3.2), which converts an incident gamma-ray photon into several thousand visible photons, where the photon number is proportional to gamma-ray energy (Sample Problem 12.2). (The basic mechanism underlying this conversion process is described in Fig. 11.13.) Some alternative scintillation materials, including those used for PET imaging, are listed in Table 12.2.

The thousands of visible photons that are produced by the scintillator are detected by a subset of PMTs in a detector array consisting of ~40– 90 PMTs (Fig. 12.9). (See Section 9.1.1 for a discussion of the operation of a PMT.) The spread of signal over multiple detector elements enables more precise computation of the (x/y) coordinates of the initiating gamma ray (the scintillation event). We address this issue next.

12.3.3 Coordinate Computation

A key step in image production is estimating the positions (x_s, y_s) of scintillation events, which are markers for decay events, from the evoked PMT

SAMPLE PROBLEM 12.2 SCINTILLATION DETECTION

Roughly one UV or visible photon is emitted from a sodium iodide scintillation crystal for each 25 eV of energy deposited by an incident gamma ray. Calculate the number of UV or visible photons produced by the scintillator when activated by absorption of a 140-keV gamma ray.

SOLUTION

The number of UV or visible photons emitted is equal to the energy of the gamma ray divided by the energy required to generate a photon:

$$\text{Number of UV or visible photons per gamma ray} = \frac{140{,}000 \text{ eV}}{25 \text{ eV}} = 5600$$

This calculation verifies the statement in the text that a gamma ray generates several thousand photons in the scintillator.

TABLE 12.2

Key Attributes of Some Prominent, Medically Relevant Scintillators

Scintillator	Light Yield (Photons per MeV)	Emission Wavelength	Decay Time (ns)	Uses
NaI(Tl)	38,000	415	230	Gamma camera/SPECT
CsI(Tl)	54,000	550	1000	SPECT
Bismuth germanate	9000	480	300	PET
Lutetium oxyorthosilicate (Ce)	27,000	420	40	PET
Lutetium perovskites (Ce)	10,000	365	17	PET

Source: Data from Lecoq, P. 2016. Development of new scintillators for medical applications. *Nucl. Instrum. Methods Phys. Res. A.* 809:130–139. https://doi.org/10.1016/j.nima.2015.08.041.

signals (Fig. 12.10). The estimation process is based on the fact that the number of visible photons impacting a particular PMT depends on its distance from the scintillation site (Fig. 12.10). The PMT closest to the site subtends the largest solid angle relative to the scintillation event and registers the largest signal from the (isotropically) emitted visible photons. Signal strength falls off progressively with distance.

The largest PMT signal alone could be used to localize events, but then resolution would be unacceptably limited by PMT (in essence pixel) size. To avoid this limitation, single events instead are localized using signals from multiple PMTs. One approach is based on weighting PMT signals, P_i, based on their strength using a linear combination of signals that exceed some threshold. One option is the analog of a centroid (center-of-mass) calculation using PMT (center) coordinates (x_i, y_i):

$$x_s = \frac{\sum_i x_i P_i}{\sum_i P_i} \tag{12.5a}$$

$$y_s = \frac{\sum_i y_i P_i}{\sum_i P_i} \tag{12.5b}$$

The sum of the PMT signals, which is a measure of total gamma energy, also is computed as

$$Z = \sum_i P_i \tag{12.5c}$$

The symbol Z stands for Z-pulse. The Z-pulse is used to reduce image-degrading effects of scattered and background radiation, as described next. It also is the normalizing denominator in Eqs. (12.5a) and (12.5b).

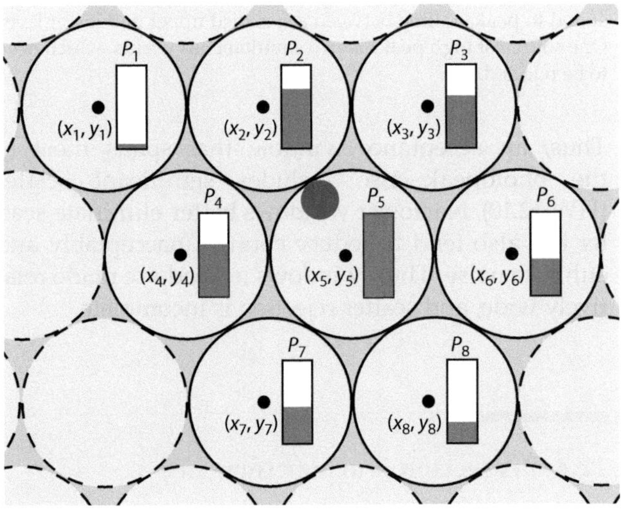

FIGURE 12.10
Signals generated in the PMTs in a gamma camera depend on distance from the scintillation site (blue dot). These signals are used to calculate event coordinates using linear combinations, such as those described in Eqs. (12.5a) and (12.5b). This method localizes signals to within a few mm.

12.3.4 Pulse Height Analysis

The scintillator responds to several types of events (Fig. 12.11). Ideal events arise when a gamma ray escapes the patient without scattering, passes through the collimator, and deposits all of its energy in the scintillator via a PE interaction. This type of event provides correct position data. Nonideal events arise when a gamma ray scatters in the patient or in the detector, or when a ray penetrates septae of the collimator. If nonideal events are counted, they lead to mispositioned data that enhance background and degrade contrast.

In X-ray transmission imaging, degradative effects of scattered photons are reduced using grids (Fig. 11.21). However, this approach is not effective in nuclear emission imaging because gamma rays have no intrinsic, well-defined direction of origin;

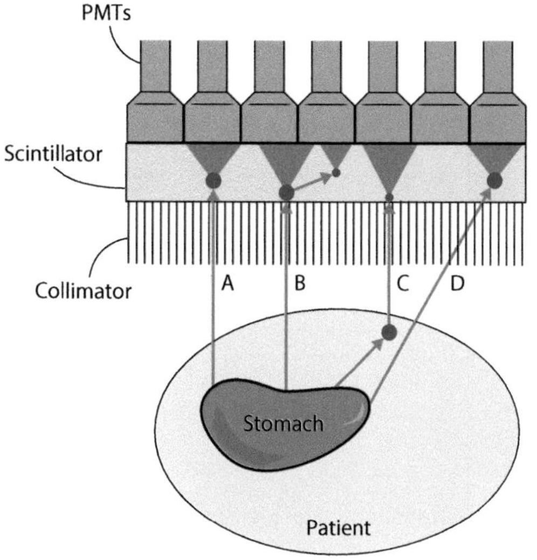

FIGURE 12.11
Types of events detected by a scintillation crystal. Event A is ideal, whereas events B (detector scatter), C (patient scatter), and D (septal penetration) are nonideal. More complicated events also are possible. Event type B, which first scatters in the detector and then interacts a second time and deposits the remaining energy in the detector, is indistinguishable from event type A based on energy discrimination. Thus, B will lead to an event that is mispositioned. Event types C and D also may be accepted and degrade image quality (e.g., if the energy loss for event C due to patient scatter is small).

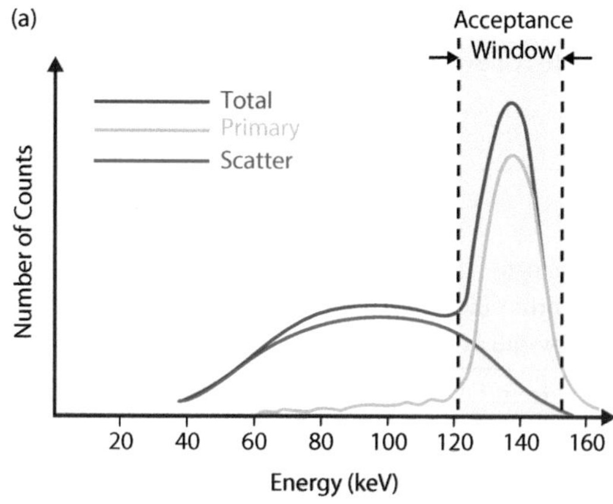

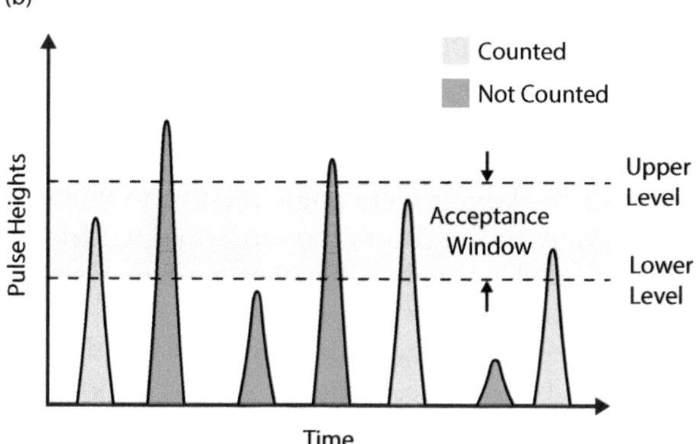

FIGURE 12.12
Essence of how energy discrimination is used to reduce scatter and background in radionuclide imaging. The spectrum shown in Panel (a) is for Tc-99*m*, which emits a gamma ray with energy 140 keV. The PHA accepts only events that generate a Z-pulse that is within ~20% of the photopeak value. The acceptance window has nonzero width due to effects that include nonuniformities in scintillator and PMT response. In Panel (b), accepted events correspond to peaks falling between a specified upper and lower level. One source of high peaks is two simultaneous events, which need to be rejected.

thus, collimators do not preferentially reduce the scatter fraction in the image. In planar nuclear imaging, rejection of scattered and background photons is based instead on energy discrimination (e.g., on the lower energy of scattered rays).

Energy discrimination is implemented using the Z-pulse (Eq. [12.5c]) and a pulse-height analyzer (PHA). The operation of a PHA can be understood by examining the spectrum of Z-pulses generated when gamma rays deposit energy in a scintillator. Fig. 12.12 shows an exemplary spectrum for Tc-99*m* under patient imaging conditions. The spectrum has a high-energy, 140-keV photopeak of finite width, which is generated by gamma rays that deposit all of their energy in the scintillator (e.g., via a PE interaction). The spectrum also has a lower energy band generated by a spectrum of effects that includes patient-scattered photons. The PHA aids in distinguishing unscattered photons from lower energy scattered photons by accepting/counting only Z-pulses with a height in a relatively narrow range around the expected photopeak (Fig. 12.12). All other pulses are rejected. If the Z-pulse is accepted by the PHA, the counts at a given pixel location in the image matrix are updated by one.

Notably, scatter rejection using a PHA is only partially effective. One reason is that energy changes associated with scattering can be relatively small. Thus, an acceptance window that spans most of the photopeak also includes significant scatter (HW 12.10). Narrower windows better eliminate scatter but also tend to reduce counts unacceptably and enhance noise. Thus, windows instead are made relatively wide, and scatter rejection is incomplete.

12.4 Projection Nuclear Imaging

We already have discussed several aspects of planar scintigram production, notably the roles played by collimators, scintillators, PMTs, and PHAs. In what follows, we discuss additional relevant image acquisition details, as well as image quality and image quantification.

12.4.1 Modern Camera Systems

Planar scintigrams, like X-ray projection images, suffer from structural overlap. Thus, several views collected from different angles often are required to make an accurate diagnosis. To meet this need, modern nuclear imaging systems typically are equipped with two large FOV gamma cameras attached to a movable gantry. These systems are capable of collecting two views, such as anterior and posterior views, simultaneously, and also are capable of generating multiple views for tomographic nuclear imaging (Fig. 12.13). Modern systems also include a movable bed to facilitate whole-body imaging.

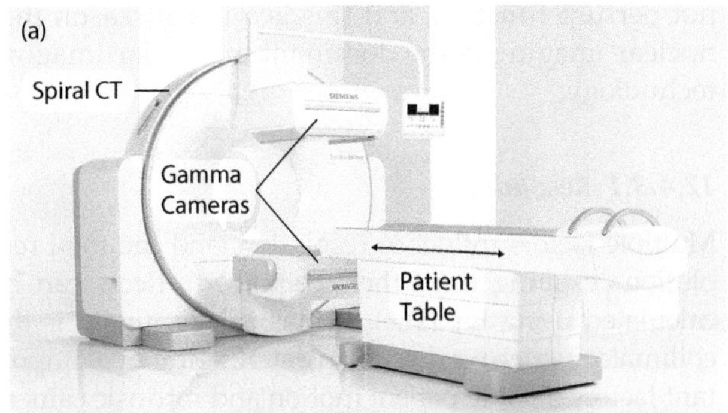

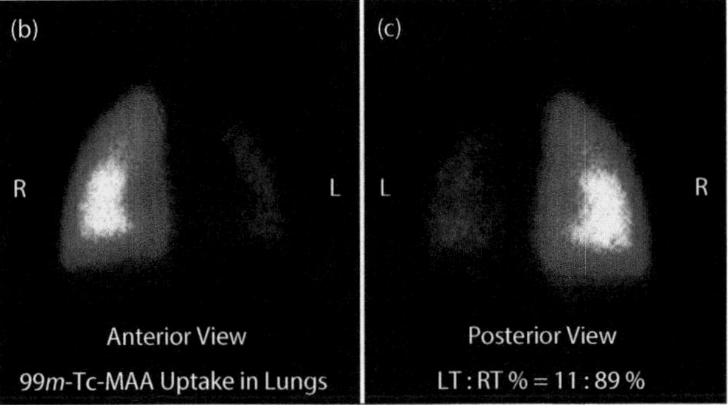

FIGURE 12.13
Dual-headed nuclear imaging systems. Panel (a) shows a modern nuclear imaging system with two gamma cameras and a spiral CT scanner. Panels (b) and (c) show anterior and posterior images generated by a dual-headed camera during a lung perfusion scan using Tc-99*m*-labeled macroaggregated albumin (MAA). The scan revealed an absence of left lung perfusion. (Panel (a) courtesy of Siemens Healthcare GmbH. Panels (b) and (c) adapted with permission from Gandhi, S.J., et al. 2013. Tc-99*m* macro aggregated albumin scintigraphy – indications other than pulmonary embolism: a pictorial essay. *Indian J. Nucl Med.* 28:152–162. DOI: 10.4103/0972-3919.119546.)

12.4.2 Production of a Digital Image

To generate a digital image, nuclear decays are assigned to positions/pixels in a computer matrix using a coordinate assignment strategy, such as that outlined in Section 12.3.3. Typical matrix dimensions are 64 × 64, 128 × 128, or 256 × 256, although larger matrices are used for whole-body imaging. An image of the distribution of the radionuclide then is generated from the histogram of events at each location (Fig. 12.14). Frequently, images are smoothed using a convolution filter.

12.4.3 Image Quality

We now turn to the important topic of image quality and address resolution and contrast. An important take-home message from what follows is that nuclear images often have good contrast because the signal is set against a relatively dark background. On the other hand, resolution is relatively poor.

Another important message is that nuclear imaging can be exceptionally sensitive, in part because background is low. For example, PET tracers can be used at picomolar concentrations, whereas prominent MRI and CT contrast agents often are used at micromolar to millimolar concentrations. High sensitivity means that tracers can be administered at doses that do

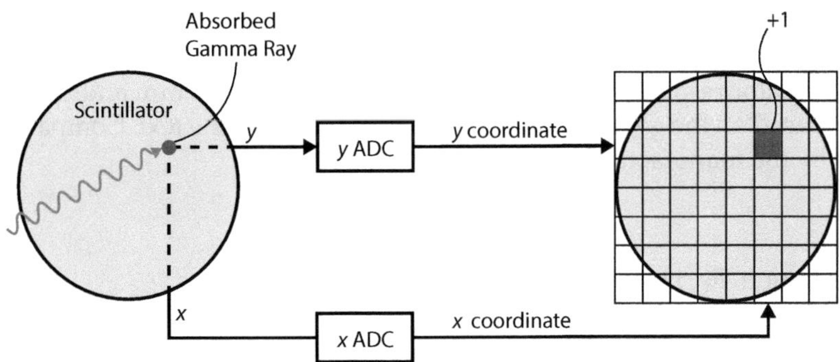

FIGURE 12.14
Schematic representation of "static, frame-mode" image acquisition. The image matrix initially is zeroed. During acquisition, the matrix is updated by acquiring counts for a defined time interval or until a defined number of total counts are collected. Coordinates and energies are computed from PMT signals that are collected in a pulse-counting mode. In this cartoon, we follow convention, and use x and y coordinates for the two-dimensional image matrix. However, when two-dimensional projections are collected, one of the dimensions is along the z direction of the patient.

not perturb function, and this is a major reason that nuclear imaging is the dominant molecular imaging technology.

12.4.3.1 Resolution

Multiple factors influence resolution, and the total resolution resulting from their combined effects can be calculated using Eq. (11.6). A major determinant is the collimator, which we discuss in detail. Additional important factors include patient motion and intrinsic camera resolution, which is a property of the detection system that neglects blurring effects caused by the collimator.

There are several ways to quantify collimator resolution, R_C. One of the most rigorous, which we adopt here, is to measure the FWHM of the image of a point source of radioactivity (the PSF) at a distance z from a collimator.

To quantify the width of the PSF, we analyze the scenario in Fig. 12.15, which shows the span of gamma-ray trajectories from a point source that are able to pass through a parallel hole collimator with septa of length L and holes of diameter D. The FWHM $\approx R_C$ of the PSF can be calculated by noting that the triangles SEJ and $S'E'J'$ are similar, giving

$$\frac{R_C}{z+L+c} = \frac{D}{L}$$

$$R_C = \frac{D}{L}(z+L+c) = D\left(1+\frac{z+c}{L}\right) \tag{12.6}$$

In Eq. (12.6), c is the scintillator thickness.

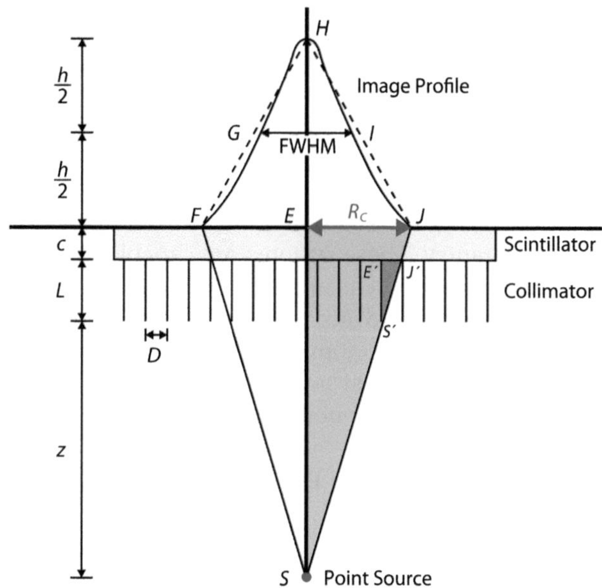

FIGURE 12.15
Geometry and notation used in the derivation of Eq. (12.6). Note that FWHM $\approx R_C$ because triangles HIG and HJF are similar.

Equation (12.6) has several important implications. These include the fact that resolution worsens with increasing distance of the source from the collimator, as well as with increasing collimator hole diameter and scintillator thickness. On the other hand, resolution improves with increasing septal length. A typical collimator resolution for a source that is 10 cm deep is ~5 mm, and a typical overall resolution at 10 cm is ~6–7 mm (Sample Problem 12.3). These numbers underscore that resolution in nuclear imaging is relatively poor.

SAMPLE PROBLEM 12.3 GAMMA CAMERA RESOLUTION

A camera manufacturer reports the following parameters for its gamma camera.

L (mm)	D (mm)	c (mm)	FWHM @ 0 mm	FWHM @ 100 mm
58.4	1.78	19	2.1 mm	5.4 mm

Calculate the collimator resolution for the two object distances, $z = 0$ mm and $z = 100$ mm, provided by the manufacturer, based on the geometrical principles presented in the text. Compare the results with the resolutions stated by the manufacturer.

SOLUTION

The collimator resolution, R_C, is given by Eq. (12.6):

$$R_{C,\, z=0\text{ mm}} = D\left(1+\frac{z+c}{L}\right) = 1.78\text{ mm}\left(1+\frac{0\text{ mm}+19\text{ mm}}{58.4\text{ mm}}\right) = 2.36\text{ mm}$$

$$R_{C,\, z=100\text{ mm}} = 1.78\text{ mm}\left(1+\frac{100\text{ mm}+19\text{ mm}}{58.4\text{ mm}}\right) = 5.4\text{ mm}$$

The calculated and stated values are in close agreement, especially for the larger object distance. The calculated and stated values both confirm that resolution worsens as object distance increases.

12.4.3.2 Contrast

Multiple factors also influence contrast in nuclear imaging. A significant source of contrast degradation is relatively uniform background radioactivity. Background can arise from incomplete rejection of scattered photons (Section 12.3.4), insufficient shielding, and a suboptimal radiopharmaceutical (e.g., one that does not have a strong affinity for the lesion relative to background).

An additional important source of background is signal that arises from regions above and below the ROI (similar to fluorescence). This problem is most significant in planar images. Tomographic slices are less prone to contrast degradation due to overlying and underlying radioactivity, improving the visibility of low-contrast lesions.

To quantify contrast, and the degradative effects of background, we define the contrast, C_L, of a lesion relative to normal tissue in terms of their respective radioactive signals, R_L and R_N, as

$$C_L = \frac{\Delta R}{R_N} = \frac{R_L - R_N}{R_N} \tag{12.7}$$

Uniform background effects can be modeled mathematically by adding a constant background signal, R_B, to each term in Eq. (12.7). This yields C_L^B

$$\begin{aligned} C_L^B &= \frac{R_L + R_B - (R_N + R_B)}{R_N + R_B} \\ &= \frac{R_L - R_N}{R_N\left(1 + \frac{R_B}{R_N}\right)} = \frac{C_L}{1 + \frac{R_B}{R_N}} \end{aligned} \tag{12.8}$$

Equation (12.8) is very similar to Eq. (11.11) and reveals that background reduces contrast. The magnitude of the effect is explored in Sample Problem 12.4.

Statistical noise also can markedly degrade visibility, especially for low-contrast lesions. Statistical variation in signal strength is a significant problem in nuclear imaging because the photon detection efficiency is very low. For example, >99.9% of emitted photons go undetected due to processes that include movement away from the detector, absorption before exiting the patient, and collimator effects. The effects of statistical noise on visibility are discussed in Box 12.1 and are the focus of HW 12.11. Degradative effects of noise also are illustrated in Figs. 9.9 and 9.10.

12.4.4 Image Quantification

For some applications, it is desirable to obtain quantitative data, such as relative signal strengths, from planar images. This goal is complicated by the fact that counts and resolution vary with structure depth. Counts vary with depth because gamma-ray photons are attenuated as they traverse matter, similar to X-rays. Resolution varies with depth because, as the source of activity gets farther from the collimator, photons can pass through a broader range of collimator holes, and the signal is less well localized. This latter issue was addressed in Section 12.4.3.1.

Simple corrective measures that minimize resolution degradation with depth include collecting views in which the structure of interest is as close as possible to the collimator. In the case of the kidney, this is a posterior view. Moreover, more sophisticated

SAMPLE PROBLEM 12.4 EFFECT OF BACKGROUND ON CONTRAST

Calculate and compare the contrast of a lesion in the absence and presence of a uniform background. Assume that the signal is 40% above background (i.e., $R_L = 1.4R$ and $R_N = R_B = R$).

SOLUTION

Equations (12.7) and (12.8) describe contrast in the absence and presence of background, respectively. Inserting the given values into Eq. (12.7) yields

$$C_L = \frac{1.4R - R}{R} = 0.4$$

Similarly, inserting values into Eq. (12.8) yields

$$C_L^B = \frac{C_L}{1 + \frac{R_B}{R_N}} = \frac{0.4}{1 + \frac{R}{R}} = 0.2$$

The effect of background in this case is to reduce contrast by a factor of two.

BOX 12.1 CONTRAST-TO-NOISE RATIO (CNR) IN PROJECTION IMAGING

To quantify the effect of statistical noise on visibility, consider an image showing a circular lesion of radius r_L and contrast, C_L, amidst a background count rate per unit area of R_B (in counts/s/cm^2). The noise is governed by Poisson statistics (Section 9.2). Thus, the coefficient of variation, COV, the ratio of the standard deviation of the noise to its mean in a background region of radius r_L, which is a measure of noise contrast, is given by (HW 9.3)

$$\text{COV} = \frac{1}{\sqrt{N}} = \frac{1}{\sqrt{R_B \pi r_L^2 t}}$$

Albert Rose (1973) studied the visual ability of humans to identify small lesions (objects) with contrast, C_L, in the presence of noise-generated contrast variations. Rose found that this identification can be accomplished accurately when the contrast-to-noise ratio, CNR, exceeds ~5, where the CNR is defined as the ratio of the lesion-to-noise contrast:

$$\text{CNR} = \frac{|C_L|}{\text{COV}} = \frac{|C_L|}{\left(1/\sqrt{R_B \pi r_L^2 t}\right)} = |C_L| \times r_L \times \sqrt{\pi \times \text{ID}} \qquad \text{(B12.1.1)}$$

In Eq. (B12.1.1), "ID" = $R_B t$ is the "information density" of the background noise in counts/cm^2. The Rose criterion implies that the minimum lesion contrast, $C_L^{\min}$, required for detectability is given by

$$\left|C_L^{\min}\right| \times r_L \times \sqrt{\pi \times \text{ID}} \approx 5 \qquad \text{(B12.1.2)}$$

Equation (B12.1.2) shows that $C_L^{\min}$ is inversely proportional to lesion size. In other words, the contrast must be larger for a small lesion than a large lesion.

camera systems are capable of mapping body contours. A special gantry then follows the contour and keeps the collimator as close to the patient as possible (Fig. 12.16).

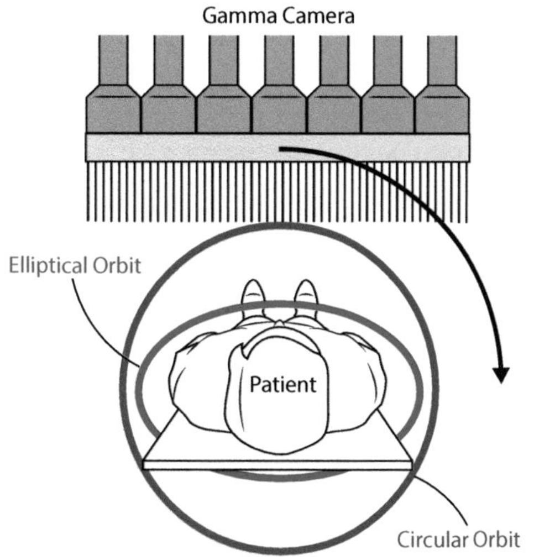

FIGURE 12.16
Body contouring is used to minimize depth-dependent problems. Some imaging systems are equipped with an infrared beam that outlines body contours and a sophisticated gantry that can follow the contoured trajectory. Here, an elliptical orbit keeps the camera closer to the body than a circular orbit.

A more theoretical corrective approach is to collect anterior and posterior views using a dual-headed camera and then compute the geometric mean (GM) of the views. The GM of two numbers is the square root of their product. Under conditions where the linear attenuation coefficient in the body is roughly constant (e.g., the brain or abdomen), the GM yields depth-independent counts and resolution.

To verify count invariance with depth, we consider the scenario in Fig. 12.17. A point source of radioactivity is located at a distance x from an anterior camera and at a distance $T-x$ from a posterior camera in a region of the body with thickness, T. Attenuation of gamma-ray photons with distance is governed by

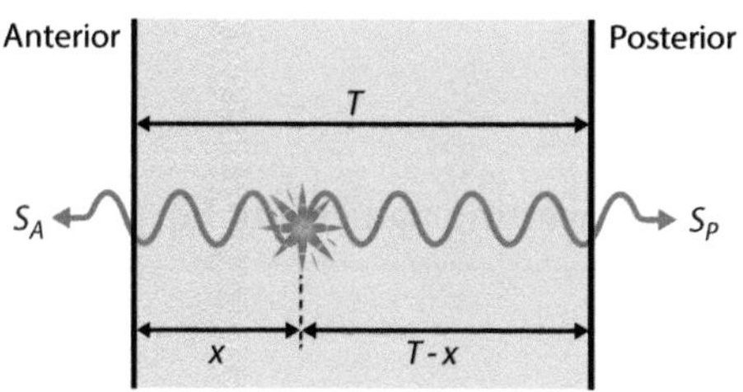

FIGURE 12.17
Geometry and notation used in the derivation of Eq. (12.9).

Eq. (11.4). Thus, we can write the signals in the anterior, S_A, and posterior, S_P, views as

$$S_A \propto A_0 e^{-\mu x}$$

$$S_P \propto A_0 e^{-\mu(T-x)}$$

where μ is the (assumed constant) linear attenuation coefficient in the body, and A_0 is the activity at the source. We now compute the GM of these signals, $S_{GM} = (S_A S_B)^{1/2}$, to obtain

$$S_{GM} \propto \left(A_0 e^{-\mu x} A_0 e^{-\mu(T-x)}\right)^{1/2} = A_0 e^{-\mu T/2} \tag{12.9}$$

Equation (12.9) verifies the assertion that the signal computed as a GM is independent of the depth of the source, x, when μ is constant (Fig. 12.18). Equation (12.9) also shows that (given T) the activity at the source, A_0, can be estimated as

$$A_0 \propto S_{GM} e^{+\mu T/2}$$

Depth, x, is best determined tomographically but also can be estimated using anterior and posterior signals and the expression (HW 12.13)

$$x = \frac{1}{2}\left[T + \frac{1}{\mu}\ln\frac{S_P}{S_A}\right]$$

Computation of the GM has proven useful in assessing relative left-side/right-side renal function with radionuclide imaging when the two kidneys are at different depths. The GM also is useful in restoration schemes employed in SPECT because spatial resolution, as well as counts, are nearly depth independent (Cherry et al. 2012).

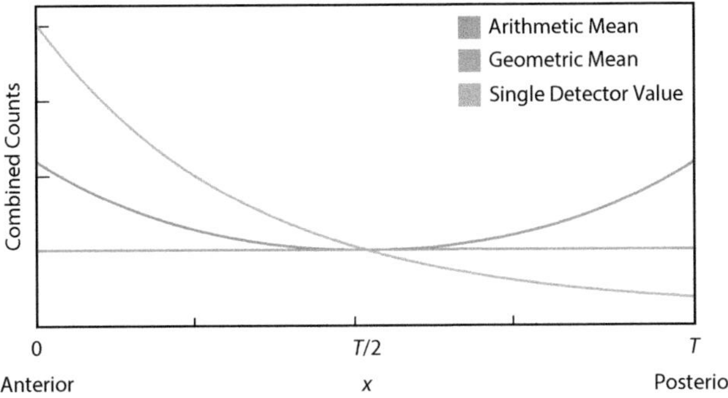

FIGURE 12.18
Opposing projections and the geometric mean can be used to generate depth-independent signal strength. The graph compares the (unnormalized) variation in counts from a point at depth x monitored using the geometric, and arithmetic, mean of the anterior and posterior views. Data presented in these graphs were obtained using Eq. (12.9) and the analogous result for the arithmetic mean, $S_{AM} \propto \frac{1}{2}\left[A_0 e^{-\mu x} + A_0 e^{-\mu(T-x)}\right]$. Homework Problem 12.14 deals with attributes of the geometric mean for two point sources.

12.5 Emission Computed Tomography

ECT produces computer-generated cross-sectional and three-dimensional images that have several advantages over planar scintigrams. For example, emission tomographic images do not suffer from the activity superposition present in planar images. Moreover, they have better contrast (Fig. 12.19), better spatial localization of lesions, and are better suited to quantitative analysis. On the other hand, SPECT images reconstructed using FBP often have lower spatial resolution than planar images because the detectors generally are farther from the source of activity. In addition, reconstructed images are extensively smoothed to reduce noise, and this lowers their spatial resolution. In contrast, the spatial resolution of SPECT images reconstructed using iterative methods can exceed that of planar images.

ECT is, in many respects, similar to X-ray CT, but there are noteworthy differences. For example, in ECT, the goal is to reconstruct the distribution of activity in a patient from projections generated by emission photons. In contrast, in X-ray CT, the goal is to reconstruct the distribution of linear attenuation coefficients, μ, from projections generated by transmitted photons (Section 11.5.4). There also are differences in implementation of reconstruction. We discuss this issue in Section 12.6.

ECT usually is sub-divided into two techniques – SPECT and PET. The goal in both techniques is to

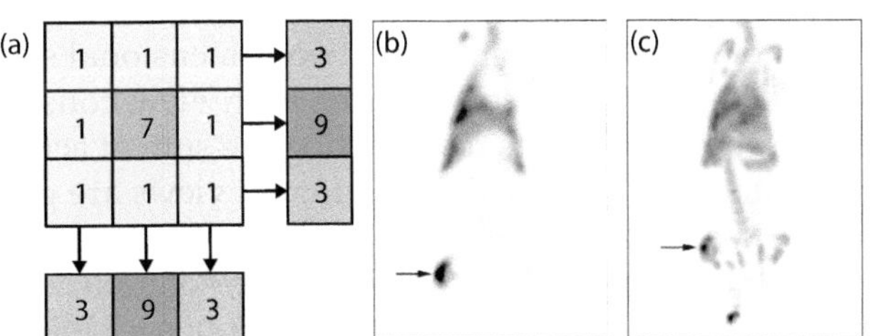

FIGURE 12.19
Comparison of contrast in projection and tomographic images. Panel (a) shows a simple example of contrast degradation in projection imaging. For the case shown, a 7:1 contrast ratio in an object becomes a 3:1 contrast ratio in a projection. In a similar spirit, Panel (b) shows relatively poor contrast in a projection image of an ^{111}In-leukocyte-labeled site of bone infection (an osteomyelitic lesion) in a juvenile minipig. Indium-111 (^{111}In) is a radioisotope that emits gamma rays with energies of 171 keV and 245 keV during its decay. Panel (c) shows markedly better contrast in a tomographic slice of the same lesion. This increased contrast, which results from elimination of activity from underlying and overlying structures, is the main advantage of the tomographic approach (SPECT) that was used. (Panels [b] and [c] adapted from Afzelius, P., et al. 2017. Post mortem leukocyte scintigraphy in juvenile pigs with experimentally induced osteomyelitis. *Contrast Media Mol. Imaging*. 3603929. DOI: 10.1155/2017/3603929. Open access.)

collect a set of projection images that permit reconstruction of the distribution of activity in the patient. We first discuss SPECT, which is less expensive than PET, but which also generates images with lower resolution. Resolution in SPECT typically ranges from 10 mm to 20 mm, whereas resolution in PET can be less than 5 mm.

12.5.1 SPECT

As the name suggests, SPECT uses radiopharmaceuticals, commonly labeled with Tc-99*m*, that generate a single photon per decay event. In contrast, PET uses radiopharmaceuticals that (indirectly) generate two photons per decay event.

12.5.1.1 Data Acquisition

SPECT imaging systems include the scintillator, collimator(s), and PMTs discussed in Section 12.3.2 and can include one, or several, gamma cameras. One advantage of systems with multiple cameras is that several angular projections can be acquired simultaneously. The primary disadvantage of systems with more than one camera is higher cost. Despite this negative, many modern nuclear imaging systems include two or more cameras and are designed for both planar and SPECT imaging, as shown in Fig. 12.13.

Similar to X-ray CT, image acquisition in SPECT involves the collection of many (e.g., 64–128) one-dimensional views (projections) of radionuclide distribution. (For simplicity, we restrict the discussion to radiation emitted from a thin two-dimensional slice orthogonal to the detector face.) Each view consists of a set of data points representing the sum of activities along LORs (Fig. 12.20). Different views are generated by rotating the camera(s) around the patient. If attenuation were negligible, and resolution were not distance dependent, views from opposite sides of the patient would be redundant, and data could be collected over just a 180° arc. However, these effects are significant, and thus data generally are collected over 360°.

Again, similar to CT, cross-sectional and three-dimensional SPECT images are reconstructed from multiple views using approaches that include FBP and iterative approaches. We will discuss reconstruction in Section 12.6.

12.5.1.2 Data Correction

Under ideal conditions, each of the points in a SPECT view would represent a sum of activities from the sample along an LOR defined by the collimator

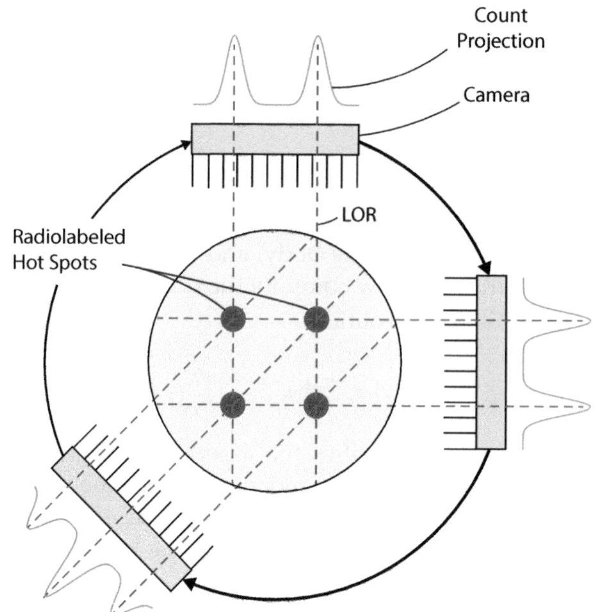

FIGURE 12.20
Schematic of data collection and image reconstruction in SPECT. The schematic shows three projections (the blue curves) obtained by summing activities along each of many LORs defined by the collimator. The camera rotates completely around the patient to facilitate collection of many projections, which are used to reconstruct the object. Although de-emphasized in the cartoon, the camera collects two-dimensional projections. Thus, reconstruction generates multiple axial slices and the associated three-dimensional object.

(Fig. 12.20). Unfortunately, several effects cause deviation from ideality. Notable among these is attenuation, which we encountered in Section 12.4.4 and which is again the focus here (Fig. 12.21). The reader interested in additional information on nonidealities is referred to the texts by Bushberg et al. (2012) and Cherry et al (2012).

Attenuation-induced loss of signal needs to be corrected to quantify SPECT images. For example, if attenuation is left uncorrected, activity closer to the patient surface that is adjacent to the camera is emphasized at the expense of deeper activity. One simple corrective option is to use the GM (Section 12.4.4). An alternative is to measure attenuation with a dual-mode imaging system that performs both emission and transmission measurements. The transmission data are reconstructed to obtain a map of attenuation coefficients, and the resulting $\mu(x,y)$ are incorporated into an iterative reconstruction algorithm to account for signal loss due to attenuation (Section 12.6).

Dual-mode imaging systems have evolved considerably since their introduction in the 1980s. Early systems used an external radionuclide source to generate gamma rays that, like X-rays, were transmitted

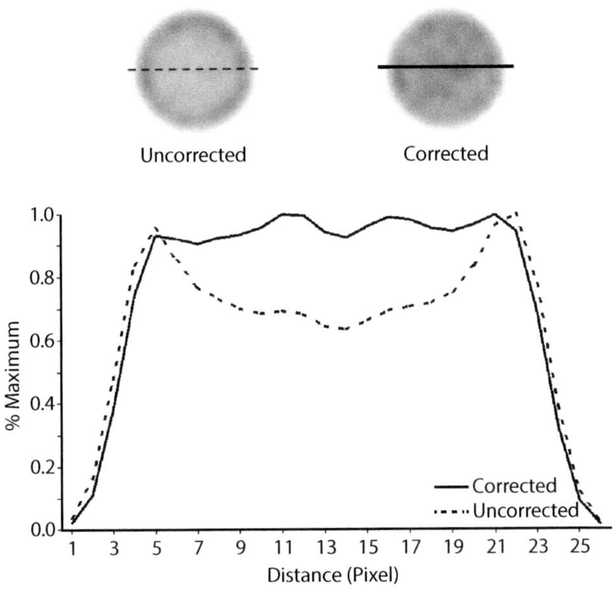

FIGURE 12.21
Attenuation correction improves the quality of SPECT reconstructions. The top panel shows uncorrected and corrected SPECT images of a phantom *uniformly* filled with indium-111. The bottom panel shows corresponding data quantification using line profiles. (Adapted with permission from Römer, W., et al. 2005. Attenuation correction of SPECT images based on separately performed CT: effect on the measurement of regional uptake values. *Nuklearmedizin* 44[1]:20–28. DOI: 10.1267/nukl05010020.)

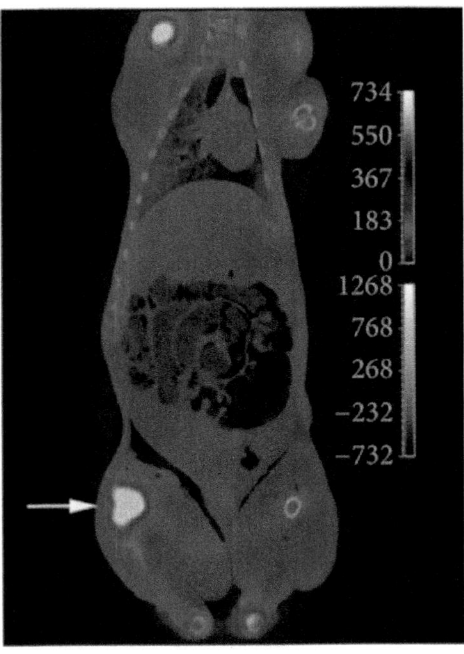

FIGURE 12.22
Dual-mode radionuclide/CT imaging improves lesion localization. The image is a fused SPECT/CT image obtained from the minipig in Fig. 12.19. (Adapted with permission from Afzelius, P., et al. 2017. Post mortem leukocyte scintigraphy in juvenile pigs with experimentally induced osteomyelitis. *Contrast Media Mol. Imaging*. 3603929. DOI: 10.1155/2017/3603929. Open access.)

through the patient. A commonly used radionuclide was Gadolinium-153, which emits 97- and 103-keV photons that are distinguishable from emission photons based on differences in energy. Current, state-of-the-art dual-mode systems are designed for SPECT/CT. In addition to generating data for attenuation correction, SPECT/CT systems, and PET/CT systems, can be used to obtain high-resolution anatomical images that are accurately aligned with radionuclide images of function. Superimposed functional and anatomical images generated using SPECT/CT and PET/CT systems have found major clinical success. For example, they have markedly improved ability to localize lesions in organs and injuries in bones (Fig. 12.22).

12.5.2 PET

Unlike SPECT, PET uses radiopharmaceuticals, commonly labeled with ^{18}F, that decay and generate a positron. Individual positrons typically traverse short distances in the body and then are annihilated via interaction with an electron. Each annihilation yields two 511-keV photons that ideally move in antiparallel directions and arrive, nearly simultaneously, at two detectors, as mentioned in Section 12.2.2.2 (Fig. 12.23a). The associated detection circuitry registers a coincidence event if certain energy and timing conditions are met (Fig. 12.23b).

12.5.2.1 Data Acquisition

Like SPECT, PET detectors include a scintillator and PMTs. However, in PET, the detectors typically are organized as "blocks," which are arranged in stationary rings (Fig. 12.24). In addition, PET uses different scintillators. State-of-the-art PET scintillators include lutetium compounds activated with cerium, which are chosen for their rapid light emission, to facilitate coincidence detection, as well as for their high μ for 511-keV photons (Table 12.2). Modern PET systems almost always are combined PET/CT devices.

Unlike SPECT, collimator grids are not required in PET because their function, which is to define an LOR, is fulfilled by coincidence detection. Specifically, coincidence detection assigns the LOR to the line connecting the two detectors that register the events (Fig. 12.23a). The absence of collimator grids, which markedly reduce signal and cause resolution to vary with depth (Eq. [12.6]), is a major advantage for PET, enhancing sensitivity by two to three orders of magnitude. This is one reason that PET images generally

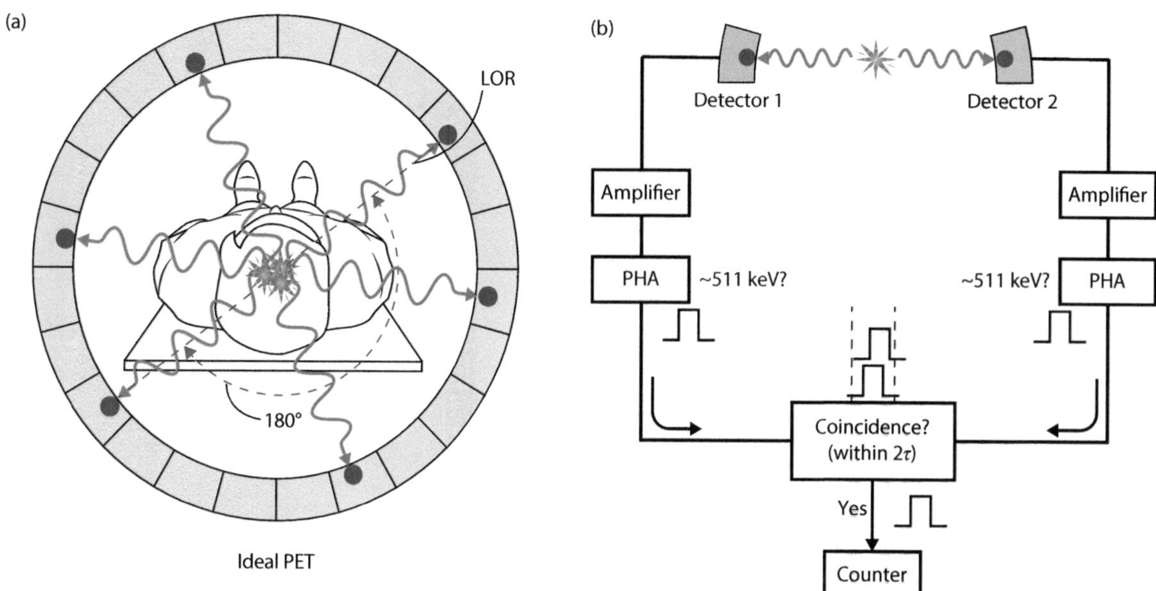

FIGURE 12.23
PET is based on coincidence detection. Panel (a) shows ideal coincidence detection in PET, which neglects effects that include the finite range of the positron in the body and noncollinearity. Regions of activity correspond to the intersection of multiple LORs. Panel (b) shows the essence of electronic coincidence detection. Energy signals from the detectors are sent to a PHA and are accepted if their energy falls within a window around 511 keV. A coincidence event is recorded if signals from two detectors overlap (i.e., they both occur within a time interval 2τ, where τ is the pulse width). For fast scintillators, this time acceptance window is set to be ~5 ns.

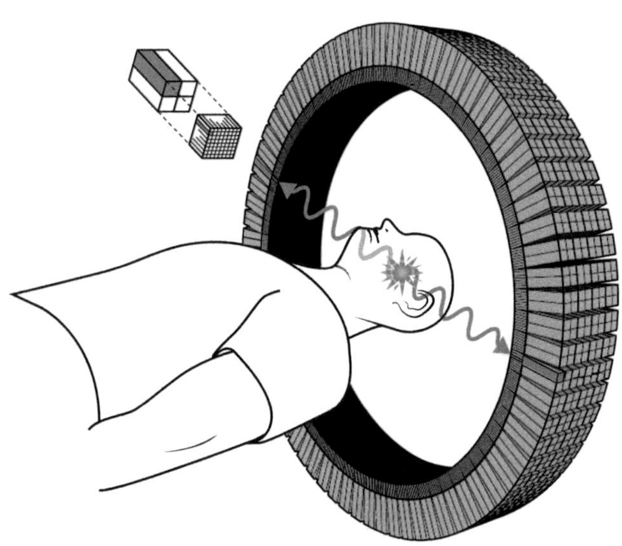

FIGURE 12.24
Schematic of a PET detector system. The enlargement on the left shows a block detector, which consists of a scintillator in contact with four PMTs arranged in a 2 × 2 array. Cuts in the scintillator divide it into smaller detector elements (typically numbering 64 per block). The cuts confine light using opaque reflective etching. Cut depth varies in a manner that allows events to be localized to an element using linear combinations of PMT signals. Unlike in SPECT, the detectors typically do not rotate. (Adapted from a work by Jens Maus.)

are of higher quality than SPECT images. Projections at all angles also can be acquired simultaneously in PET (Fig. 12.25), facilitating dynamic studies of fast processes.

Another distinct attribute of PET is the possibility of implementing "time-of-flight" (TOF) detection, which uses differences in arrival times to refine the location of the decay event along the LOR. Identifying decay positions is the goal of reconstruction; thus, highly accurate TOF detection is an alternative approach that could supplant reconstruction.

To understand the basic idea underlying TOF detection, consider the scenario in Fig. 12.26, which shows two photons generated by annihilation at distances x_1 and x_2 from the detectors that register their arrival. If the events are detected at times t_1 and t_2 after annihilation, we have

$$x_1 = ct_1 \text{ and } x_2 = ct_2$$

$$|x_2 - x_1| = c\,|t_2 - t_1| \quad \Rightarrow \quad \Delta x = c\Delta t \qquad (12.10)$$

In Eq. (12.10), $\Delta x = |x_2 - x_1|$, $\Delta t = |t_2 - t_1|$, and $c = 3 \times 10^8$ m/s = 30 cm/ns is the speed of light (because the refractive index of gamma rays and X-rays in soft

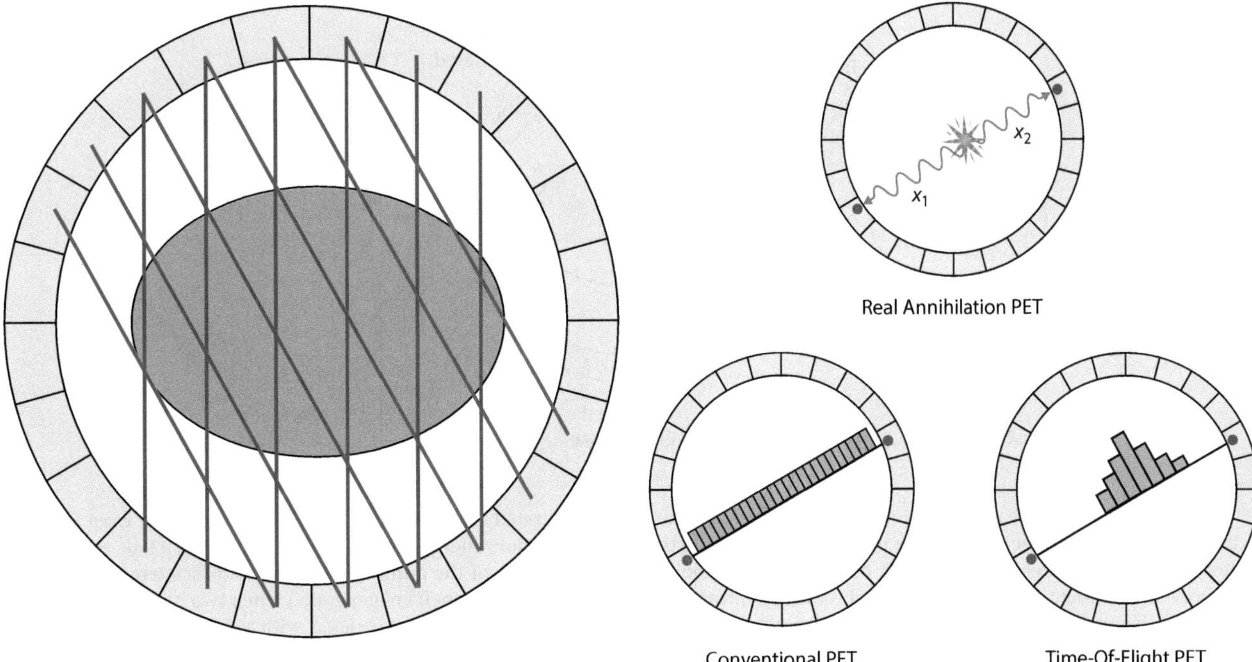

FIGURE 12.25
The 360° ring of detectors used in PET allows different projection views to be collected simultaneously rather than requiring that detectors be rotated for each view. Here, the blue and red lines show two of the many projection angles that can be collected at the same time.

FIGURE 12.26
TOF detection improves event localization in PET. TOF detection enhances localization by generating a nonconstant probability distribution along the LOR. In contrast, conventional PET yields a constant probability distribution.

tissue is very close to unity). Defining d to be the distance of the decay from the center of the LOR yields

$$d = \left| x_2 - \frac{x_2 + x_1}{2} \right| = \left| \frac{x_2 - x_1}{2} \right| = c\frac{\Delta t}{2} \qquad (12.11)$$

Equation (12.11) correctly predicts that $\Delta t = 0$ when the decay is at the center of the LOR (i.e., when $d = 0$). Most importantly, Eq. (12.11) reveals that TOF detection can, in principle, be used to localize a decay event to a point along the LOR.

Unfortunately, current technology, particularly timing resolution, has not advanced to the stage where TOF data can replace reconstruction (Sample Problem 12.5). Specifically, coincidence timing resolution at best is a few tenths of a nanosecond due to limitations on scintillator rise time (the time it takes the scintillation pulse intensity to increase before decay). However, TOF data can be used to develop a probability distribution for the decay event along the LOR, and this information is enough to produce significant improvements in image quality (Fig. 12.26). As a consequence, TOF-PET/CT systems are increasing in number.

SAMPLE PROBLEM 12.5 TIME-OF-FLIGHT PET

Show that a (currently unattainable) timing resolution as fine as 100 ps in TOF PET will yield a positional resolution of approximately 1.5 cm. Notably, 1.5 cm is still much larger than the resolution of PET. Thus, the current value of TOF PET is improved SNR associated with refining event positions.

SOLUTION

Equation (12.11) describes the relationship between position and TOF differences. Inserting the stated temporal resolution and the speed of light $c = 30$ cm/ns into Eq. (12.11) yields

$$d = c\frac{\Delta t}{2} = 30 \text{ cm/ns} \times \frac{0.10 \text{ ns}}{2} = 1.5 \text{ cm}$$

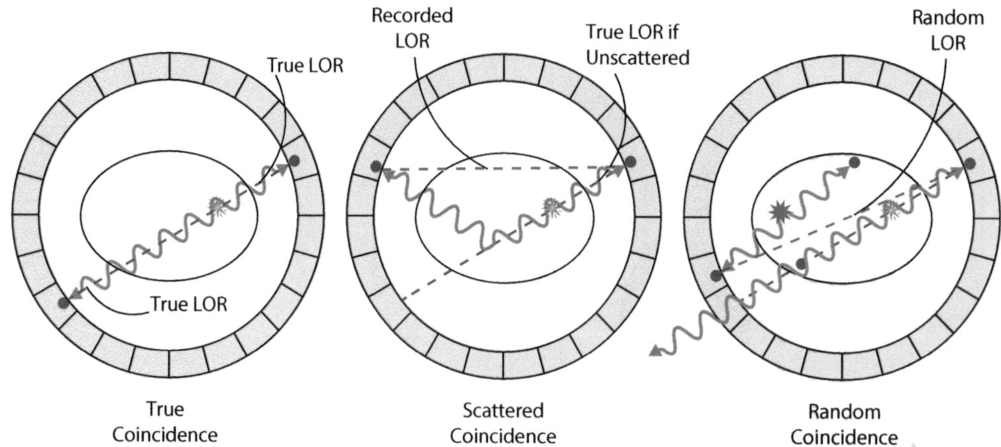

FIGURE12.27
Three types of events are detected in PET. True events (left panel) are generated when two photons, which originated from the same decay, are detected in the absence of a process like scattering that alters direction/energy. The LOR associated with a true event generates useful data. Scatter events (middle panel) are generated when one, or both, of the annihilation photons is scattered in the body; these events generate a misplaced LOR that reduces contrast. Random events (right panel) are generated when two photons from different annihilations are detected nearly simultaneously (only one from each decay); the LOR associated with a random event contains no relevant information.

12.5.2.2 Data Correction

Coincidence detection in PET also necessitates data correction. The PET coincidence circuitry counts prompt events corresponding to the detection of two photons within a short time window, and these include "true" events, as well as scatter and random events, as shown in Fig. 12.27. Both scatter and random events are sources of noise and serve to degrade image contrast. Thus, each LOR of PET data needs to be corrected for their presence as well as for attenuation and other effects (HW 12.16). In the interests of brevity, we mention a few aspects of corrective approaches.

Correction for random events is relatively straightforward. One method, which is based on use of a delayed time window, is outlined in Fig. 12.28. To implement the method, the computer checks for cases where two detectors generate signals that differ in time by ~5–10 ns (the coincidence time window). The computer tabulates these "prompt" events, updating the memory by one for the LOR associated with the detector pair. In addition, the computer tabulates delayed events. These correspond to cases where the second signal occurs in a time window that is temporally delayed (e.g., 50–60 ns). Prompt and delayed events include the same number of random coincidences because the rate at which uncorrelated photons hit the detectors is not altered by the delay. In contrast, true events are uniquely prompt, and thus the delayed data for an LOR are subtracted from the prompt data to remove random coincidences for the detector pair (LOR).

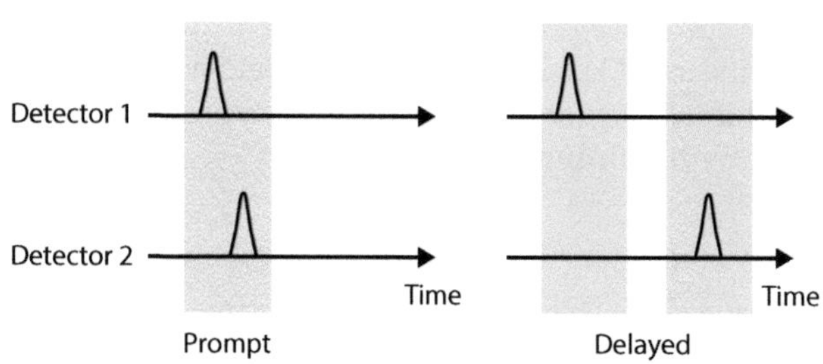

FIGURE 12.28
A delayed window scheme can be used to correct for random events in PET. In particular, prompt events can be true or random, while delayed events are always random.

A simple approximate method of attenuation correction in PET, which parallels our discussion of the GM in Section 12.4.4, is described in Box 12.2. The method also is illustrated in HW 12.16.

More sophisticated attenuation correction can be implemented using a map of attenuation coefficients obtained from a PET/CT system and incorporating the CT data into an iterative reconstruction, as mentioned briefly in Section 12.5.1.2. The CT data also can be used for scatter correction because attenuation coefficients are overwhelmingly dominated by contributions from Compton scattering at 511 keV (Section 11.5.2).

12.5.3 Comparison of SPECT and PET

We close this section with a comparison of SPECT and PET (Table 12.3).

BOX 12.2 ATTENUATION CORRECTION IN PET

In this box, we describe a simple method of attenuation correction in PET and, in the process, derive an expression for the probability that both annihilation photons will escape the body.

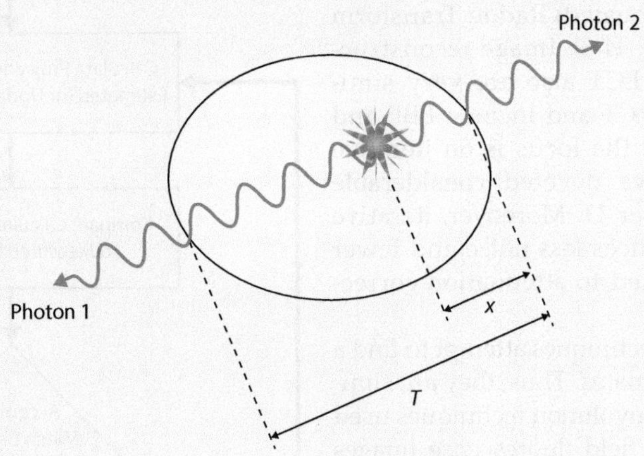

To set up the notation, let "p_j" represent the probability that annihilation photon "j" will escape a uniform attenuator of thickness "T." It then follows that two annihilation photons have escape probabilities given by (see the figure above)

$$p_2 = e^{-\mu x} \text{ and } p_1 = e^{-\mu(T-x)}$$

where x is the distance of the decay event from one surface in the patient. The probability that both photons will escape and generate a coincidence event is the product ($p_1 \times p_2$) because the two events are independent. Thus, we have

$$p_{\text{coincidence}} = e^{-\mu x} e^{-\mu(T-x)} = e^{-\mu T} \quad \text{(B12.2.1)}$$

Equation (B12.2.1) quantifies the degree to which attenuation reduces the detection of annihilation events (e.g., to 5% of events). To correct for this effect, we need to multiply the detected counts by the inverse of Eq. (B12.2.1). Thus, the attenuation correction factor (ACF) for a given LOR between detectors i and j is

$$\text{ACF}_{ij} = e^{\mu T_{ij}} \quad \text{(B12.2.2)}$$

where T_{ij} is the tissue thickness along the LOR between the detectors (HW 12.16).

TABLE 12.3

Comparison of SPECT and PET

	SPECT	PET
No. of photons	One	Two
Tracer type	Gamma emitter	Positron emitter
Predominant tracer	Tc-99*m*	^{18}F
Resolution	10 mm–20 mm	5 mm (or less)
Relative sensitivity	Lower (due to collimator)	Much higher
Relative cost	Lower	Higher
Multi-hole collimator	Yes	No – replaced by coincidence detection
Camera	Rotates	Stationary ring
Predominant application	Cardiac imaging	Brain imaging
Time-of-flight correction	No	Yes
Dual-tracer imaging	Yes	Yes

12.6 Reconstruction

The reconstruction problem in ECT is similar to that in X-ray CT but involves the inversion of an attenuated (instead of an unattenuated) Radon Transform in the case of SPECT (Box 11.3). Image reconstruction approaches used in ECT also are very similar to those used in X-ray CT and include FBP and iterative techniques. Here the focus is on iterative reconstruction because we devoted considerable attention to FBP in Chapter 11. Moreover, iterative reconstruction often produces less noise and fewer artifacts and is better suited to attenuation correction in ECT.

Iterative reconstruction techniques attempt to find a solution via successive estimates. Thus, they are similar in spirit to iterative deconvolution techniques used to remove blur from wide field fluorescence images (Section 6.5). The basic idea underlying iterative reconstruction is to use the current "estimate" of activity to compute estimated projection data. The estimated and measured projection data are compared, and their difference is used to update the activity estimate. Iteration can start with an arbitrary first estimate, consisting of zeroes or ones, or an informed first estimate (Fig. 12.29).

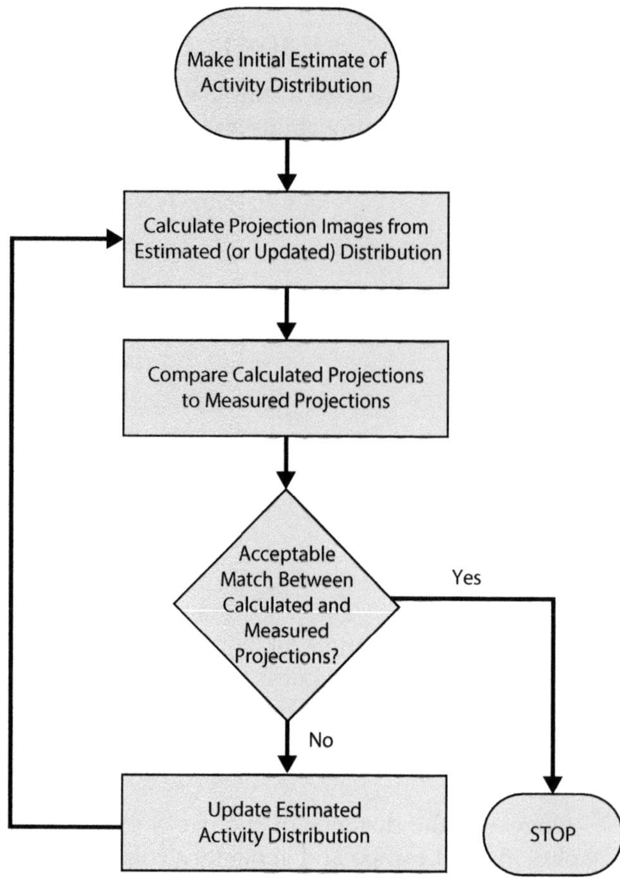

FIGURE 12.29
Flow chart summary of iterative reconstruction. (Adapted from Bushberg et al. 2012.)

Several different iterative approaches are used in nuclear medicine. Here, we illustrate the essence of iterative reconstruction by focusing on an example using an additive update, which is written mathematically as

$$a_{ji}^{k+1} = a_{ji}^{k} + \frac{P_i - \sum_{j=1}^{N} a_{ji}^{k}}{N} \tag{12.12}$$

In Eq. (12.12), the superscript k is an iteration index, the subscript j is a pixel index, the subscript i is a ray index, N is the number of pixels along ray i, a_{ji}^{k+1} and a_{ji}^{k} are the new and current activity estimates, and P_i is the measured sum along ray i (the total measured counts along ray i). Equation (12.12) implements correction in an intuitively reasonable way. If the estimated sum, $\Sigma_{j=1}^{N} a_{ji}^{k}$, is too high, the activity estimate, a_{ji}^{k}, is lowered by adding a negative correction term. Conversely, if the estimated sum is too low, the activity estimate is raised by adding a positive correction term.

To develop a feel for the mathematics embodied in Eq. (12.12), it is useful to work through a simple example. To this end, consider the 2 × 2 activity matrix shown in Fig. 12.30. The upper matrix shows the correct activity distribution, which we wish to reconstruct. (In real imaging scenarios, the answer is unknown. To acknowledge this, the numbers in the initial matrix are suppressed by making them nearly invisible.) The six numbers on the periphery of this matrix show measured projection data along the LORs defined by the direction of the nearby arrows. In this simple, idealized example, use of the six measured ray sums and Eq. (12.12) yields an exact reconstruction of the activity distribution, as described in the legend. In realistic cases involving large, noisy data sets, results from iterative reconstruction are approximate.

12.7 Applications

We close with a tabular summary of some key medical applications of radionuclide imaging to augment, and reiterate, the applications mentioned throughout this chapter (Fig. 12.31). The total number of radionuclide applications is quite large, since diagnostic

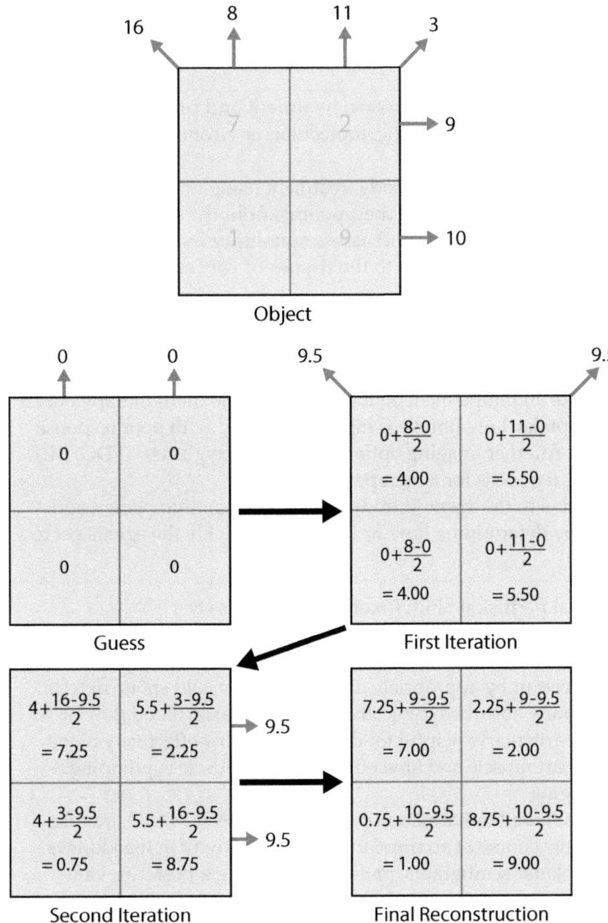

FIGURE 12.30
Simple example of iterative reconstruction. The matrix at the top is the object, and the six numbers on its periphery show associated measured projection data along the LORs defined by the arrows. The "Guess" matrix contains the initial estimate or guess. All of its elements are set to zero. The two numbers on its periphery show calculated projection data along the two vertical LORs defined by the direction of the adjacent arrows. These calculated projection data and the associated measured projection data, 8 and 11, were used in conjunction with Eq. (12.12) to obtain a first updated activity matrix (First Iteration). Further iterations were performed, first using data along the two diagonal LORs (Second Iteration), and then using data along the two horizontal LORs (Final Reconstruction). The Final Reconstruction is a perfect replica of the object for this very simple case.

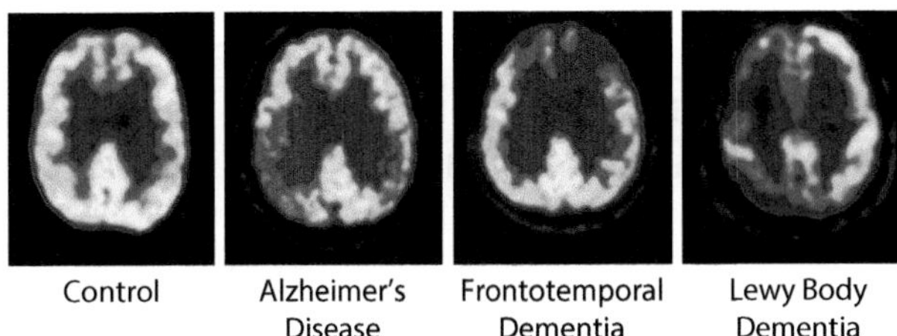

FIGURE 12.31
PET images of normal (control) and impaired brains obtained using the glucose analog FDG. Tracer distribution is lower in impaired brains engaged in reduced energy consumption. Cooler (bluer) colors denote lower intensities and thus lower activities. (Adapted from Ong, K. T. 2018. Neuroimaging in dementia. *OBM Geriatr.* 2(3). DOI: 10.21926/obm.geriatr.1803011. Attribution 4.0 International [CC BY 4.0].)

radiopharmaceuticals have been developed for use with most organs and many physiological functions. Radiopharmaceuticals can be administered intravenously, orally, and by inhalation, depending on the intended target(s), and are excreted in urine, feces, and exhaled air. While radionuclide procedures sound scary, and do carry health risks, they can involve less radiation than a CT scan (Table 12.4).

TABLE 12.4
Risk from Nuclear Imaging Procedures[a]

	Procedure	Effective Dose (mSv)
Bone	Tc-99*m* MDP	6.3
Brain	^{18}F-FDG	14.1
Heart	Tc-99*m* sestamibi (1-day protocol)	9.4
Lung Perfusion	Tc-99*m* MAA	2.0
Renal	Tc-99*m* MAG3	2.6
Thyroid scan	^{123}I	1.9
Tumor	^{18}F-FDG	14.1

Source: Data from Metter et al. 2008. Effective doses in radiology and diagnostic nuclear medicine: a catalog. *Radiology* 248:254–263. DOI: 10.1148/radiol.2481071451.

[a] See Box 11.1 for information on effective dose, the unit mSv and its relation to cancer risk, and typical exposure from background radiation.

Medical Applications of Radionuclide Imaging	
Select Applications	**Select Aspects of Implementation**
Radionuclide theranostics (often a component of personalized medicine; see below)	Use radiopharmaceuticals for both therapeutic and diagnostic (i.e., for theranostic) purposes. Label the therapeutic radiopharmaceutical with a beta emitter to limit the penetration of destructive radiation to a few millimeters and generally spare nearby healthy tissue. Label the diagnostic radiopharmaceutical, which typically is used to detect some molecular target, with a gamma or positron emitter for imaging purposes. ***Example*** – treatment of thyroid disease (cancer or hyperthyroidism) Introduce iodine-131 (^{131}I) into the bloodstream and allow accumulation in the thyroid. Collect planar or SPECT images using the gamma rays emitted by ^{131}I, which comprise ~10% of the radiation. Destroy tissue using the beta particles emitted by ^{131}I, which comprise ~90% of the radiation.

(Continued)

Medical Applications of Radionuclide Imaging	
Select Applications	**Select Aspects of Implementation**
Radionuclide-based personalized medicine	Develop a personalized approach for disease treatment and prevention based on knowledge of an individual's genetic, molecular, environmental, and lifestyle factors. ***Example 1*** – ascertain suitability for antibody treatment using immuno-PET Add a suitable radiolabel to a potential therapeutic antibody. Introduce labeled agent into candidate patient and assess suitability by using PET to determine whether the agent localizes to the tissues of concern while avoiding unintended targets. ***Example 2*** – ascertain suitability for endocrine therapy for breast cancer Use radiolabeled estradiol (fluoroestradiol/FES) and PET to evaluate the presence of estrogen receptors in breast cancers. More specifically, use FES PET to quantify FES uptake and implement endocrine (e.g., tamoxifen) therapy only if there is sufficient uptake. Low uptake is highly correlated with poor response to endocrine therapy. Another imaging option is fluorodeoxyglucose (FDG) PET. ***Example 3*** – fix dosing regimens for a therapeutic agent Add a suitable radiolabel to the agent. Introduce labeled agent into patient and fix dosing regimens by determining how well and how quickly the agents get to their targets.
Assess function of organs, including heart, kidney, liver, gall bladder, thyroid, lung, and brain (see below)	***Example 1*** – myocardial perfusion SPECT scan/nuclear stress test Inject a suitable radiopharmaceutical, such as Tc-99*m* sestamibi. Collect a series of images to assess blood flow to the heart muscle during stress, using SPECT. Induce stress by exercise or by appropriate medicines. Note cold spots/defects where the tracer is poorly absorbed, typically due to decreased blood flow or damage. This test is particularly helpful for determining how effectively blood passes through the heart muscle and how effectively the muscle is pumping. ***Example 2*** – renal imaging Inject a bolus of a suitable radiopharmaceutical, such as Tc-99*m*-MAG3, into the blood. Monitor the time course of accumulation of radioactivity in the kidneys by taking a series of planar scintigrams, preferentially using a posterior view that minimizes gamma ray travel through the body. Correlate the rate of accumulation with kidney function (e.g., slow or asymmetric accumulation may indicate impaired function, such as a blockage).
Identify tumors, stage cancer, and assess response to therapy	Inject a suitable radiopharmaceutical. One option is FDG – a metabolic reporter that accumulates in rapidly dividing tumor cells. Another is fluorodeoxythymidine – a marker for thymidine kinase activity, which is correlated with cell proliferation. Map the distribution of radioactivity, for example, using PET or combined PET/CT. Note hotspots, which may indicate the presence and location of primary tumors and metastases. Repeat the procedure following therapy to assess the therapy's effectiveness. Additional radiopharmaceuticals have been developed to localize, and treat, specific cancers (such as radioactive iodine for imaging thyroid cancer; see above).
Scan bones for pathologies, including stress fractures, nonmalignant bone lesions, inflammation and infections, and cancer metastases	Inject a suitable radiopharmaceutical, such as Tc-99*m*-methylene diphosphonate (MDP), which binds to bone at sites of active bone mineralization. Map the accumulation of MDP by taking anterior and posterior planar scintigrams or, if better contrast is desired, using SPECT. Cancer metastases in bone also can be identified using metabolic reporters and PET (see above), but at greater cost.
Assess brain disorders, including dementia, clogged blood vessels, seizures, epilepsy, and head injuries	***Example 1*** – brain lesions Inject a radiopharmaceutical that cannot cross the blood-brain barrier (BBB), such as Tc-99*m* glucoheptonate. Map the distribution of radioactivity, for example, using SPECT. Hot spots may indicate a breakdown in the BBB caused by lesions, such as tumors, hemorrhages, regions of inflammation, and other pathologies. ***Example 2*** – dementia Inject a metabolic reporter, such as FDG, that passes through the BBB. Map distribution using a suitable technique, such as PET. Unexpected cold spots may indicate diminished or otherwise altered brain activity (Fig. 12.31).

Suggested Reading and Additional Resources

Abdulla, S. and Clarke, C. 2011 – 2022. Radiology Café: FRCR lecture notes: Molecular imaging. https://www.radiologycafe.com/radiology-trainees/frcr-physics-notes/molecular-imaging

Bushberg, J.T., Seibert, J.A., Leidholdt, E.M., and Boone, J.M. 2012. *The essential physics of medical imaging (3rd Edition)*. Wolters Kluwer | Lippincott Williams & Wilkins, Philadelphia, PA. Chapters 18 and 19.

Cherry, S.R., Sorenson, J.A., and Phelps, M.E. (Eds.) 2012. *Physics in nuclear medicine (4th Edition)*. Saunders, Philadelphia, PA.

De Deene, Y. 2019. Teaching the principles of X-ray CT and SPECT using optical CT, glowsticks and a scaled anthropomorphic phantom. *J. Phys: Conf. Ser.* 1305:1–6. DOI: 10.1088/1742-6596/1305/1/012058.

Kane, S.A. and Gelman, B.A. 2020. *Introduction to physics in modern medicine* (3rd Edition). CRC Press, Boca Raton, FL. Chapter 6.

Rose, A. 1973. *Vision: Human and electronic*. Plenum Press, New York-London. Pages 21–23.

Sprawls, P. 1995. Radioactivity. In *The physical principles of medical imaging*. http://www.sprawls.org/ppmi2/RADIOACT/

HOMEWORK PROBLEMS

12.1 [Activity] The continuous loss of radioactivity due to radioactive decay means that dosages must be adjusted to account for losses between when initial activity is measured, when samples are administered, and when imaging is performed. A sample labeled with ^{18}F is prepared at 8:15 a.m. The activity of the sample is 25 $\times 10^6$ Bq. The sample is administered to a patient at 8:55 a.m., and a PET scan is performed at 10:30 a.m. (after allowing time for the radionuclide to reach sites of activity). How much activity did the sample have when it was administered? How much activity did the sample have when imaging was performed? Note: $\tau_{1/2}$ = 110 minutes for ^{18}F.

12.2 [Nuclear Decay – Radium] Marie and Pierre Curie successfully isolated radioactive radium salts approximately 120 years ago. One of the component isotopes was radium-226, which has a decay constant of 4.29×10^{-4}/year.

a. Compute the half-life of radium-226 in years.
b. What fraction of the radium-226 isolated by the Curies remains today?

12.3 [Nuclear Decay – Carbon Dating] Throughout its lifetime, an organism accumulates $^{14}_{6}$C, a radioactive isotope of $^{12}_{6}$C, such that the activity *at death* is 0.25 Bq/g of carbon. The activity decays exponentially after death because uptake of carbon ceases. Assume that you find a bone containing two grams of carbon with an activity of 0.20 Bq. What is the age of the bone? Hint: The half-life of $^{14}_{6}$C is 5730 years.

12.4 [Radioactive Decay – Half-Life 1] Show that the half-life for radioactive decay, $\tau_{½}$, and the probability of radioactive decay, λ, are inversely related according to the following expression:

$$\tau_{½} = \frac{\ln(2)}{\lambda} = \frac{0.693}{\lambda}$$

12.5 [Radioactive Decay – Half-Life 2] Use Eq. (12.2), and the definition of half-life, to show that the number of radioactive nuclei remaining after the elapse of "n" half-lives is given by

$$N\left(n\tau_{½}\right) = \left(\frac{1}{2}\right)^n N_0$$

12.6 [Effective Half-Life 1] Signal strength in radionuclide imaging is a function of the number of radionuclides in the tissue or body of interest. This strength falls as radionuclides decay. However, the strength also falls as radionuclides are eliminated from the body. The combined process follows an exponential decay characterized by an effective half-life that is a function of the radioactive half-life and the biological half-life:

$$\frac{1}{\tau_{\text{eff}}} = \frac{1}{\tau_{\text{radio}}} + \frac{1}{\tau_{\text{biol}}}$$

The radioactive half-life does not depend on environment (under physiological conditions). However, the biological half-life, or clearance rate, will depend on the pharmaceutical with which the radionuclide is associated and other conditions. Calculate the effective half-life of the following common radionuclides (half-lives in days):

Radionuclide	τ_{radio}	τ_{biol}	τ_{eff}
^{3}H	4500	12	
^{32}P	14.3	1155	
^{90}Sr	11,000	18,000	
Tc-99*m*	0.25	1	

12.7 [Effective Half-Life 2] Prove the equation for the effective half-life in HW 12.6. Hint: Assume that the activity decays as the product of two exponentials, one with the radioactive half-life, the other with the biological half-life.

12.8 *[Cyclotron 1] Consider a particle with charge q and mass m moving in a uniform magnetic field $\vec{B}$. Assume that the particle moves in a plane perpendicular to $\vec{B}$.

a. Prove that the particle moves in a circular orbit of radius $R = mv/qB$.
b. Show that the period T of the circular motion is $2\pi m/qB$. This period is termed the cyclotron period.

Hint: If unfamiliar, look up "Lorentz force" in a general introductory physics text or on the Internet.

12.9 *[Cyclotron 2] Cyclotrons were briefly discussed in Section 12.3.1 because of their role in radionuclide production.

a. Show that the kinetic energy, E_k, of a particle leaving a cyclotron is equal to $q^2B^2R^2/(2m)$, where R is the maximum radius of the Dees.
b. Consider a cyclotron with $B = 1.0$ Tesla and maximum radius $R = 0.6$ m. Show that the cyclotron frequency $1/T$ (HW 12.8) for a proton is 15.2 MHz and that E_k is 17.2 MeV.

12.10 [Scatter Rejection] The energy of a Compton-scattered photon, E_s, is given by

$$E_s = \frac{E_0}{1 + \frac{E_0}{511 \text{ keV}}(1 - \cos\theta)}$$

where E_0 is the incident energy in keV and θ is the photon scattering angle relative to the direction of incidence. Use this result to show that a low-energy PHA window of 130 keV should reject photons that scatter in the patient through an angle greater than ~44° when $E_0 = 140$ keV.

12.11 [Contrast-to-Noise Ratio] Radiopharmaceutical HS selectively accumulates in a lesion of interest, at 8× the level in surrounding tissue, creating a hotspot that generates contrast. Radiopharmaceutical CS is selectively excluded from a similar lesion, at 1/8 the level in surrounding tissue, generating a coldspot that creates contrast. Assume that each lesion is 2 cm thick, has a 2-cm diameter, and is embedded in a region of total thickness 15 cm. Also assume that the count rate in the surrounding tissue is 10 cpm/cm^2 per cm thickness of tissue for both HS and CS, and that the imaging time is 1 minute.

a. Calculate the contrast of each lesion relative to the surrounding (normal) tissue. Which radiopharmaceutical provides the best contrast, given that the relative accumulation ratio for both is the same: 8:1?
b. Calculate the CNR for each radiopharmaceutical. Which radiopharmaceutical would you choose to image the lesion, based on the Rose criterion? Hint: See Box 12.1.

12.12 [Attenuation Estimation] This problem is directed at developing a quantitative appreciation for the magnitude of attenuation effects. To this end, consider 140-keV photons, which have an attenuation coefficient for soft tissue, $\mu = 0.15$/cm. Compute the photon fraction transmitted by 5 cm of soft tissue and by 15 cm of soft tissue.

12.13 [Depth Estimation] Prove the depth equation in Section 12.4.4, which reads

$$x = \frac{1}{2}\left[T + \frac{1}{\mu}\ln\frac{S_P}{S_A}\right]$$

12.14 [Geometric Mean – Two Point Sources] Show that the GM of the signal from two point sources lying along a line between two detectors (see figure) is given by

$$S_{GM} = \left[(A_1^2 + A_2^2)e^{-\mu T} + A_1A_2e^{-\mu(x_1+y_2)} + A_1A_2e^{-\mu(x_2+y_1)}\right]^{1/2}$$

In this equation, A_1 and A_2 are unattenuated activities. The important take-home message is that the GM is no longer location independent. Hint: See Section 12.4.4.

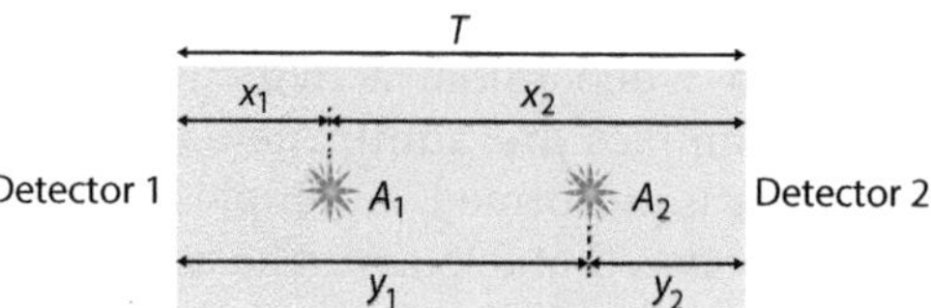

12.15 [Time-of-Flight PET] TOF detection, based on measured differences in gamma-ray arrival times, may ultimately supplant reconstruction. Show that to determine the location of an annihilation event to within 1 cm would require being able to measure a time difference of 0.067 ns. Such precision in timing currently is not possible, but it is possible to use TOF PET to refine decay locations and improve reconstructions.

12.16 [PET] Attenuation presents a special issue in PET relative to SPECT (and X-ray imaging). Specifically, to register an event, both PET photons must escape the body.

a. Show that the probability that both photons escape a uniformly attenuating object is given by the following expression:

$$P = e^{-\mu T}$$

where μ is the attenuation coefficient, and T is the thickness of the body between the two coincidence detectors. Significantly, the probability is independent of where the gamma rays are created along the line of flight.

b. Calculate the probability that both 511-keV annihilation photons escape soft tissue, assuming that the body thickness along the LOR is $T = 20$ cm, the mass attenuation coefficient is $\mu/\rho = 0.095\ \text{cm}^2/\text{g}$, and the density of soft tissue is $\rho = 1.0\ \text{g/cm}^3$.

c. Calculate the attenuation correction for this LOR. Hint: See Box 12.2.

12.17 [Scintillation – PET] Scintillators are used to convert X-rays and gamma rays into visible photons for detection. The stopping power of a scintillator is described by the same mathematics that describes attenuation by tissue and lead shielding, namely, Eq. (11.4). Preferred scintillators depend on technique. A common scintillator for PET scans is bismuth germanate (BGO). The scintillator must be thick enough to stop a reasonable percentage of incident photons, but not so thick that the location of the X-ray or gamma ray is obscured. Calculate the thickness of BGO required to stop 90% of 511-keV gamma rays. The attenuation coefficient is 0.96/cm.

ACTIVITIES

A12.1 SPECT

In the spirit of Activity A11.1, the purpose of this activity is to alert readers to

1. Educationally oriented devices that are capable of reconstructing two- and three-dimensional images of *emitting* specimens from projections collected at many angles.
2. Possible experiments and activities in the absence of detail.

The DeskCAT™ multi-slice optical CT scanner described in Activity A11.1 also can be used to investigate emission imaging. To this end, the manufacturer provides a fluorescent specimen, as shown.[2] There also are non-commercial options, such as the scanner and chemiluminescent specimens described by De Deene (2019).

Both scanners allow students to collect projection images and generate reconstructions of emitting specimens. The DeskCAT system also allows students to augment an emission scan with a transmission scan and use the latter to correct the emission data.

[2] Image courtesy of Modus Medical Devices: www.modusqa.com/products/deskcat-optical-ct-scanner-for-biophysics-education/.

13

Magnetic Resonance Imaging

This chapter covers an unusual and exceptionally powerful technique – magnetic resonance imaging (MRI). MRI differs from the other methods that we have encountered, which use "ray trajectories" to achieve localization. This difference allows MRI to bypass the diffraction-dictated link between wavelength and resolution and thereby generate high (submillimeter) resolution images despite using EM radiation with wavelengths on the order of meters!

MRI is intricate, as well as unusual, and has foundations in concepts from magnetism and quantum mechanics that have not yet arisen in the text. Thus, this chapter includes, of necessity, a very succinct review of select topics from these areas of physics.[1]

13.1 Essence of the Technique

An image is created by correlating a signal attribute (e.g., a ray trajectory or an echo time) with the location of the source. What makes MRI unusual is its use of signal frequencies to encode and map the location of the sources. In MRI, the sources typically are hydrogen nuclei (protons) from water in living tissue.

"Frequency encoding" is based on a resonance phenomenon exhibited by protons residing in a static magnetic field of strength B, typically specified using the unit tesla (T). The protons will undergo transitions between quantum energy states when exposed to an oscillating radiofrequency (RF) magnetic field, but only if the frequency of the oscillating field satisfies the resonance condition $\nu = \gamma B$ (γ is a property of the proton).[2,3] This link between ν and B means that spatially separated protons will absorb differently, and thus be distinguishable, if they are exposed to a spatially varying magnetic field.

Image production using frequency encoding and spatially varying magnetic fields was first demonstrated in 1973 by Paul Lauterbur, who won the Nobel Prize in 2003 (in conjunction with Peter Mansfield) for his contributions to the development of MRI. It is worthwhile briefly reviewing this seminal experiment, which is shown schematically in Fig. 13.1. Lauterbur collected absorption spectra from a sample consisting of two water-filled tubes residing in a one-dimensional magnetic field gradient. In each spectrum, absorption frequency encoded proton location along the direction of field variation, and absorption intensity at a given frequency was a measure of the "number" of protons absorbing at that frequency/position (Fig. 13.1b). Thus, each spectrum was a one-dimensional map (a projection) of proton distribution along the direction of the gradient.

Lauterbur collected different views of distribution (by varying the direction of the gradient) and then reconstructed a cross-sectional image of the object based on back projection, similar to CT. Lauterbur's projection-based approach is no longer used, but frequency encoding and field gradients remain a key feature of MRI.

MRI has many strengths. MRI is very safe, due to the use of low-energy RF radiation, and is capable of evaluating both anatomy and function. MRI can generate slices at any orientation and can generate contrast using parameters, such as signal decay times, that are much more tissue-specific than proton density. This latter capability is a source of exceptional soft tissue contrast, and thus MRI commonly is used to study abdominal organs and soft tissue (e.g., cartilage and ligament) damage and to identify tumors. One application in which MRI particularly excels relative to competing modalities, notably CT, is imaging both function and abnormalities of the brain. In these cases, the superior contrast detail of MRI makes abnormalities more visible. Limitations of MRI include the high cost of the equipment, generally slow image acquisition and an associated need for patients to remain motionless, and incompatibility with pacemakers and metal implants. Exposure to high levels of RF radiation also has the potential to induce patient heating.

[1] If time is limited, the material in Sections 13.1–13.6 can be used as the basis for a more introductory treatment of MRI.

[2] For the moment, we ignore the difference between γ and $\bar{\gamma} = \gamma/2\pi$ (Section 13.5.2).

[3] Electrons will similarly undergo transitions between quantum states, but typically this requires application of an oscillatory *microwave* magnetic field. Microwaves are less able to penetrate tissue and would deposit too much energy in the patient. In part for this reason, clinical MRI derives its signal from nuclei. Theoretically, any nucleus with an odd number of protons and/or neutrons (e.g., ^{13}C with its one "extra neutron") can be used in MRI; however, the proton is emphasized because it generates by far the largest MRI signal.

DOI: 10.1201/b22076-15

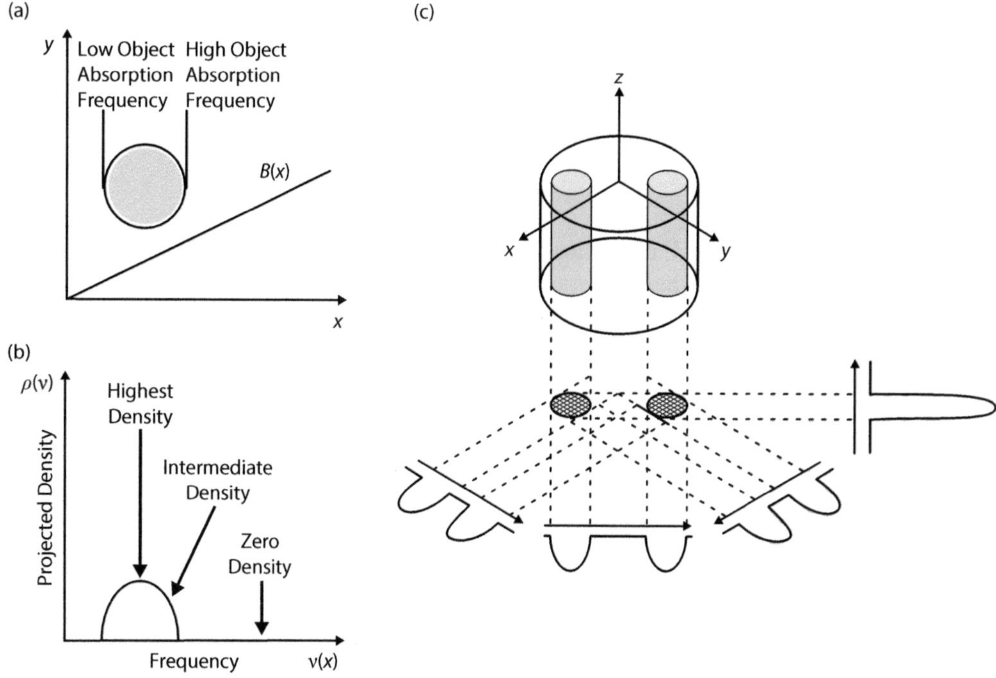

FIGURE 13.1
Schematic illustration of an early method of image production based on frequency encoding. Panel (a) shows a cross-section of a cylindrical object embedded in a magnetic field that varies linearly along the x direction. Thus, absorption frequency $\nu(x) = \gamma B(x)$ also varies linearly with x. Panel (b) shows an absorption spectrum obtained from the cylindrical object in the presence of the gradient. The spectrum yields a projection of proton density along the gradient because a one-dimensional gradient does not distinguish protons along directions orthogonal to the gradient. (In the absence of a gradient, the spectrum in Panel (b) would be a sharp peak at the identical resonance frequency of all protons.) Panel (c) illustrates reconstruction of a cross-sectional image of two cylindrical water-filled objects based on the collection of multiple projections generated using gradients with different directions, followed by back projection. (Adapted with permission from Lauterbur, P.C. 1973. *Nature* 242:190–191.)

13.2 Essential Basics from Classical Magnetism and Quantum Mechanics

The MRI signal arises because the proton behaves like a tiny dipole magnet and has an associated "magnetic moment," as discussed below. More specifically, in the main MRI magnetic field, the magnetic moment of the proton has access to two quantum states that differ in energy. The preference for protons to adopt the lower energy state results in a population bias that generates a net macroscopic magnetic moment. In MRI, this equilibrium configuration is perturbed by pulses of RF radiation, and the subsequent evolution of the non-equilibrium state yields the MRI signal.

13.2.1 Magnetic Dipole Moments and Angular Momentum

MRI signals are generated by the nuclear magnetic dipole moments, $\vec{\mu}$, of magnetically active nuclei, notably hydrogen. To cultivate intuition about $\vec{\mu}$, consider a compass needle. A compass needle has a preferred orientation in the earth's magnetic field due to the tendency of its magnetic dipole moment vector to align with the earth's field and thereby achieve a low potential energy configuration (Fig. 13.2). On the atomic scale, the magnetic dipole moment of a nucleus behaves similarly, tending to align with an externally imposed field and thereby minimize energy.

The mechanism underlying $\vec{\mu}$'s "quest for alignment" is shown in the enlargement in Fig. 13.2. The enlargement emphasizes that the magnetic field lines outside of a magnet run from its north pole to its south pole, whereas $\vec{\mu}$ points from south pole to north pole. Thus, the alignment of $\vec{\mu}$ with the field is driven by the attraction of opposite poles.

What is the origin of the magnetic moment of the proton? One historically important possibility is that $\vec{\mu}$ arises from the classical link between moving charge (i.e., current) and magnetism. Specifically, the proton can be envisioned as a spinning (moving) charge, which generates a magnetic field and an associated magnetic moment (Fig. 13.3). The spinning motion also creates angular momentum, $\vec{L}$, which is the rotational analog of linear momentum. If this

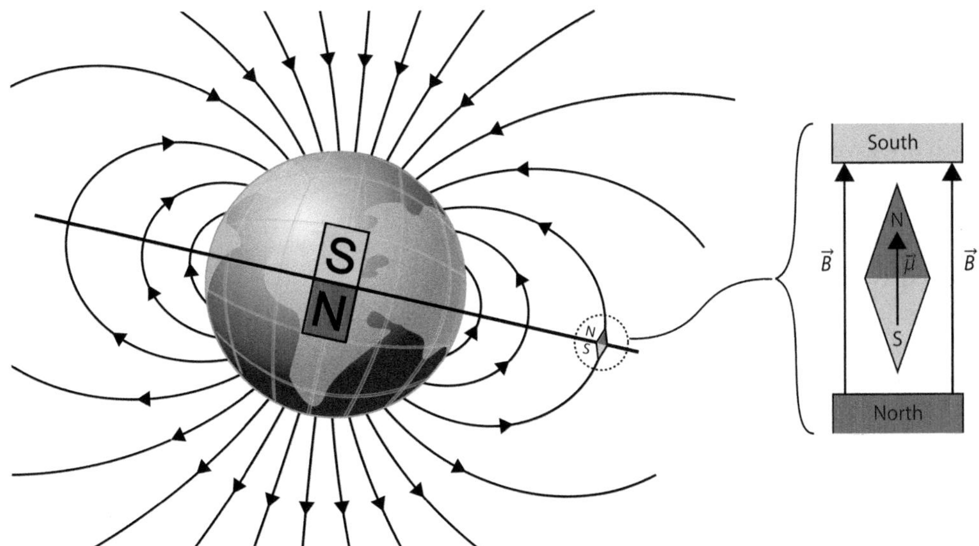

FIGURE 13.2
Behavior of a magnetic dipole moment in an external magnetic field. The dipole moment tends to align parallel to the field to achieve a low-energy configuration. This tendency reflects the attraction of opposite poles, as shown in the enlargement. Interestingly, the earth's north pole is a south magnetic pole. (Left image adapted from "Earth's Magnetic Field," Wikimedia Commons. Creative Commons CC0 1.0 Universal Public Domain Dedication.)

model is adopted, it can be shown that $\vec{\mu}$ and $\vec{L}$ are related by (HW 13.1)

$$\vec{\mu} = \frac{q}{2m}\vec{L} \tag{13.1}$$

In Eq. (13.1), q and m are charge and mass, respectively.

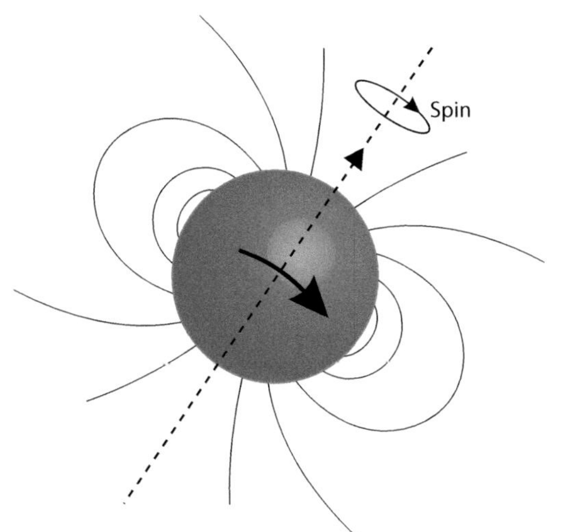

FIGURE 13.3
Classical representation of a spinning charge and its associated magnetic field and magnetic moment. When applied to the magnetism of microscopic particles, this picture is not correct, although it is conceptually useful. One obvious flaw is that the electron is a point particle, and points do not spin. Another flaw is that the neutron is neutral and thus cannot be modeled as a classical spinning charge, yet it has a magnetic moment.

When applied to subatomic structures like the proton, Eq. (13.1) has been found experimentally to be qualitatively, but not quantitatively, correct. Thus, the spinning-charge model is flawed. The correct version of Eq. (13.1) is obtained by invoking a more abstract idea from quantum mechanics, which is that magnetism of microscopic particles is intrinsic, like charge and mass.

13.2.2 Quantum Spin Angular Momentum and Magnetic Moments

In a major departure from classical physics, quantum mechanics stipulates that microscopic particles, like the electron and nucleons, have an intrinsic spin angular momentum that is the source of an intrinsic magnetic moment. Spin angular momentum and the magnetic moment are related by

$$\vec{\mu} = \gamma\vec{I} \tag{13.2}$$

In Eq. (13.2), spin angular momentum is represented using the notation $\vec{I}$, and $\gamma = gq/2m$ is known as the gyromagnetic ratio. The "g" factor varies for different particles. For the proton, $g_p \sim 5.6$. Equation (13.2) differs from Eq. (13.1) due to the presence of the g factor.

Quantum physics diverges even more significantly from classical physics in predicting that measured properties of $\vec{I}$, and thus $\vec{\mu}$, can assume only specific (discrete) values. In the interest of brevity, we restrict our discussion of quantization of spin angular momentum to "spin one-half particles," notably the proton. We also focus on I_z (the z component of $\vec{I}$)

because the main external magnetic field conventionally is along the z axis in MRI. In this case, there are two possible quantized values for I_z, namely, $I_z = \pm\hbar/2$, where $\hbar = h/2\pi$. There thus also are two allowed values of μ_z, which are given by

$$\mu_z = \gamma I_z = \pm\tfrac{1}{2}\gamma\hbar \tag{13.3}$$

We will invoke Eq. (13.3) when we discuss energy.

13.2.3 Energy of a Magnetic Dipole in a Magnetic Field

In MRI, the patient is placed in a strong external magnetic field, $\vec{B}$, and this generates an energy, E, associated with the interaction between protons and the field. The energy is given by the expression

$$E = -\vec{\mu}\cdot\vec{B} \tag{13.4}$$

In Eq. (13.4), the symbol "·" denotes a vector dot product.[4] The minus sign in Eq. (13.4) ensures that the preferred, aligned configuration (Fig. 13.2) has the lowest energy.

Quantum mechanics constrains the energy in Eq. (13.4) to assume only a few discrete values, and this critically impacts the MRI signal. To determine the allowed energies, we follow convention and place the external field along z. With this choice, Eq. (13.4) reduces to

$$E = -\mu_z B_z \tag{13.5}$$

We also restrict our analysis to hydrogen nuclei. In this case, there are two possible values for μ_z, given by Eq. (13.3), and two associated energy levels, which are

$$E = \pm\tfrac{1}{2}\gamma\hbar B_z \tag{13.6}$$

By convention, B_z is referred to as B_0, so we will use the symbol B_0 for the strength of the external field in what follows (Fig. 13.4).

RF radiation will induce transitions between the two energy states specified in Eq. (13.6) and depicted in Fig. 13.4, and this is the basis of signal generation in MRI mentioned in Section 13.1.

13.2.4 Torque on a Magnetic Dipole in a Magnetic Field

A magnetic moment in a magnetic field experiences a torque given by

$$\vec{\tau} = \vec{\mu}\times\vec{B} \tag{13.7}$$

[4] The dot product of two vectors $\vec{A}$ and $\vec{B}$ is the scalar $\vec{A}\cdot\vec{B} = A_xB_x + A_yB_y + A_zB_z$.

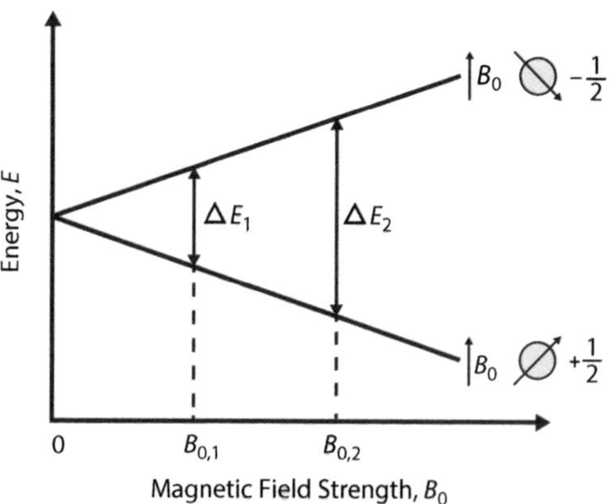

FIGURE 13.4
Schematic representation of the two quantum spin states of the proton and their energies in an external field. The energy varies linearly with magnetic field strength, as shown. Spin angular momentum has no classical analog and arises, in the case of the proton, from the quarks and gluons that reside within it.

In Eq. (13.7), the symbol "×" denotes a vector cross product.[5] Thus, the direction of the torque is given by the "right-hand rule." The torque will induce rotation of the angular momentum $\vec{L}$ as stipulated by

$$\vec{\tau} = \frac{d\vec{L}}{dt} \tag{13.8}$$

Moreover, since $\vec{\mu}$ is proportional to $\vec{L}$ (Eq. [13.1]), the torque will induce a rotation of $\vec{\mu}$. Equations (13.7) and (13.8) figure prominently in the derivation of the Bloch equations (Section 13.7 and Box 13.3).

13.3 Nuclear Moments and the Strength of the MRI Signal

Strong MRI signals arise from abundant nuclei that possess a large magnetic moment. Which nuclei fit these criteria?

Some potentially useful nuclei are shown in Table 13.1. Many, but not all, isotopes can be observed by MRI. This is because protons and neutrons tend to pair up (proton with proton and neutron with neutron) so that their spins cancel. Thus, nuclei that contain an even

[5] The cross product of two vectors $\vec{A}$ and $\vec{B}$ is a new vector, $\vec{A}\times\vec{B}$. The magnitude of the vector cross product is given by $|\vec{A}\times\vec{B}| = AB\sin\theta$, where θ is the angle between the vectors. The direction of the vector cross product is perpendicular to the plane containing $\vec{A}$ and $\vec{B}$. The correct perpendicular direction (of the two options) is obtained by placing $\vec{A}$ and $\vec{B}$ tail-to-tail and using the fingers of the right hand to rotate $\vec{A}$ into $\vec{B}$ through the smaller angle. The thumb points in the direction for the vector cross product. This is known as the right-hand rule.

TABLE 13.1

Key Attributes of Magnetically Active Nuclei

Nucleus	Spin	Gyromagnetic Ratio (MHz/T)	Relative Sensitivity	Abundance %
^{1}H	1/2	42.58	1.000	99.99
^{3}He	1/2	−32.43	0.442	0.0001
^{13}C	1/2	10.71	0.016	1.108
^{19}F	1/2	40.06	0.833	100.0
^{23}Na	3/2	11.26	0.083	100.0
^{31}P	1/2	17.24	0.066	100.0

Source: Data from http://mriquestions.com/other-nuclei.html.

number of protons and neutrons have no net spin, and no net $\vec{\mu}$, and are not visible in MRI.

In contrast, nuclei with an odd number of protons and/or neutrons are potentially useful. Among these is the hydrogen nucleus, which has a very large magnetic moment. This attribute, coupled with high abundance (nearly 100%), accounts for the strong MRI signal generated by protons and for the dominance of proton MRI.

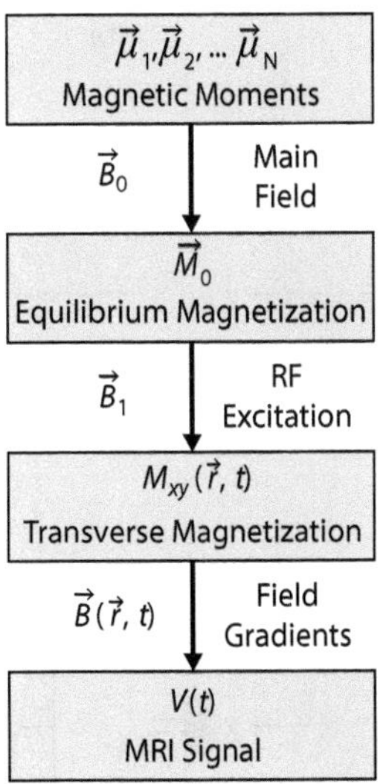

FIGURE 13.5
Key steps in an MRI experiment. See text for details.

13.4 Overview of an MRI Experiment

MRI is most accurately analyzed using concepts from quantum mechanics. However, many aspects of the technique also can be described using concepts from classical physics,[6] and at certain key junctures the classical approach is more accessible. Thus, we will adopt a hybrid approach, using quantum and/or classical ideas where they are most appropriate.

We start with an intuitive, big picture view of MRI. The goal is to establish a foundation for the more rigorous discussion that follows, primarily utilizing concepts from classical physics.

Fig. 13.5 is a block diagram illustrating the key elements of an MRI experiment. The patient is placed in a strong main magnetic field, $\vec{B}_0$, which is oriented along the z axis. The long axis of the patient also is along z. As a consequence, protons have access to two quantum states with distinct energies, with the lower energy state corresponding to the configuration where the z component of $\vec{\mu}$ is oriented parallel to the main field. The disparity in energy between the states is small relative to thermal energies, so the bias in orientation is tiny. Nevertheless, the high proton density in the body leads to a measurable (macroscopic) net equilibrium "magnetization" (magnetic moment per unit volume), $\vec{M}_0$, along z.

To generate a detectable, image-encoding signal, $\vec{M}_0$ is perturbed by briefly exposing the patient to an on-resonance RF magnetic field, $\vec{B}_1$ (an RF "pulse") that lies in the xy plane (Figs. 13.5 and 13.6). From a classical perspective, a torque from the RF field rotates the net magnetization away from its lowest energy, equilibrium configuration along z, and energy absorbed from the field moves the system into an excited state. The RF pulse *terminates* with the creation of purely "transverse" magnetization, $\vec{M}_{xy}$, which then rotates in the xy plane at the resonance frequency under the influence of the torque, $\vec{M}_{xy} \times \vec{B}_0$, created by the main field (Fig. 13.6c).[7] While rotating in the xy plane, $\vec{M}_{xy}$ induces a voltage oscillating at the resonance frequency (the detected signal) in a nearby coil.

To map proton distribution, the patient also is exposed to magnetic field gradients. Three gradients are required for imaging in three dimensions. Field gradients create a spatially dependent distribution of rotation rates for transverse magnetization because the rotation frequency, $\nu = \gamma B$, varies with position. The associated MRI signal contains a spectrum of frequencies, and analysis of its frequency content is used to produce an image.

[6] Specifically, the behavior of the observed, macroscopic magnetization vector introduced in this section is well described by classical equations (Box 13.3) that are derivable from quantum mechanics. In contrast, individual nuclei, which are not observed, must be described using quantum mechanics.

[7] In Box 13.2, we demonstrate that this classical rotation (and signal) frequency is the same as the quantum transition frequency discussed in Section 13.1. The rotation also is referred to as precession.

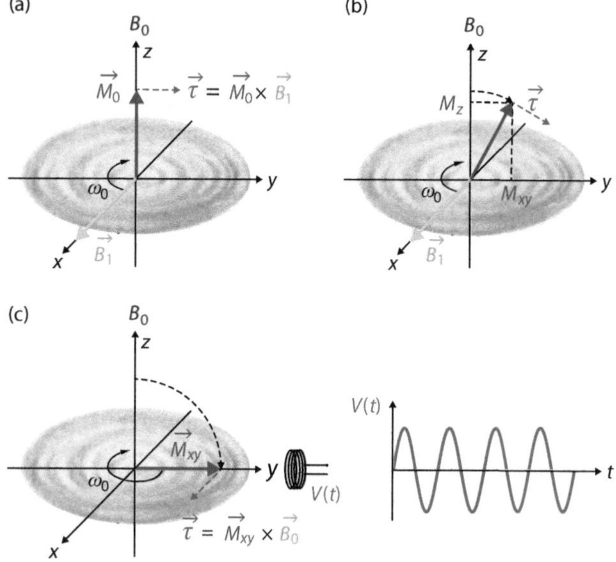

FIGURE 13.6
Diagram of basic signal generation in MRI. An RF field (a) polarized (i.e., the direction of the magnetic field) in the xy plane creates a torque that rotates the equilibrium magnetization away from the z axis, (b) generating transverse magnetization, M_{xy}, and reducing z magnetization. The decrease in M_z implies energy absorption. After completion of the RF pulse, M_{xy} rotates in the xy plane (c) and induces a voltage, $V(t)$, at the resonance frequency in a nearby coil. Despite the fact that the external field is assigned to the z axis, the field is horizontal in a clinical MRI experiment. In Panels (a) and (b), the trajectory of the magnetization during the pulse is simplified to show only rotation about the x axis due to a torque from the RF field. In reality, the magnetization simultaneously rotates about z due to a torque on its transverse component from the main field, leading to a spiral trajectory onto the xy plane (Fig. B13.3.1). In contrast, in Panel (c), rotation about z and the driving torque (dotted red arrow) are shown to illustrate the mechanism of signal generation.

We close with a list of concepts that are especially noteworthy.

1. The proton possesses a magnetic moment that tends to align along a static external magnetic field to achieve a low-energy state. For a population of protons, a slightly higher fraction exists in the low-energy state, and this leads to a net, static magnetization along the main field.
2. Static longitudinal magnetization is converted into transverse magnetization by a transient, on-resonance RF field.
3. The MRI signal is generated by rotating transverse magnetization.
4. Rotation (signal) frequencies for transverse magnetization are proportional to field strength and thus vary with position in a field gradient.
5. Resonance excitation frequencies similarly vary in a field gradient.

13.5 Signal Basics

We start by addressing the magnitude of the net equilibrium magnetization for a patient in an external field. This is followed by a discussion of the role that RF fields play in perturbing the equilibrium state and thereby initiating the process of signal generation.

13.5.1 Quantifying Equilibrium Magnetization

When the patient is placed in the external field, the z components of the magnetic moments of the protons preferentially align with the field to achieve a lower energy configuration. The result is the creation of a net population excess in the low-energy state and the associated generation of a net macroscopic magnetization, which is the vector sum of the magnetic moments of the individual protons in a unit volume.

The excess number of protons in an ROI in the low-energy state, ΔN, and the equilibrium magnetization, M_0, are given by (Box 13.1)

$$\Delta N = \frac{\hbar\gamma B_0}{k_B T}\frac{N_T}{2} \tag{13.9a}$$

$$M_0 = \frac{\Delta N \mu_z}{V} = \frac{N_T}{V}\frac{\hbar^2\gamma^2 B_0}{4k_B T} = \rho\frac{\hbar^2\gamma^2 B_0}{4k_B T} \tag{13.9b}$$

In Eqs. (13.9a) and (13.9b), k_B is Boltzmann's constant, T is the absolute temperature (in K), N_T is the total number of protons in the ROI, and ρ is the proton density. Notably, the population excess is proportional to the strength of the external field. Thus, a larger external field will generate a larger magnetization and a stronger signal.

What does Eq. (13.9a) imply about the fractional population excess, $\Delta N/N_T$, in the low-energy state? Inserting typical numbers ($T = 300$ K, $B_0 = 3$ T) yields $\Delta N/N_T = \hbar\gamma B_0/2k_B T \approx 1.6 \times 10^{-6}$. Thus, the fractional excess is very small. On the other hand, the absolute excess ΔN in an ROI (e.g., a gram of water) is large, on the order of 10^{17} (Sample Problem 13.1). The large absolute excess leads to a net magnetization that is observable in MRI (Section 13.7).

13.5.2 Quantum Analysis of RF Fields

The initial phase of signal creation in MRI involves the use of RF radiation to alter the equilibrium distribution of magnetic moments.

The quantum description of RF-induced excitation in MRI is similar to the quantum description of

BOX 13.1 QUANTIFYING EQUILIBRIUM POPULATION BIAS

The number of excess protons in the low-energy state, ΔN, which is given by Eq. (13.9a), can be derived from statistical mechanics. At thermal equilibrium, the ratio of the average number of protons in the higher energy state, N_+, to the average number of protons in the lower energy state, N_-, is given by the Boltzmann distribution:

$$\frac{N_+}{N_-} = e^{-\Delta E/k_B T} \tag{B13.1.1}$$

Here, ΔE is the energy difference between the high- and low-energy states. The remaining symbols are defined in the text. Equation (B13.1.1) shows that the average number of protons in the high-energy state will always be less than the average number of protons in the low-energy state. The energy difference can be calculated from Eq. (13.6):

$$\Delta E = E_+ - E_- = \tfrac{1}{2}\gamma\hbar B_0 - \left(-\tfrac{1}{2}\gamma\hbar B_0\right) = \gamma\hbar B_0 \tag{B13.1.2}$$

The energy differences encountered in MRI are very small relative to thermal energies, $k_B T$ (HW 13.2). Therefore, Eq. (B13.1.1) can be simplified by replacing the exponential with the first two terms of its Taylor series expansion:

$$\frac{N_+}{N_-} \approx 1 - \frac{\Delta E}{k_B T} \tag{B13.1.3}$$

Equation (13.9a) can now be derived by computing the difference between N_+ and N_- using Eq. (B13.1.3) and replacing ΔE with the expression in Eq. (B13.1.2):

$$\Delta N = N_- - N_+ = N_- - \left(N_- - \frac{\Delta E}{k_B T} N_-\right) = \frac{\Delta E N_-}{k_B T} = \frac{\hbar\gamma B_0}{k_B T}\frac{N_T}{2}$$

Here, we have made use of the fact that ΔN is small, meaning that $N_- \approx N_+ \approx \tfrac{1}{2}N_T$, to replace N_- with $\tfrac{1}{2}N_T$.

SAMPLE PROBLEM 13.1 NUMBER OF EXCESS PROTONS IN LOW-ENERGY SPIN STATE

Calculate the number of excess protons in the low-energy state in one gram of water in a 3 T magnetic field. Assume $T = 300$ K. How will this number change if the magnetic field is doubled?

SOLUTION

The number of excess protons, ΔN, in the low-energy state can be calculated using Eq. (13.9a). In this simple sample, the free protons all correspond to hydrogen atoms in the water molecules. The molecular weight of water is 18 g/mole. Therefore, 1 gram of water contains 1/18 mole of water molecules and 1/9 mole of protons (since there are two hydrogen atoms per water molecule). The total number of protons is $N_T = (1/9)N_A = (1/9)(6.022\times10^{23}) = 6.68\times10^{22}$, where N_A is Avogadro's number. The number of excess protons can now be calculated:

$$\Delta N = \frac{\hbar\gamma B_0}{k_B T}\frac{N_T}{2} = \frac{h\bar{\gamma} B_0}{k_B T}\frac{N_T}{2} = \frac{\left(6.63\times10^{-34}\,\text{J·s}\right)\left(42.58\times10^{6}/\text{s·T}\right)(3\ \text{T})}{\left(1.38\times10^{-23}\ \text{J/K}\right)(300\ \text{K})}\frac{6.68\times10^{22}}{2} = 6.8\times10^{17}$$

Here $\bar{\gamma} = \gamma/2\pi$ (Section 13.5.2). Since biological tissue is mostly water, this number is the approximate value in medical MRI. The number of excess protons in the low-energy state will double if the magnetic field is doubled, because the number is directly proportional to the magnetic field strength.

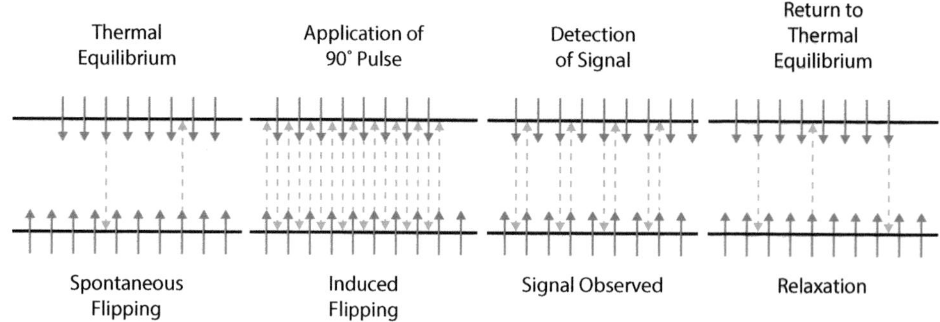

FIGURE 13.7
Quantum picture of the creation of an MRI signal. Before application of the RF field, there is a slight excess of protons in the low-energy state (10 versus 8 in this schematic). Application of a 90° RF pulse induces protons to undergo transitions into the higher energy state and equalizes the populations. After the field is turned off, the input energy is released. However, the emission photons that are released during the downward energy transitions are not the primary source of the voltage signal or of energy release. Instead, the voltage primarily is determined by a non-propagating magnetic "near" field, and most energy release is nonradiative. The atypically long range of near-field effects is due to the long wavelength of RF radiation.

fluorescence excitation (Fig. 13.7 and Section 6.1.2). The primary difference is that magnetic moments are nearly evenly distributed between the low- and high-energy states before excitation, whereas all fluorophores initially are in the lowest electronic state because electronic energy spacings are much larger than $k_B T$.

Incident RF radiation can promote low-energy protons into the high-energy state if the photon energy is on resonance (i.e., matched to the energy difference between states). In light of Eq. (13.6), the required photon energy is

$$\begin{aligned} E_{\text{photon}} &= h\nu = \hbar\omega = E_{\text{high}} - E_{\text{low}} \\ &= \frac{1}{2}\gamma\hbar B_0 - \left(-\frac{1}{2}\gamma\hbar B_0\right) = \gamma\hbar B_0 \end{aligned} \tag{13.10}$$

Equation (13.10) shows that the required (angular) frequency, ω_L, is given by

$$\omega_L = \gamma B_0 \tag{13.11a}$$

or, equivalently,

$$\nu_L = \frac{\omega_L}{2\pi} = \frac{\gamma}{2\pi} B_0 = \bar{\gamma} B_0 \tag{13.11b}$$

In Eqs. (13.11a) and (13.11b), we have inserted a subscript "L" because the resonance frequency generally is called the Larmor frequency. Equations (13.11a) and (13.11b) are the most important relationships in MRI. Their most notable feature is the (linear) dependence of frequency on field strength. This attribute will play a critical role in spatial mapping (Section 13.6).

We have derived Eq. (13.11a) using quantum mechanical and spectroscopic arguments. As a complement, we re-derive this equation in Box 13.2 using classical torque arguments. The take-home message from the box is that transverse magnetization in a longitudinal field of strength B will experience (from a classical perspective) a torque that induces rotation at $\omega_L = \gamma B$. This, in turn, creates an MRI signal consisting of a voltage oscillating at ω_L (Fig. 13.6).

Equation (13.11b) reveals that transitions between energy states in MRI are induced by radiation in the RF region of the EM spectrum. To verify this, we note that the Larmor frequency of hydrogen can be computed using $\bar{\gamma} = \gamma/2\pi = 42.58$ MHz/T (HW 13.3). Thus, $\nu_L = 85$ MHz (megahertz) for protons in an external field of strength 2 T. Analogous results for other nuclei are the focus of HW 13.2.

When a pulse of RF radiation at ν_L is used to promote low-energy protons into the higher energy state, the equilibrium population difference is reduced, and a nonequilibrium configuration is generated (Fig. 13.7). Timescales associated with the return to equilibrium play a key role in MRI.

13.6 An Introduction to Image Production and Quality

What follows is a discussion of three pivotal topics: spatial encoding, spatial resolution, and contrast. Contrast is tackled last because this allows us to defer discussion of intricate MRI "pulse sequences" and to start with a shorter, more introductory treatment of image production and quality.

In the introductory treatment, contrast is implicitly determined by the net magnetization in a voxel,

BOX 13.2 EQUIVALENCE OF QUANTUM TRANSITION AND CLASSICAL PRECESSION FREQUENCIES

The purpose of this box is to establish an important connection between the quantum and classical descriptions of MRI. In particular, the goal is to prove that transverse magnetization rotates classically at the quantum transition frequency, ω_0.

Figs. 13.6 and B13.2.1 emphasize key aspects of the classical view of signal generation. Their important message is that the MRI signal is created by transverse magnetization $\vec{M}_{xy}$ rotating under the influence of a torque created by an external field of magnitude B_0 along z. The equation that describes this motion is

$$\frac{d\vec{M}_{xy}}{dt} = \gamma \vec{M}_{xy} \times \vec{B}_0 \quad \text{(B13.2.1)}$$

where the symbol "×" represents a vector cross product (Section 13.2.4).

The geometric implications of Eq. (B13.2.1) are illustrated in Fig. B13.2.1. In an infinitesimal time interval, dt, the transverse magnetization vector rotates through an infinitesimal angle, $d\varphi$ (measured in radians). The associated change in magnetization, $d\vec{M}_{xy}$, is perpendicular to both $\vec{M}_{xy}$ and $\vec{B}_0$, as dictated by the right-hand rule. Thus, $\vec{M}_{xy}$ changes in direction but not in magnitude, tracing out a circular orbit of radius M_{xy} at angular frequency $d\varphi/dt$.

Computation of $d\varphi/dt$ is based on two expressions for the magnitude, dM_{xy}, of the vector $d\vec{M}_{xy}$. First, Fig. B13.2.1 and the definition of the radian yield

$$dM_{xy} = M_{xy} d\varphi$$

Second, Eq. (B13.2.1) and the definition of the cross product yield

$$dM_{xy} = \gamma M_{xy} B_0 \sin 90 dt = \gamma M_{xy} B_0 dt$$

Equating these two expressions for dM_{xy} yields the desired result:

$$\frac{d\varphi}{dt} = \gamma B_0 = \omega_0$$

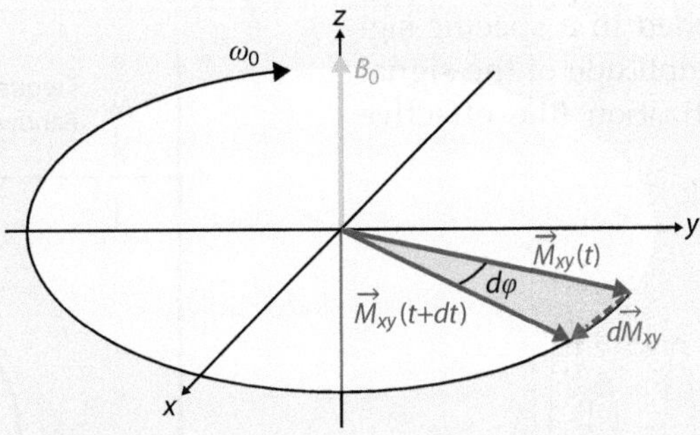

FIGURE B13.2.1
Vector diagram showing precessing magnetization. See text for details.

which is proportional to proton density (Eq. [13.9b]). Unfortunately, as mentioned in Section 13.1, density is a relatively poor contrast parameter; thus, to do full justice to contrast will require addressing the complex topic of the time dependence of the MRI signal.

13.6.1 Spatial Mapping

Spatial mapping is achieved by introducing additional (position-dependent) magnetic fields into MRI. The utility of spatially varying fields stems from the fact that the Larmor frequency is directly proportional to magnetic field strength, i.e., $\omega_L = \gamma B$ (Eq. [13.11a]). Thus, imposition of a spatially varying field will cause the signal frequency, ω_L, to vary with (and be a signature of) position. This is the key concept underlying spatial mapping.

In what follows, we discuss the process of encoding position using frequency in detail. For illustrative purposes, we start with the comparatively straightforward case of one-dimensional imaging before tackling two- and three-dimensional imaging.

13.6.1.1 One-Dimensional Imaging

One-dimensional encoding (e.g., along x) can be achieved by imposing a linear magnetic field gradient (with strength G_x in T/m) that adds to the static magnetic field; the net field is $B(x) = B_0 + G_x x$. The direction of the gradient field is along z, like the main field; what varies along x is the strength of the field (Fig. 13.8). The function of the gradient is to establish a one-to-one correspondence between frequency and x position given by

$$\omega_L(x) = \gamma B(x) = \gamma (B_0 + G_x x) = \omega_0 + \gamma G_x x \qquad (13.12)$$

In Eq. (13.12), we have used the notation $\omega_0 = \gamma B_0$ to represent the Larmor frequency in the absence of a gradient (or in the presence of a gradient at $x = 0$).

Equation (13.12) shows that structures residing at the higher end of the gradient generate signal (broadcast) at higher frequency, whereas those at the lower end of the gradient broadcast at lower frequency. In other words, the presence of the gradient creates a signal containing a spectrum of frequencies, with structure location along x encoded in a specific signal frequency (Fig. 13.8). The amplitude of the signal reveals the amount of magnetization (the effective proton density) at each position x.

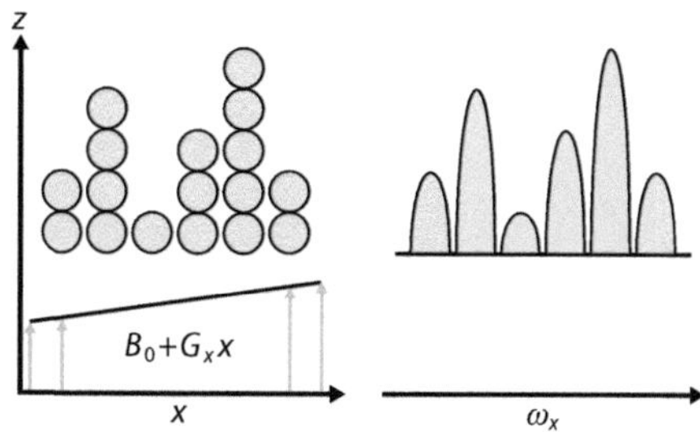

FIGURE 13.8
Generation of a one-dimensional image using a one-dimensional gradient. Structures residing at the high and low ends of the gradients create signals at high and low frequencies, respectively. Note that while the strength of the gradient field changes along the x direction, the field is directed along z, as depicted by the vertical arrows.

13.6.1.2 Two-Dimensional Imaging

One field gradient will generate a one-dimensional (projection) image; see Figs. 13.1 and 13.8. In contrast, three field gradients are required to generate a three-dimensional image or to generate a sectional (two-dimensional) image corresponding to a particular slice in the third dimension. The gradients are applied in a well-defined temporal sequence; they involve slice selection followed by encoding within the slice.

The first step in generating a sectional image is to "select a slice" (Fig. 13.9). This process involves simultaneous application of RF radiation and a field that varies spatially along what we assume is the z axis. In reality, the slice selection gradient could lie along any of the three axes, or along an oblique axis using a combination of gradients, and this versatility is a major strength of MRI (HW 13.4). The RF radiation contains a small range of frequencies, $\Delta\nu$, about a center frequency, ν_c. As a consequence, the Larmor condition is satisfied only for protons in a thin slice along the gradient, with the slice location and width determined by ν_c and $\Delta\nu$, respectively (HW 13.5 and 13.6). Only protons in the slice are excited by the RF field and rotated into the xy plane to generate signal. Thus, the signal arises only from protons in this slice. The determinants of slice thickness are discussed in Section 13.6.3.

The second and third steps involve mapping signals to positions within the slice. This process is more intricate and involves what is known as phase and frequency encoding.

Phase encoding is implemented first, typically by imposing a gradient along what is defined to be the

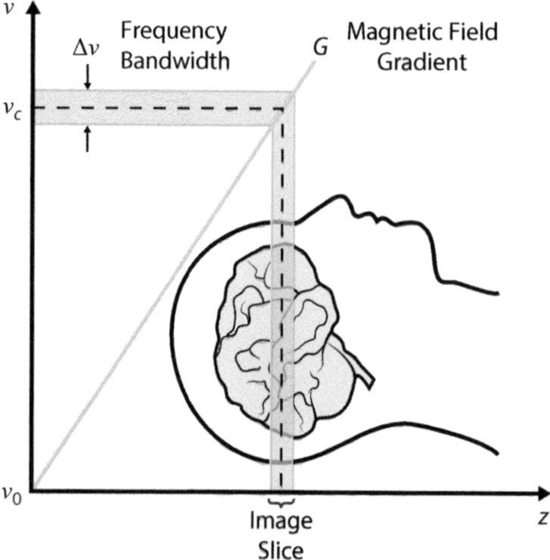

FIGURE 13.9
Schematic representation of slice selection. A gradient is imposed along the z axis, and an RF pulse of narrow bandwidth is applied simultaneously. As a consequence, the resonance excitation condition is satisfied for only a small slice within the patient.

y axis. The y gradient is imposed for a short time, τ_{PE}, and then is turned off before signal readout. During application of this second gradient field, transverse magnetization at different y locations rotates with different angular speeds (Section 13.4). Thus, at the end of the gradient, transverse magnetization at different y locations will have different phase angles that can be used to map signals (Fig. 13.10). (By analogy, runners with different speeds would similarly traverse different arc lengths on a track.) The position-dependent phase angle is given by $\theta_L(y) = \omega_L(y)\tau_{PE}$ (HW 13.7).

After phase encoding, frequency encoding is implemented along the remaining axis (x). This process is the analog of the mapping procedure described for one-dimensional imaging. The x gradient is imposed while the signal simultaneously is read out, and thus magnetizations at different positions generate RF signals with different frequencies that are used as a basis for mapping.

Monitoring a one-dimensional signal (a time-dependent voltage) just once is insufficient to map proton density throughout the slice in two dimensions. Thus, the excitation/phase encoding/frequency encoding sequence is repeated many times with different strengths of the phase encoding gradient (Sections 13.8 and 13.9). The need for repetition is significant because it slows data acquisition.

Together, phase and frequency encoding produce a unique signature for the RF signal from each tissue voxel, so they are distinguishable (Fig. 13.11). In contrast, two-dimensional localization cannot be achieved by applying two gradients simultaneously during readout because this will simply generate a net gradient along an oblique direction.

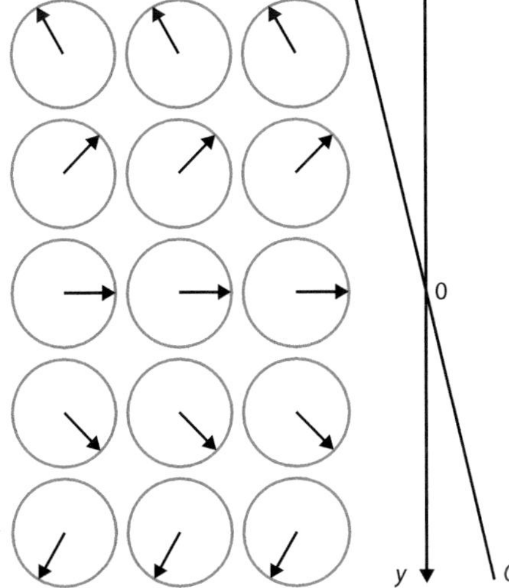

FIGURE 13.10
Schematic representation of phase encoding. The field strength at $y = 0$ (the null) is B_0 (Eq. [13.12]). Magnetization at negative y positions is exposed to a field that is slightly weaker than B_0 and thus rotates more slowly than magnetization at the null. Lagging is depicted as a counterclockwise rotation relative to the null. Similarly, magnetization at positive y positions rotates more rapidly than magnetization at the null. Advancing is depicted as a clockwise rotation relative to the null.

13.6.2 Image Reconstruction via Fourier Transformation

Like CT and ECT images, MRI images are reconstructed from raw data that are stored in a matrix (e.g., with dimensions 256 × 256). However, one distinguishing feature of image reconstruction in MRI is that the process currently is implemented via Fourier transformation of the raw data. In this section, we discuss the basis of Fourier image reconstruction in MRI conceptually for the relatively simple case of imaging in one dimension. A more rigorous treatment, also extended to imaging in two dimensions, is presented in Section 13.9.

Fourier transformation has arisen several times in the text (Boxes 4.3 and 8.1 and Figs. 8.2 and 9.25). As a brief reminder, the method decomposes a signal into its frequency spectrum, i.e., the set of frequencies that contribute to the signal and their amplitudes and phases. Given this, consider an MRI signal (a processed voltage), $S(t)$, collected in the presence of a magnetic field that varies along x. The spectrum of $S(t)$ generated by a one-dimensional FT will contain a number of frequencies generated by magnetization rotating at different angular speeds, ω, in the gradient (Section 13.4 and Activity A13.1). The amplitude

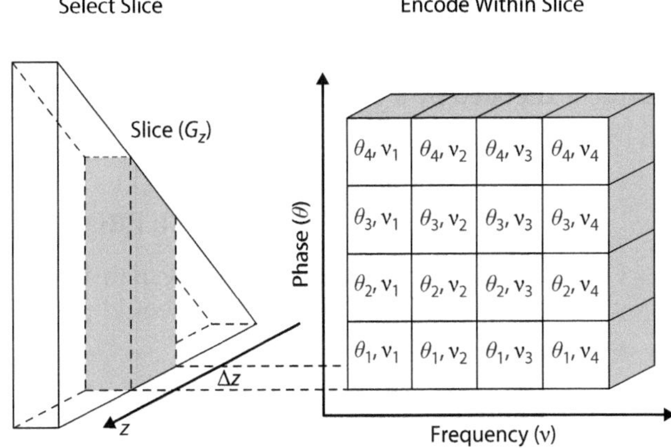

FIGURE 13.11
Schematic depiction of the spatial mapping strategy used in MRI. Unlike in microscopy, ultrasound, CT, and ECT, the MRI signal in the detection "coil" arises from the entire irradiated area, and thus encoding must be used to map signals between object and image space. Phase and frequency encoding generate a unique "address" for the signal from each voxel. Image reconstruction decomposes the aggregate signal into contributions from individual voxels, which are delivered to the corresponding pixel in the image matrix.

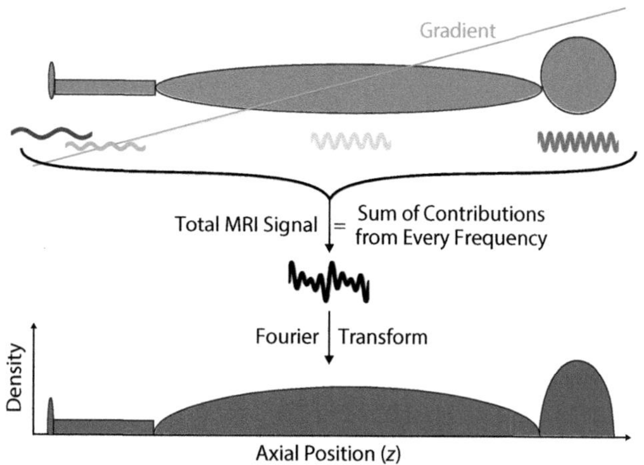

FIGURE 13.12
Fourier transformation is used to generate an image from the time-dependent MRI signal. The magnetic field gradient ensures that magnetization at different positions in the object can be distinguished based on distinct signal frequencies. Fourier transformation of the total signal yields the weighting of each frequency, which is equivalent to the weighting associated with each position, as exemplified in Eq. (13.12).

at each frequency point in the spectrum will be determined by the amount of magnetization with that specific frequency. Thus, the spectrum of $S(t)$ will yield an image consisting of a one-dimensional map of magnetization (or equivalently density) as a function of frequency (Fig. 13.12). This can be converted into an image consisting of a map of density as a function of position using Eq. (13.12).

13.6.3 Resolution

Most commonly, the resolution of an imaging technique is dictated by a physical effect, like diffraction. In contrast, the in-plane resolution in MRI typically is pixel limited. This means that the smallest resolvable detail has dimensions on the order of twice the pixel size, as dictated by the sampling theorem; see Section 9.3.

Pixel size in MRI is determined by image acquisition parameters, such as FOV and the number of data points collected (i.e., the dimensions of the image matrix). To establish the connection, we use the notation FOV_x and FOV_y for the FOV (e.g., in mm) along the x and y axes, and N_x and N_y for the number of frequency- and phase-encoding data points along these axes. Then, we have the following expressions for the pixel sizes, Δx and Δy:

$$\Delta x = \frac{FOV_x}{N_x} \tag{13.13a}$$

$$\Delta y = \frac{FOV_y}{N_y} \tag{13.13b}$$

These equations imply that increasing the number of data points, for a fixed FOV, will increase resolution. The same effect can be obtained by decreasing the FOV for a fixed number of data points, and without paying the penalty of increased data acquisition time. Typical in-plane resolutions are <1 mm (Sample Problem 13.2) and can be <0.25 mm (using microMRI on small animals or small body structures). Ultimately, resolution is limited by the weak signal that comes with very small pixel size and by molecular diffusion, which allows protons to move between pixels during a scan. An alternative version of Eq. (13.13) that emphasizes a link between FOV and acquisition parameters from frequency space is the focus of HW 13.8.

SAMPLE PROBLEM 13.2 MRI PIXEL SIZE

Calculate the pixel size of an MRI system having a 150-mm FOV when both frequency and phase are sampled with 256 data points. How would pixel size change if frequency were sampled with a different number of points than phase?

SOLUTION

The x and y pixel sizes, Δx and Δy, are given by Eqs. (13.13a) and (13.13b):

$$\Delta x = \frac{FOV_x}{N_x} = \frac{150 \text{ mm}}{256} = 0.59 \text{ mm/pixel}$$

The x and y pixel sizes in this case are identical, because the FOV and the number of data points in x and y are identical. This approach generates isotropic pixels. However, if the FOV or number of data points in x and y were different from one another, the pixels would be anisotropic (i.e., rectangular instead of square).

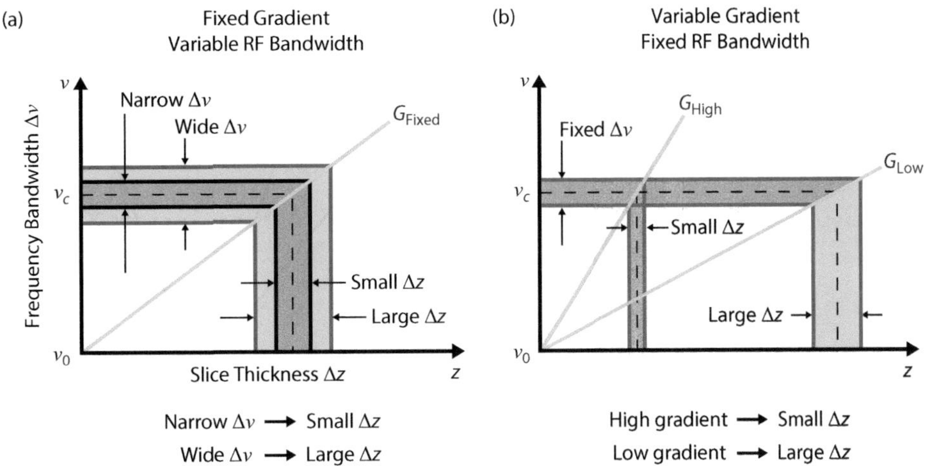

FIGURE 13.13
The bandwidth of the RF pulse (a) and the strength of the z gradient (b) affect resolution along the slice selection direct.

Resolution along the z axis is determined by voxel width, Δz, where

$$\Delta z = \text{slice thickness} \tag{13.13c}$$

A thin slice will lead to good resolution along z. To generate thinner slices, the gradient can be made steeper or the bandwidth of the exciting RF pulse can be reduced (Fig. 13.13). Resolution along z typically is the poorest of the three dimensions. A good z resolution is ~1 mm (HW 13.5).

13.7 A More Advanced Analysis

One of the primary strengths of MRI is its exquisite soft tissue contrast, which is due in large part to encoding brightness based on timescales over which perturbed tissues return to equilibrium. In this extended treatment, we discuss the rather complex topic of "relaxation time-weighted" contrast and the closely linked issue of the time dependence of the MRI signal. Our analysis is founded in classical physics. The motivation for pursuing a classical analysis is that it cultivates intuition while being less complex.

It is instructive to start with a classical description of the equilibrium behavior of individual protons in an external magnetic field. This will set the stage for a classical analysis of the origin of the net equilibrium magnetization, $\vec{M}_0$, and its manipulation by an RF pulse to create the MRI signal.

Individual protons possess a magnetic moment and therefore experience a torque created by the main field (Eq. [13.7]). As a consequence, each proton behaves like a spinning top that precesses under the influence of the torque created by the gravitational field. Specifically, protons precess about the main magnetic field at the Larmor frequency (Fig. 13.14a and HW 13.10). The net equilibrium magnetization, $\vec{M}_0$, arises from the net effect of the separate moments of each individual proton precessing in the parallel and antiparallel orientations (Fig. 13.14b).

From a classical perspective, the MRI signal is created by tipping the net equilibrium magnetization, $\vec{M}_0$, away from the main magnetic field. Tipping is implemented by applying an oscillating RF field, $\vec{B}_1(t)$, in the xy plane, which exerts a torque on $\vec{M}_0$ (Eq. [13.7] and Fig. 13.6). The RF field is roughly one hundred-thousandth as strong as the main field and thus does not alter the Larmor frequency appreciably.

The pulse of RF radiation can be used to rotate the equilibrium magnetization completely into the xy plane and thereby create a net nonequilibrium transverse magnetization, $\vec{M}_{xy}$. The required duration, t, for such a "90° tipping pulse" is given by (HW 13.12)

$$\gamma B_1 t = \frac{\pi}{2} \tag{13.14}$$

After its creation by the 90° pulse, $\vec{M}_{xy}$ rotates in the xy plane at the Larmor frequency (Box 13.2), and a receiver coil oriented in the xy plane detects the presence of $\vec{M}_{xy}$, and generates the MRI signal, based on the law of electromagnetic induction (Fig. 13.6). In the presence of a magnetic field gradient, the precession frequency for transverse magnetization will vary with position, creating a voltage signal containing a spectrum of frequencies.

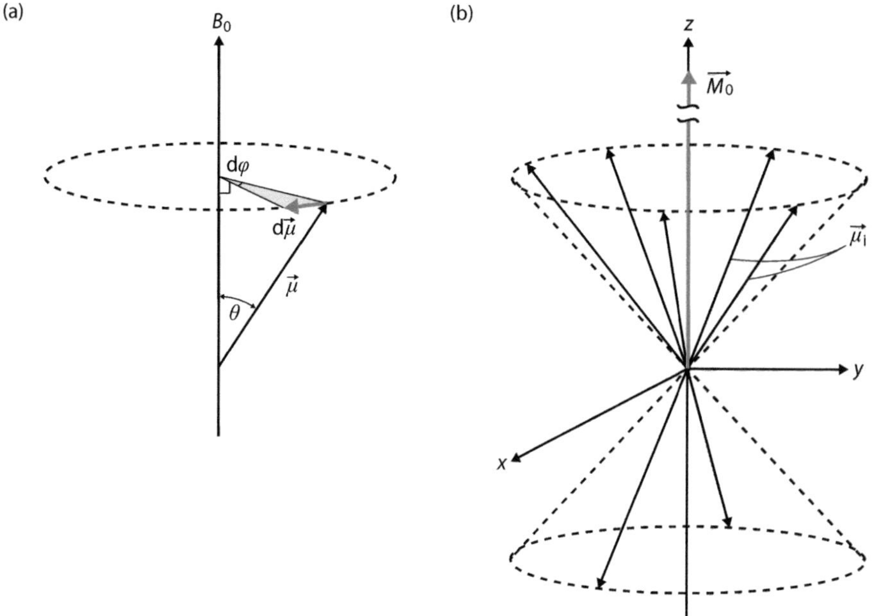

FIGURE 13.14
Behavior of individual protons in an external field. Like a gyroscope, individual protons (a) precess around the external field (at an angle $\theta = 54.7°$ – HW 13.11). A collection of protons (b) creates a net equilibrium magnetization that is generated from the slight (here five versus two) excess that is precessing in the parallel orientation.

13.7.1 Relaxation Mechanisms

Nonequilibrium configurations are unstable, and thus the transverse magnetization created by the RF field exists transiently. Specifically, after the 90° pulse, M_{xy} is large because a subset of spins is phase coherent,[8] pointing asymmetrically in the same direction in the transverse plane (Fig. 13.15a).[9] However, coherence is lost with time, and M_{xy} decays (relaxes) back to its equilibrium value, which is zero. Relaxation of transverse magnetization occurs over a timescale, T_2, that is tens to hundreds of milliseconds (Fig. 13.15). After the 90° pulse, the longitudinal magnetization also relaxes (from zero) back to its equilibrium value M_0, although typically this recovery process occurs over a timescale, T_1, that is much longer than T_2 (Fig. 13.15). Thus, the MRI signal usually decays before M_z returns to its equilibrium value.

13.7.1.1 Qualitative Analysis of Relaxation Effects

Relaxation of M_z back to its equilibrium value occurs when the energy that was input by the on-resonance RF field is lost to the environment. This process requires a stimulus (HW 13.14), and a primary source of the required stimulus is on-resonance fluctuating fields in the sample that have a component in the xy plane. In particular, a proton can be stimulated to release energy by on-resonance fields arising from a nearby magnetic moment, e.g., a nearby proton.

Molecular motion, particularly proton tumbling, is a key source of the required fluctuating fields. Thus, systems with the most extensive motional tumbling near ν_L will undergo the most rapid longitudinal relaxation (Fig. 13.16). Significant tumbling at ν_L is not present in free fluids or "solids" at clinical field strengths, and thus T_1 is long. In contrast, partially bound water (e.g., associated with the surface of macromolecules), which is found in many body tissues, undergoes significant tumbling near ν_L. This causes T_1 to shorten, but to a tissue-specific extent. In upcoming sections, we will discuss how the tissue specificity of the longitudinal recovery time can be used to generate contrast.

[8] A rigorous treatment of the presence of transverse spin coherence at the end of the RF pulse is beyond the scope of this text. Instead, we present an intuitive rationalization. Before application of the pulse, the spin system is asymmetric, possessing a slight bias in orientation along the longitudinal direction. The RF pulse rotates this skewed distribution into the transverse plane, turning longitudinal bias into transverse bias (phase coherence), with the same spins generating the two kinds of asymmetry. Succinctly stated, the RF pulse rotates the complete spin distribution without disrupting its order.

[9] In MRI, the words proton and spin typically are used interchangeably. Moreover, observable quantities consist of ensembles of tens of thousands of nearby protons that all precess at the same macroscopic frequency. Such a set is referred to as a spin "isochromat," to reinforce the fact that the color (or frequency) of the protons is the same. Thus, in Fig. 13.15, and several others, the arrows represent spin isochromats, which are referred to simply as spins.

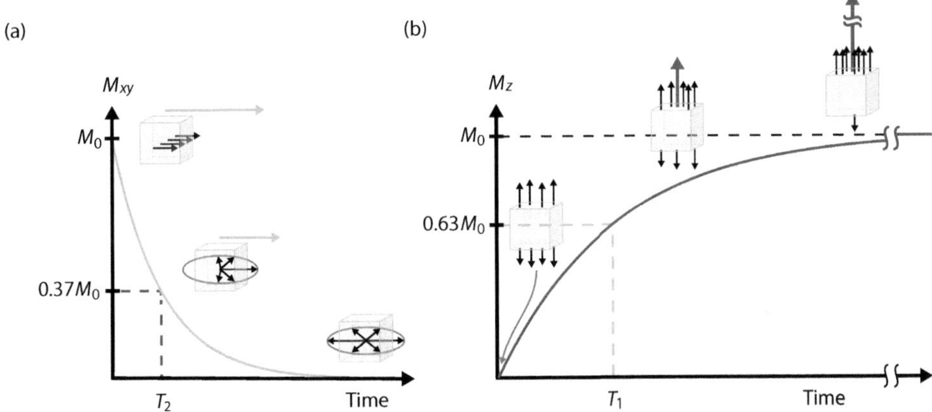

FIGURE 13.15
Exponential relaxation of transverse (a) and longitudinal (b) components of the net magnetization. Cubes next to the curves reveal the temporal behavior of individual spins, and the net transverse and longitudinal vectors (light and dark blue), although the spin bias is greatly exaggerated. Longitudinal relaxation leads to loss of proton energy to the environment, whereas transverse relaxation involves only energy exchange between protons without any net loss to the environment. Finally, transverse magnetization typically relaxes much more quickly than longitudinal relaxation; thus, the timescales in Panels (a) and (b) can differ significantly.

Relaxation of M_{xy} is driven by motion at ν_L as well as by "low-frequency" motion, and thus $T_2 \leq T_1$. The effect of low-frequency motion is to create nearly static fields that combine with the main field and cause the local field to differ slightly from B_0. These transient variations in the local field lead to a variation in spin precession frequency (Eq. [13.11]). As a consequence, spins lose the coherence created by the RF field, and M_{xy} decays over a timescale that is structure specific. For example, a watery component like cerebrospinal fluid (CSF) or blood is relatively lacking in low-frequency motion, and thus decay of M_{xy} is comparatively slow (Fig. 13.16). In contrast, cortical bone is highly constrained, and decay of M_{xy} is very rapid.

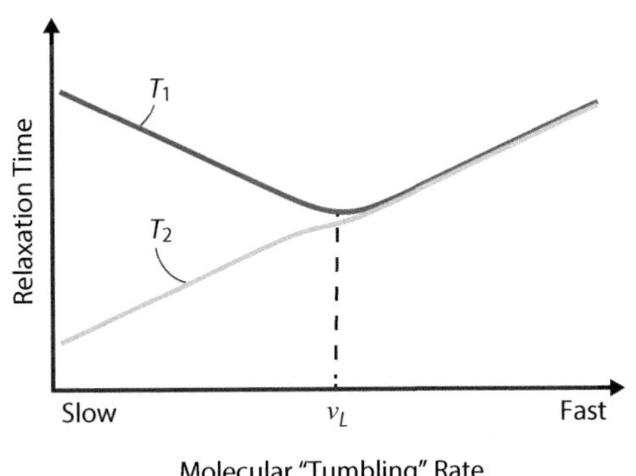

FIGURE 13.16
Molecular dependence of the longitudinal (T_1) and transverse (T_2) relaxation times. In free fluids, molecular tumbling on average is very fast; few protons tumble near ν_L, and T_1 is large. At the other extreme, in solids, tumbling on average is too slow to induce longitudinal relaxation, and so T_1 again is large. In marked contrast, in solids, T_2 is small. This difference arises because T_2 relaxation, but not T_1 relaxation, *additionally* is driven by low-frequency, fluctuating fields which are relatively extensive in solids (but absent in liquids). As a consequence, T_2 is very short in solids, making it difficult to conduct MRI studies of structures like cortical bone. (The low proton density of cortical bone is another impediment.) T_1 varies significantly with field strength, whereas T_2 is relatively independent of field strength. Curves were generated from theoretical expressions for T_1 and T_2 and are plotted on a log-log scale.

Field inhomogeneity also arises from imperfections in the main magnet. Moreover, this *extrinsic* effect typically dominates the decay of transverse magnetization, as discussed next. In many applications of MRI, pulse sequences, which we will encounter later, are used to circumvent the extrinsic effects, and then purely intrinsic effects generate contrast. "Gradient-echo" imaging is a notable exception.

13.7.1.2 Quantifying Transverse Relaxation

In the absence of extrinsic field inhomogeneity, decay of M_{xy} after a 90° pulse occurs exponentially with a time constant, T_2, which is known as the spin-spin or transverse relaxation time (Fig. 13.15). Specifically, the temporal dependence is (HW 13.15)

$$M_{xy} = M_0 e^{-t/T_2} \qquad (13.15)$$

Equation (13.15) correctly predicts that $M_{xy}(0) = M_0$ and $M_{xy}(\infty) = 0$. In writing Eq. (13.15), we have focused on the temporal dependence that gives rise to relaxation and ignored the time dependence associated with the ongoing rotation of M_{xy} at ω_0 (Box 13.3 and HW 13.15).

More generally, M_{xy} decays with a (shorter) time constant T_2^*, which is an effective transverse relaxation

TABLE 13.2

Relaxation Time and Proton Density Data for Brain Tissue and Blood

Field Strength (T)	Tissue	T_1 (ms)	T_2 (ms)	T_2^* (ms)	Proton Density (gm/cm³)
1.5	White matter	510	67	n.a.	0.61
	Gray matter	760	77	69	0.69
	Arterial blood	1441	290	55	0.72
	CSF	2650	280	n.a.	1.0
3.0	White matter	1080	70	50	0.61
	Gray matter	1820	100	50	0.69
	Arterial blood	1932	275	46	0.72
	CSF	3817	1442	n.a.	1.0

Source: Data from MacIntosh, B. J., et al. 2013. Magnetic resonance imaging to visualize stroke and characterize stroke recovery: a review. *Front. Neurol.* 4:60. DOI: 10.3389/fneur.2013.00060.

time that also incorporates extrinsic field inhomogeneity. Some typical values for T_2 and T_2^* for brain components are given in Table 13.2.

13.7.1.3 Quantifying Longitudinal Relaxation

Immediately after a 90° pulse, M_z is zero. Then M_z recovers exponentially to its equilibrium value, M_0, with a time constant T_1, which is known as the spin-lattice or longitudinal relaxation time (Fig. 13.15). Specifically, the temporal dependence is given by (Box 13.3 and HW 13.15)

$$M_z(t) = M_0\left(1 - e^{-t/T_1}\right) \tag{13.16}$$

Equation (13.16) correctly predicts that $M_z(0) = 0$ and $M_z(\infty) = M_0$. Typically, $T_1 \sim 5 - 10 \times T_2$ (Table 13.2).

BOX 13.3 THE BLOCH EQUATIONS

Equations (13.15) and (13.16), and many related results, can be derived from the Bloch equations, which describe the temporal dependence of magnetization during excitation and relaxation. The starting point in developing the Bloch equations is Eq. (13.8), which states that the time derivative of angular momentum, $\vec{L}$, is torque, $\vec{\tau}$:

$$\frac{d\vec{L}}{dt} = \vec{\tau} \tag{B13.3.1}$$

Invoking Eq. (13.7) and a classical version of Eq. (13.2), $\vec{\mu} = \gamma\vec{L}$, yields

$$\frac{d\vec{\mu}}{dt} = \gamma\vec{\mu} \times \vec{B} \tag{B13.3.2}$$

Equation (B13.3.2) also applies to the net magnetization $\vec{M}$ because it is the sum of individual magnetic moments. Writing the cross product and net magnetization in component form yields

$$\frac{dM_x}{dt} = \gamma\left(M_y B_z - M_z B_y\right) \tag{B13.3.3a}$$

$$\frac{dM_y}{dt} = \gamma\left(M_z B_x - M_x B_z\right) \tag{B13.3.3b}$$

$$\frac{dM_z}{dt} = \gamma\left(M_x B_y - M_y B_x\right) \tag{B13.3.3c}$$

Felix Bloch augmented Eqs. (B13.3.3a)–(B13.3.3c) with a phenomenological description of relaxation using standard, first-order differential equations. The augmented equations read

$$\frac{dM_x}{dt} = \gamma\left(M_y B_z - M_z B_y\right) - \frac{M_x}{T_2} \tag{B13.3.4a}$$

$$\frac{dM_y}{dt} = \gamma\left(M_z B_x - M_x B_z\right) - \frac{M_y}{T_2} \tag{B13.3.4b}$$

$$\frac{dM_z}{dt} = \gamma\left(M_x B_y - M_y B_x\right) - \frac{M_z - M_0}{T_1} \tag{B13.3.4c}$$

Here, M_0 is the equilibrium magnetization. To proceed further, we assume that the magnetic field has a z component generated by the external field and x and y components that are generated by an on-resonance RF pulse rotating in the clockwise direction. This implies

$$B_x = B_1 \cos(\omega_0 t) \quad B_y = -B_1 \sin(\omega_0 t) \quad B_z = B_0 \tag{B13.3.5}$$

Equations (B13.3.4a)–(B13.3.4c) can be solved for some special cases. One case of interest is the trajectory of the magnetization during the application of the short RF pulse, when the relaxation terms can be neglected. It can be shown by direct substitution into Eqs. (B13.3.4a)–(B13.3.4c) that, in the absence of the relaxation terms, the solution is (HW 13.15)

$$\begin{aligned} M_x &= M_0 \sin(\omega_1 t)\sin(\omega_0 t) \\ M_y &= M_0 \sin(\omega_1 t)\cos(\omega_0 t) \\ M_z &= M_0 \cos(\omega_1 t) \end{aligned} \tag{B13.3.6}$$

In Eq. (B13.3.6), $\omega_1 = \gamma B_1$. The magnetization spirals down into the xy plane on the surface of a sphere, as shown in Fig. B13.3.1.

The Bloch equations also can be used to demonstrate that, after the RF pulse (when $B_1 = 0$), the transverse magnetization oscillates in the xy plane at ω_0 while relaxing back to zero with time constant T_2. In contrast, the longitudinal magnetization recovers exponentially with time constant T_1 (HW 13.15).

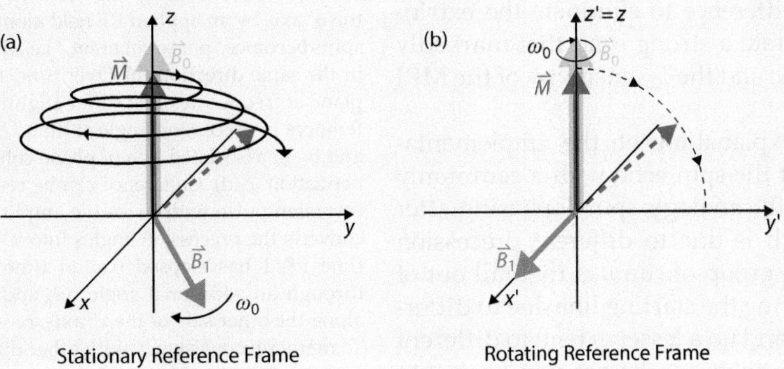

FIGURE B13.3.1
Temporal behavior of the magnetization during a 90° RF pulse. The magnetization spirals down (Panel (a)) into the xy plane as predicted by Eq. (B13.3.6). (Relaxation after the 90° RF pulse resembles a "corkscrew.") However, the rotational motion at the Larmor frequency often is not shown (Panel (b) and Fig. 13.6). This is equivalent to viewing the effects of the pulse in a reference frame that rotates at the Larmor frequency. A 90° pulse leads to the largest signal for a single excitation.

13.7.2 An Introduction to the Spin Echo

Image acquisition in MRI involves the use of sequences of RF pulses, gradient pulses, and delays between them. These are known as "pulse sequences." Insight into one reason for the use of pulse sequences can be obtained by examining the data in Table 13.2, which show the sensitivity of relaxation times and proton density to variations in biological structure. Notably, T_1 and T_2 vary by a factor of 5–10 for different biological structures, whereas proton density varies by less than a factor of two. This suggests that contrast will be enhanced if the amplitude of the MRI signal reflects relaxation times instead of density. Pulse sequences that emphasize relaxation disparities achieve this goal, but, as we will see, they are intricate.

Another reason for the use of pulse sequences is that relaxation of the transverse magnetization tends to be dominated by spatial variations in the main magnetic field. Thus, intrinsic relaxation effects, which are tightly linked to tissue type, are masked by extrinsic effects (Fig. 13.17). In this section, we address how to circumvent the rapid T_2^* decay associated with intrinsic effects and extrinsic field inhomogeneity through the use of a spin echo. In the next section, we discuss

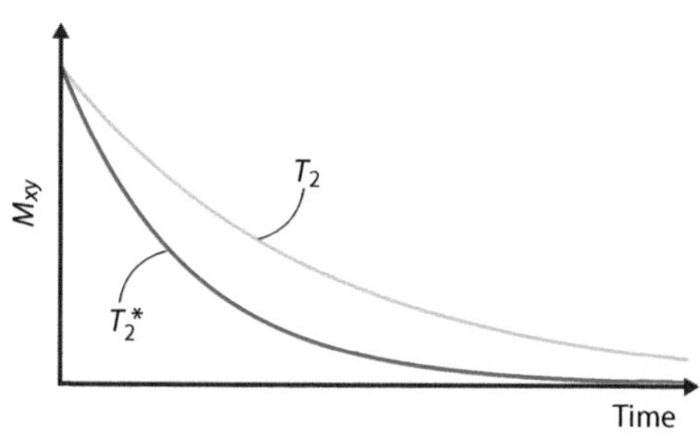

FIGURE 13.17
Transverse magnetization decays rapidly with a time constant T_2^*, which incorporates intrinsic and extrinsic effects.

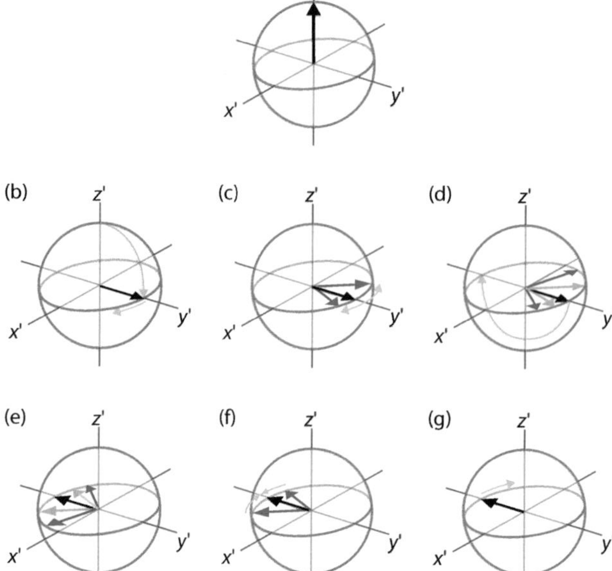

FIGURE 13.18
Schematic diagram of the creation of a spin echo. The net magnetization, which is initially aligned with the z' axis (a), is rotated onto the y' axis by an applied RF field along the x' axis, and a subset of spins becomes "phase coherent," i.e., the associated moments point in the same direction (b). Over time, the spins precess in the $x'y'$ plane at frequencies that differ slightly (by $\Delta\omega$). This leads to differences in precession angle at time $TE/2$ given by $\Delta\theta = \Delta\omega \times TE/2$ and to an associated loss of phase coherence and transverse magnetization (c–d). Coherence can be re-established using a 180° (π) inversion pulse with twice the amplitude of the 90° pulse, which converts the precession angles into $\pi - \Delta\theta$ (e). After an additional time $TE/2$ has elapsed (i.e., at time TE), each spin will rotate through an additional angle $\Delta\theta$, and thus all spins will realign along the other side of the y' axis, re-establishing coherence (f–g). To maintain consistency with other diagrams and animations, the events are depicted in a "primed" frame that rotates at the Larmor frequency, so that the overall Larmor precession is suppressed. (Adapted with permission from Dr. Aaron G. Filler.)

how a more involved spin-echo pulse sequence can be used to generate contrast based on relaxations times as well as density.

In Section 13.7.1.2, we mentioned that spins rapidly lose synchrony after a 90° pulse due to intrinsic effects as well as extrinsic effects, like main field inhomogeneity (Fig. 13.17). The latter, less anatomically relevant, effect is time independent and thus reversible, whereas the intrinsic effects are not. The spin echo exploits this difference to eliminate the extrinsic effects and generate a strong echo that markedly extends the timescale and the T_2 sensitivity of the MRI signal (Fig. 13.18).

We begin our explanation of the implementation and outcome of the spin echo with a commonly applied analogy. In the analogy, spin dephasing after the 90° pulse, which is due to different precession rates, is likened to a group of runners that fall out of synchrony after leaving the starting line due to different running speeds and (to a lesser extent) to different fatigue rates. Fast runners can be brought back into approximate synchrony with slow runners if the runners all reverse their direction of motion at the same time. After direction reversal, the faster runners are behind the slow, but they catch up and arrive back at the starting line together with the slower runners. Some lack of synchrony remains due to the different fatigue rates.

In the spirit of the racer analogy, the outcome of a spin echo is to bring dephasing spins back into synchrony at a time TE, known as the echo time. The spin echo consists of a standard 90° RF pulse followed, at a time $TE/2$ later, by a 180° inversion (refocusing) pulse (Fig. 13.18). As a consequence, nuclei reconverge and produce a transverse magnetization that again is large. At time TE, the signal (an echo) is measured.

The refocusing pulse reverses loss of nuclear synchrony that arises due to time-invariant field inhomogeneity but does not reverse loss of nuclear synchrony arising from time-varying tissue-specific T_2 processes, which are analogous to different fatigue rates for the runners. Thus, the echo is somewhat weaker than the signal generated by the first 90° pulse due to effects of T_2, and the envelope of a sequence consisting of repetitive 180° pulses follows a decay dictated by T_2 instead of T_2^* (Fig. 13.19). More complex pulse sequences that are the foundation for versatile, relaxation-weighted contrast are our next topic.

13.7.3 Pulse Sequences

Contrast generation in MRI is achieved using relatively intricate pulse sequences. Complex pulse sequences are used to create contrast based on tissue-specific variations in relaxation times and to generate spatial localization using frequency and phase encoding. We will discuss a dominant data acquisition sequence, which is a two-dimensional spin-echo

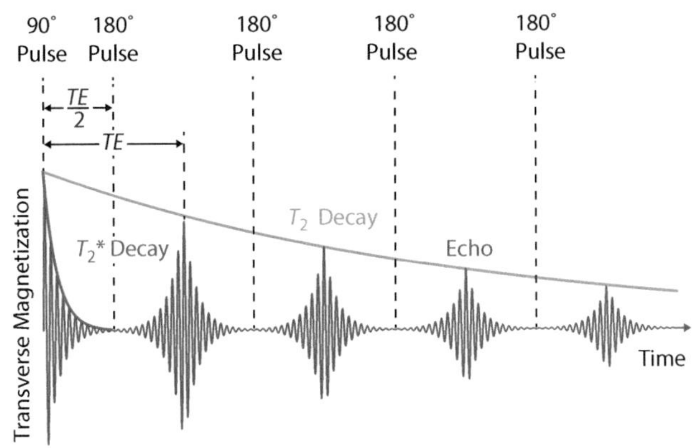

FIGURE 13.19
Echo amplitude in a multiple spin-echo sequence decays with a time course determined by T_2. Double spin-echo sequences were commonly used until relatively recently.

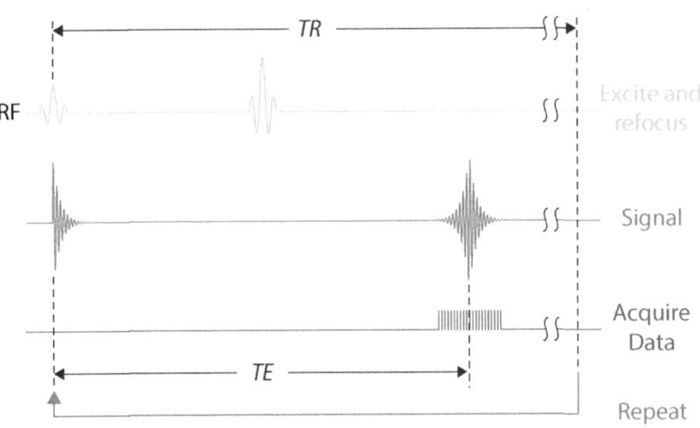

FIGURE 13.20
Simplified spin-echo pulse sequence, in the absence of the spatial encoding steps.

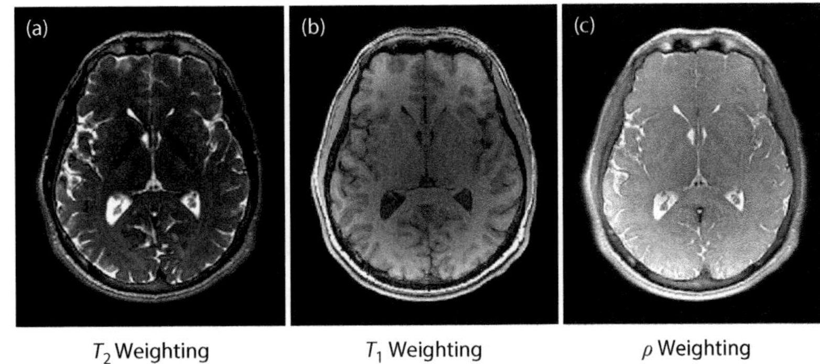

FIGURE 13.21
Comparison of (a) T_2, (b) T_1, and (c) proton density-weighted images of the brain. The bright region in the center of the T_2-weighted image is CSF. (Adapted with permission from Kopanoglu, E., et al. 2020. Simultaneous use of individual and joint regularization terms in compressive sensing: joint reconstruction of multi-channel multi-contrast MRI acquisitions. *NMR Biomed.* 33:4e4247. https://doi.org/10.1002/nbm.4247.)

approach. Readers interested in other sequences are referred to the texts by Bushberg et al. (2012) and Brown et al. (2014).

For the moment, we put aside discussion of the steps in spin-echo imaging that achieve localization and focus on the steps that generate contrast. This simplified sequence is shown schematically in Fig. 13.20. The sequence uses two RF pulses and two associated times – *TE* and *TR* – and can generate images with contrast that is weighted by ρ, T_2, or T_1 (Fig. 13.21).

The first step in the sequence is to use a 90° and a 180° RF pulse to create a spin echo at time *TE*, where data are sampled. After a repetition time, *TR*, has elapsed, a second cycle of 90° and 180° RF pulses is applied, and generally this cycle of pulses and sampling is repeated many times, at intervals *TR*.

The amplitude of the second, and subsequent, echoes, which is proportional to "$M_{xy}(TE)$," can be determined by mapping the time dependence of both the transverse and longitudinal magnetizations during the pulse cycle.[10] This involves taking into account the effect of the 90° pulse and of the 180° pulse. If it is assumed that $TR \gg T_2^*$ and that $TR \gg TE$, the result is (Box 13.4)

$$\text{Signal Amplitude}(x,y) \propto M_{xy}(TE) \propto \rho(x,y)\left(1-e^{-TR/T_1}\right)e^{-TE/T_2} \tag{13.17}$$

Generating relaxation-weighted contrast involves manipulating *TR* and *TE* to generate tissue-specific variation in the signal amplitude given by Eq. (13.17). Sample Problem 13.3 illustrates this idea in a very concrete context, and the reader is strongly encouraged to work through this sample problem before proceeding with the more general discussion in the main text.

[10] A more accurate notation, valid for the second pulse, would be $M_{xy}(TR + TE)$ because readout during the second pulse is at the time $TR + TE$.

SAMPLE PROBLEM 13.3 SPIN ECHO AMPLITUDES

The following table presents exemplary values of proton ρ, T_1, and T_2 for white matter and cerebrospinal fluid (CSF):

Tissue	Proton $\rho(g/cm^3)$	T_1 (ms)	T_2 (ms)
White matter	0.61	510	67
CSF	1.00	2650	280

(a) Calculate the ratio of the proton densities of white matter and CSF. (b) Calculate the spin-echo signal amplitudes for white matter and CSF, keeping track of each term so that their relative importance in determining signal disparities can be assessed. Assume $TR = 400$ ms and $TE = 5$ ms. (c) Which tissue will be brighter? (d) Which relaxation time is dominating the amplitude difference? (This question addresses the type of weighting that is being used – Table 13.3.)

SOLUTION

a. The ratio of proton densities for white matter and CSF are $\rho_{WM}/\rho_{CSF} = 0.61/1.00 = 0.61$.

b. The spin-echo signal amplitude is given by Eq. (13.17). The amplitudes for white matter and CSF can be calculated using values in the table:

$$SA_{WM} \propto \rho\left(1-e^{-TR/T_1}\right)e^{-TE/T_2} = 0.61\ \frac{g}{cm^3}\left(1-e^{-400\text{ ms}/510\text{ ms}}\right)\left(e^{-5\text{ ms}/67\text{ ms}}\right) = 0.61\times 0.544\times 0.928 = 0.308$$

$$SA_{CSF} \propto \rho\left(1-e^{-TR/T_1}\right)e^{-TE/T_2} = 1.00\ \frac{g}{cm^3}\left(1-e^{-400\text{ ms}/2650\text{ ms}}\right)\left(e^{-5\text{ ms}/280\text{ ms}}\right) = 1.00\times 0.140\times 0.982 = 0.138$$

Notice that the major determinant of the signal difference is the second term, which differs by nearly a factor of four. The other two terms are relatively similar. Thus, the second term is dictating the difference.

c. White matter will be brighter, because it has a larger amplitude.

d. The T_1 term is dominating the amplitude difference, so this is a T_1-weighted protocol.

Note: The white matter and CSF data analyzed in this sample problem correspond to the data plotted in Fig. 13.22.

We first analyze how to create T_2-weighted images. In this case, the goal is to minimize T_1-related differences between tissues. Examination of Eq. (13.17) reveals that this can be accomplished mathematically if the term $\left(1-e^{-TR/T_1}\right)$ is approximately unity, and this requires a long TR. More conceptually, a long TR will minimize the effect of T_1 because all tissues will have a largely recovered longitudinal magnetization when the next 90° pulse is applied (Fig. 13.22). If TE also is (relatively) long, ~60–150 ms (Table 13.3), tissues with a shorter T_2 will generate a weak echo because their transverse magnetization will largely have decayed. Examples are liver and muscle with $T_2 \sim 40-45$ ms and white and gray matter with $T_2 \sim 70-80$ ms. In contrast, tissues with a longer T_2 will generate a strong echo. An example is CSF with $T_2 \sim 280$ ms. Thus, T_2-weighting can be achieved with a long TR and a relatively long TE (Table 13.3). This method yields high-contrast images, with tissues like CSF appearing bright (Fig. 13.21).

In contrast, if the pulse sequence is implemented with a relatively short TR and a short TE, T_1-weighting is achieved. Mathematically, a short TE minimizes T_2 disparities in Eq. (13.17) because all tissues are described by $e^{-TE/T_2} \cong 1$. Conceptually, a short TE minimizes T_2 effects because all tissues similarly retain their transverse magnetization during echo

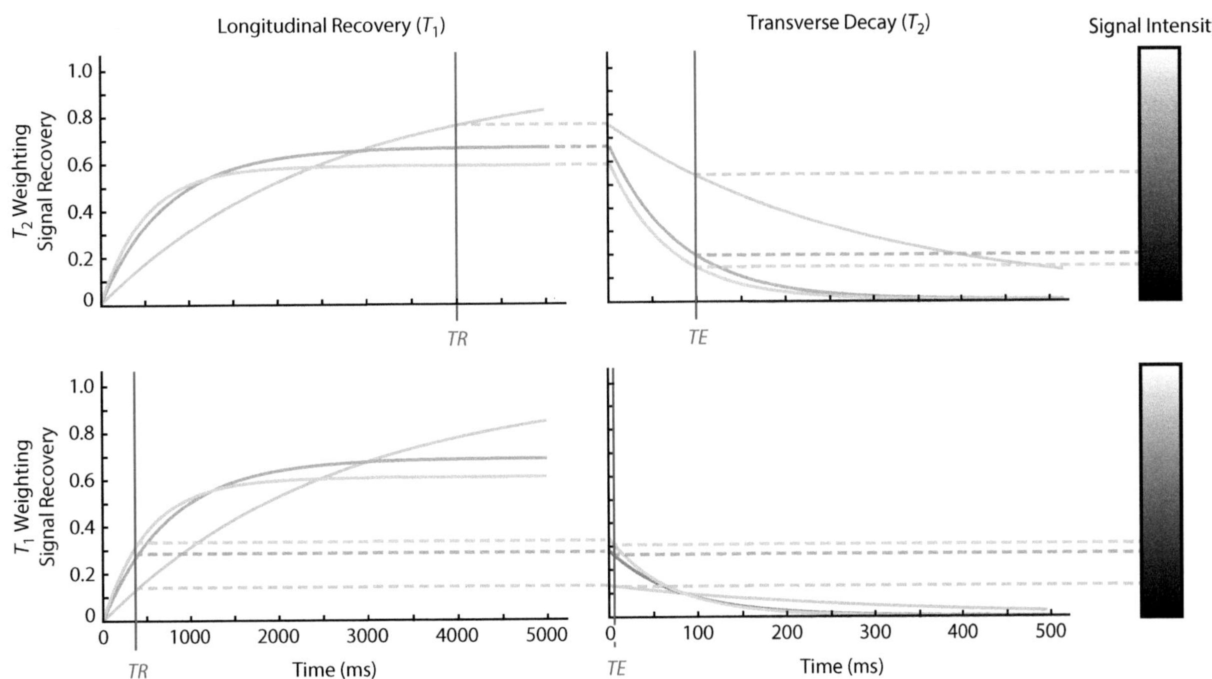

FIGURE 13.22
Graphical illustration of the generation of T_2-weighted and T_1-weighted contrast. The former is accomplished by selection of a long *TR* that de-emphasizes differences in longitudinal recovery together with a relatively long *TE* that emphasizes effects of transverse decay. The latter is accomplished by selection of a relatively short *TR* that emphasizes differences in longitudinal recovery together with a short *TE* that de-emphasizes effects of transverse decay. Curves were generated for white matter (green), gray matter (gray), and CSF (blue).

measurement (Fig. 13.22). In addition, a relatively short *TR*, ~400–600 ms, emphasizes T_1 differences during the return to equilibrium. For example, fat, with its short T_1 = 250 ms, will generate a large signal because its longitudinal magnetization will recover during *TR* and then subsequently generate large transverse magnetization. Muscle, with its intermediate T_1 = 900 ms, will generate an intermediate signal, and CSF with its long T_1 ~2600 ms, will generate a weak signal (Figs. 13.21 and 13.22). Finally, density-weighting can be achieved using a long *TR* and a short *TE*, which minimizes effects of both T_1 and T_2.

The important take-home message is that MRI signals can be made to reflect effects of ρ, T_1, or T_2 with judicious choices of *TR* and *TE*, as summarized in Table 13.3. In addition, tissues will be encoded with different relative degrees of brightness in images with different types of contrast weighting. This versatility is a major reason that MRI images have such incomparable soft tissue contrast.

TABLE 13.3
Time Delays (*TR*/*TE*) Used for Generating Different Forms of Contrast

Parameter	T_1 Contrast	Proton Density Contrast	T_2 Contrast
TR (ms)	400–600	2000–4000	2000–4000
TE (ms)	5–30	5–30	60–150

13.8 Summary of Two-Dimensional Spin-Echo Imaging

The complexity of MRI image acquisition justifies a summary description. In brief the steps are

1. The patient is placed in a strong, static, external magnetic field.
2. A pulse of RF radiation with narrow bandwidth is applied together with a slice-selection gradient. Protons in a small slice are excited, and transverse magnetization is generated.
3. A phase-encoding gradient is applied transiently. This introduces a phase difference between the protons along the *y* direction, which will generate data that fill in one "row" of the data matrix (Fig. 13.11).

BOX 13.4 DERIVATION OF MRI SPIN-ECHO SIGNAL AMPLITUDE

Here we present a succinct derivation of the spin-echo signal amplitude, which is given in Eq. (13.17). Additional detail can be found in Brown et al. (2014).

The signal amplitude, $\mathrm{SA}(x,y)$, is proportional to the transverse magnetization, M_{xy}, at readout (during the echo). Thus, our task is to determine the temporal evolution of the transverse magnetization during the pulse sequence and then to evaluate M_{xy} at $t = TE$. We also need to determine the temporal evolution of the longitudinal magnetization because, at the start of each cycle, M_z gets tipped by the 90° RF pulse and is converted into M_{xy} (Fig. 13.6).

Before presenting the full argument, which is somewhat subtle, we illustrate the ideas by analyzing the first cycle and via a semi-rigorous analysis of follow-up cycles.

The first cycle involves application of a 90° RF pulse to the equilibrium configuration ($M_z = M_0$ and $M_{xy} = 0$). At the end of this first pulse, the magnetization components are

$$M_z = 0 \tag{B13.4.1a}$$

$$M_{xy} = M_0 \tag{B13.4.1b}$$

The effect of the 180° refocusing pulse is to cause M_{xy} at time TE (the echo height) to differ exponentially from the "initial value," M_0, by only the effect of T_2 relaxation. Thus, for the first cycle (Eq. [13.15] and Fig. 13.19)

$$M_{xy}(TE) = M_0 e^{-TE/T_2} \tag{B13.4.2}$$

Equation (B13.4.2) generally does not hold for subsequent cycles because the longitudinal magnetization does not fully recover unless the repetition time $TR \gtrsim 5T_1$. Thus, the transverse magnetization created by 90° RF pulses in cycles two and onward generally is less than that given in Eq. (B13.4.1b), and so too is the signal at the associated echo time.

The key issue for cycles two and onward is determining the value of M_z at the end of the previous cycle. This value for M_z will (like M_0 for the first cycle) get converted into transverse magnetization and then evolve at the echo by analogy with Eq. (B13.4.2). The take-home message from the rigorous analysis below is that the longitudinal magnetization at the end of all cycles is the same as it is at the end of the first cycle, as long as two simplifying assumptions hold. These are that $TR \gg T_2^*$ and $TR \gg TE$, as stated in the main text. In this limit (Eq. [13.16])

$$M_z(TR^-) = M_0\left(1 - e^{-TR/T_1}\right) \tag{B13.4.3}$$

where the symbol TR^- is notation for a time just before TR *for all cycles*. Thus, the signal amplitude (SA) for cycles two and onward is given by

$$\mathrm{SA}(x,y) \propto M_{xy}(TE) \propto M_0(x,y)\left(1 - e^{-TR/T_1}\right)e^{-TE/T_2} \tag{B13.4.4}$$

where $M_0(x,y) \propto \rho(x,y)$.

To make the analysis more rigorous, we need to address the discontinuous change in M_z induced by the 180° RF refocusing pulse at $\tau_{180} = TE/2$. Specifically, the 180° pulse causes twice the rotation of a 90° pulse and thus inverts M_z. To model this effect requires a generalized version of Eq. (13.16) that reads (HW 13.16)

$$M_z(t) = M_z(t_0)e^{-(t-t_0)/T_1} + M_0\left(1 - e^{-(t-t_0)/T_1}\right) \tag{B13.4.5}$$

where t_0 is an arbitrary starting time, and M_0 is the equilibrium magnetization. In light of Eq. (B13.4.5), the evolution of M_z before and after the first 180° pulse is given by the expressions (Eq. [B13.4.1a])

$$M_z(t) = \begin{cases} M_0\left(1-e^{-t/T_1}\right) & 0<t<\tau_{180} \\ M_z\left(\tau_{180}^{+}\right)e^{-(t-\tau_{180})/T_1} + M_0\left(1-e^{-(t-\tau_{180})/T_1}\right) & \tau_{180}<t<TR \end{cases} \quad \text{(B13.4.6)}$$

The symbol τ_{180}^{+} is notation for a time just after $\tau_{180} = TE/2$. Invoking the inverting effect of the 180° pulse, we have

$$M_z\left(\tau_{180}^{+}\right) = -M_0\left(1-e^{-\tau_{180}/T_1}\right) \quad \text{(B13.4.7)}$$

Inserting Eq. (B13.4.7) into Eq. (B13.4.6) yields

$$M_z(t) = M_0\left(1-2e^{-(t-\tau_{180})/T_1} + e^{-t/T_1}\right) \quad \tau_{180}<t<TR \quad \text{(B13.4.8)}$$

At this point, we invoke the simplifying assumption $TR \gg \tau_{180}$ to obtain

$$M_z\left(TR^{-}\right) = M_0\left(1-2e^{-(TR-\tau_{180})/T_1} + e^{-TR/T_1}\right) \cong M_0\left(1-e^{-TR/T_1}\right) \quad TR \gg \tau_{180} \quad \text{(B13.4.9)}$$

As stated above, Eq. (B13.4.9) also is valid for cycle two and onward if $TR \gg T_2^*$. To see why this is true, note that Eqs. (B13.4.5)–(B13.4.9) will be valid for a cycle in the sequence as long as the longitudinal magnetization is zero at the end of the 90° pulse for that cycle (as it is for the first cycle – see Eq. [B13.4.1a]). This requirement is met for all cycles if $TR \gg T_2^*$ because, in this limit, all transverse magnetization has decayed before application of each 90° RF pulse. Thus, all 90° pulses in the sequence start, like the first, with no transverse magnetization to get tipped into the z direction. As a result, all 90° pulses in the sequence end, like the first, with no longitudinal magnetization.

4. A rephasing (echo) pulse is applied at time $TE/2$, together with the slice-selection gradient, to regenerate proton synchrony and enhance the MRI signal.
5. A frequency-encoding gradient is applied along x in conjunction with data acquisition during echo evolution and decay, which occurs at time TE.
6. Data are collected for multiple repetition times, TR. For each TR, the phase-encoding gradient is altered to yield new phase data and thereby fill out the data matrix.
7. The image is computed via Fourier transformation of the MRI signal, with the choice of TR and TE determining the signal amplitude and thus image contrast.
8. Steps 2–7 can be repeated for additional slices.

These steps also are depicted visually in a full spin-echo pulse sequence diagram, which is an augmented version of the diagram in Fig. 13.20 (Fig. 13.23).

13.9 K-Space and the MRI Imaging Equation [Optional]

This section contains a very brief introduction to the rigorous mathematical foundations of image production in two-dimensional MRI. The discussion is heavily based in Fourier analysis and can be omitted if desired. Most of the ideas that follow already have been discussed at a more conceptual level in previous sections.

The goal of two-dimensional MRI is to convert the time-dependent voltage induced in the receiver coil

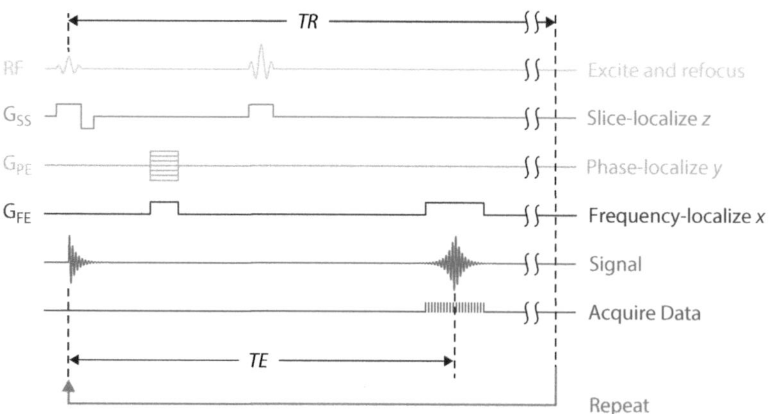

FIGURE 13.23
Exemplary full spin-echo pulse sequence, including the spatial encoding steps. One phase is collected for each echo, and the phase is incremented for different repetitions by changing the gradient. This is depicted in the schematic using peaks of different height along the phase-encoding axis. The half-area, reverse-polarity gradient that follows the 90° RF pulse serves to reverse de-phasing that occurs during the pulse. This sequence really is a combined gradient echo/spin echo, but nevertheless typically is called a spin echo.

into the in-plane distribution of protons, $\rho(x,y)$, in the selected slice. A rigorous treatment of this signal detection and analysis problem requires starting with Faraday's Law of Induction and is beyond the scope of this text. Fortunately, the final result assumes a relatively simple form, which reads

$$S(t;\tau_{PE}) \propto \iint \rho(x,y) e^{-i(\gamma G_x x t + \gamma G_y y \tau_{PE})} dxdy \quad (13.18)$$

In Eq. (13.18), $S(t;\tau_{PE})$ is the detected signal at time t for a given phase-encoding duration, τ_{PE}, and G_x and G_y are gradient strengths. For simplicity, we have suppressed the position-invariant contrast terms. The reader interested in a full treatment is referred to the text by Brown et al. (2014).

To rewrite Eq. (13.18) in a form that is more suggestive of a spatial FT, we make a change of notation that involves defining two spatial frequencies as

$$k_x = \bar{\gamma} G_x t \quad (13.19a)$$

$$k_y = \bar{\gamma} G_y \tau_{PE} \quad (13.19b)$$

With this notational change, Eq. (13.18) becomes

$$S(k_x,k_y) \propto \iint \rho(x,y) e^{-2\pi i(k_x x + k_y y)} dxdy \quad (13.20)$$

Two attributes of Eq. (13.20) are of particular importance. First, the MRI signal can be regarded as a function of the two spatial frequencies (k_x,k_y) instead of as a function of (t,τ_{PE}). Thus, signal readout can be viewed as generating data that fill in a matrix in the spatial frequency domain (K-Space). Second, the MRI signal and the density function constitute an FT pair. Thus, once the K-Space matrix is filled, the density function can be computed by inverse Fourier transformation of the MRI signal as follows:

$$\rho(x,y) \propto FT^{-1}\{S(k_x,k_y)\} = \iint S(k_x,k_y) e^{2\pi i(k_x x + k_y y)} dk_x dk_y \quad (13.21)$$

Equation (13.21) constitutes the rigorous basis for statements made earlier, notably that the image is generated via Fourier transformation of the MRI signal (HW 13.19).

Examination of Eqs. (13.19a) and (13.19b) shows that signal readout as a function of t for a fixed τ_{PE} generates a complete set of k_x data but only a single value of k_y. To circumvent this problem, phase-encoding is repeated multiple times with different values of G_y, and a complete set of k_y data are generated that fill out the K-Space matrix.

Representative K-Space data, and the associated image obtained by two-dimensional Fourier transformation, are shown in Fig. 13.24a. Also shown in the figure are results obtained when the K-Space data are subjected to the process of spatial filtering, which has arisen several times in the text (Figs. 4.9, B4.3.3, and 9.25). If peripheral K-Space data are removed via spatial filtering and the central raw data are used for the reconstruction, the image contains the large-scale features in the object, but it is blurry and lacks detail. Conversely, if central data are removed and only peripheral data used, the image emphasizes edges and fine detail. Experiments of this type are easily performed using Fiji™ (Activity A13.2).

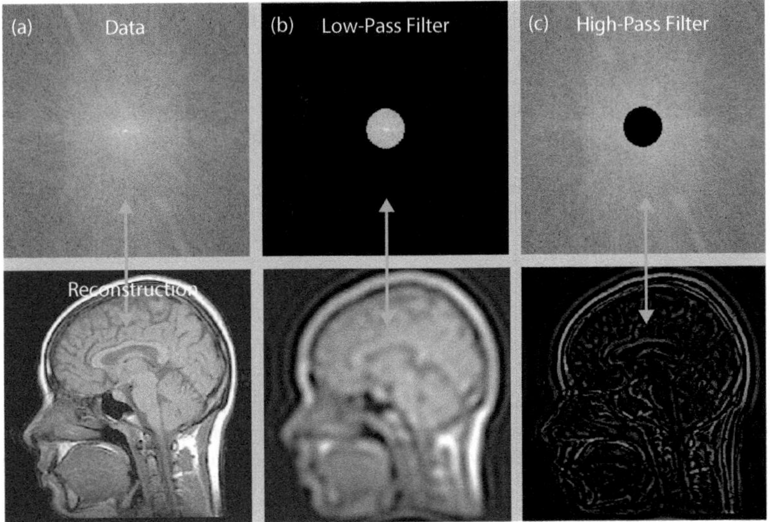

FIGURE 13.24
Exemplary MRI signal data and effects of selective use of data on the reconstructed image. The signal data (shown as a magnitude representation) and the associated reconstruction are shown in Panel (a).[11] Use of only central data generates a low-resolution image (b), and use of only peripheral data generates a fine-detail-biased image (c). Note that there is no one-to-one correspondence between raw and reconstructed data locations. The image in Panel (c) was gamma scaled (Section 9.4.1.1.3) to improve visibility. (Full MRI image from National Aeronautics and Space Administration (NASA); remaining images by authors.)

13.10 Applications

We close with a tabular summary of some key medical applications of MRI to augment, and reiterate, the applications mentioned throughout this chapter. Some applications overlap with those of CT and US, and thus the choice of MRI over these alternatives often depends on the goal of the examination, as well as issues such as cost and speed. Different pulse sequences and weightings can be used to provide additional forms of contrast, leading to techniques such as diffusion-weighted MRI, perfusion-weighted MRI, (magnetic) susceptibility-weighted MRI, phase-contrast MRI, and functional MRI, among others. In addition, contrast agents, such as gadolinium chelates, may be used to alter relaxation rates and therefore contrast in T_1- or T_2-weighted images.

Medical Applications of MRI

Select Applications	**Select Aspects of Implementation**
Detection of brain and nervous system abnormalities, including tumors, strokes, epilepsy, and dementia	***Gadolinium*** may be used to enhance contrast. For example, gadolinium can accumulate in a brain tumor after intravenous injection and passage through a tumor-associated leak in the blood-brain barrier. Gadolinium will shorten the longitudinal relaxation time and cause the tumor to appear bright on a T_1-weighted image. ***T_2-weighted images*** can be used to detect increased water content that is associated with many brain abnormalities. ***Diffusion-weighted*** MRI may be used to detect stroke.
MRI angiography – visualization of the vascular system to detect, diagnose, and aid treatment of stroke and diseases of the heart and blood vessels	***Contrast-enhanced angiography*** involves intravenous injection of a "pulse" of gadolinium-based contrast agent. Vessels exhibit a concentration-dependent increase in brightness on T_1-weighted images. Multiple images can be acquired to visualize passage of the pulse through the vascular system. ***Time-of-flight (TOF) angiography*** relies on entry of "unexcited" protons into the imaging slice due to blood influx from an orthogonal direction. The unexcited protons have full longitudinal magnetization and thus appear effectively "relaxed." As a result, T_1 effectively is reduced. TOF MRI does not require a contrast agent, which is a strength relative to CT, especially for patients that are allergic to contrast agents and/or have poor kidney function.

(Continued)

[11] Although not stressed in this chapter, raw MRI data are, in general, complex numbers, characterized by both a magnitude and a phase. For simplicity, only the magnitude is typically displayed.

Medical Applications of MRI

Select Applications	Select Aspects of Implementation
MRI-guided focused ultrasound to destroy tissue	Uses MRI to image an area of interest while US is focused on, and used to destroy, a region of concern (e.g., a tumor) within that area. The focused US is different from that used in imaging and typically is applied continuously rather than in pulses. The US heats a small region at the center of focus, while leaving surrounding areas unaffected, much like a magnifying glass focusing sunlight to a hotspot. MRI often is used for imaging instead of US because of its greater sensitivity.
Diagnosis of soft tissue sports-related injuries, including tendinitis, bursitis, meniscal tears, ligamentous injuries, and joint effusions	Relies on MRI's sensitivity to subtle changes in the water content of tissue, its ability to probe different planes, and its exceptional anatomic detail and contrast.
Organ imaging in two and three dimensions, including visualizing organ shape	May involve the use of a contrast agent. An example of a liver-specific contrast agent is the manganese-based compound, mangafodipir trisodium, which undergoes specific uptake due to chemical similarity to Vitamin B_6. May also involve oblique-plane imaging.
Real-time imaging, including visualizing the beating heart, monitoring minimally invasive surgical procedures, studying joint kinetics, and analyzing speech	Involves the use of fast techniques. These include iterative reconstruction to bypass time-consuming acquisition of K-Space data and gradient echo pulse sequences that involve small flip angles and a short *TR*.
Functional MRI – mapping brain regions associated with critical functions, including speaking, moving, and vision	Detects differences in signal intensity based on blood oxygenation level dependent (BOLD) contrast. BOLD contrast arises from two effects. One is the paramagnetism of deoxyhemoglobin, which increases field disturbances and shortens transverse relaxation times. The other is enhanced blood flow in regions experiencing neural activation, which leads to a small reduction in levels of deoxyhemoglobin in blood and to an associated increase in transverse relaxation times. BOLD contrast is generated using T_2/T_2^*-weighted images in which neurally active regions appear brighter.
Planning surgical procedures on the brain	May involve the use of functional MRI to map a patient's functional anatomy and thereby plan safe and effective neurosurgery.

Suggested Reading and Additional Resources

Abdulla, S. and Clarke, C. 2011 – 2022. Radiology Café: FRCR lecture notes: MR imaging. https://www.radiologycafe.com/radiology-trainees/frcr-physics-notes/mr-imaging

Brown, R.W., Cheng, Y.-C.N., Haacke, E.M., Thompson, M.R., and Venkatesan, R. 2014. *Magnetic resonance imaging: physical principles and sequence design (2nd Edition)*. Wiley-Blackwell, Hoboken, NJ.

Bushberg, J.T., Seibert, J.A., Leidholdt, E.M., and Boone, J.M. 2012. *The essential physics of medical imaging (3rd Edition)*. Wolters Kluwer | Lippincott Williams & Wilkins, Philadelphia, PA. Chapters 12 and 13.

Callaghan, P.T. 1991. *Principles of nuclear magnetic resonance microscopy*. Clarendon Press, Oxford, England.

Cooley, C.Z., Stockmann, J.P., Witzel, T., LaPierre, C., Mareyam, A., Jia, F., Zaitsev, M., Wenhui, Y., Zheng, W., Stang, P., Scott, G., Adalsteinsson, E., White, J.K., and Wald, L.L. 2020. Design and implementation of a low-cost, tabletop MRI scanner for education and research prototyping. *J. Magn. Reson*. 310:1–10.

Elster, A.D. Questions and answers in MRI. http://mriquestions.com/complete-list-of-questions.html

Kane, S.A. and Gelman, B.A. 2020. *Introduction to physics in modern medicine (3rd Edition)*. CRC Press, Boca Raton, FL. Chapter 8.

McRobbie, D.W., Moore, E.A., Graves, M.J., and Prince, M.R. 2017. *MRI from picture to proton (3rd Edition)*. Cambridge University Press, Cambridge, England.

Sprawls, P. 2000. *Magnetic resonance imaging: principles, methods, and techniques*. Medical Physics Publishing, Madison, WI. (ISBN 0-944838-97-9).

HOMEWORK PROBLEMS

13.1 *[Classical Relationship between Magnetic Moment and Angular Momentum] Equation (13.1) is a broadly valid classical relationship between angular momentum and the magnetic moment. Prove that the equation is valid for the case of a charged particle undergoing circular motion. The proof for a spinning charge is a little more complex because it involves integration. Hint: Recall that $\vec{L} = \vec{r} \times \vec{p}$ and that $\vec{\mu} = IA\hat{n}$, where I is current,

A is area, and $\hat{n}$ is a unit vector perpendicular to the plane of the circle.

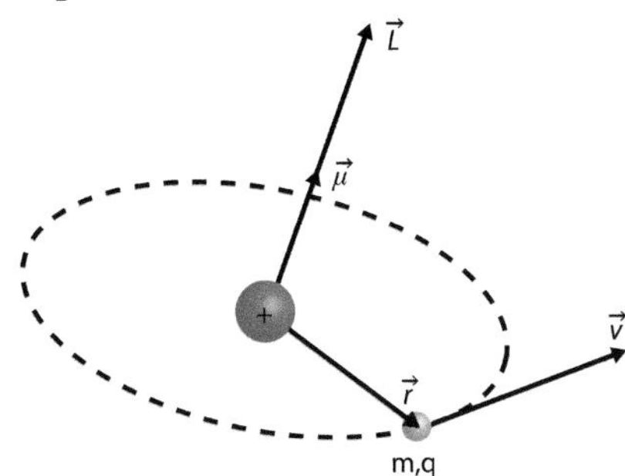

13.2 [MRI Energies] A scientist wants to perform an MRI study using a state-of-the-art 8.0-T research MRI instrument.

a. What are the resonance frequencies for ^1H and ^{31}P with this system? Hint: See Table 13.1.

b. What are the corresponding photon energies (in eV)? Compare these energies to typical chemical bond energies, such as those in the body, and typical diagnostic X-ray and gamma-ray energies, such as those encountered in Chapters 11 and 12. What are the medical implications of any differences?

c. Also compare the photon energies to typical thermal energies, $k_B T$. What are the implications of this result for the equilibrium population bias? Assume $T = 300\ K$.

13.3 [Gyromagnetic Ratio for Hydrogen] The gyromagnetic ratio of a proton is given by $\gamma = \frac{+e}{2m} g$, where $g = 5.586$.

a. Show that $\gamma = 2.675 \times 10^8$ rad/(s · T).

b. The result in part (a) can be used to compute an angular Larmor frequency, ω_L (Eq. [13.11]). More commonly, the gyromagnetic ratio is quoted in a form, "gamma bar," that is appropriate for computation of the Larmor frequency ν_L in Hz. Show that $\bar{\gamma} = 42.58$ MHz/T, where the bar is meant to serve as a reminder that this value will yield a resonance frequency in Hz.

13.4 [Slice Selection] Explain, using words and a drawing, how the magnetic field slice-selection gradient should be oriented to image coronal and sagittal cross-sections.

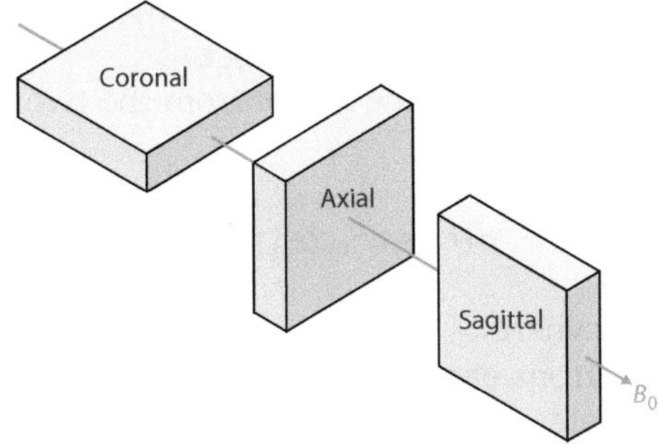

13.5 [Characterization of Field Gradient 1] The magnetic field in a 1.50-T MRI scanner varies by 10 mT/m along the slice-section gradient direction. The resonance frequency at $z = 0$ (isocenter) is 63.8700 MHz.

a. Assume that a range of frequencies 63.9000 ± 0.0002 MHz is used to excite protons in the patient. What is the thickness of the excited slice?

b. Determine the z location of the slice.

c. Assume that the main field varies by 20 parts per million over a standard MRI FOV (150 mm). How much variation in the Larmor frequency does this field inhomogeneity create?

13.6 [Characterization of Field Gradient 2] A 3.00-T MRI instrument has a 150-mm FOV. The instrument is set to generate a 6 mT change in the field that varies linearly across the FOV. The null (zero) point of the gradient is in the middle of the FOV.

a. Calculate the strength of the field gradient in Tesla per mm and in Hz per mm.

b. Calculate the minimum, null, and maximum magnetic field strengths. What are the associated Larmor frequencies?

c. What is the net frequency range across the FOV? By what percentage does the frequency vary across the FOV?

13.7 [Phase Encoding] Assume that two spins are separated by 10 mm and that a 10 mT/m gradient is switched on for 5 ms. What phase difference does the gradient create between the spins?

13.8 *[Effect of K-Space Sampling on FOV] Prove an important link between sampling in K-Space and FOV, which reads for (the x direction)

$$FOV_x = \frac{1}{\Delta k_x}$$

13.9 [MRI Resolution] The spatial resolution of most imaging techniques is wavelength limited. For example, the resolution of standard optical microscopy is determined by the wavelength of visible light. Similarly, the resolution of US imaging is limited by the wavelength of the US. However, these considerations do not apply to MRI. What wavelengths of radiation are used for MRI? Explain why the spatial resolution of MRI is much smaller than the wavelength.

13.10 *[Derivation of the Precession Frequency] Use the notation and the geometry in Fig. 13.14, in conjunction with Eqs. (13.7) and (13.8), to show that a magnetic moment in a field of strength B_0 will precess at a frequency

$$\omega_0 = \frac{d\varphi}{dt} = -\gamma B_0$$

The minus sign serves as reminder that the precession is clockwise (when $\gamma > 0$), whereas counterclockwise rotations typically are taken to be positive.

13.11 [Spin Orientation in an External Field] Prove the result in the legend to Fig. 13.14, which states that $\vec{\mu}$ makes an angle $\theta = 54.7°$ with respect to the z axis. Hint: The allowed z components of $\vec{\mu}$ are given in Eq. (13.3). The magnitude of $\vec{\mu}$ is given by $\mu = \sqrt{3}/2\gamma\hbar$.

13.12 [Attributes of 90° RF Pulses] Assume that the amplitude of an RF pulse is $B_1 = 10\ \mu T$ (microTesla). For this case, what is the duration of a "90°" tipping pulse?

13.13 [Contrast – Relaxation Times] Why does MRI base contrast on differences in relaxation times, such as T_1 and T_2, instead of on differences in proton density?

13.14 [Spontaneous Emission in MRI] Energy loss in MRI occurs predominantly via stimulated emission that is driven by coupling of the spins to the environment. To get a feel for the importance of stimulated emission, calculate the probability of spontaneous emission, which is given by

$$P = \frac{\mu_0 \omega_0^3 \gamma^2 \hbar}{6\pi c^3}$$

where $\mu_0 = 1.26 \times 10^{-6}\ \text{T} \cdot \text{m/A}$ and c is the speed of light. Assume that the main field strength is 1.0 T. Comment on the significance of your answer.

13.15 *[The Bloch Equations] The Bloch equations are discussed in Box 13.3.

a. Demonstrate by direct substitution that the solutions given in Eq. (B13.3.6) satisfy the Bloch equations, Eqs. (B13.3.4a)–(B13.3.4c), during the RF pulse and in the absence of the relaxation terms. You also should verify that the postulated solution satisfies the appropriate initial conditions.

b. Demonstrate by direct substitution that the solutions given below satisfy the Bloch equations immediately after a 90° pulse that tilts the magnetization onto the y axis. Hint: The RF terms are not present after the pulse. You also should verify that the postulated solution satisfies the appropriate initial conditions immediately after the pulse.

$$M_x(t) = M_0 \sin(\omega_0 t) e^{-t/T_2}$$
$$M_y(t) = M_0 \cos(\omega_0 t) e^{-t/T_2}$$
$$M_z(t) = M_0 \left[1 - e^{-t/T_1}\right]$$

13.16 *[The Bloch Equations] Derive Eq. (B13.4.5) in Box 13.4.

13.17 [Contrast – Spin Echo] Equation (13.17) describes how the spin-echo amplitude depends on experimental parameters (TR and TE) and tissue parameters (T_1 and T_2). Assume, for a given experiment, that TR and TE are 2000 ms and 70 ms, respectively.

a. Calculate the spin-echo amplitude for arterial blood and white matter at 3 T, using values from Table 13.2.
b. Which component will be bright and which will be dark?
c. What kind of weighting is being used (Table 13.3)? Justify your answer!

13.18 *[Contrast – Spin-Echo Optimal TE] The absolute contrast between two tissues A and B generating signals S_A and S_B is defined as

$$C_{ABS} = |S_A - S_B|$$

Assume that the two tissues have similar longitudinal relaxation times and densities and that their signals are given by Eq. (13.17). Show that the absolute contrast between the tissues is maximal for TE given by

$$TE = \frac{\ln\left(\frac{T_{2B}}{T_{2A}}\right) T_{2A} T_{2B}}{T_{2B} - T_{2A}}$$

13.19 *[Fourier Analysis and MRI] Consider two spins located at positions $\pm z_0$ that reside in a linear field gradient along z.

a. Show that the signal from the two spins is given by

$$S(t) = 2S_0 \cos(\gamma G_z t z_0)$$

where S_0 is the magnitude of the signal from each spin. Hint: Each spin generates a signal of the form $e^{i\theta}$.

b. Compute the Fourier transform of the signal and show that it yields the density:

$$FT^{-1}\left(S(t)\right) = \rho(z) = S_0\left(\delta(z - z_0) + \delta(z + z_0)\right)$$

where δ is the Dirac delta function. This simple example confirms the general concepts developed in the main text.

ACTIVITIES[12]

A13.1 Simulating MRI Fourier Data

This activity provides an opportunity to explore frequency encoding and Fourier decomposition of a time-dependent signal using the program Microsoft Excel®. It may be necessary to load the analysis ToolPak (e.g., following the directions at https://support.office.com/en-us/article/load-the-analysis-toolpak-in-excel-6a63e598-cd6d-42e3-9317-6b40ba1a66b4). The instructions provided below should be approximately correct for current versions of Excel.

We have provided a sample signal (shown below), which consists of the superposition of sine terms with frequencies of 4, 12, and 20 Hz and relative amplitudes of 2, 5, and 3, respectively.

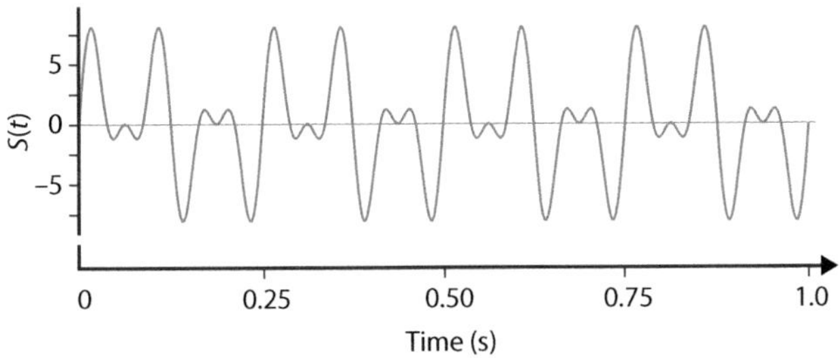

A. Open the Excel file A13.1.xlsx.

B. Compute the FT of the signal (column B) by going to Tools > Data Analysis > Fourier Analysis and then highlighting all 1024 elements in column B.

C. Give Excel a 1024-column destination (e.g., column E labeled FT) for the (complex-valued) Fourier amplitudes.

D. Compute the relative "size" of the complex Fourier amplitudes for frequencies ranging from 0 to 30 Hz by computing their moduli. This can be accomplished using the function "IMABS" and Edit > Fill > Down.

E. Generate a graph of the modulus versus frequency from 0 to 30 Hz. Did the FT accurately determine the frequencies and weightings of the sine waves that were used to generate the signal?

F. In MRI, field gradients create a one-to-one correspondence between frequency and position, and thus the FT in essence is an image. To illustrate the idea, assume, for simplicity, that frequency in Hz is mapped into a corresponding x position where $x = \nu/10$. Also assume that the amplitude of the FT is proportional to the number of spins at that location. Use Excel to generate the resulting one-dimensional image.

G. If desired, use Excel to superpose three sine waves with the specified frequencies and amplitudes and verify that this superposition generates the signal shown in the figure.

A13.2 Spatial Filtering of MRI Data

This activity explores Fourier filtering. It is similar in spirit to the Fourier filtering component of Activity A9.2. You will use the open source Fiji™ (ImageJ™) image processing package. Download instructions are provided in Activity A6.1.

A. Open A13.2.tif, which is the proton density-weighted image in Fig. 13.21.

B. Compute the FFT of the image (Process > FFT > FFT).

C. Scale the background of the FFT image so that it is completely black. To do this, click the "Color Picker" icon on the Fiji toolbar. A color palette should appear, with rectangular foreground (F) and background (B) icons at the bottom. Click on the background icon, which will bring up a window for Background Color. Set R = G = B = 0.

D. Filter the FFT image using a circular mask. First, choose Process > FFT > Make Circular Selection. Next, choose a radius of 30 pixels. Finally, remove the frequency components in the circle using Edit > Cut (the circle should go black).

E. Reconstruct the image from the filtered transform using Process > FFT > Inverse FFT. Save both the filtered image and the filter (flatten the filter first) and explain the change in appearance of the image in terms of the nature of the filter.

F. Repeat steps (B) to (E) but instead set the background to white, which will retain only frequencies in the circular section.

[12] Activity files are available for download in the online Support Materials.

A13.3 Benchtop MRI

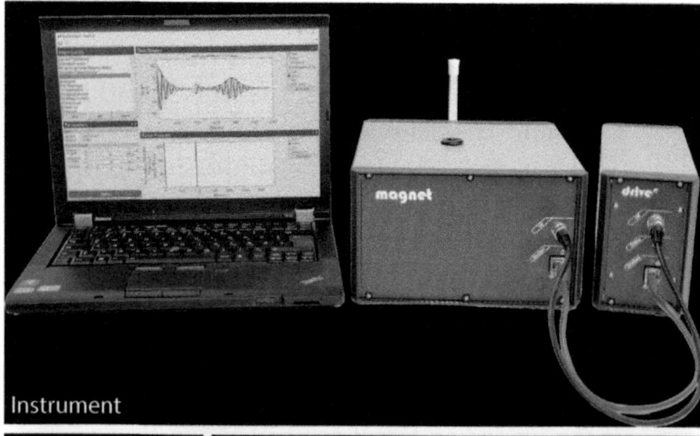

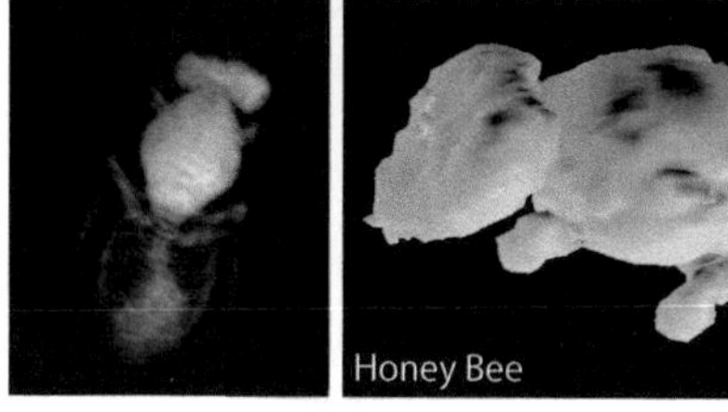

The concepts presented in this chapter can be implemented and reinforced via a variety of accessible experimental activities. Here we alert readers to

1. Educationally oriented devices that are capable of performing real MRI, and closely related nuclear magnetic resonance (NMR), experiments.
2. Possible experiments and activities in the absence of detail.

One commercial option, pictured here[13] and used at Lewis & Clark College, is the portable*Lab* benchtop MRI training system developed by Pure Devices GmbH (Rimpar, Germany). Two other commercial options are the EduMR desktop NMR and MRI system, developed by Niumag Corporation (Suzhou, China), and the Terranova-MRI system, developed by Magritek Ltd (Wellington, New Zealand); the latter is demonstrated in some educationally oriented videos featuring the renowned physicist Sir Paul Callaghan at https://magritek.com/resources/videos/. There also are several noncommercial options that are less expensive but nontrivial to construct. A comprehensive source is the publication by Cooley et al. (2020).

These instruments allow students to conduct a broad spectrum of enlightening and engaging experiments that are remarkably similar to those conducted with full-sized clinical instruments (differing primarily by sample size). Examples include

1. Measurement of the Larmor frequency.
2. Measurement of relaxation times.
3. Creation and manipulation of spin echoes.
4. Imaging with different types of weighting (e.g., ρ, T_1, and T_2).

[13] Images courtesy of Stefan Wintzheimer, Pure Devices GmbH.

Appendices

Appendix A: List of Abbreviations

This appendix lists abbreviations used in the text, excluding those used only once or twice in the immediate vicinity of their definition.

Abbreviation	Description
ACF	Attenuation correction factor
ADC	Analog-to-digital converter
ADU	Analog-to-digital unit
AHE	Adaptive histogram equalization
A-mode	Amplitude mode (US)
BFP	Back focal plane
B-mode	Brightness mode (US)
C	Contrast
CCD	Charge-coupled device
CEUS	Contrast-enhanced ultrasound
CLAHE	Contrast-limited adaptive histogram equalization
CMOS	Complementary metal-oxide semiconductor
CNR	Contrast-to-noise ratio
CSF	Cerebrospinal fluid
CT	Computed tomography
D	Diopter
dB	Decibel
DCV	Dense-core vesicle
DF	Darkfield
DIC	Differential interference contrast
DR	Dynamic range
ECT	Emission computed tomography
EGFP	Enhanced green fluorescent protein
EM	Electromagnetic (e.g., EM radiation)
EMCCD	Electron-multiplying charge-coupled device
EUS	Endoscopic ultrasound
FAIM	Fluorescence anisotropy imaging microscopy
FBP	Filtered back projection
FCS	Fluorescence correlation spectroscopy
FDG	Fluorodeoxyglucose
FF	Flat field
FFP	Front focal plane
FFT	Fast Fourier transform
FLIM	Fluorescence lifetime imaging microscopy
FOV	Field of view
FPD	Flat panel detector
FRAP	Fluorescence recovery after photobleaching
FRET	Fluorescence resonance energy transfer
FWHM	Full width at half maximum
FT	Fourier transform
GFP	Green fluorescent protein

Abbreviation	Description
GM	Geometric mean
HSI	Hue saturation intensity
HVL	Half-value layer
HVT	Half-value thickness
HU	Hounsfield units
HW	Homework
IR	Infrared (e.g., IR radiation)
kVp	Peak kilovoltage
LCE	Local contrast enhancement
LED	Light emitting diode
LOR	Line of response
LSCM	Laser scanning confocal microscopy
LSFM	Light sheet fluorescence microscopy
LUT	Look-up table
M	Magnification
MAA	Macroaggregated albumin
M-mode	Motion mode (ultrasound)
MOS	Metal-oxide-semiconductor
MRI	Magnetic resonance imaging
MSCT	Multi-slice CT
MTF	Modulation transfer function
NA	Numerical aperture
ND	Neutral density
OD	Optical density
OPD	Optical path difference
OPL	Optical path length
OR-PAM	Optical-resolution photoacoustic microscopy
PALM	Photoactivated localization microscopy
PAM	Photoacoustic microscopy
PE	Photoelectric effect
PET	Positron emission tomography
PHA	Pulse-height analyzer
PMT	Photomultiplier tube
PRF	Pulse repetition frequency (US)
PRP	Pulse repetition period (US)
PSF	Point spread function
RBC	Red blood cell
RF	Radiofrequency
RGB	Red green blue
ROI	Region of interest
SBP	Simple back projection

Abbreviation	Description
sCMOS	Scientific complementary metal-oxide semiconductor
SDCM	Spinning disk confocal microscopy
SI	Structured illumination
SIM	Structured illumination microscopy
SNR	Signal-to-noise ratio
SPECT	Single-photon emission computed tomography
SPIM	Selective plane illumination microscopy
SPL	Spatial pulse length
SPT	Single-particle tracking
SR-SIM	Super-resolution structured illumination
SSCT	Single-slice CT
STED	Stimulated emission depletion
STORM	Stochastic optical reconstruction microscopy
TE	Echo time (MRI)
TEE	Transesophageal echocardiogram
TGC	Time gain compensation
TFT	Thin film transistor
TIR	Total internal reflection
TIRF	Total internal reflection fluorescence
TIRFM	Total internal reflection fluorescence microscopy
TOF	Time of flight
TR	Repetition time (MRI)
TTE	Transthoracic echocardiogram
US	Ultrasound
UV	Ultraviolet (e.g., UV radiation)

Appendix B: Fundamental Constants

Symbol	Name	Value
c	Speed of light in a vacuum	3.00×10^8 m/s
e	Elementary charge	1.60×10^{-19} C
γ	Gyromagnetic ratio (proton)	2.675×10^8 rad/(s·T)
$\bar{\gamma} = \gamma/2\pi$		42.58 MHz/T
h	Planck's constant	6.63×10^{-34} J·s
$\hbar \equiv h/2\pi$	Reduced Planck's constant	1.05×10^{-34} J·s
k_B	Boltzmann constant	1.38×10^{-23} J/K
m_e	Electron rest mass	9.11×10^{-31} kg 0.511 MeV/c^2
m_p	Proton rest mass	1.67×10^{-27} kg 938.3 MeV/c^2
N_A	Avogadro's number	6.022×10^{23}

Appendix C: Units

Abbreviation	Full Name	Description
Bq	Becquerel	Radioactivity – one decay per second
C	Coulomb	Electric charge
Ci	Curie	Radioactivity – 3.7×10^{10} Bq
eV	Electron volt	Energy
HU	Hounsfield unit	Radiodensity (grayscale)
Hz	Hertz	Frequency – one cycle per second
J	Joule	Energy
K	Kelvin	Temperature
kg	Kilogram	Mass
m	Meter	Distance
Rayl	Rayleigh	Acoustic impedance
s	Second	Time
T	Tesla	Magnetic field strength
V	Volt	Electrical potential

Appendix D: Conversions

1 electron volt	=	1.602×10^{-19} Joules
1 kiloelectron volt	=	1.602×10^{-16} Joules
1 Joule	=	6.24×10^{18} electron volts
2π radians	=	360 degrees
1 degree	=	0.01745 radians

Appendix E: Mathematical Relations

This appendix provides a brief summary of mathematical expressions that appear throughout the text and homework.

Trigonometry

$$\sin\theta = \frac{\text{Opposite}}{\text{Hypotenuse}}$$

$$\cos\theta = \frac{\text{Adjacent}}{\text{Hypotenuse}}$$

$$\tan\theta = \frac{\text{Opposite}}{\text{Adjacent}}$$

Similar Triangles

$$\frac{h}{H} = \frac{c}{C} = \frac{b}{B} = \frac{a}{A}$$

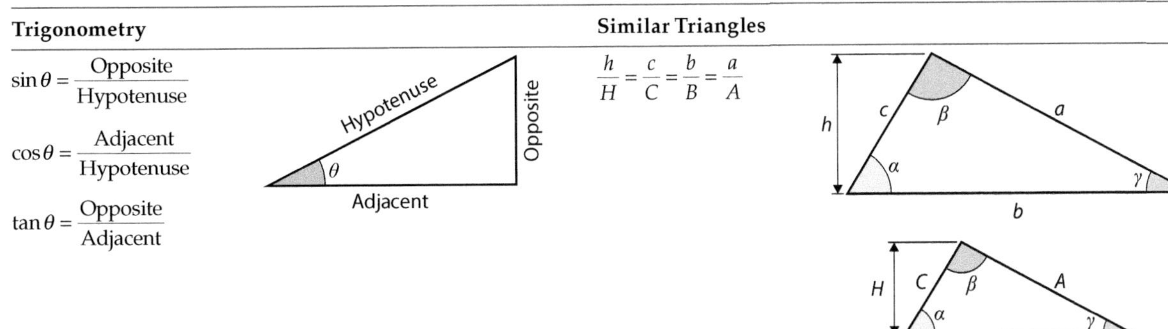

Trigonometric Identities

$$\sin^2\theta + \cos^2\theta = 1$$

$$\sin(\theta \pm \varphi) = \sin\theta\cos\varphi \pm \cos\theta\sin\varphi$$

$$\sin 2\theta = 2\sin\theta\cos\theta$$

$$\cos(\theta \pm \varphi) = \cos\theta\cos\varphi \mp \sin\theta\sin\varphi$$

$$\sin\theta + \sin\varphi = 2\cos\left(\frac{\theta - \varphi}{2}\right)\sin\left(\frac{\theta + \varphi}{2}\right)$$

$$\sin(\theta + \pi) = -\sin\theta$$

$$\cos\left(\theta \pm \frac{\pi}{2}\right) = \mp\sin\theta$$

$$\sin^2\left(\frac{\theta}{2}\right) = \frac{1}{2}(1 - \cos\theta)$$

Approximations/Taylors Series

$\sin\theta \simeq \tan\theta \simeq \theta$ (small θ in radians)

$\cos\theta \simeq 1$ (small θ in radians)

$e^x \simeq 1 - x$ (small x)

Euler Relations

$$e^{i\theta} = \cos\theta + i\sin\theta$$

$$\sin\theta = \frac{e^{i\theta} - e^{-i\theta}}{2i}$$

$$\cos\theta = \frac{e^{i\theta} + e^{-i\theta}}{2}$$

Log and Exponential

$$\ln\left(e^x\right) = x$$

$$e^{\ln x} = x$$

$$e^x \times e^y = e^{x+y}$$

$$\left(e^x\right)^y = e^{xy}$$

$$\log(10^x) = x$$

$$10^{\log x} = x$$

Derivatives and Integrals

$$\frac{d}{dx}(\sin ax) = a\cos ax$$

$$\frac{d}{dx}(\cos ax) = -a\sin ax$$

$$\frac{d}{dx}\left(e^{ax}\right) = ae^{ax}$$

$$\int \frac{dx}{x} = \ln|x|$$

Index

Note: Locators in *italics* represent figures and **bold** indicate tables in the text. Locators followed by 'b' refer to boxes

E

F

R

S

9781032328195